Lecture Notes in Computer Science 8561

Commenced Publication in 1973
Founding and Former Series Editors:
Gerhard Goos, Juris Hartmanis, and Jan van Leeuwen

T0183493

Carsten Sinz   Uwe Egly (Eds.)

Theory and Applications of Satisfiability Testing – SAT 2014

17th International Conference
Held as Part of the Vienna Summer of Logic, VSL 2014
Vienna, Austria, July 14-17, 2014
Proceedings

 Springer

Volume Editors

Carsten Sinz
Karlsruher Institut für Technologie (KIT)
Am Fasanengarten 5, 76131 Karlsruhe, Germany
E-mail: carsten.sinz@kit.edu

Uwe Egly
TU Wien
Favoritenstraße 9-11, 1040 Wien, Austria
E-mail: uwe@kr.tuwien.ac.at

ISSN 0302-9743   e-ISSN 1611-3349
ISBN 978-3-319-09283-6   e-ISBN 978-3-319-09284-3
DOI 10.1007/978-3-319-09284-3
Springer Cham Heidelberg New York Dordrecht London

Library of Congress Control Number: 2014943561

LNCS Sublibrary: SL 1 – Theoretical Computer Science and General Issues

Typesetting: Camera-ready by author, data conversion by Scientific Publishing Services, Chennai, India

Printed on acid-free paper

Springer is part of Springer Science+Business Media (www.springer.com)

Foreword

In the summer of 2014, Vienna hosted the largest scientific conference in the history of logic. The Vienna Summer of Logic (VSL, http://vsl2014.at) consisted of twelve large conferences and 82 workshops, attracting more than 2000 researchers from all over the world. This unique event was organized by the Kurt Gödel Society and took place at Vienna University of Technology during July 9 to 24, 2014, under the auspices of the Federal President of the Republic of Austria, Dr. Heinz Fischer.

The conferences and workshops dealt with the main theme, logic, from three important angles: logic in computer science, mathematical logic, and logic in artificial intelligence. They naturally gave rise to respective streams gathering the following meetings:

Logic in Computer Science / Federated Logic Conference (FLoC)

- 26th International Conference on Computer Aided Verification (CAV)
- 27th IEEE Computer Security Foundations Symposium (CSF)
- 30th International Conference on Logic Programming (ICLP)
- 7th International Joint Conference on Automated Reasoning (IJCAR)
- 5th Conference on Interactive Theorem Proving (ITP)
- Joint meeting of the 23rd EACSL Annual Conference on Computer Science Logic (CSL) and the 29th ACM/IEEE Symposium on Logic in Computer Science (LICS)
- 25th International Conference on Rewriting Techniques and Applications (RTA) joint with the 12th International Conference on Typed Lambda Calculi and Applications (TLCA)
- 17th International Conference on Theory and Applications of Satisfiability Testing (SAT)
- 76 FLoC Workshops
- FLoC Olympic Games (System Competitions)

Mathematical Logic

- Logic Colloquium 2014 (LC)
- Logic, Algebra and Truth Degrees 2014 (LATD)
- Compositional Meaning in Logic (GeTFun 2.0)
- The Infinity Workshop (INFINITY)
- Workshop on Logic and Games (LG)
- Kurt Gödel Fellowship Competition

Logic in Artificial Intelligence

- 14th International Conference on Principles of Knowledge Representation and Reasoning (KR)
- 27th International Workshop on Description Logics (DL)
- 15th International Workshop on Non-Monotonic Reasoning (NMR)
- 6th International Workshop on Knowledge Representation for Health Care 2014 (KR4HC)

The VSL keynote talks which were directed to all participants were given by Franz Baader (Technische Universität Dresden), Edmund Clarke (Carnegie Mellon University), Christos Papadimitriou (University of California, Berkeley) and Alex Wilkie (University of Manchester); Dana Scott (Carnegie Mellon University) spoke in the opening session. Since the Vienna Summer of Logic contained more than a hundred invited talks, it would not be feasible to list them here.

The program of the Vienna Summer of Logic was very rich, including not only scientific talks, poster sessions and panels, but also two distinctive events. One was the award ceremony of the Kurt Gödel Research Prize Fellowship Competition, in which the Kurt Gödel Society awarded three research fellowship prizes endowed with 100.000 Euro each to the winners. This was the third edition of the competition, themed Logical Mind: Connecting Foundations and Technology this year.

The 1st FLoC Olympic Games formed the other distinctive event and were hosted by the Federated Logic Conference (FLoC) 2014. Intended as a new FLoC element, the Games brought together 12 established logic solver competitions by different research communities. In addition to the competitions, the Olympic Games facilitated the exchange of expertise between communities, and increased the visibility and impact of state-of-the-art solver technology. The winners in the competition categories were honored with Kurt Gödel medals at the FLoC Olympic Games award ceremonies.

Organizing an event like the Vienna Summer of Logic was a challenge. We are indebted to numerous people whose enormous efforts were essential in making this vision become reality. With so many colleagues and friends working with us, we are unable to list them individually here. Nevertheless, as representatives of the three streams of VSL, we would like to particularly express our gratitude to all people who helped to make this event a success: the sponsors and the Honorary Committee; the Organization Committee and

Preface

This volume contains the papers presented at the 17th International Conference on Theory and Applications of Satisfiability Testing (SAT 2014) held during July 14–17, 2014, in Vienna, Austria. SAT 2014 was part of the Federated Logic Conference (FLoC) 2014 and the Vienna Summer of Logic (VSL) and was hosted by the Vienna University of Technology.

The International Conference on Theory and Applications of Satisfiability Testing (SAT) is the primary annual meeting for researchers focusing on the theory and applications of the propositional satisfiability problem, broadly construed: Besides plain propositional satisfiability, it includes Boolean optimization (including MaxSAT and Pseudo-Boolean, PB, constraints), Quantified Boolean Formulas (QBF), Satisfiability Modulo Theories (SMT), and Constraint Programming (CP) for problems with clear connections to propositional reasoning. Many hard combinatorial problems can be tackled using SAT-based techniques, including problems that arise in formal verification, artificial intelligence, operations research, biology, cryptology, data mining, machine learning, mathematics, etc. Indeed, the theoretical and practical advances in SAT research over the past 20 years have contributed to making SAT technology an indispensable tool in various domains.

SAT 2014 welcomed scientific contributions addressing different aspects of SAT, including (but not restricted to) theoretical advances (including exact algorithms, proof complexity, and other complexity issues), practical search algorithms, knowledge compilation, implementation-level details of SAT solvers and SAT-based systems, problem encodings and reformulations, applications, as well as case studies and reports on insightful findings based on rigorous experimentation.

A total of 78 papers were submitted to SAT 2014, distributed into 51 regular papers, 15 short papers, and 12 tool papers. Three regular paper submissions were found by the Program Committee to be out of scope for the conference (based on guidelines in the Call for Papers), and were returned without review. The 75 remaining paper submissions were assigned for review to at least four Program Committee members and their selected external reviewers. Continuing the procedure initiated in SAT 2012, the review process included an author-response period, during which the authors of submitted papers were given the opportunity to respond to the initial reviews for their submissions. For reaching final decisions, a Program Committee discussion period followed the author-response period. This year, external reviewers supporting the Program Committee were also invited to participate directly in the discussions for the papers they reviewed. In the end, the Program Committee decided to accept 21 regular papers, seven short papers, and four tool papers.

In addition to presentations of the accepted papers, the scientific program of SAT 2014 included two invited talks:

- Leonardo de Moura (Microsoft Research, USA):
  "A Model-Constructing Satisfiability Calculus"
- Jakob Nordström (KTH Royal Institute of Technology, Sweden):
  "A (Biased) Proof Complexity Survey for SAT Practitioners"

Two additional keynote talks were held jointly with other conferences of the Vienna Summer of Logic:

- Christos Papadimitriou (University of California, Berkeley, USA):
  "Computational Ideas and the Theory of Evolution"
- Alex Wilkie (University of Manchester, UK):
  "The Theory and Applications of o-Minimal Structures"

Moreover, there was a VSL opening speech given by Dana Scott from Carnegie Mellon University, USA.

SAT 2014, together with the other conferences of the Vienna Summer of Logic, hosted various associated events: 14 workshops, held on July 12/13 and 18/19 were (co-)affiliated with SAT 2014:

- FLoC Workshop on Proof Complexity (PC 2014)
  Organizers: Olaf Beyersdorff, Jan Johannsen
- 5th Pragmatics of SAT Workshop (POS 2014)
  Organizers: Daniel Le Berre, Allen Van Gelder
- Second International Workshop on Quantified Boolean Formulas (QBF 2014)
  Organizers: Charles Jordan, Florian Lonsing, Martina Seidl
- All About Proofs, Proofs for All (APPA 2014)
  Organizers: David Delahaye, Bruno Woltzenlogel Paleo
- International Joint Workshop on Implementation of Constraint and Logic Programming Systems and Logic-Based Methods in Programming Environments (CICLOPS-WLPE 2014)
  Organizers: Thomas Ströder, Terrance Swift
- 4th International Workshop on the Cross-Fertilization Between CSP and SAT (CSPSAT 2014)
  Organizers: Yael Ben-Haim, Valentin Mayer-Eichberger, Yehuda Naveh
- Higher Order Program Analysis (HOPA 2014)
  Organizer: Matthew Hague
- Interpolation: From Proofs to Applications (iPRA 2014)
  Organizers: Laura Kovacs, Georg Weissenbacher
- 4th International Workshop on Logic and Search (LaSh 2014)
  Organizers: Marc Denecker, David Mitchell, Emilia Oikarinen
- Parallel Methods for Search Optimization (ParSearchOpt 2014)
  Organizers: Philippe Codognet, Meinolf Sellmann, Guido Tack
- Second Workshop on the Parameterized Complexity of Computational Reasoning (PCCR 2014)
  Organizers: Michael R. Fellows, Serge Gaspers, Toby Walsh

- 21st RCRA International Workshop
  rithms for Solving Problems with Co
  Organizers: Toni Mancini, Marco Ma
- 12th International Workshop on Satis
  Organizers: Philipp Ruemmer, Christ
- Working Conference on Verified Softv
  (VSTTE 2014)
  Organizers: Dimitra Giannakopoulo

SAT 2014 also encompassed five compet

- SAT Competition 2014
  Organizers: Anton Belov, Daniel Die
- Configurable SAT Solver Challenge
  dauer, Sam Bayless, Holger Hoos, K
- MaxSAT Evaluation
  Organizers: Josep Argelich, Chu Mi
- QBF Gallery
  Organizers: Charles Jordan, Martin
- SMT-COMP 2014 (also affiliated
  David Deharbe, Tjark Weber

Moreover, an *SAT/SMT Summer*
Pascal Fontaine, Dejan Jovanović, and
days before SAT 2014, during July 10-

We would like to thank everyone
success. First and foremost we would l
Committee and the additional externa
work, without which it would not h
such an outstanding conference progr
who submitted their work for our con
chair Armin Biere, vice chair John Fr
their help and advice in organizationa
chair Inês Lynce and the competitic
work, and all the organizers of the S
Special thanks go to the organizers o
Baaz, Helmut Veith, and Moshe Var
the various conferences. The EasyC
assistance in coordinating the subm
assembly of these proceedings. Tha
Gelder, the co-chairs of SAT 2013,
We also thank the local organizatior
of local organization.

Finally, we gratefully thank the
Association, and Intel for financial

July 2014

Conference Organization

Program Committee Chairs

Uwe Egly   Vienna University of Technology, Austria
Carsten Sinz   Karlsruhe Institute of Technology, Germany

Workshops Chair

Inês Lynce   Technical University of Lisbon, Portugal

Competitions Chair

Laurent Simon   University of Bordeaux, France

Steering Committee

Matti Järvisalo   University of Helsinki, Finland
Allen Van Gelder   University of California at Santa Cruz, USA
Allesandro Cimatti   Fondazione Bruno Kessler, Italy
Roberto Sebastiani   University of Trento, Italy
Karem Sakallah   University of Michigan, USA
Laurent Simon   University of Bordeaux, France
Ofer Strichman   Technion, Israel
Stefan Szeider   Vienna University of Technology, Austria

Program Committee

Gilles Audemard   Artois University, France
Fahiem Bacchus   University of Toronto, Canada
Anton Belov   University College Dublin, Ireland
Olaf Beyersdorff   University of Leeds, UK
Armin Biere   Johannes Kepler University Linz, Austria
Alessandro Cimatti   Fondazione Bruno Kessler, Italy
Nadia Creignou   University of Aix-Marseille, France
John Franco   University of Cincinnati, USA
Enrico Giunchiglia   University of Genoa, Italy
Youssef Hamadi   Microsoft Research Cambridge, UK

Marijn Heule	University of Texas at Austin, USA
Holger H. Hoos	University of British Columbia, Canada
Matti Järvisalo	University of Helsinki, Finland
Hans Kleine Büning	University of Paderborn, Germany
Oliver Kullmann	Swansea University, UK
Daniel Le Berre	Artois University, France
Chu Min Li	University of Picardie Jules Verne, France
Florian Lonsing	Vienna University of Technology, Austria
Inês Lynce	Technical University of Lisbon, Portugal
Panagiotis Manolios	Northeastern University, USA
Norbert Manthey	TU Dresden, Germany
Joao Marques-Silva	University College Dublin, Ireland
Alexander Nadel	Intel Haifa, Israel
Jakob Nordström	KTH Royal Institute of Technology, Sweden
Albert Oliveras	Technical University of Catalonia, Spain
Jussi Rintanen	Aalto University, Finland
Lakhdar Sais	Artois University, France
Karem Sakallah	University of Michigan, USA
Horst Samulowitz	IBM T.J. Watson Research Center, USA
Tobias Schubert	Albert Ludwigs University of Freiburg, Germany
Roberto Sebastiani	University of Trento, Italy
Martina Seidl	Johannes Kepler University Linz, Austria
Bart Selman	Cornell University, USA
Stefan Szeider	Vienna University of Technology, Austria
Jacobo Torán	University of Ulm, Germany
Allen Van Gelder	University of California at Santa Cruz, USA
Xishun Zhao	Sun Yat-sen University, China

Additional Reviewers

Asín Achá, Roberto Javier	Compton, Kevin
Austrin, Per	De Haan, Ronald
Avellaneda, Florent	Fichte, Johannes Klaus
Bayless, Sam	Franzén, Anders
Beame, Paul	Frédéric, Olive
Bonacina, Ilario	Fröhlich, Andreas
Bordeaux, Lucas	Gebser, Martin
Bova, Simone	Goldberg, Eugene
Bruttomesso, Roberto	Habet, Djamal
Bueno, Denis	Hajiaghayi, Mohammadtaghi
Buss, Sam	Hoessen, Benoît
Cai, Shaowei	Hsu, Eric
Cao, Weiwei	Huang, Sangxia
Chen, Zhenyu	Ignatiev, Alexey

Jabbour, Said
Janota, Mikolas
Jordan, Charles
Katsirelos, George
Korhonen, Janne H.
Kovásznai, Gergely
Kulikov, Alexander
Lagniez, Jean Marie
Lallouet, Arnaud
Lauria, Massimo
Lettmann, Theodor
Lindauer, Marius
Manquinho, Vasco
Martin, Barnaby
Martins, Ruben
Meier, Arne
Miksa, Mladen
Morgado, Antonio
Oikarinen, Emilia
Papavasileiou, Vasilis

Previti, Alessandro
Ramanujan, M.S.
Reimer, Sven
Rollini, Simone Fulvio
Ryvchin, Vadim
Sabharwal, Ashish
Schaerf, Andrea
Shen, Yuping
Simon, Laurent
Slivovsky, Friedrich
Suter, Philippe
Tomasi, Silvia
Vinyals, Marc
Vizel, Yakir
Vollmer, Heribert
Wieringa, Siert
Wintersteiger, Christoph M.
Yue, Weiya
Zanuttini, Bruno

Invited Talks
(Abstracts)

A Model-Constructing Satisfiability Calculus

Leonardo de Moura[1] and Dejan Jovanović [2]

[1] Microsoft Research
[2] SRI International

Abstract. Considering the theoretical hardness of SAT, the astonishing adeptness of SAT solvers when attacking practical problems has changed the way we perceive the limits of algorithmic reasoning. Modern SAT solvers are based on the idea of *conflict driven clause learning* (CDCL). The CDCL algorithm is a combination of an explicit backtracking search for a satisfying assignment complemented with a deduction system based on Boolean resolution. In this combination, the worst-case complexity of both components is circumvented by the components guiding and focusing each other. The generalization of the SAT problem into the first-order domain is called satisfiability modulo theories (SMT). The common way to solve an SMT problem is to employ a SAT solver to enumerate the assignment of the Boolean abstraction of the formula. The candidate Boolean assignment is then either confirmed or refuted by a *decision procedure* dedicated to reasoning about conjunctions of theory-specific constraints. This framework is commonly called DPLL(T) and is employed by most of the SMT solvers today. Although DPLL(T) at its core relies on a CDCL SAT solver, this SAT solver is only used as a black-box. This can be seen as an advantage since the advances in SAT easily transfer to performance improvements in SMT. On the other hand, in the last few years the idea of direct model construction complemented with conflict resolution has been successfully generalized to fragments of SMT dealing with theories such as linear real arithmetic, linear integer arithmetic, nonlinear arithmetic, and floating-point. All these procedures, although quite effective in their corresponding first-order domains, have not seen a more widespread acceptance due to their limitations in purely Boolean reasoning and incompatibility with DPLL(T). In this talk we describe a *model-constructing satisfiability calculus* (MCSAT) that encompasses all the decision procedures above, including the decision procedures aimed at DPLL(T), while resolving the limitations mentioned above. The MCSAT framework extends DPLL(T) by allowing assignments of variables to concrete values, while relaxing the restriction that decisions, propagations, and explanations of conflicts must be in term of existing atoms.

A (Biased) Proof Complexity Survey for SAT Practitioners

Jakob Nordström

School of Computer Science and Communication
KTH Royal Institute of Technology
SE-100 44 Stockholm, Sweden

Abstract. This talk is intended as a selective survey of proof complexity, focusing on some comparatively weak proof systems that are of particular interest in connection with SAT solving. We will review resolution, polynomial calculus, and cutting planes (related to conflict-driven clause learning, Gröbner basis computations, and pseudo-Boolean solvers, respectively) and some proof complexity measures that have been studied for these proof systems. We will also briefly discuss if and how these proof complexity measures could provide insights into SAT solver performance.

Table of Contents

Proof Complexity

Parallel and Incremental (Q)SAT

Applications

Structure

Simplification and Solving

Analysis

Tool Papers

A (Biased) Proof Complexity Survey
for SAT Practitioners

Jakob Nordström

School of Computer Science and Communication
KTH Royal Institute of Technology
SE-100 44 Stockholm, Sweden

Abstract. This talk is intended as a selective survey of proof complexity, focusing on some comparatively weak proof systems that are of particular interest in connection with SAT solving. We will review resolution, polynomial calculus, and cutting planes (related to conflict-driven clause learning, Gröbner basis computations, and pseudo-Boolean solvers, respectively) and some proof complexity measures that have been studied for these proof systems. We will also briefly discuss if and how these proof complexity measures could provide insights into SAT solver performance.

Proof complexity studies how hard it is to find succinct certificates for the unsatisfiability of formulas in conjunctive normal form (CNF), i.e., proofs that formulas always evaluate to false under any truth value assignment, where these proofs should be efficiently verifiable. It is generally believed that there cannot exist a proof system where such proofs can always be chosen of size at most polynomial in the formula size. If this belief could be proven correct, it would follow that $NP \neq coNP$, and hence $P \neq NP$, and this was the original reason research in proof complexity was initiated by Cook and Reckhow [18]. However, the goal of separating P and NP in this way remains very distant.

Another, perhaps more recent, motivation for proof complexity is the connection to applied SAT solving. Any algorithm for deciding SAT defines a proof system in the sense that the execution trace on an unsatisfiable instance is itself a polynomial-time verifiable witness (often referred to as a *refutation* rather than a *proof*). In the other direction, most SAT solvers in effect search for proofs in systems studied in proof complexity, and upper and lower bounds for these proof systems hence give information about the potential and limitations of such SAT solvers.

In addition to running time, an important concern in SAT solving is memory consumption. In proof complexity, time and memory are modelled by *proof size* and *proof space*. It therefore seems interesting to understand these two complexity measures and how they are related to each other, and such a study reveals intriguing connections that are also of intrinsic interest to proof complexity. In this context, it is natural to concentrate on comparatively weak proof systems that are, or could plausibly be, used as a basis for SAT solvers. This talk will focus on such proof systems, and the purpose of these notes is to summarize the main points. Readers interested in more details can refer to, e.g, the survey [31].

C. Sinz and U. Egly (Eds.): SAT 2014, LNCS 8561, pp. 1–6, 2014.

1   Resolution

The proof system *resolution* [13] lies at the foundation of state-of-the-art SAT solvers based on conflict-driven clause learning (CDCL) [5,28,30]. In resolution, one derives new clauses from the clauses of the original CNF formula until an explicit contradiction is reached. Haken [24] proved the first (sub)exponential lower bound on proof size (measured as the number of clauses in a proof), and truly exponential lower bounds—i.e., bounds $\exp(\Omega(n))$ in the size n of the formula—were later established in [16,33].

The study of space in resolution was initiated by Esteban and Torán [20], measuring the space of a proof (informally) as the maximum number of clauses needing to be kept in memory during proof verification. Alekhnovich et al. [1] later extended the concept of space to a more general setting, including other proof systems. The (clause) space measure can be shown to be at most linear in the formula size, and matching lower bounds were proven in [1,8,20].

Ben-Sasson and Wigderson [11] instead focused on *width*, measured as the size of largest clause in a proof. It is easy to show that upper bounds on width imply upper bounds on size. More interestingly, [11] established the converse that strong enough lower bounds on width imply strong lower bounds on size, and used this to rederive essentially all known size lower bounds in terms of width. The relation between size and width was elucidated further in [4,15].

Atserias and Dalmau [3] proved that width also yields lower bounds on space[1] and that all previous space lower bounds could be obtained in this way. This demonstrates that width plays a key role in understanding both size and space. It should be noted, however, that in contrast to the relation between width and size the connection between width and space does not go in both directions, and an essentially optimal separation of the two measures was obtained in [9].

Regarding the connections between size and space, it follows from [3] that formulas of low space complexity also have short proofs. For the subsystem of *tree-like resolution*, where each line in the proof can only be used once, [20] showed that size upper bounds also imply space upper bounds, but for general resolution [9] established that this is false in the strongest possible sense. There have also been strong size-space trade-offs proven in [6,7,10].

The most comprehensive study to date of the question if and how hardness with respect to these complexity measures for resolution is correlated with actual hardness as measured by CDCL running time would seem to be [27], but it seems fair to say that the results so far are somewhat inconclusive.

2   Polynomial Calculus

Resolution can be extended with algebraic reasoning to form the stronger proof system *polynomial calculus (PC)* as defined in [1,17],[2] which corresponds to

---

[1] Note that this relation is nontrivial since space is measured as the number of *clauses*.

[2] We will be slightly sloppy in these notes and will not distinguish between polynomial calculus (PC) [17] and the slightly more general proof system polynomial calculus

Gröbner basis computations. In a PC proof, clauses are interpreted as multilinear polynomials (expanded out to sums of monomials), and one derives contradiction by showing that these polynomials have no common root. Intriguingly, while proof complexity-theoretic results seem to hold out the promise that SAT solvers based on polynomial calculus could be orders of magnitude faster than CDCL, such algebraic solvers have so far failed to be truly competitive (except for limited "hybrid versions" that incorporate reasoning in terms of linear equations into CDCL solvers).

Proof size in polynomial calculus is measured as the total number of monomials in a proof and the analogue of resolution space is the number of monomials needed simultaneously in memory during proof verification. Clause width in resolution translates into polynomial degree in PC. While size, space and width in resolution are fairly well understood, our understanding of the corresponding complexity measures in PC is more limited.

Impagliazzo et al. [26] showed that strong degree lower bounds imply strong size lower bounds. This is a parallel to the size-width relation for resolution in [11] discussed above, and in fact [11] can be seen as a translation of the bound in [26] from PC to resolution. This size-degree relation has been used to prove exponential lower bounds on size in a number of papers, with [2] perhaps providing the most general setting.

The first lower bounds on space were reported in [1], but only sublinear bounds and only for formulas of unbounded width. The first space lower bounds for k-CNF formulas were presented in [22], and asymptotically optimal (linear) lower bounds were finally proven by Bonacina and Galesi [14]. However, there are several formula families with high resolution space complexity for which the PC space complexity still remains unknown.

Regarding the relation between space and degree, it is open whether degree is a lower bound for space (which would be the analogue of what holds in resolution), but some limited results in this direction were proven in [21]. The same paper also established that the two measures can be separated in the sense that there are formulas of minimal (i.e., constant) degree complexity requiring maximal (i.e., linear) space.

As to size versus space in PC, it is open whether small space complexity implies small size complexity, but [21] showed that small size does not imply small space, just as for resolution. Strong size-space trade-offs have been shown in [7], essentially extending the results for resolution in [6,10] but with slightly weaker parameters.

3 Cutting Planes

In the proof system *cutting planes (CP)* [19] clauses of a CNF formula are translated to linear inequalities and the formula is refuted by showing that the

---

resolution (PCR) [1], using the term "polynomial calculus" to refer to both. PC is the proof system that is actually used in practice, but PCR is often more natural to work with in the context of proof complexity.

polytope defined by these inequalities does not have any zero-one integer points (corresponding to satisfying assignments). As is the case for polynomial calculus, cutting planes is exponentially stronger than resolution viewed as a proof system, but we are not aware of any efficient implementations of cutting planes-based SAT solvers that are truly competitive with CDCL solvers on CNF inputs in general (although as shown in [12,29] there are fairly natural formulas for which one can observe exponential gains in performance also in practice).

Cutting planes is much less well understood than both resolution and polynomial calculus. For proof size there is only one superpolynomial lower bound proven by Pudlák [32], but this result relies on a very specific technique that works only for formulas with a very particular structure. It remains a major challenge in proof complexity to prove lower bounds for other formulas such as random k-CNF formulas or so-called Tseitin formulas.

It is natural to define the *line space* of a CP proof to be the maximal number of linear inequalities that need to be kept in memory simultaneously during the proof. Just as for monomial space in polynomial calculus, line space in cutting planes is easily seen to be a generalization of clause space in resolution and is hence upper bounded by the clause space complexity. As far as we are aware, however, no lower bounds are known for CP space. Also, it should perhaps be noted that there does not seem to exist any generalization of width/degree for cutting planes with interesting connections to size or space.

Given the state of knowledge regarding proof size and space, maybe it is not too surprising that we also do not know much about size-space trade-offs. The recent papers [23,25] developed new techniques for this problem by making a connection between size-space trade-offs and communication complexity, and used this connection to show results that could be interpreted as circumstantial evidence that similar trade-off results as for resolution could be expected to hold also for cutting planes. However, so far all that has been proven using the approach in [23,25] are conditional space lower bounds, i.e., space lower bounds that seem likely to hold unconditionally, but which can so far be established only for cutting planes proofs of polynomial size.

References

1. Alekhnovich, M., Ben-Sasson, E., Razborov, A.A., Wigderson, A.: Space complexity in propositional calculus. SIAM Journal on Computing 31(4), 1184–1211 (2002), preliminary version appeared in STOC 2000
2. Alekhnovich, M., Razborov, A.A.: Lower bounds for polynomial calculus: Non-binomial case. Proceedings of the Steklov Institute of Mathematics 242, 18–35 (2003), http://people.cs.uchicago.edu/razborov/files/misha.pdf, Preliminary version appeared in FOCS 2001
3. Atserias, A., Dalmau, V.: A combinatorial characterization of resolution width. Journal of Computer and System Sciences 74(3), 323–334 (2008), preliminary version appeared in CCC 2003
4. Atserias, A., Lauria, M., Nordström, J.: Narrow proofs be maximally long. In: Proceedings of the 29th Annual IEEE Conference on Computational Complexity (CCC 2014) (to appear, Jun 2014)

5. Bayardo Jr., R.J., Schrag, R.: Using CSP look-back techniques to solve real-world SAT instances. In: Proceedings of the 14th National Conference on Artificial Intelligence (AAAI 1997), pp. 203–208 (July 1997)
6. Beame, P., Beck, C., Impagliazzo, R.: Time-space tradeoffs in resolution: Superpolynomial lower bounds for superlinear space. In: Proceedings of the 44th Annual ACM Symposium on Theory of Computing (STOC 2012), pp. 213–232 (May 2012)
7. Beck, C., Nordström, J., Tang, B.: Some trade-off results for polynomial calculus. In: Proceedings of the 45th Annual ACM Symposium on Theory of Computing (STOC 2013), pp. 813–822 (May 2013)
8. Ben-Sasson, E., Galesi, N.: Space complexity of random formulae in resolution. Random Structures and Algorithms 23(1), 92–109 (2003), preliminary version appeared in CCC 2001
9. Ben-Sasson, E., Nordström, J.: Short proofs be spacious: An optimal separation of space and length in resolution. In: Proceedings of the 49th Annual IEEE Symposium on Foundations of Computer Science (FOCS 2008), pp. 709–718 (October 2008)
10. Ben-Sasson, E., Nordström, J.: Understanding space in proof complexity: Separations and trade-offs via substitutions. In: Proceedings of the 2nd Symposium on Innovations in Computer Science (ICS 2011), pp. 401–416 (January 2011), full-length version available at http://eccc.hpi-web.de/report/2010/125/
11. Ben-Sasson, E., Wigderson, A.: Short proofs are narrow—resolution made simple. Journal of the ACM 48(2), 149–169 (2001), preliminary version appeared in STOC 1999
12. Biere, A., Berre, D.L., Lonca, E., Manthey, N.: Detecting cardinality constraints in CNF. In: Proceedings of the 17th International Conference on Theory and Applications of Satisfiability Testing (SAT 2014) (to appear, July 2014)
13. Blake, A.: Canonical Expressions in Boolean Algebra. Ph.D. thesis. University of Chicago (1937)
14. Bonacina, I., Galesi, N.: Pseudo-partitions, transversality and locality: A combinatorial characterization for the space measure in algebraic proof systems. In: Proceedings of the 4th Conference on Innovations in Theoretical Computer Science (ITCS 2013), pp. 455–472 (January 2013)
15. Bonet, M.L., Galesi, N.: Optimality of size-width tradeoffs for resolution. Computational Complexity 10(4), 261–276 (2001), preliminary version appeared in FOCS 1999
16. Chvátal, V., Szemerédi, E.: Many hard examples for resolution. Journal of the ACM 35(4), 759–768 (1988)
17. Clegg, M., Edmonds, J., Impagliazzo, R.: Using the Groebner basis algorithm to find proofs of unsatisfiability. In: Proceedings of the 28th Annual ACM Symposium on Theory of Computing (STOC 1996), pp. 174–183 (May 1996)
18. Cook, S.A., Reckhow, R.: The relative efficiency of propositional proof systems. Journal of Symbolic Logic 44(1), 36–50 (1979)
19. Cook, W., Coullard, C.R., Turn, G.: On the complexity of cutting-plane proofs. Discrete Applied Mathematics 18(1), 25–38 (1987)
20. Esteban, J.L., Torn, J.: Space bounds for resolution. Information and Computation 171(1), 84–97 (2001), preliminary versions of these results appeared in STACS 1999 and CSL 1999
21. Filmus, Y., Lauria, M., Mikša, M., Nordström, J., Vinyals, M.: Towards an understanding of polynomial calculus: New separations and lower bounds (extended abstract). In: Fomin, F.V., Freivalds, R., Kwiatkowska, M., Peleg, D. (eds.) ICALP 2013, Part I. LNCS, vol. 7965, pp. 437–448. Springer, Heidelberg (2013)

22. Filmus, Y., Lauria, M., Nordström, J., Thapen, N., Ron-Zewi, N.: Space complexity in polynomial calculus (extended abstract). In: Proceedings of the 27th Annual IEEE Conference on Computational Complexity (CCC 2012), pp. 334–344 (June 2012)
23. Gs, M., Pitassi, T.: Communication lower bounds via critical block sensitivity. In: Proceedings of the 46th Annual ACM Symposium on Theory of Computing (STOC 2014) (to appear, May 2014)
24. Haken, A.: The intractability of resolution. Theoretical Computer Science 39(2-3), 297–308 (1985)
25. Huynh, T., Nordström, J.: On the virtue of succinct proofs: Amplifying communication complexity hardness to time-space trade-offs in proof complexity (extended abstract). In: Proceedings of the 44th Annual ACM Symposium on Theory of Computing (STOC 2012), pp. 233–248 (May 2012)
26. Impagliazzo, R., Pudlák, P., Sgall, J.: Lower bounds for the polynomial calculus and the Gröbner basis algorithm. Computational Complexity 8(2), 127–144 (1999)
27. Järvisalo, M., Matsliah, A., Nordström, J., Živný, S.: Relating proof complexity measures and practical hardness of SAT. In: Milano, M. (ed.) CP 2012. LNCS, vol. 7514, pp. 316–331. Springer, Heidelberg (2012)
28. Marques-Silva, J.P., Sakallah, K.A.: GRASP—a new search algorithm for satisfiability. In: Proceedings of the IEEE/ACM International Conference on Computer-Aided Design (ICCAD 1996), pp. 220–227 (November 1996)
29. Mikša, M., Nordström, J.: Long proofs of (seemingly) simple formulas. In: Proceedings of the 17th International Conference on Theory and Applications of Satisfiability Testing (SAT 2014) (to appear, July 2014)
30. Moskewicz, M.W., Madigan, C.F., Zhao, Y., Zhang, L., Malik, S.: Chaff: Engineering an efficient SAT solver. In: Proceedings of the 38th Design Automation Conference (DAC 2001), pp. 530–535 (June 2001)
31. Nordström, J.: Pebble games, proof complexity and time-space trade-offs. Logical Methods in Computer Science 9, 15:1–15:63 (2013)
32. Pudlák, P.: Lower bounds for resolution and cutting plane proofs and monotone computations. Journal of Symbolic Logic 62(3), 981–998 (1997)
33. Urquhart, A.: Hard examples for resolution. Journal of the ACM 34(1), 209–219 (1987)

Cores in Core Based MaxSat Algorithms: An Analysis

Fahiem Bacchus and Nina Narodytska

Department of Computer Science, University of Toronto,
Toronto, Ontario, Canada, M5S 3H5
{fbacchus,ninan}@cs.toronto.edu

Abstract. A number of MAXSAT algorithms are based on the idea of generating unsatisfiable cores. A common approach is to use these cores to construct cardinality (or pseudo-boolean) constraints that are then added to the formula. Each iteration extracts a core of the modified formula that now contains cardinality constraints. Hence, the cores generated are not just cores of the original formula, they are cores of more complicated formulas. The effectiveness of core based algorithms for MAXSAT is strongly affected by the structure of the cores of the original formula. Hence it is natural to ask the question: how are the cores found by these algorithms related to the cores of the original formula? In this paper we provide a formal characterization of this relationship. Our characterization allows us to identify a possible inefficiency in these algorithms. Hence, finding ways to address it may lead to performance improvements in these state-of-the-art MAXSAT algorithms.

1 Introduction

MAXSAT is an optimization version of SAT in which the problem is to find a truth assignment that satisfies a maximum weight of clauses. In its most general form, a MAXSAT problem is expressed as a CNF formula partitioned into *hard* and *soft* clauses. Associated with each soft clause c_i is a numeric weight, w_i, and with each set of soft clauses S a cost, $cost(S)$, equal to the sum of the weights of the clauses in S. Various restricted versions of MAXSAT have also been studied [12]. Like SAT many practical problems can be encoded as MAXSAT formulas making the development of efficient MAXSAT solvers an important research problem.

There are a variety of algorithmic approaches to solving MAXSAT including solvers based on branch and bound, e.g., [10,13], solvers based on conversion to integer linear programs [5], solvers based on hybrid SAT and MIPs approaches [8], and core based solvers that use cardinality constraints, e.g., [15,4].

Core based solvers solve MAXSAT by solving a sequence of SAT problems using the cores returned by these SAT solving episodes to construct the next SAT problem. The performance of such solvers seems to depend on the structure of the cores of the original MAXSAT formula. For example, these solvers are quite successful when there are a large number of hard clauses which tends to reduce the size of the cores that must be dealt with. Similarly, these solvers do not work well on random problems and it is known that some types of random problems contain large cores [7]. However, achieving a clearer understanding of this relationship remains an open research problem.

C. Sinz and U. Egly (Eds.): SAT 2014, LNCS 8561, pp. 7–15, 2014.

In this paper we point out that core based solvers using cardinality constraints generate cores of a more complicated formula than the original MAXSAT formula. The cores they generate are cores of the MAXSAT formula augmented by cardinality constraints. We show that there is a precise relationship between the cores they generate and cores of the original MAXSAT formula. Our results could potentially help in obtaining a deeper understanding of how the structure of the cores of the MAXSAT instance affects the performance of this class of MAXSAT algorithms. More concretely, however, our results allow us to identify a possible source of inefficiency in such solvers. Developing techniques for removing this inefficiency thus becomes one way of potentially improving these solvers.

2 Background

A MAXSAT instance $\mathcal{F}$ is expressed as a CNF formula that is partitioned into two subsets of clauses $hard(\mathcal{F})$ and $soft(\mathcal{F})$. Note that in this paper we do not consider $\mathcal{F}$ to be a multi-set of clauses: multiple copies of a hard clause can be discarded, and multiple copies of a soft clause replaced with one copy with weight equal to the sum of the weights of the copies.

Definition 1 (Cost). *Each clause c_i in $soft(\mathcal{F})$ has an associated weight w_i. For any set of soft clauses $A \subseteq soft(\mathcal{F})$ we say that $cost(A) = \sum_{c_i \in A} w_i$, i.e., the cost of A is the sum of the weights of its soft clauses.*

Definition 2 (Solutions of $\mathcal{F}$). *A solution of $\mathcal{F}$ is a truth assignment π to the variables of $\mathcal{F}$ such that $\pi \models hard(\mathcal{F})$. The cost of a solution $cost(\pi)$ is the sum of the weights of the soft clauses it falsifies: $cost(\pi) = \sum_{\pi \not\models c_i} w_i$. The MAXSAT problem is to find a solution of $\mathcal{F}$ of minimum (optimal) cost.*

In this paper we assume that $hard(\mathcal{F})$ is satisfiable, i.e., solutions of $\mathcal{F}$ exist.

Definition 3 (Cores of $\mathcal{F}$). *A core κ of $\mathcal{F}$ is a subset of $soft(\mathcal{F})$ such that $\kappa \wedge hard(\mathcal{F})$ is unsatisfiable. Let $Cores(\mathcal{F})$ be the set of all cores of $\mathcal{F}$.*

We observe that for any core κ of $\mathcal{F}$ and solution π of $\mathcal{F}$, π must falsify at least one clause of κ. Furthermore, if κ is a core of $\mathcal{F}$ then any set of soft clauses A, that is a superset of κ, $A \supseteq \kappa$, is also a core.

3 The Fu and Malik Algorithm

To illustrate the type of MAXSAT algorithms under consideration we first describe one of the original core based MAXSAT algorithms due to Fu and Malik [9].

Fu & Malik works on restricted MAXSAT problems in which every soft clause has unit weight. These are called partial MAXSAT problems in the MAXSAT literature.

The algorithm executes a series of iterations, with the i-th iteration operating on the CNF formula $\mathcal{F}^i$, and the first iteration operating on the input MAXSAT formula, i.e., $\mathcal{F}^0 = \mathcal{F}$. Each iteration performs the following steps:

1. A SAT solver is called on $\mathcal{F}^i$. Note that the SAT solver ignores clause weights, regarding both $hard(\mathcal{F}^i)$ and $soft(\mathcal{F}^i)$ as ordinary clauses.

2. If $\mathcal{F}^i$ is satisfiable, then the satisfying truth assignment, restricted to the variables of $\mathcal{F}$ is an optimal MAXSAT solution.
3. Else $\mathcal{F}^i$ is unsatisfiable and we obtain a core κ from the SAT solver. Now the algorithm constructs the next formula $\mathcal{F}^{i+1}$ in two steps:
   (a) For every soft clause $c \in \mathcal{F}^i$ such that $c \in \kappa$ we add to c a literal b which is the positive literal of a brand new blocking variable (b-variable). Thus in $\mathcal{F}^{i+1}$ the clause c becomes the new clause $(c \vee b)$. This new clause $(c \vee b)$ is a soft clause of $\mathcal{F}^{i+1}$.
   (b) We add to $\mathcal{F}^i$ a new set of hard clauses encoding the cardinality constraint that the sum of the above newly added b-variables is equal to one. These new clauses are hard clauses of $\mathcal{F}^{i+1}$.

The added cardinality constraint allows one and only one soft clause of the discovered core to be relaxed by setting its b-variable to *true*. Each iteration installs an additional cardinality constraint which permits one more clause to be relaxed. Eventually $\mathcal{F}^i$, for some i, permits the relaxation of a sufficient number of clauses to achieve satisfaction.

One important point to notice is that if the SAT solver finds $\mathcal{F}^i$ to be unsatisfiable, then the core it returns is not a core of the original MAXSAT formula $\mathcal{F}$, it is a core of the relaxed formula $\mathcal{F}^i$. Since we want to increase our understanding of how algorithms like Fu & Malik are affected by the core structure of the MAXSAT formula, it becomes important to understand how the cores of $\mathcal{F}^i$, generated by the algorithm, are related to the cores of the original MAXSAT formula $\mathcal{F}$.

4   Cardinality Constraints

Now we present a general formulation of the problem we are addressing. This formulation is applicable not only to the Fu & Malik algorithm, but also to other core guided algorithms exploiting cardinality constraints like WPM1 [2] and WPM2 [3]. Our results are also applicable to the lower bounding phase of core guided algorithms that exploit binary search [15]. However, our results do not directly apply to iterative MAXSAT solvers, e.g., [6,11].

In general core guided algorithms impose linear inequalities or equalities over the blocking variables. These linear constraints are usually encoded into CNF and added to the formula. In some cases these constraints are handled directly without conversion to CNF, e.g., [1,14]. But even in these cases the constraints serve to restrict the satisfying models of the formula, so they are in effect "added" to the formula.

For convenience, we will call all such constraints *cardinality constraints*, although some of them are actually pseudo-boolean contraints. One important restriction of the analysis provided in this paper is that the cardinality contraints can only mention the b-variables and perhaps some other auxiliary variables. In particular, the cardinality constraints cannot mention any of the variables of $\mathcal{F}$. To the best of our knowledge, this restriction is satisfied by all existing core guided algorithms.

The cardinality constraints allow various sets of soft clauses to be "turned off" or blocked by allowing various combinations of the b-variables to be set to true. Since the b-variables appear only positively in the soft clauses of $\mathcal{F}^i$ each true b-variable satisfies some soft clause making it impossible for that clause to contribute to unsatisfiability (the cardinality constraints do not affect $hard(\mathcal{F})$).

As seen in the previous section, every time a cardinality constraint is added the soft clauses of the current formula $\mathcal{F}^i$ are modified. Current algorithms use two types of modifications to the clauses in $soft(\mathcal{F}^i)$:

Adding a b-variable: This involves replacing $c \in soft(\mathcal{F}^i)$ by $c \vee b$ where b is the positive literal of a new b-variable. Since $c \in \mathcal{F}^i$ it might already contain some b-variables added in previous iterations.[1]

Cloning: This involves adding a duplicate c' of a clause $c \in soft(\mathcal{F}^i)$ where c' contains a new b-variable: $c' = (c \vee b)$.[2] The clone c' is given a weight w and w is subtracted from c's weight. Since $c \in F^i$, it might be that c is itself a clone added in a previous iteration. Thus an original soft clause of $\mathcal{F}$ might be split into multiple clones, each with its own sequence of b-variables. The total sum of the weights of all these clones is always equal to the weight of the original soft clause.

Let $\mathbf{card}^i$ be the set of cardinality constraints that have been added up to iteration i of a MAXSAT algorithm, $soft(\mathcal{F}^i)$ be the corresponding modified set of soft clauses, and $\mathcal{B}^i$ be the set of all b-variables in $\mathcal{F}^i$.

Definition 4 (Solutions of $\mathbf{card}^i$). *A truth assignment β to all of the variables of $\mathcal{B}^i$ that satisfies the cardinality constraints in $\mathbf{card}^i$ is called a **solution** of $\mathbf{card}^i$. The set of all solutions of $\mathbf{card}^i$ is denoted by $\mathbf{soln}(\mathbf{card}^i)$.*

4.1 Residues and Reductions

A solution of $\mathbf{card}^i$, β, relaxes various clauses of $soft(\mathcal{F}^i)$. Each clauses $c^i \in soft(F^i)$ is an original soft clause $c \in soft(\mathcal{F})$ disjoined with some b-variables. If β makes any of these b-variables true then c^i is in effect removed from the formula: β relaxes c^i. If c^i is not relaxed by β then it is reduced to the original soft clause c by β: β values every b-variable so if none of the b-variables in c^i are made true, then they must all be made false, in effect removing them from c^i. These two stages of reduction by β—the removal or relaxation of soft clauses and the reduction of the remaining soft clauses to original soft clauses—are important in characterizing the relationship between the cores of $\mathcal{F}^i$ and those of $\mathcal{F}$. These two stages or reduction are formalized in our definitions of *Residues* and *Reductions*.

Definition 5 (Residues). *Let β be a solution of $\mathbf{card}^i$ ($\beta \in \mathbf{soln}(\mathbf{card}^i)$) and let A^i be a subset of $soft(\mathcal{F}^i)$. The **residue** of A^i induced by β, denoted by $A^i \Downarrow \beta$, is the subset of A^i formed by removing all clauses satisfied by β:*

$$A^i \Downarrow \beta = A^i - \{c^i | c^i \in A^i \text{ and } \beta \models c^i\}.$$

Note that in a residue, $A^i \Downarrow \beta$, the clauses of A^i not satisfied by β are unchanged. Thus $A^i \Downarrow \beta$ is a subset of A^i which in turn is a subset of $soft(\mathcal{F}^i)$. Thus a residue is subset of F^i.

---

[1] Some algorithms like WPM2 add at most one b-variable to a soft clause, others like Fu & Malik can add multiple b-variables to a clause.

[2] This type of modification is used in the WPM1 algorithm to deal with weighted MAXSAT.

The next stage of reduction is achieved with the standard notion of the **reduction** of a set of clauses by a truth assignment.

Definition 6 (Reduction). *Let $\beta \in \mathbf{soln}(\mathbf{card}^i)$ and $A^i \subseteq soft(\mathcal{F}^i)$. The **reduction** of A^i induced by β, denoted $A^i|_\beta$ is the new set of clauses formed by (a) removing all clauses satisfied by β from A^i, (b) removing all literals falsified by β from the remaining clauses, and (c) removing all duplicate clauses and setting the weight of the remaining clauses to their original weights in $\mathcal{F}$.*

In the three steps to compute a reduction we see that step (a) is the same as forming the residue—removing all satisfied clauses. Step (b) reduces the soft clauses to original soft clauses of $\mathcal{F}$—as noted above, all remaining b-variables in the clauses after step (a) must be falsified by β and thus will be removed by step (b). However, step (b) does not quite produce soft clauses of $\mathcal{F}$ as the weights of these clauses might differ from the weights they had in $\mathcal{F}$ (due to cloning). This is fixed by step (c) which removes all duplicate clauses (due to cloning) and resets the weights back to the original weights. Thus it can be observed that the reduction of A^i is a subset of the original MAXSAT formula $\mathcal{F}$.

We make a few observations about residues and reductions. Let A and B be any subsets of $soft(\mathcal{F}^i)$ and let β and β' be any two solutions of $\mathbf{card}^i$.

1. $A \subseteq B$ implies $(A \Downarrow \beta) \subseteq (B \Downarrow \beta)$ and $(A|_\beta) \subseteq (B|_\beta)$.
2. $A \Downarrow \beta \subseteq soft(\mathcal{F}^i)$ while $A|_\beta \subseteq soft(\mathcal{F})$.
3. $A \Downarrow \beta = A \Downarrow \beta'$ implies $A|_\beta = A|_{\beta'}$.
4. $(A \Downarrow \beta)|_\beta = A|_\beta$, although sometimes we will use the notation $(A \Downarrow \beta)|_\beta$ as this more clearly indicates that reduction has two stages.
5. When the clauses of $soft(\mathcal{F}^i)$ have more than one b-variable it can be the case that $soft(F^i) \Downarrow \beta = soft(F^i) \Downarrow \beta'$ even when $\beta \neq \beta'$.

Example 1. Consider formula $\mathcal{F}^i$ with soft clauses c_1, c_2 and c_3 (as well as other hard clauses). Say a run of Fu & Malik discovers the sequence of cores (specified as clause indicies) $\kappa_1 = \{1,2\}$, $\kappa_2 = \{2,3\}$, and $\kappa_3 = \{1,2,3\}$. Using b-variables with a superscript to indicate the core number and a subscript to indicate the clause number, $soft(\mathcal{F}^3) \supset \{(c_1, b_1^1, b_1^3), (c_2, b_2^1, b_2^2, b_2^3), (c_3, b_3^2, b_3^3)\}$, and $\mathbf{card}^3 = \{ CNF(b_1^1 + b_2^1 = 1), CNF(b_2^2 + b_3^2 = 1), CNF(b_1^3 + b_2^3 + b_3^3 = 1)\}$.

There are 12 different solutions to $\mathbf{card}^3$. However if we compute the residue of $A^3 = \{(c_1, b_1^1, b_1^3), (c_2, b_2^1, b_2^2, b_2^3), (c_3, b_3^2, b_3^3)\}$ with respect to these 12 solutions we obtain only 5 different residues: $\{(c_3, b_3^2, b_3^3)\}$, $\{(c_2, b_2^1, b_2^2, b_2^3)\}$, $\{(c_1, b_1^1, b_1^3), (c_3, b_3^2, b_3^3)\}$, $\{(c_1, b_1^1, b_1^3)\}$, and $\{\}$.

4.2 The Relationship between $Cores(\mathcal{F}^i)$ and $Cores(\mathcal{F})$

We can now present the paper's main result: a formalization of the relationship between the cores of $\mathcal{F}^i$ and the cores of the original MAXSAT formula $\mathcal{F}$. The next theorem shows that each core of $\mathcal{F}^i$ corresponds to a union of many cores of $\mathcal{F}$, and that every solution of $\mathbf{card}^i$ adds a core of $\mathcal{F}$ to this union.

Theorem 1. $\kappa^i \in Cores(\mathcal{F}^i)$ *if and only if*

$$\kappa^i = \bigcup_{\beta \,\in\, \mathbf{soln}(\mathbf{card}^i)} \kappa^\beta \textbf{ where } \kappa^\beta \subseteq (soft(\mathcal{F}^i) \Downarrow \beta) \textbf{ and } \kappa^\beta|_\beta \in Cores(\mathcal{F})$$

Each κ^β in this union is a set of soft clauses of $\mathcal{F}^i$ that remain after removing all clauses satisfied by β (i.e., $\kappa^\beta \subseteq soft(\mathcal{F}^i) \Downarrow \beta$), such that its reduction by β ($\kappa^\beta|_\beta$) is a core of $\mathcal{F}$.

Proof. Note that $hard(\mathcal{F}^i) = hard(\mathcal{F}) \wedge \mathbf{card}^i$. Thus a core of $\mathcal{F}^i$ is a subset of $soft(\mathcal{F}^i)$ that together with $hard(\mathcal{F}) \wedge \mathbf{card}^i$ is unsatisfiable.

First we show that if κ^i is a core of $\mathcal{F}^i$ then it is a union of sets κ^β satisfying the stated conditions. For any $\beta \in \mathbf{soln}(\mathbf{card}^i)$ let $\kappa^\beta = (\kappa^i \Downarrow \beta)$. Then we observe that (a) since $\kappa^i \subseteq soft(\mathcal{F}^i)$ then $\kappa^\beta = \kappa^i \Downarrow \beta \subseteq (soft(\mathcal{F}^i) \Downarrow \beta)$ (by observation 1 above), and (b) $\kappa^\beta|_\beta \in Cores(\mathcal{F})$. To see that (b) holds we observe that if $\kappa^\beta|_\beta$ is not a core of $\mathcal{F}$ then there exists a truth assignment π to the variables of $\mathcal{F}$ such that $\pi \models hard(\mathcal{F}) \wedge \kappa^\beta|_\beta$. Since κ^β is a residue induced by β, no clause of κ^β is satisfied by β (the residue operation removes all clauses satisfied by β). Therefore, we have that $\pi \models hard(\mathcal{F}) \wedge \kappa^\beta$ even though π does not assign a value to any b-variable. Then $\langle \beta, \pi \rangle \models hard(\mathcal{F}) \wedge \mathbf{card}^i \wedge \kappa^i$ since β satisfies all clauses in $\kappa^i - \kappa^\beta$ (these were the clauses removed from κ^i when taking its residue with respect to β because they were satisfied by β) and all clauses in $\mathbf{card}^i$, while π satisfies κ^β and $hard(\mathcal{F})$. That is, if (b) does not hold we obtain a contradiction of the premise that κ^i is a core of $\mathcal{F}^i$. Since this argument holds for every $\beta \in \mathbf{soln}(\mathbf{card}^i)$ we see that $\kappa^i = \cup\, \kappa^\beta$.

Second, we show that $\cup\, \kappa^\beta$ is a core of $\mathcal{F}^i$. Say that it is not. Then there exists a truth assignment $\langle \gamma, \pi \rangle$ that satisfies $hard(\mathcal{F}) \wedge \mathbf{card}^i \wedge (\cup\, \kappa^\beta)$, where γ assigns the variables in $\mathcal{B}^i$ and satisfies $\mathbf{card}^i$ while π assigns all of the other variables. Since $\gamma \in \mathbf{soln}(\mathbf{card}^i)$ we have that $\kappa^\gamma \subseteq (\cup\, \kappa^\beta)$. Furthermore, since $\kappa^\gamma \subseteq (soft(\mathcal{F}^i) \Downarrow \gamma)$ no clause in κ^γ is satisfied by γ; therefore, all literals of $\mathcal{B}^i$ in κ^γ must be falsified by γ. Since $\langle \gamma, \pi \rangle \models \kappa^\gamma$ and we must also have that $\langle \gamma, \pi \rangle \models \kappa^\gamma|_\gamma$, and since $\kappa^\gamma|_\gamma$ has no variables of $\mathcal{B}^i$, we must have that $\pi \models \kappa^\gamma|_\gamma$. This, however, is a contradiction as $\pi \models hard(\mathcal{F})$ and $\kappa^\gamma|_\gamma$ is a core of $\mathcal{F}$.

5   Residue Subsumption in $\mathbf{card}^i$

Theorem 1 allows us to identify a potential inefficiency of MAXSAT algorithms that compute cores after adding cardinality constraints.

Definition 7 (Residue Subsumption). *Let β and β' be two solutions of $\mathbf{card}^i$. We say that β residue subsumes β' if (1) $\beta \neq \beta'$ and (2) $soft(\mathcal{F}^i) \Downarrow \beta \subseteq soft(\mathcal{F}^i) \Downarrow \beta'$.*

Residue subsumption means that β relaxes (satisfies) all or more of the soft clauses of $\mathcal{F}^i$ that are relaxed by β'.

Example 2. Continuing with Example 1, we observed that there are only 5 different residues of A^3 generated by the 12 different solutions to $\mathbf{card}^3$: $\{(c_3, b_3^2, b_3^3)\}$, $\{(c_2, b_2^1, b_2^2, b_2^3)\}$, $\{(c_1, b_1^1, b_1^3), (c_3, b_3^2, b_3^3)\}$, $\{(c_1, b_1^1, b_1^3)\}$, and $\{\}$. The empty residue $\{\}$ is generated by three solutions of $\mathbf{card}^3$ one of which is β which sets b_1^1, b_2^2 and b_3^3 to true. Hence, β residue subsumes *all* other solutions to $\mathbf{card}^i$.

Our next result shows that when computing a core of $\mathcal{F}^i$ it is possible to ignore residue subsumed solutions of $\mathbf{card}^i$.

Proposition 1. *Let $RS \subset \mathbf{soln}(\mathbf{card}^i)$ be a set of solutions such that for all $\rho' \in RS$ there exists a $\rho \in (\mathbf{soln}(\mathbf{card}^i) - RS)$ such that ρ residue subsumes ρ'. Then any*

$$\kappa^i = \bigcup_{\beta \in (\mathbf{soln}(\mathbf{card}^i)-RS)} \kappa^\beta \text{ where } \kappa^\beta \subseteq (soft(\mathcal{F}^i) \Downarrow \beta) \text{ and } \kappa^\beta|_\beta \in Cores(\mathcal{F})$$

is a core of F^i.

Proof. Let

$$\kappa^+ = \bigcup_{\beta \in \mathbf{soln}(\mathbf{card}^i)} \kappa^\beta \text{ where } \kappa^\beta \subseteq (soft(\mathcal{F}^i) \Downarrow \beta) \text{ and } \kappa^\beta|_\beta \in Cores(\mathcal{F}).$$

By Theorem 1 κ^+ is a core of $\mathcal{F}^i$. Let $\rho' \in RS$ be residue subsumed by $\rho \in (\mathbf{soln}(\mathbf{card}^i) - RS)$. κ^+ is a union of sets including sets $\kappa^{\rho'}$ and κ^ρ both of which satisfy the above conditions.

By substituting $\kappa^{\rho'}$ by κ^ρ for each $\rho' \in RS$ we see that κ^+ becomes equal to κ^i and thus κ^i must also be a core of $\mathcal{F}^i$ if this substitution is valid. To show that the substitution is valid we must show that κ^ρ satisfies the two conditions required for $\kappa^{\rho'}$. First, $\kappa^\rho \subseteq (soft(F^i) \Downarrow \rho) \subseteq (soft(\mathcal{F}^i) \Downarrow \rho')$. Second, since κ^ρ is a subset of both $soft(\mathcal{F}^i) \Downarrow \rho$, and $soft(\mathcal{F}^i) \Downarrow \rho'$, neither ρ not ρ' satisfy any clauses of κ^ρ. Furthermore, both ρ and ρ' assign all b-variables so all b-variables left in κ^ρ must be falsified by both ρ and ρ'. This means that $\kappa^\rho|_{\rho'} = \kappa^\rho|_\rho$ and $\kappa^\rho|_\rho$ is already known to be a core of $\mathcal{F}^i$.

Theorem 1 shows that when the SAT solver computes a core κ^i of $\mathcal{F}^i$ it must refute all solutions of $\mathbf{card}^i$. The SAT solver might not need to refute two residue subsuming solutions ρ and ρ' separately—it might be able to find a single conflict that eliminates both candidate solutions. Nevertheless, it is possible that the solver ends up constructing two separate and different refutations of ρ and ρ', eventually unioning them into a refutation of $\mathcal{F}^i$. This can make the core extracted from the refutation larger, and also requires more time for the SAT solver. From Prop. 1 it can be seen that the solver need only refute ρ, finding the required κ^ρ. Adding κ^ρ to the core suffices to refute both ρ and ρ'.

For the Fu & Malik algorithm it has previously been noted that many symmetries exist over the introduced b-variables [1], and symmetry breaking constraints can be introduced over the b-variables to remove some of these symmetries. These symmetry breaking constraints serve to reduce the set of solutions $\mathbf{soln}(\mathbf{card}^i)$. Residue subsumption offers a more general way achieving this result. In particular, Lemma 15 of [4] introduces an ordering (weight) over all solutions in $\mathbf{soln}(\mathbf{card}^i)$ and shows that the introduced symmetry breaking constraints block solutions that are residue subsumed by other higher weight solutions. In fact, the introduced symmetries define a mapping between solutions that induce *equivalent* residues of $soft(\mathcal{F}^i)$, i.e., solutions ρ and ρ' where each residue subsumes the other. Residue subsumption as we have defined it here is more general than symmetry breaking as we don't need ρ and ρ' to subsume each other, we only require that ρ subsumes ρ'.

Example 3. Continuing with Example 2. Given that β which sets b_1^1, b_2^2 and b_3^3 to true residue subsumes all other solutions to $\mathbf{card}^3$, Prop. 1 shows that the SAT solver need only check if $\mathcal{F}^3$ is satisfiable under β. This makes intuitive sense, we have found 3 cores over 3 soft clauses, indicating that they all need to be relaxed. Fu & Malik would have the SAT solver refute all 12 solutions to $\mathbf{card}^3$. Note also that any form of symmetry reduction restricted to blocking solutions with equivalent residues, would still have to check at least 5 solutions.

5.1 Exploiting Residue Subsumption

There are two issues that arise when trying to exploit residue subsumption in core based solvers.

The first issue is that the correctness of the algorithms used in these solvers relies on certain properties of the formulas $\mathcal{F}^i$ constructed at each iteration. For example, a common way of proving the correctness of these algorithms is to prove that the original MAXSAT formula $\mathcal{F}$ is MaxSat reducible to each $\mathcal{F}^i$ [4]. If we alter $\mathbf{card}^i$ so as to block residue subsumed solutions, this would change $\mathcal{F}^i$ and we would have to verify that the new formula continues to satisfy the properties required of it by each algorithm. Unfortunately, proving these properties can be quite intricate.

However, this first issue is easily resolved. Let $\mathcal{F}^{i+}$ be a modification of $\mathcal{F}^i$ that blocks some residue subsumed solutions of $\mathbf{card}^i$. Applying Theorem 1 to $\mathcal{F}^{i+}$ we see that any core of $\mathcal{F}^{i+}$ returned by the SAT solver is a union over the solutions of $\mathbf{card}^i$ that have not been blocked in $\mathcal{F}^{i+}$. Then Prop. 1 shows that the returned core of $\mathcal{F}^{i+}$ is in fact also a core of $\mathcal{F}^i$. That is, the SAT solver will still return cores of $\mathcal{F}^i$ even if it has been modified to block residue subsumed solutions of $\mathbf{card}^i$. Furthermore, if the SAT solver returns a satisfying assignment this assignment must also satisfy $\mathcal{F}^i$ as it satisfies a more constrained version of $\mathcal{F}^i$.

Core based algorithms use the SAT solver as a black-box, expecting it to return a solution or core of $\mathcal{F}^i$, and as explained above modifying the solver so as to block residue subsumed solutions of $\mathbf{card}^i$ does not impact this functionality. Thus, any core based algorithm can exploit residue subsumption without affecting its correctness.

The second issue is more difficult to resolve, and remains an open research question. This is the issue of modifying the SAT solver so as to efficiently block residue subsumed solutions of $\mathbf{card}^i$. Potentially, extra clauses could be added to $\mathbf{card}^i$, an SMT-like theory could be consulted during search, or some modification could be made to the solver's search. How best to accomplish this is a problem we are continuing to work on.

6 Conclusion

In this paper we have presented a formal characterization of the relationship between the cores computed by core based MAXSAT algorithms using cardinality constraints and cores of the original MAXSAT formula. Our main result allowed us to identify a condition, residue subsumption, that could potentially be used to improve these algorithms.

References

1. Ansótegui, C., Bonet, M.L., Gabàs, J., Levy, J.: Improving SAT-based weighted MaxSAT solvers. In: Milano, M. (ed.) CP 2012. LNCS, vol. 7514, pp. 86–101. Springer, Heidelberg (2012)
2. Ansótegui, C., Bonet, M.L., Levy, J.: Solving (weighted) partial MaxSAT through satisfiability testing. In: Kullmann, O. (ed.) SAT 2009. LNCS, vol. 5584, pp. 427–440. Springer, Heidelberg (2009)
3. Ansótegui, C., Bonet, M.L., Levy, J.: A new algorithm for weighted partial MaxSAT. In: Proceedings of the AAAI National Conference, AAAI (2010)
4. Ansótegui, C., Bonet, M.L., Levy, J.: SAT-based MaxSAT algorithms. Artificial Intelligence 196, 77–105 (2013)
5. Ansótegui, C., Gabàs, J.: Solving (weighted) partial MaxSAT with ILP. In: International Conference on Integration of AI and OR Techniques in Constraint Programming for Combinatorial Optimization Problems (CPAIOR), pp. 403–409 (2013)
6. Berre, D.L., Parrain, A.: The sat4j library, release 2.2. Journal on Satisfiability, Boolean Modeling and Computation (JSAT) 7(2-3), 6–59 (2010)
7. Chvátal, V., Reed, B.A.: Mick gets some (the odds are on his side). In: Symposium on Foundations of Computer Science (FOCS). pp. 620–627 (1992)
8. Davies, J., Bacchus, F.: Postponing optimization to speed up MaxSAT solving. In: Schulte, C. (ed.) CP 2013. LNCS, vol. 8124, pp. 247–262. Springer, Heidelberg (2013)
9. Fu, Z., Malik, S.: On solving the partial max-sat problem. In: Biere, A., Gomes, C.P. (eds.) SAT 2006. LNCS, vol. 4121, pp. 252–265. Springer, Heidelberg (2006)
10. Heras, F., Larrosa, J., Oliveras, A.: MiniMaxSAT: An efficient weighted Max-SAT solver. Journal of Artificial Intelligence Research (JAIR) 31, 1–32 (2008)
11. Koshimura, M., Zhang, T., Fujita, H., Hasegawa, R.: QMaxSAT: A partial Max-SAT solver. Journal on Satisfiability, Boolean Modeling and Computation (JSAT) 8(1/2), 95–100 (2012)
12. Li, C.M., Manyà, F.: MaxSAT, hard and soft constraints. In: Biere, A., Heule, M., van Maaren, H., Walsh, T. (eds.) Handbook of Satisfiability, Frontiers in Artificial Intelligence and Applications, vol. 185, pp. 613–631. IOS Press (2009)
13. Li, C.M., Manyà, F., Mohamedou, N.O., Planes, J.: Resolution-based lower bounds in MaxSAT. Constraints 15(4), 456–484 (2010)
14. Manquinho, V., Marques-Silva, J., Planes, J.: Algorithms for weighted boolean optimization. In: Kullmann, O. (ed.) SAT 2009. LNCS, vol. 5584, pp. 495–508. Springer, Heidelberg (2009)
15. Morgado, A., Heras, F., Marques-Silva, J.: Improvements to core-guided binary search for MaxSAT. In: Cimatti, A., Sebastiani, R. (eds.) SAT 2012. LNCS, vol. 7317, pp. 284–297. Springer, Heidelberg (2012)

Solving MaxSAT and #SAT on Structured CNF Formulas

Sigve Hortemo Sæther, Jan Arne Telle, and Martin Vatshelle

Department of Informatics, University of Bergen, Norway

Abstract. In this paper we propose a structural parameter of CNF formulas and use it to identify instances of weighted MaxSAT and #SAT that can be solved in polynomial time. Given a CNF formula we say that a set of clauses is projection satisfiable if there is some complete assignment satisfying these clauses only. Let the **ps**-value of the formula be the number of projection satisfiable sets of clauses. Applying the notion of branch decompositions to CNF formulas and using **ps**-value as cut function, we define the **ps**-width of a formula. For a formula given with a decomposition of polynomial **ps**-width we show dynamic programming algorithms solving weighted MaxSAT and #SAT in polynomial time. Combining with results of 'Belmonte and Vatshelle, Graph classes with structured neighborhoods and algorithmic applications, THEOR. COMPUT. SCI. 511: 54-65 (2013)' we get polynomial-time algorithms solving weighted MaxSAT and #SAT for some classes of structured CNF formulas. For example, we get $\mathcal{O}(m^2(m+n)s)$ algorithms for formulas F of m clauses and n variables and total size s, if F has a linear ordering of the variables and clauses such that for any variable x occurring in clause C, if x appears before C then any variable between them also occurs in C, and if C appears before x then x occurs also in any clause between them. Note that the class of incidence graphs of such formulas do not have bounded clique-width.

1 Introduction

Given a CNF formula, propositional model counting (#SAT) is the problem of computing the number of satisfying assignments, and maximum satisfiability (MaxSAT) is the problem of determining the maximum number of clauses that can be satisfied by some assignment. Both problems are significantly harder than simply deciding if a satisfying assignment exists. #SAT is #P-hard [11] even when restricted to Horn 2-CNF formulas, and to monotone 2-CNF formulas [22]. MaxSAT is NP-hard even when restricted to Horn 2-CNF formulas [15], and to 2-CNF formulas where each variable appears at most 3 times [20]. Both problems become tractable under certain structural restrictions obtained by bounding width parameters of graphs associated with formulas, see for example [9,10,23,25]. For earlier work on width decompositions in this setting see e.g. [8,1]. The work we present here is inspired by the recent results of Paulusma

C. Sinz and U. Egly (Eds.): SAT 2014, LNCS 8561, pp. 16–31, 2014.
© Springer International Publishing Switzerland 2014

et al [18] and Slivovsky and Szeider [24] showing that #SAT is solvable in polynomial time when the incidence graph $I(F)$ of the input formula F has bounded modular treewidth, and more strongly, bounded symmetric clique-width.

We extend these results in several ways. We give algorithms for both #SAT and MAXSAT, and also weighted MAXSAT, finding the maximum weight of satisfiable clauses, given a set of weighted clauses. We introduce the parameter ps-width, and express the runtime of our algorithms as a function of ps-width.

Theorem 3. *Given a formula F over n variables and m clauses and of total size s, and a decomposition of F of* ps-*width k, we solve #SAT, and weighted* MAXSAT *in time $\mathcal{O}(k^3 s(m + n))$.*

Thus, given a decomposition having a ps-width k that is *polynomially-bounded* in the number of variables n and clauses m of the formula, we get polynomial-time algorithms. These are dynamic programming algorithms similar to the one given for #SAT in [24], but we believe that the ps-width parameter is a better measure of the inherent runtime bottleneck of #SAT and MAXSAT when using this type of dynamic programming. The essential combinatorial result enabling this improvement is Lemma 5 of this paper. The algorithm of [24] solves #SAT in time $(n+m)^{\mathcal{O}(w)}$ for w being the symmetric clique-width of the decomposition, and is thus a polynomial-time algorithm if given a decomposition with *constantly bounded* w. The result of Theorem 3 encompasses this, since we show via the concept of MIM-width [26], that any formula with constantly bounded symmetric clique-width also has polynomially bounded ps-width.

We show that a relatively rich class of formulas, including classes of unbounded clique-width, have polynomially bounded ps-width. This is shown using the concept of MIM-width of graphs, introduced in the thesis of Vatshelle [26]. See Figure 1. In particular, this holds for classes of formulas having incidence graphs that can be represented as intersection graphs of certain objects, like interval graphs [2]. We prove this also for bigraph bipartizations of these graphs, which are obtained by imposing a bipartition on the vertex set and keeping only edges between the partition classes. Some such bigraph bipartizations have been studied previously, in particular the interval bigraphs. The interval bigraphs contain all bipartite permutation graphs, and these latter graphs have been shown to have unbounded clique-width [4].

By combining an alternative definition of interval bigraphs [13] with a fast recognition algorithm [17,19] we arrive at the following. Say that a CNF formula F has an interval ordering if there exists a linear ordering of variables and clauses such that for any variable x occurring in clause C, if x appears before C then any variable between them also occurs in C, and if C appears before x then x occurs also in any clause between them.

Theorem 10. *Given a CNF formula F over n variables and m clauses and of total size s, we can in time $\mathcal{O}((m + n)s)$ decide if F has an interval ordering (yes iff $I(F)$ is an interval bigraph), and if yes we solve #SAT and weighted* MAXSAT *with a runtime of $\mathcal{O}(m^2(m + n)s)$.*

The algorithms of Theorem 10 may be of interest for practical applications, as there are no big hidden constants in the runtimes.

Our paper is organized as follows. In Section 2 we give formal definitions. We will be using a type of decomposition that originates in the theory of graphs and matroids where it is known as branch decomposition, see [12,21]. The standard approach is to apply this type of decomposition to the incidence graph of a formula, and evaluate its width using as cut function a graph parameter, as done in [24]. The cut function we will use is not a graph parameter, but rather the ps-value of a formula, being the number of distinct subsets of clauses that are satisfied by some complete assignment. We thus prefer to apply the decomposition directly to the formula and not to its incidence graph, although the translation between the two will be straightforward. We define cuts of formulas and ps-width of a formula. Note that a formula can have ps-value exponential and ps-width polynomial. In Section 3 we present dynamic programming algorithms that given a formula and a decomposition solves #SAT and weighted MaxSAT, proving Theorem 3. In Section 4 we investigate classes of formulas having decompositions of low ps-width, basically proving the correctness of the hierarchy presented in Figure 1. In Section 5 we consider formulas having an interval ordering and prove Theorem 10. We end in Section 5 with some open problems.

2   Framework

A *literal* is a propositional *variable* or a negated variable, x or $\neg x$, a *clause* is a set of literals, and a *formula* is a multiset of clauses. For a formula F, $\mathtt{cla}(F)$ denotes the clauses in F. For a clause C, $\mathtt{lit}(C)$ denotes the set of literals in C and $\mathtt{var}(C)$ denotes the variables of the literals in $\mathtt{lit}(C)$. For a (multi-) set S of variables and clauses, $\mathtt{var}(S)$ denotes the variables of S and $\mathtt{cla}(S)$ denotes the clauses. All sets of clauses mentioned in this paper are multisets. For a formula F, $\mathtt{var}(F)$ denotes the union $\bigcup_{C \in \mathtt{cla}(F)} \mathtt{var}(C)$. For a set X of variables, an *assignment* of X is a function $\tau : X \to \{0, 1\}$. For a literal ℓ, we define $\tau(\ell)$ to be $1 - \tau(\mathtt{var}(\ell))$ if ℓ is a negated variable ($\ell = \neg x$ for some variable x) and to be $\tau(\mathtt{var})$ otherwise ($\ell = x$ for some variable x). A clause C is said to be *satisfied* by an assignment τ if there exists at least one literal $\ell \in \mathtt{lit}(C)$ so that $\tau(\ell) = 1$. All clauses an assignment τ do not satisfy are said to be *unsatisfied* by τ. We notice that this means an empty clause will be unsatisfied by all assignments. A formula is satisfied by an assignment τ if τ satisfies all clauses in $\mathtt{cla}(F)$.

The problem #SAT, given a formula F, asks how many distinct assignments of $\mathtt{var}(F)$ satisfy F. The optimization problem weighted MaxSAT, given a formula F and weight function $w : \mathtt{cla}(F) \to \mathbb{N}$, asks what assignment τ of $\mathtt{var}(F)$ maximizes $\sum_C w(C)$ for all $C \in \mathtt{cla}(F)$ satisfied by τ. The problem MaxSAT is weighted MaxSAT where all clauses have weight one. When given a CNF formula F, we use s to denote the total size of F. More precisely, the total size of F is $s = |\mathtt{cla}(F)| + \sum_{C \in \mathtt{cla}(F)} |\mathtt{lit}(C)|$. For weighted MaxSAT, we

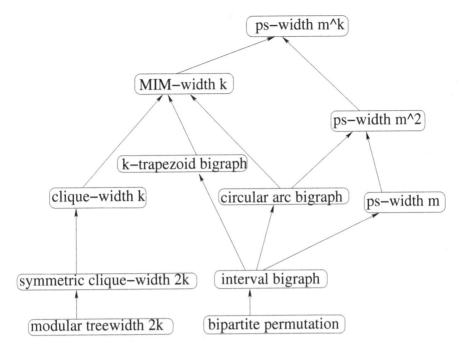

Fig. 1. A hierarchy of structural parameters and classes of bipartite graphs, where k is a constant and F a CNF formula having m clauses. An arc from P to Q means 'any formula (or incidence graph of a formula) that has a decomposition of type P, also has a decomposition of type Q'. The lack of an arc means that no such relation holds, i.e. this is a Hasse diagram.

assume the sum of all the weights are at most $2^{O(\mathtt{cla}(F))}$, and thus we can do summation on the weights in time linear in $\mathtt{cla}(F)$.

For a set A, with elements from a universe U we denote by $\overline{A}$ the elements in $U \setminus A$, as the universe is usually given by the context.

2.1   Cut of a Formula

In this paper, we will solve MaxSAT and #SAT by the use of dynamic programming. We will be using a divide and conquer technique where we solve the problem on smaller subformulas of the original formula F and then combine the solutions to each of these smaller formulas to form a solution to the entire formula F. Note however, that the solutions found for a subformula will depend on the interaction between the subformula and the remainder of the formula. We use the following notation for subformulas.

For a clause C and set X of variables, by $C|_X$ we denote the clause $\{\ell \in C : \mathtt{var}(\ell) \in X\}$. We say $C|_X$ is the clause C *induced* by X. Unless otherwise specified, all clauses mentioned in this paper is from the set $\mathtt{cla}(F)$ (e.g., if we write $C|_x \in \mathtt{cla}(F')$, we still assume $C \in \mathtt{cla}(F)$). For a formula F and subsets

$C \subseteq \mathtt{cla}(F)$ and $X \subseteq \mathtt{var}(F)$, we say the subformula $F_{C,X}$ of F *induced* by C and X is the formula consisting of the clauses $\{C_i|_X : C_i \in C\}$. That is, $F_{C,X}$ is the formula we get by removing all clauses not in C followed by removing each literal that consists of a variable not in X. For a set C of clauses, we denote by $C|_X$ the set $\{C|_X : C \in C\}$. As with a clause, for an assignment τ over a set X of variables, we say the assignment τ *induced* by $X' \subseteq X$ is the assignment $\tau|_{X'}$ where the domain is restricted to X'.

For a formula F and sets $C \subseteq \mathtt{cla}(F)$, $X \subseteq \mathtt{var}(F)$, and $S = C \cup X$, we call S a *cut* of F and note that it breaks F into four subformulas $F_{C,X}$, $F_{\overline{C},X}$, $F_{C,\overline{X}}$, and $F_{\overline{C},\overline{X}}$. See Figure 2. One important fact we may observe from this definition is that a clause C in F is satisfied by an assignment τ of $\mathtt{var}(F)$, if and only if C (induced by X or $\overline{X}$) is satisfied by τ in at least one of the formulas of any cut of F.

2.2   Projection Satisfiable Sets and ps-value of a Formula

For a formula F and assignment τ of all the variables in $\mathtt{var}(F)$, we denote by $\mathtt{sat}(F, \tau)$ the set $C \subseteq \mathtt{cla}(F)$ so that each clause in C is satisfied by τ, and each clause not in C is unsatisfied by τ. If for a set $C \subseteq \mathtt{cla}(F)$ we have $\mathtt{sat}(F, \tau) = C$ for some τ over $\mathtt{var}(F)$, then C is known as a *projection* (see e.g. [16,24]) and we say C is *projection satisfiable* in F. We denote by $\mathtt{PS}(F)$ the family of all projection satisfiable sets in F. That is,

$$\mathtt{PS}(F) = \{\mathtt{sat}(F, \tau) : \tau \text{ is an assignment of } \mathtt{var}(F)\}.$$

The cardinality of this set, $\mathtt{PS}(F)$, is referred to as the $\mathtt{ps}$-value of F.

2.3   The ps-width of a Formula

We define a *branch decomposition* of a formula F to be a pair (T, δ) where T is a rooted binary tree and δ is a bijective function from the leaves of T to the clauses and variables of F. If all the non-leaf nodes (also referred to as *internal* nodes) of T induce a path, we say that (T, δ) is a *linear* branch decomposition. For a non-leaf node v of T, we denote by $\delta(v)$ the set $\{\delta(l) : l \text{ is a leaf in the subtree rooted in } v\}$. Based on this, we say that the decomposition (T, δ) of formula F induces certain cuts of F, namely the cuts defined by $\delta(v)$ for each node v in T.

For a formula F and branch decomposition (T, δ), for each node v in T, by F_v we denote the formula induced by the clauses in $\mathtt{cla}(F) \setminus \delta(v)$ and the variables in $\delta(v)$, and by $F_{\overline{v}}$ we denote the formula on the complement sets; i.e. the clauses in $\delta(v)$ and the variables in $\mathtt{var}(F) \setminus \delta(v)$. In other words, if $\delta(v) = C \cup X$ with $C \subseteq \mathtt{cla}(F)$ and $X \subseteq \mathtt{var}(F)$ then $F_v = F_{\overline{C},X}$ and $F_{\overline{v}} = F_{C,\overline{X}}$. To simply the notation, we will for a node v in a branch decomposition and a set C of clauses denote by $C|_v$ the set $C|_{\mathtt{var}(F_v)}$. We define the $\mathtt{ps}$-*value* of the cut $\delta(v)$ to be

$$\mathtt{ps}(\delta(v)) = \max\{|PS(F_v)|, |PS(F_{\overline{v}})|\}$$

We define the **ps**-*width* of a branch decomposition to be

$$\texttt{psw}(T, \delta) = \max\{\texttt{ps}(\delta(v)) : v \text{ is a node of } T\}$$

We define the **ps**-*width* of a formula F to be

$$\texttt{psw}(F) = \min\{\texttt{psw}(T, \delta) : (T, \delta) \text{ is a branch decompositions of } F\}$$

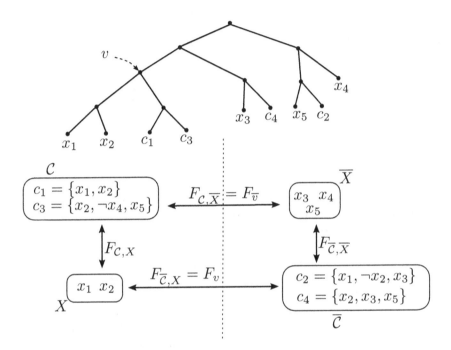

Fig. 2. On top is a branch decomposition of a formula F with $\texttt{var}(F) = \{x_1, x_2, x_3, x_4, x_5\}$ and the 4 clauses $\texttt{cla}(F) = \{c_1, c_2, c_3, c_4\}$ as given in the boxes. The node v of the tree defines the cut $\delta(v) = \mathcal{C} \uplus X$ where $\mathcal{C} = \{c_1, c_3\}$ and $X = \{x_1, x_2\}$. On the bottom is an illustration of the 4 subformulas defined by this cut. For example, $F_{\overline{\mathcal{C}}, X} = \{\{x_1, \neg x_2\}, \{x_2\}\}$ and $F_{\mathcal{C}, \overline{X}} = \{\emptyset, \{\neg x_4, x_5\}\}$. We have $F_v = F_{\overline{\mathcal{C}}, X}$ and $F_{\overline{v}} = F_{\mathcal{C}, \overline{X}}$ with projection satisfiable sets of clauses $PS(F_v) = \{\{c_2|_v\}, \{c_4|_v\}, \{c_2|_v, c_4|_v\}\}$ and $PS(F_{\overline{v}}) = \{\emptyset, \{c_3|_{\overline{v}}\}\}$ and the **ps**-value of this cut is $\texttt{ps}(\delta(v)) = \max\{|PS(F_v)|, |PS(F_{\overline{v}})|\} = 3$.

Note that the **ps**-value of a cut is a symmetric function. That is, the **ps**-value of cut S equals the **ps**-value of the cut $\overline{S}$. See Figure 2 for an example.

3   Dynamic Programming for MaxSAT and #SAT

Given a branch decomposition (T, δ) of a CNF formula F over n variables and m clauses and of total size s, we will give algorithms that solve MaxSAT and

#SAT on F in time $\mathcal{O}(\mathtt{psw}(T,\delta)^3 s(m+n))$. Our algorithms are strongly inspired by the algorithm of [24], but in order to achieve a runtime polynomial in ps-width, and also to solve MAXSAT, we make some changes.

In a pre-processing step we will need the following which, for each node v in T computes the sets $\mathtt{PS}(F_v)$ and $\mathtt{PS}(F_{\overline{v}})$.

Theorem 1. *Given a CNF formula F of n variables and m clauses with a branch decomposition (T,δ) of $\mathtt{ps}$-width k, we can in time $\mathcal{O}(k^2 \log(k) m (m+n))$ compute the sets $\mathtt{PS}(F_v)$ and $\mathtt{PS}(F_{\overline{v}})$ for each v in T.*

Proof. We notice that for a node v in T with children c_1 and c_2, we can express $\mathtt{PS}(F_v)$ as

$$\mathtt{PS}(F_v) = \left\{ (C_1 \cup C_2)|_v \cap \mathtt{cla}(F_v) : \begin{array}{l} C_1|_{c_1} \in \mathtt{PS}(F_{c_1}), \text{ and} \\ C_2|_{c_2} \in \mathtt{PS}(F_{c_2}) \end{array} \right\} .$$

Similarly, for sibling s and parent p of v in T, the set $\mathtt{PS}(F_{\overline{v}})$ can be expressed as

$$\mathtt{PS}(F_{\overline{v}}) = \left\{ (C_p \cup C_s)|_{\overline{v}} \cap \mathtt{cla}(F_{\overline{v}}) : \begin{array}{l} C_p|_{\overline{p}} \in \mathtt{PS}(F_{\overline{p}}), \text{ and} \\ C_s|_s \in \mathtt{PS}(F_s) \end{array} \right\} .$$

By transforming these recursive expressions into a dynamic programming algorithm, as done in Procedure 1 and Procedure 2 below, we are able to calculate all the desired sets as long as we can compute the sets for the base cases $\mathtt{PS}(F_l)$ when l is a leaf of T, and $\mathtt{PS}(F_{\overline{r}})$ for the root r of T. However, these formulas contain at most one variable, and thus we can easily construct their set of projection satisfiable clauses in linear amount of time for each of the formulas. For the rest of the formulas, we construct the formulas using Procedure 1 and Procedure 2. As there are at most twice as many nodes in T as there are clauses and variables in F, the procedures will run at most $\mathcal{O}(|\mathtt{cla}(F)| + |\mathtt{var}(F)|)$ times. In each run of the algorithms, we iterate through at most k^2 pairs of projection satisfiable sets, and do a constant number of set operations that might take $\mathcal{O}(|\mathtt{cla}(F)|)$ time each. Then we sort the list of at most k^2 sets of clauses. When we sort, we can expect the runtime of comparing two elements to spend time linear in $|\mathtt{cla}(F)|$, so the total runtime for sorting L and deleting duplicates takes at most $\mathcal{O}(k^2 \log(k)|\mathtt{cla}(F)|)$ time. This results in a total runtime of $\mathcal{O}(k^2 \log(k)|\mathtt{cla}(F)|(|\mathtt{cla}(F)| + |\mathtt{var}(F)|))$ for all the nodes of T combined.   □

We first give the algorithm for MAXSAT and then briefly describe the changes necessary for solving weighted MAXSAT and #SAT.

Our algorithm relies on the following binary relation, $\leq$, on assignments τ and τ' related to a cut $S = \mathcal{C} \cup X$ with $\mathcal{C} \subseteq \mathtt{cla}(F)$, $X \subseteq \mathtt{var}(F)$. For $C'|_{\overline{X}} \in \mathtt{PS}(F_{\mathcal{C},\overline{X}})$, we define $\tau' \leq_S^{\mathcal{C}'} \tau$ if it holds that $|\mathtt{sat}(F,\tau') \setminus \mathcal{C}'| \leq |\mathtt{sat}(F,\tau) \setminus \mathcal{C}'|$. Note that for each cut $S = \mathcal{C} \cup X$ and each $C'|_{\overline{X}} \in \mathtt{PS}(F_{\mathcal{C},\overline{X}})$ this gives a total preorder (transitive, reflexive and total) on assignments. The largest elements of this total preorder will be important for our algorithm, as they satisfy the maximum number of clauses under the given restrictions.

Procedure 1: Generating $\mathsf{PS}(F_v)$
input: $\mathsf{PS}(F_{c_1})$ and $\mathsf{PS}(F_{c_2})$ for children c_1 and c_2 of v in branch decomposition
output: $\mathsf{PS}(F_v)$
$L \leftarrow$ empty list of projection satisfiable clause-sets   **for** each $(C_1\|_{c_1}, C_2\|_{c_2}) \in \mathsf{PS}(F_{c_1}) \times \mathsf{PS}(F_{c_2})$ **do**   add $(C_1 \cup C_2)\|_v \cap \mathtt{cla}(F_v)$ to L sort L lexicographically by what clauses each element contains   remove duplicates in L by looking only at consecutive elements   **return** L

Procedure 2: Generating $\mathsf{PS}(F_{\overline{v}})$
input: $\mathsf{PS}(F_s)$ and $\mathsf{PS}(F_{\overline{p}})$ for sibling s and parent p of v in branch decomposition
output: $\mathsf{PS}(F_{\overline{v}})$
$L \leftarrow$ empty list of projection satisfiable clause-sets   **for** each $(C_s\|_s, C_p\|_{\overline{p}}) \in \mathsf{PS}(F_s) \times \mathsf{PS}(F_{\overline{p}})$ **do**   add $(C_s \cup C_p)\|_{\overline{v}} \cap \mathtt{cla}(F_{\overline{v}})$ to L sort L lexicographically by what clauses each element contains   remove duplicates in L by looking only at consecutive elements   **return** L

Given (T, δ) of a formula F our dynamic programming algorithm for MAXSAT will generate, for each node v in T, a table $\mathtt{Tab}_v$ indexed by pairs of $\{(C_1, C_2) : C_1\|_v \in \mathsf{PS}(F_v), C_2\|_{\overline{v}} \in \mathsf{PS}(F_{\overline{v}})\}$. For projection satisfiable sets $C_v\|_v \in \mathsf{PS}(F_v)$ and $C_{\overline{v}}\|_{\overline{v}} \in \mathsf{PS}(F_{\overline{v}})$ the contents of the table at this index $\mathtt{Tab}_v(C_v, C_{\overline{v}})$ should be an assignment $\tau : \mathtt{var}(\delta(v)) \to \{0, 1\}$ satisfying the following constraint:

$$\mathtt{Tab}_v(C_v, C_{\overline{v}}) = \tau \text{ such that } \mathtt{sat}(F_v, \tau) = C_v\|_v \text{ and } \tau' \leq^{C_{\overline{v}}}_{\delta(v)} \tau \text{ for any}$$
$$\tau' : \mathtt{var}(\delta(v)) \to \{0, 1\} \text{ having } \mathtt{sat}(F_v, \tau') = C_v\|_v \tag{1}$$

Let us give some intuition for this constraint. Our algorithm uses the technique of 'expectation from the outside' introduced in [5,6]. The partial assignment τ to variables in $\mathtt{var}(\delta(v))$ stored at $\mathtt{Tab}_v(C_v, C_{\overline{v}})$ will be combined with partial assignments to variables in $\mathtt{var}(F) \setminus \mathtt{var}(\delta(v))$ satisfying $C_{\overline{v}}$. These latter partial assignments constitute 'the expectation from the outside'. Constraint (1) implies that τ, being a largest element of the total preorder, will be a best combination with this expectation from the outside since it satisfies the maximum number of remaining clauses.

By bottom-up dynamic programming along the tree T we compute the tables of each node of T. For a leaf l in T, generating $\mathtt{Tab}_l$ can be done easily in linear time since the formula F_v contains at most one variable. For an internal node v of T, with children c_1, c_2, we compute $\mathtt{Tab}_v$ by the algorithm described in Procedure 3. There are 3 tables involved in this update, one at each child and one at the parent. A pair of entries, one from each child table, may lead to an update of an entry in the parent table. Each table entry is indexed by a pair, thus there are 6 indices involved in a single potential update. A trick first introduced in [6] allows us to loop over triples of indices and for each triple compute the remaining 3 indices forming the 6-tuple involved in the update, thereby reducing the runtime.

Procedure 3: Computing $\mathtt{Tab}_v$ for inner node v with children c_1, c_2

input: $\mathtt{Tab}_{c_1}$, $\mathtt{Tab}_{c_2}$
output: $\mathtt{Tab}_v$

1. initialize $\mathtt{Tab}_v : \mathrm{PS}(F_v) \times \mathrm{PS}(F_{\bar{v}}) \to \{\mathtt{unassigned}\}$ // *dummy entries*
2. **for each** $(C_{c_1}|_{c_1}, C_{c_2}|_{c_2}, C_{\bar{v}}|_{\bar{v}})$ in $\mathrm{PS}(F_{c_1}) \times \mathrm{PS}(F_{c_2}) \times \mathrm{PS}(F_{\bar{v}})$ **do**
3. $\quad C_{\overline{c_1}} \leftarrow (C_{c_2} \cup C_{\bar{v}}) \cap \delta(c_1)$
4. $\quad C_{\overline{c_2}} \leftarrow (C_{c_1} \cup C_{\bar{v}}) \cap \delta(c_2)$
5. $\quad C_v \leftarrow (C_{c_1} \cup C_{c_2}) \setminus \delta(v)$
6. $\quad \tau \quad\leftarrow \mathtt{Tab}_{c_1}(C_{c_1}, C_{\overline{c_1}}) \uplus \mathtt{Tab}_{c_2}(C_{c_2}, C_{\overline{c_2}})$
7. $\quad \tau' \leftarrow \mathtt{Tab}_v(C_v, C_{\bar{v}})$
8. $\quad$ **if** $\tau' = \mathtt{unassigned}$ or $\tau' \leq^{C_{\bar{v}}}_{\delta(v)} \tau$ **then** $\mathtt{Tab}_v(C_v, C_{\bar{v}}) \leftarrow \tau$
9. **return** $\mathtt{Tab}_v$

Lemma 2. *For a CNF formula F of total size s and an inner node v, of a branch decomposition (T, δ) of ps-width k, Procedure 3 computes $\mathtt{Tab}_v$ satisfying Constraint (1) in time $\mathcal{O}(k^3 s)$.*

Proof. We assume $\mathtt{Tab}_{c_1}$ and $\mathtt{Tab}_{c_2}$ satisfy Constraint (1). Procedure 3 loops over all triples in $\mathrm{PS}(F_{c_1}) \times \mathrm{PS}(F_{c_2}) \times \mathrm{PS}(F_{\bar{v}})$. From the definition of ps-width of (T, δ) there are at most k^3 such triples. Each operation inside an iteration of the loop take $\mathcal{O}(s)$ time and there is a constant number of such operations. Thus the runtime is $\mathcal{O}(k^3 s)$.

To show that the output $\mathtt{Tab}_v$ of Procedure 3 satisfies Constraint (1), we will prove that for any $C|_v \in \mathrm{PS}(F_v)$ and $C'|_{\bar{v}} \in \mathrm{PS}(F_{\bar{v}})$ the value of $\mathtt{Tab}_v(C, C')$ satisfies Constraint (1). That is, we will assure that the content of $\mathtt{Tab}_v(C, C')$ is an assignment τ so that $\mathtt{sat}(F_v, \tau) = C|_v$ and for all other assignments τ' over $\mathtt{var}(\delta(v))$ so that $\mathtt{sat}(F_v, \tau') = C|_v$, we have $\tau' \leq^{C'}_{\delta(v)} \tau$.

Let us assume for contradiction, that $\mathtt{Tab}_v(C, C')$ contains an assignment τ but there exists an assignment τ' over $\mathtt{var}(\delta(v))$ so that $\mathtt{sat}(F_v, \tau') = C|_v$, and we do not have $\tau' \leq^{C'}_{\delta(v)} \tau$. As τ is put into $\mathtt{Tab}_v(C, C')$ only if it is an assignment over $\mathtt{var}(\delta(v))$ and $\mathtt{sat}(F_v, \tau) = C|_v$. So, what we need to show to prove that $\mathtt{Tab}_v$ is correct is that in fact $\tau' \leq^{C'}_{\delta(v)} \tau$:

First, we notice that τ' consist of assignments $\tau'_1 = \tau'|_{\mathtt{var}(\delta(c_1))}$ and $\tau'_2 = \tau'|_{\mathtt{var}(\delta(c_2))}$ where τ'_1 is over the variables in $\mathtt{var}(\delta(c_1))$ and τ'_2 is over $\mathtt{var}(\delta(c_2))$. Let $C_1|_{c_1} = \mathtt{sat}(F_{c_1}, \tau'_1)$ and $C_2|_{c_2} = \mathtt{sat}(F_{c_2}, \tau'_2)$ and let $C'_1 = (C_2 \cup C') \cap \delta(c_1)$ and $C'_2 = (C_1 \cup C') \cap \delta(c_2)$. By how $\mathtt{Tab}_{c_1}$ and $\mathtt{Tab}_{c_2}$ is defined, we know for the assignment τ_1 in $\mathtt{Tab}_{c_1}(C_1, C'_1)$ and τ_2 in $\mathtt{Tab}_{c_2}(C_2, C'_2)$, we have $\tau'_1 \leq^{C'_1}_{\delta(c_1)} \tau_1$ and $\tau'_2 \leq^{C'_2}_{\delta(c_2)} \tau_2$. From our definition of the total preorder $\leq$ for assignments, we can deduce that $\tau'_1 \uplus \tau'_2 \leq^{C'}_{\delta(v)} \tau_1 \uplus \tau_2$;

$$|\mathtt{sat}(F, \tau_1' \uplus \tau_2') \setminus C'|$$
$$= |\mathtt{sat}(F, \tau_1') \setminus C_1'| - |C_1 \cap C'| + |\mathtt{sat}(F, \tau_2') \setminus C_2'| - |C_2 \cap C'| - |C_1 \cap C_2|$$
$$\le |\mathtt{sat}(F, \tau_1) \setminus C_1'| - |C_1 \cap C'| + |\mathtt{sat}(F, \tau_2) \setminus C_2'| - |C_2 \cap C'| - |C_1 \cap C_2|$$
$$= |\mathtt{sat}(F, \tau_1 \uplus \tau_2) \setminus C'| .$$

However, since $\tau_1 \uplus \tau_2$ at the iteration of the triple $(C_1|_{c_1}, C_2|_{c_2}, C'|_{\overline{v}})$ in fact is considered by the algorithm to be set as $\mathtt{Tab}_v(C, C')$, it must be the case that $\tau_1 \uplus \tau_2 \le_{\delta(v)}^{C'} \tau$. As $\le_{\delta(v)}^{C'}$ clearly is a transitive relation, we conclude that $\tau' \le_{\delta(v)}^{C'} \tau$. □

Theorem 3. *Given a formula F over n variables and m clauses and of total size s, and a branch decomposition (T, δ) of F of ps-width k, we solve MaxSAT, #SAT, and weighted MaxSAT in time $\mathcal{O}(k^3 s(m + n))$.*

Proof. To solve MaxSAT, we first compute $\mathtt{Tab}_r$ for the root node r of T. This requires that we first compute $\mathtt{PS}(F_v)$ and $\mathtt{PS}(F_{\overline{v}})$ for all nodes v of T, and then, in a bottom up manner, compute $\mathtt{Tab}_v$ for each of the $\mathcal{O}(m + n)$ nodes in T. The former part we can do in $\mathcal{O}(k^3 s(m + n))$ time by Theorem 1, and the latter part we do in the same amount of time by Lemma 2.

At the root r of T we have $\delta(r) = \mathtt{var}(F) \cup \mathtt{cla}(F)$. Thus $F_r = \emptyset$ and $F_{\overline{r}}$ contains only empty clauses, so that $PS(F_r) \times PS(F_{\overline{r}})$ contains only $(\emptyset, \emptyset)$. By Constraint (1) and the definition of the $\le$ total preorder on assignments, the assignment τ stored in $\mathtt{Tab}_r(\emptyset, \emptyset)$ is an assignment of $\mathtt{var}(F)$ maximizing $|\mathtt{sat}(F, \tau)|$, the number of clauses satisfied, and hence is a solution to MaxSAT.

For a weight function $w : \mathtt{cla}(F) \to \mathbb{N}$, by redefining $\tau_1 \le_A^B \tau_2$ to mean $w(\mathtt{sat}(F, \tau_1) \setminus B) \le w(\mathtt{sat}(F, \tau_2) \setminus B)$ both for the definition of $\mathtt{Tab}$ and for Procedure 3, we are able to solve the more general problem weighted MaxSAT in the same way.

For the problem #SAT, we care only about assignments satisfying all the clauses of F, and we want to decide the number of distinct assignments doing so. This requires a few alterations. Firstly, alter the definition of the contents of $\mathtt{Tab}_v(C, C')$ in Constraint (1) to be the number of assignments τ over $\mathtt{var}(\delta(v))$ where $\mathtt{sat}(F_v, \tau) = C|_v$ and $\mathtt{cla}(\delta(v)) \setminus C' \subseteq \mathtt{sat}(F, \tau)$. Secondly, when computing $\mathtt{Tab}_l$ for the leaves l of T, we set each of the entries of $\mathtt{Tab}_l$ to either zero, one, or two, according to the definition. Thirdly, we alter the algorithm to compute $\mathtt{Tab}_v$ (Procedure 3) for inner nodes. We initialize $\mathtt{Tab}_v(C, C')$ to be zero at the start of the algorithm, and substitute lines 6, 7 and 8 of Procedure 3 by the following line which increases the table value by the product of the table values at the children

$$\mathtt{Tab}_v(C_v, C_{\overline{v}}) \leftarrow \mathtt{Tab}_v(C_v, C_{\overline{v}}) + \mathtt{Tab}_{c_1}(C_{c_1}, C_{\overline{c_1}}) \cdot \mathtt{Tab}_{c_2}(C_{c_2}, C_{\overline{c_2}})$$

This will satisfy our new constraint of $\mathtt{Tab}_v$ for internal nodes v of T. The value of $\mathtt{Tab}_r(\emptyset, \emptyset)$ at the root r of T will be exactly the number of distinct assignments satisfying all clauses of F. □

The bottleneck giving the cubic factor k^3 in the runtime of Theorem 3 is the number triples in $\mathsf{PS}(F_{\overline{v}}) \times \mathsf{PS}(F_{c_1}) \times \mathsf{PS}(F_{c_2})$ for any node v with children c_1 and c_2. When (T, δ) is a linear branch decomposition, it is always the case that either c_1 or c_2 is a leaf of T. In this case either $|\mathsf{PS}(F_{c_1})|$ or $|\mathsf{PS}(F_{c_2})|$ is a constant. Therefore, for linear branch decompositions $\mathsf{PS}(F_{\overline{v}}) \times \mathsf{PS}(F_{c_1}) \times \mathsf{PS}(F_{c_2})$ will contain no more than $\mathcal{O}(k^2)$ triples. Thus we can reduce the runtime of the algorithm by a factor of k.

Theorem 4. *Given a formula F over n variables and m clauses and of total size s, and a linear branch decomposition (T, δ) of F of ps-width k, we solve #SAT, MaxSAT, and weighted MaxSAT in time $\mathcal{O}(k^2 s(m + n))$.*

4   CNF Formulas of Polynomial ps-width

In this section we investigate classes of CNF formulas having decompositions with ps-width polynomially bounded in the total size s of the formula. In particular, we show that this holds whenever the incidence graph of the formula has constant MIM-width (maximum induced matching-width). We also show that a large class of bipartite graphs, using what we call bigraph bipartizations, have constant MIM-width.

Let us start by defining bigraph bipartizations. For a graph G and subset of vertices $A \subseteq V(G)$ the bipartite graph $G[A, \overline{A}]$ is the subgraph of G containing all edges of G with exactly one endpoint in A. We call $G[A, \overline{A}]$ a bigraph bipartization of G, note that G has a bigraph bipartization for each subset of vertices. For a graph class X define the class of X bigraphs as the bipartite graphs H for which there exists $G \in X$ such that H is isomorphic to a bigraph bipartization of G. For example, H is an interval bigraph if there is some interval graph G and some $A \subseteq V(G)$ with H isomorphic to $G[A, \overline{A}]$.

To establish the connection to MIM-width we need to look at induced matchings in the incidence graph of a formula. The incidence graph of a formula F is the bipartite graph $I(F)$ having a vertex for each clause and variable, with variable x adjacent to any clause C in which it occurs. An induced matching in a graph is a subset M of edges with the property that any edge of the graph is incident to at most one edge in M. In other words, for any 3 vertices a, b, c, if ab is an edge in M and bc is an edge then there does not exist an edge cd in M. The number of edges in M is called the size of the induced matching. The following result provides an upper bound on the ps-value of a formula in terms of the maximum size of an induced matching of its incidence graph.

Lemma 5. *Let F be a CNF formula and let k be the maximum size of an induced matching in $I(F)$. We then have $|\mathsf{PS}(F)| \leq |\mathtt{cla}(F)|^k$.*

Proof. Let $\mathcal{C} \in \mathsf{PS}(F)$ and $\mathcal{C}_f = \mathtt{cla}(F) \setminus \mathcal{C}$. Thus, there exists a complete assignment τ such that the clauses not satisfied by τ are $\mathcal{C}_f = \mathtt{cla}(F) \setminus \mathtt{sat}(F, \tau)$. Since every variable in $\mathtt{var}(F)$ appears in some clause of F this means that $\tau|_{\mathtt{var}(\mathcal{C}_f)}$ is the unique assignment of the variables in $\mathtt{var}(\mathcal{C}_f)$ which do not satisfy any clause of C_f. Let $C'_f \subseteq C_f$ be an inclusion minimal set such that $\mathtt{var}(C_f) = \mathtt{var}(C'_f)$, hence $\tau|_{\mathtt{var}(C_f)}$ is also the unique assignment of the variables in $\mathtt{var}(C_f)$ which do not satisfy any clause of C'_f. An upper bound on the number of different such minimal C'_f, over all $C \in \mathtt{PS}(F)$, will give an upper bound on $|\mathtt{PS}(F)|$. For every $C \in C'_f$ there is a variable v_C appearing in C and no other clause of C'_f, otherwise C'_f would not be minimal. Note that we have an induced matching M of $I(F)$ containing all such edges v_C, C. By assumption, the induced matching M can have at most k edges and hence $|C'_f| \leq k$. There are at most $|\mathtt{cla}(F)|^k$ sets of at most k clauses and the lemma follows. $\qquad\square$

In order to lift this result on the $\mathtt{ps}$-value of F, i.e $|\mathtt{PS}(F)|$, to the $\mathtt{ps}$-width of F, we use MIM-width of the incidence graph $I(F)$, which is defined using branch decompositions of graphs. A branch decomposition of the formula F, as defined in Section 2, can also be seen as a branch decomposition of the incidence graph $I(F)$. Nevertheless, for completeness, we formally define branch decompositions of graphs and MIM-width.

A branch decomposition of a graph G is a pair (T, δ) where T is a rooted binary tree and δ a bijection between the leaf set of T and the vertex set of G. For a node w of T let the subset of $V(G)$ in bijection δ with the leaves of the subtree of T rooted at w be denoted by V_w. We say the decomposition defines the cut $(V_w, \overline{V_w})$. The MIM-value of a cut $(V_w, \overline{V_w})$ is the size of a maximum induced matching of $G[V_w, \overline{V_w}]$. The MIM-width of (T, δ) is the maximum MIM-value over all cuts $(V_w, \overline{V_w})$ defined by a node w of T. The MIM-width of graph G, denoted $mimw(G)$, is the minimum MIM-width over all branch decompositions (T, δ) of G. As before a *linear branch decomposition* is a branch decomposition where inner nodes of the underlying tree induces a path.

We now give an upper bound on the $\mathtt{ps}$-value of a formula in terms of the MIM-width of any graph G such that the incidence graph of the formula is a bigraph bipartization of G.

Theorem 6. *Let F be a CNF formula of m clauses, G a graph, and (T, δ_G) a (linear) branch decomposition of G of MIM-width k. If for a subset $A \subseteq V(G)$ the graph $G[A, \overline{A}]$ is isomorphic to $I(F)$, then we can in linear time produce a (linear) branch decomposition (T, δ_F) of F having $\mathtt{ps}$-width at most m^k.*

Proof. Since each variable and clause in F has a corresponding node in $I(F)$, and each node in $I(F)$ has a corresponding node in G, by defining δ_F to be the function mapping each leaf l of T to the variable or clause in F corresponding to the node $\delta_G(l)$, (T, δ_T) is going to be a branch decomposition of F. For any cut $(A, \overline{A})$ induced by a node of (T, δ_F), let $C \subseteq \mathtt{cla}(F)$ be the clauses corresponding to vertices in A and $X \subseteq \mathtt{var}(F)$ the variables corresponding to vertices in A. The cut $S = C \cup X$ of F defines the two formulas $F_{C,\overline{X}}$ and $F_{\overline{C},X}$, and it holds that $I(F_{C,\overline{X}})$ and $I(F_{\overline{C},X})$ are induced subgraphs of $G[A, \overline{A}]$ and hence by Lemma 5, we have $|\mathtt{PS}(F_{C,\overline{X}})| \leq |\mathtt{cla}(F)|^{\mathtt{mim}(A)}$, and likewise we have $|\mathtt{PS}(F_{\overline{C},X})| \leq |\mathtt{cla}(F)|^{\mathtt{mim}(A)}$. Since the $\mathtt{ps}$-width of the decomposition is the maximum $\mathtt{ps}$-value of each cut, the theorem follows. $\qquad\square$

Note that by taking $G = I(F)$ and $A = \mathtt{cla}(F)$ and letting (T, δ_G) be a branch decomposition of G of minimum MIM-width, we get the following weaker result.

Corollary 7. *For any CNF formula F over m clauses, the* $\mathtt{ps}$*-width of F is no larger than* $m^{\mathtt{mimw}(I(F))}$.

In his thesis, Vatshelle [26] shows that MIM-width of any graph G is at most the clique-width of G. Furthermore, the clique-width has been shown by Courcelle [7] to be at most twice the symmetric clique-width. Thus, we can conclude that MIM-width is bounded on any graph class with a bound on the symmetric clique-width, in accordance with Figure 1.

Many classes of graphs have intersection models, meaning that they can be represented as intersection graphs of certain objects, i.e. each vertex is associated with an object and two vertices are adjacent iff their objects intersect. The objects used to define intersection graphs usually consist of geometrical objects such as lines, circles or polygons. Many well known classes of intersection graphs have constant MIM-width, as in the following which lists only a subset of the classes proven to have such bounds in [2,26].

Theorem 8 ([2,26]). *Let G be a graph. If G is a:*
interval graph then $\mathtt{mimw}(G) \leq 1$.
circular arc graph then $\mathtt{mimw}(G) \leq 2$.
k-trapezoid graph then $\mathtt{mimw}(G) \leq k$.
Moreover there exist linear decompositions satisfying the bound.

Let us briefly mention the definition of these graph classes. A graph is an interval graph if it has an intersection model consisting of intervals of the real line. A graph is a circular arc graph if it has an intersection model consisting of arcs of a circle. To build a k-trapezoid we start with k parallel line segments $(s_1, e_1), (s_2, e_2), ..., (s_k, e_k)$ and add two non-intersecting paths s and e by joining s_i to s_{i+1} and e_i to e_{i+1} respectively by straight lines for each $i \in \{1, ..., k-1\}$. The polygon defined by s and e and the two line segments $(s_1, e_1), (s_k, e_k)$ forms a k-trapezoid. A graph is a k-trapezoid graph if it has an intersection model consisting of k-trapezoids. See [3] for information about graph classes and their containment relations. Combining Theorems 6 and 8 we get the following.

Corollary 9. *Let F be a CNF formula containing m clauses. If $I(F)$ is a:*
interval bigraph then $\mathtt{psw}(F) \leq m$.
circular arc bigraph then $\mathtt{psw}(F) \leq m^2$.
k-trapezoid bigraph then $\mathtt{psw}(F) \leq m^k$.
Moreover there exist linear decompositions satisfying the bound.

5   Interval Bigraphs and Formulas Having Interval Orders

We will in this section show one class of formulas where we can find linear branch decompositions having $\mathtt{ps}$-width $\mathcal{O}(|\mathtt{cla}(F)|)$. Let us recall the definition of interval ordering. A CNF formula F has an interval ordering if there exists a linear ordering of variables and clauses such that for any variable x occurring in clause C, if x appears before C then any variable between them also occurs in C, and if C appears before x then x occurs also in any clause between them. By a result of Hell and Huang [13] it follows that a formula F has an interval ordering if and only if $I(F)$ is a interval bigraph.

Theorem 10. *Given a CNF formula F over n variables and m clauses and of total size s, we can in time $\mathcal{O}((m+n)s)$ decide if F has an interval ordering (yes iff $I(F)$ is an interval bigraph), and if yes we solve #SAT and weighted MAXSAT with a runtime of $\mathcal{O}(m^2(m+n)s)$.*

Proof. Using the characterization of [13] and the algorithm of [19] we can in time $\mathcal{O}((m+n)s)$ decide if F has an interval ordering and if yes, then we find it. From this interval ordering we build an interval graph G such that $I(F)$ is a bigraph bipartization of G, and construct a linear branch decomposition of G having MIM-width 1 [2]. From such a linear branch decomposition we get from Theorem 6 that we can construct another linear branch decomposition of F having **ps**-width $\mathcal{O}(m)$. We then run the algorithm of Theorem 4.   □

6   Conclusion

In this paper we have proposed a structural parameter of CNF formulas, called **ps**-width or projection-satisfiable-width. We showed that weighted MAXSAT and #SAT can be solved in polynomial time on formulas given with a decomposition of polynomially bounded **ps**-width. Using the concept of interval bigraphs we also showed a polynomial time algorithm that actually finds such a decomposition, for formulas having an interval ordering.

Could one devise such an algorithm also for the larger class of circular arc bigraphs, or maybe even for the even larger class of k-trapezoid bigraphs? In other words, is the problem of recognizing if a bipartite input graph is a circular arc bigraph, or a k-trapezoid bigraph, polynomial-time solvable?

It could be interesting to give an algorithm solving MAXSAT and/or #SAT directly on the interval ordering of a formula, rather than using the more general notion of **ps**-width as in this paper. Maybe such an algorithm could be of practical use?

Also of practical interest would be to design a heuristic algorithm which given a formula finds a decomposition of relatively low **ps**-width, as has been done for boolean-width in [14].

Finally, we hope the essential combinatorial result enabling the improvements in this paper, Lemma 5, may have other uses as well.

References

1. Bacchus, F., Dalmao, S., Pitassi, T.: Algorithms and complexity results for# sat and bayesian inference. In: Proceedings of 44th Annual IEEE Symposium on Foundations of Computer Science, pp. 340–351. IEEE (2003)
2. Belmonte, R., Vatshelle, M.: Graph classes with structured neighborhoods and algorithmic applications. Theor. Comput. Sci. 511, 54–65 (2013)

3. Brandstädt, A., Le, V.B., Spinrad, J.P.: Graph Classes: A Survey, Monographs on Discrete Mathematics and Applications, vol. 3. SIAM Society for Industrial and Applied Mathematics, Philadelphia (1999)
4. Brandstädt, A., Lozin, V.V.: On the linear structure and clique-width of bipartite permutation graphs. Ars. Comb. 67 (2003)
5. Bui-Xuan, B.M., Telle, J.A., Vatshelle, M.: H-join decomposable graphs and algorithms with runtime single exponential in rankwidth. Discrete Applied Mathematics 158(7), 809–819 (2010)
6. Bui-Xuan, B.M., Telle, J.A., Vatshelle, M.: Boolean-width of graphs. Theoretical Computer Science 412(39), 5187–5204 (2011)
7. Courcelle, B.: Clique-width of countable graphs: a compactness property. Discrete Mathematics 276(1-3), 127–148 (2004)
8. Darwiche, A.: Recursive conditioning. Artificial Intelligence 126(1), 5–41 (2001)
9. Fischer, E., Makowsky, J.A., Ravve, E.V.: Counting truth assignments of formulas of bounded tree-width or clique-width. Discrete Applied Mathematics 156(4), 511–529 (2008)
10. Ganian, R., Hlinený, P., Obdrzálek, J.: Better algorithms for satisfiability problems for formulas of bounded rank-width. Fundam. Inform. 123(1), 59–76 (2013)
11. Garey, M.R., Johnson, D.S.: Computers and Intractability: A Guide to the Theory of NP-Completeness. W.H. Freeman (1979)
12. Geelen, J.F., Gerards, B., Whittle, G.: Branch-width and well-quasi-ordering in matroids and graphs. J. Combin. Theory Ser. B 84(2), 270–290 (2002)
13. Hell, P., Huang, J.: Interval bigraphs and circular arc graphs. Journal of Graph Theory 46(4), 313–327 (2004)
14. Hvidevold, E.M., Sharmin, S., Telle, J.A., Vatshelle, M.: Finding good decompositions for dynamic programming on dense graphs. In: Marx, D., Rossmanith, P. (eds.) IPEC 2011. LNCS, vol. 7112, pp. 219–231. Springer, Heidelberg (2012)
15. Jaumard, B., Simeone, B.: On the complexity of the maximum satisfiability problem for horn formulas. Inf. Process. Lett. 26(1), 1–4 (1987)
16. Kaski, P., Koivisto, M., Nederlof, J.: Homomorphic hashing for sparse coefficient extraction. In: Thilikos, D.M., Woeginger, G.J. (eds.) IPEC 2012. LNCS, vol. 7535, pp. 147–158. Springer, Heidelberg (2012)
17. Müller, H.: Recognizing interval digraphs and interval bigraphs in polynomial time. Discrete Applied Mathematics 78(1-3), 189–205 (1997)
18. Paulusma, D., Slivovsky, F., Szeider, S.: Model counting for CNF formulas of bounded modular treewidth. In: Portier, N., Wilke, T. (eds.) STACS. LIPIcs, vol. 20, pp. 55–66. Schloss Dagstuhl - Leibniz-Zentrum fuer Informatik (2013)
19. Rafiey, A.: Recognizing interval bigraphs by forbidden patterns. CoRR abs/1211.2662 (2012)
20. Raman, V., Ravikumar, B., Rao, S.S.: A simplified NP-complete MAXSAT problem. Inf. Process. Lett. 65(1), 1–6 (1998)
21. Robertson, N., Seymour, P.D.: Graph minors X. obstructions to tree-decomposition. J. Combin. Theory Ser. B 52(2), 153–190 (1991)
22. Roth, D.: A connectionist framework for reasoning: Reasoning with examples. In: Clancey, W.J., Weld, D.S. (eds.) AAAI/IAAI, vol. 2, pp. 1256–1261. AAAI Press / The MIT Press (1996)
23. Samer, M., Szeider, S.: Algorithms for propositional model counting. J. Discrete Algorithms 8(1), 50–64 (2010)

24. Slivovsky, F., Szeider, S.: Model counting for formulas of bounded clique-width. In: Cai, L., Cheng, S.-W., Lam, T.-W. (eds.) ISAAC2013. LNCS, vol. 8283, pp. 677–687. Springer, Heidelberg (2013)
25. Szeider, S.: On fixed-parameter tractable parameterizations of SAT. In: Giunchiglia, E., Tacchella, A. (eds.) SAT 2003. LNCS, vol. 2919, pp. 188–202. Springer, Heidelberg (2004)
26. Vatshelle, M.: New width parameters of graphs. Ph.D. thesis. The University of Bergen (2012)

Solving Sparse Instances of Max SAT
via Width Reduction and Greedy Restriction

Takayuki Sakai[1], Kazuhisa Seto[2,*], and Suguru Tamaki[1,**]

[1] Kyoto University, Yoshida Honmachi, Sakyo-ku, Kyoto 606-8501, Japan
{tsakai,tamak}@kuis.kyoto-u.ac.jp
[2] Seikei University, Musashino-shi, Tokyo 180-8633, Japan
seto@st.seikei.ac.jp

Abstract. We present a moderately exponential time polynomial space algorithm for sparse instances of Max SAT. Our algorithms run in time of the form $O(2^{(1-\mu(c))n})$ for instances with n variables and cn clauses. Our deterministic and randomized algorithm achieve $\mu(c) = \Omega(\frac{1}{c^2 \log^2 c})$ and $\mu(c) = \Omega(\frac{1}{c \log^3 c})$ respectively. Previously, an exponential space deterministic algorithm with $\mu(c) = \Omega(\frac{1}{c \log c})$ was shown by Dantsin and Wolpert [SAT 2006] and a polynomial space deterministic algorithm with $\mu(c) = \Omega(\frac{1}{2^{O(c)}})$ was shown by Kulikov and Kutzkov [CSR 2007].

Our algorithms have three new features. They can handle instances with (1) weights and (2) hard constraints, and also (3) they can solve counting versions of Max SAT. Our deterministic algorithm is based on the combination of two techniques, width reduction of Schuler and greedy restriction of Santhanam. Our randomized algorithm uses random restriction instead of greedy restriction.

Keywords: Exponential time algorithm, polynomial space, weight, hard constraint, counting.

1 Introduction

In the maximum satisfiability problem (Max SAT), the task is, given a set of clauses, to find an assignment that maximizes the number of satisfied clauses, where a clause is a disjunction of literals and a literal is a Boolean variable or its negation. Max SAT is one of the most fundamental NP-hard problems. In Max ℓ-SAT, we pose a restriction on input instances that each clause contains at most ℓ literals. Max ℓ-SAT is NP-hard even when $\ell = 2$.

Given an instance of Max SAT with n variables and $m = cn$ clauses, one can solve the problem in time $O(m2^n)$. The challenge is to reduce the running time to $O(\text{poly}(m)2^{(1-\mu)n})$ for some absolute constant $\mu > 0$. This is

* This work was supported in part by MEXT KAKENHI (24106001, 24106003); JSPS KAKENHI (26730007).
** This work was supported in part by MEXT KAKENHI (24106003); JSPS KAKENHI (25240002, 26330011).

C. Sinz and U. Egly (Eds.): SAT 2014, LNCS 8561, pp. 32–47, 2014.
© Springer International Publishing Switzerland 2014

a very difficult goal to achieve since the best running time upper bound is of the form $O(\text{poly}(m)2^{(1-\frac{1}{O(\log c)})n})$ [4, 28] even for the satisfiability problem (SAT), where the task is to find an assignment that satisfies all the clauses. Therefore, it is natural to seek for an algorithm for Max SAT which runs in time $O(\text{poly}(m)2^{(1-\mu(c))n})$ for some $\mu(c) > 0$. Previously, an exponential space deterministic algorithm with $\mu(c) = \Omega(\frac{1}{c\log c})$ was shown by Dantsin and Wolpert [8] and a polynomial space deterministic algorithm with $\mu(c) = \Omega(\frac{1}{2^{O(c)}})$ was shown by Kulikov and Kutzkov [24]. In this paper, we prove the following theorems.

Theorem 1. *For instances with n variables and $m = cn$ clauses, Max SAT can be solved deterministically in time $O(\text{poly}(m)2^{(1-\mu(c))n})$ and polynomial space, where $\mu(c) = \Omega(\frac{1}{c^2\log^2 c})$.*

Theorem 2. *For instances with n variables and $m = cn$ clauses, Max SAT can be solved probabilistically in expected time $O(\text{poly}(m)2^{(1-\mu(c))n})$ and polynomial space, where $\mu(c) = \Omega(\frac{1}{c\log^3 c})$.*

Our algorithms have three new features which were not treated in [8, 24]. (1) Our algorithms can handle instances with weights. (2) Our algorithms can handle instances with hard constraints (instances of partial Max SAT), i.e., instances in which some clauses must be satisfied. (3) Our algorithms can count the number of optimal assignments. Furthermore, if we are allowed to use exponential space, our algorithms can be modified into the ones that can count the number of assignments achieving a given objective value. For more formal statements, see Theorems 4,5,7 and 8.

1.1 Related Work

The complexities of Max SAT and Max CSP (constraint satisfaction problem) has been studied with respect to several parameters [1–3, 6, 8–11, 14, 15, 19–26, 29, 30, 32]. Recall that Max CSP is a generalization of Max SAT where an instance consists of a set of arbitrary constraints instead of clauses. In Max ℓ-CSP, each constraint depends on at most ℓ variables.

We summarize the best upper bounds for each parametrization in Table 1. Here k is the objective value, i.e., the number of constraints that must be satisfied, l is the length of an instance, i.e., the sum of arities of constraints, m is the number of constraints, and n is the number of variables. We omit the factor in polynomial in k, l, m, n there. Note that for Max 2-SAT (and Max 2-CSP), the best upper bound is of the form $O(\text{poly}(m)2^{(1-\mu)n})$ for some absolute constant $\mu > 0$. However, it is not known whether $O(\text{poly}(m)2^{(1-\mu(\ell))n})$ time algorithms exist for Max ℓ-SAT when $\ell \geq 3$, where $\mu(\ell) > 0$ is a constant only depending on ℓ.

Table 1. A historical overview of upper bounds

Running time	Problem	Space	Reference
$O(2^{0.4414k})$	Max SAT	polynomial	[3]
$O(2^{0.1000l})$	Max 2-SAT	polynomial	[10]
$O(2^{0.1450l})$	Max SAT	polynomial	[1]
$O(2^{0.1583m})$	Max 2-SAT	polynomial	[9]
$O(2^{0.1901m})$	Max 2-CSP	polynomial	[9]
$O(2^{0.4057m})$	Max SAT	polynomial	[6]
$O(2^{0.7909n})$	Max 2-CSP	exponential	[21, 32]
$O(2^{(1-\alpha(\frac{m}{n}))n)}), \alpha(c) = \frac{1}{O(c\log c)}$	Max SAT	exponential	[8]
$O(2^{(1-\beta(\frac{m}{n}))n)}), \beta(c) = \frac{1}{O(2^{O(c)})}$	Max SAT	polynomial	[24]
$O(2^{(1-\gamma(\frac{m}{n}))n)}), \gamma(c) = \frac{1}{O(c^2\log^2 c)}$	Max SAT	polynomial	Theorem 1
$O(2^{(1-\delta(\frac{m}{n}))n)}), \delta(c) = \frac{1}{O(c\log^3 c)}$	Max SAT	polynomial	Theorem 2

An alternative way to parametrize MAX SAT is to ask whether at least $\tilde{m} + k$ clauses can be satisfied, where $\tilde{m}$ is an expected number of satisfied constraints by a uniformly random assignment, see, e.g., [12].

1.2   Our Technique

Our deterministic algorithm is based on the combination of two techniques, width reduction of Schuler [4, 28] and greedy restriction of Santhanam [7, 27]. Our first observation is that we can solve Max ℓ-SAT in time $O(\text{poly}(m)2^{(1-\Omega(\frac{1}{\ell^2 c^2}))n})$ by slightly modifying Santhanam's greedy restriction algorithm for De Morgan formula SAT [7, 27]. To apply the Max ℓ-SAT algorithm for general Max SAT, we adopt the technique of Schuler's width reduction [4, 28]. Briefly, width reduction is, given an instance of SAT, to produce a collection of instances of ℓ-SAT such that the original instance is satisfiable if and only if at least one of the produced instances is satisfiable. We will see that width reduction can be also applied to Max SAT instances. Once we recognize the approach of combining greedy restriction and width reduction works, the analysis of running time basically follows from the existing analysis.

Our randomized algorithm uses a simple random restriction algorithm for Max ℓ-SAT instead of the greedy restriction algorithm.

1.3   Paper Organization

In section 2, we introduce definitions and notation needed in the paper. In section 3, we present a greedy restriction algorithm for Max formula SAT and its running time analysis. The algorithm is used to solve Max ℓ-SAT. In section 4, we show a deterministic algorithm for Max SAT and its analysis. In section 5, we present an exponential space deterministic algorithm for a counting version of Max SAT and its analysis. In section 6, we show a randomized algorithm for Max ℓ-SAT and its analysis.

2   Preliminaries

We denote by $\mathbb{Z}$ the set of integers. We define $-\infty$ as $-\infty + z = z + -\infty = -\infty$ and $-\infty < z$ for any $z \in \mathbb{Z}$. ∞ is defined analogously. Let $V = \{x_1, \ldots, x_n\}$ be a set of Boolean variables. We use the value 1 to indicate Boolean 'true', and 0 'false'. The *negation* of a variable $x \in V$ is denoted by $\bar{x}$. A *literal* is either a variable or its negation. An ℓ-*constraint* is a Boolean function $\phi : \{0,1\}^\ell \to \{0,1\}$, which depends on ℓ variables of V. Note that a 0-constraint is either '0' or '1' (a constant function). An *instance* Φ of Max CSPs consists of pairs of a constraint and a weight function, i.e., $\Phi = \{(\phi_1, w_1), \ldots, (\phi_m, w_m)\}$ where each ϕ_i is a k_i-constraint and $w_i : \{0,1\} \to \{-\infty\} \cup \mathbb{Z}$. For a weight function w, we denote by $\tilde{w}$ a weight function such that $\tilde{w}(1) = w(1), \tilde{w}(0) = -\infty$. Note that a constraint with a weight of the form $w(0) = -\infty$ must be satisfied, i.e., it is a *hard* constraint. The width of Φ is $\max_i k_i$. We use the notation as $\mathrm{Val}(\Phi, a) := \sum_{i=1}^m w_i(\phi_i(a))$ and $\mathrm{Opt}(\Phi) := \max_{a \in \{0,1\}^n} \mathrm{Val}(\Phi, a)$. For an integer K, we define

$$\#\mathrm{Val}_{\geq K}(\Phi) := |\{a \in \{0,1\}^n \mid \mathrm{Val}(\Phi, a) \geq K\}|,$$
$$\#\mathrm{Opt}(\Phi) := |\{a \in \{0,1\}^n \mid \mathrm{Val}(\Phi, a) = \mathrm{Opt}(\Phi)\}|.$$

We are interested in subclasses of Max CSPs defined by restricting the type of constraints. We denote by $y_1, \ldots, y_\ell$ arbitrary literals. In *Max SAT*, each constraint must be a *clause*, i.e., a disjunction of literals of the form $y_1 \vee \cdots \vee y_\ell$. We allow 0-constraints to appear in instances of Max SAT.

3   A Greedy Restriction Algorithm for Max formula SAT

In this section, we present an algorithm for Max CSPs in which each constraint is given as a De Morgan formula. The algorithm immediately yields an algorithm for Max ℓ-SAT.

3.1   De Morgan Formula and Its Simplification

A *De Morgan formula* is a rooted binary tree in which each leaf is labeled by a literal from the set $\{x_1, \ldots, x_n, \bar{x}_1, \ldots, \bar{x}_n\}$ or a constant from $\{0,1\}$ and each internal node is labeled by $\wedge$ ("and") or $\vee$ ("or"). Given a De Morgan formula ϕ, a *subformula* of ϕ is a De Morgan formula which is a subtree in ϕ. Every De Morgan formula computes in a natural way a Boolean function from $\{0,1\}^n$ to $\{0,1\}$. The *size* of a De Morgan formula ϕ is defined to be the number of leaves in it, and it is denoted by $L(\phi)$. We denote by $\mathrm{var}(\phi)$ the set of variables which appear as literals in ϕ. The *frequency* of a variable x in ϕ is defined to be the number of leaves labeled by x or $\bar{x}$, and it is denoted by $\mathrm{freq}_\phi(x)$.

For any formula ϕ, any set of variables $\{x_{i_1}, \ldots, x_{i_k}\}$ and any constants $a_1, \ldots, a_k \in \{0,1\}$, we denote by $\phi[x_{i_1} = a_1, \ldots, x_{i_k} = a_k]$ the formula obtained from ϕ by assigning to each $x_{i_j}, \bar{x}_{i_j}$ the value $a_j, \bar{a}_j$ and applying the following procedure **Simplify**. We can similarly define $\phi[l_{i_1} = a_1, \ldots, l_{i_k} = a_k]$ for any set of literals $\{l_{i_1}, \ldots, l_{i_k}\}$. The procedure **Simplify** reduces the size of a formula by applying rules to eliminate constants and redundant literals. These are the same simplification rules used by [13, 27].

Simplify(ϕ: formula)

Repeat the following until there is no decrease in size of ϕ.

(a) If $0 \wedge \psi$ occurs as a subformula, where ψ is any formula, replace this subformula by 0.

(b) If $0 \vee \psi$ occurs as a subformula, where ψ is any formula, replace this subformula by ψ.

(c) If $1 \wedge \psi$ occurs as a subformula, where ψ is any formula, replace this subformula by ψ.

(d) If $1 \vee \psi$ occurs as a subformula, where ψ is any formula, replace this subformula by 1.

(e) If $y \vee \psi$ occurs as a subformula, where ψ is a formula and y is a literal, then replace all occurrences of y in ψ by 0 and all occurrence of $\overline{y}$ by 1.

(f) If $y \wedge \psi$ occurs as a subformula, where ψ is a formula and y is a literal, then replace all occurrences of y in ψ by 1 and all occurrence of $\overline{y}$ by 0.

Fig. 1. Simplification procedure

It is easy to see that **Simplify** runs in time polynomial in the size of ϕ and the resulting formula computes the same function as ϕ.

3.2   An Algorithm for Max Formula SAT

Max formula SAT is a subclass of Max CSPs where each constraint is given as a De Morgan formula, i.e., an instance is of the form $\Phi = \{(\phi_1, w_1), \ldots, (\phi_m, w_m)\}$ where ϕ_i is a De Morgan formula and w_i is a weight function. We use the notation as $\mathrm{var}(\Phi) := \cup_{i=1}^{m} \mathrm{var}(\phi_i)$, $\widetilde{L}(\Phi) := \sum_{i:L(\phi_i) \geq 2} L(\phi_i)$ and $\widetilde{\mathrm{freq}}_\Phi(x) := \sum_{i:L(\phi_i) \geq 2} \mathrm{freq}_{\phi_i}(x)$. For any instance Φ, any set of variables $\{x_{i_1}, \ldots, x_{i_k}\}$ and any constants $a_1, \ldots, a_k \in \{0, 1\}$, we define as $\Phi[x_{i_1} = a_1, \ldots, x_{i_k} = a_k] := \{(\phi_1', w_1), \ldots, (\phi_m', w_m)\}$ where $\phi_i' = \phi_i[x_{i_1} = a_1, \ldots, x_{i_k} = a_k]$.

We need the following lemma.

Lemma 1. *Given an instance Φ of Max formula SAT with m constraints, $\mathrm{Opt}(\Phi)$ and $\#\mathrm{Opt}(\Phi)$ can be computed in time $\mathrm{poly}(m)2^{\widetilde{L}(\Phi)}$.*

Proof. We first show the following claim.

Claim. Given an instance Φ of Max formula SAT with m constraints and $\widetilde{L}(\Phi) = 0$, $\mathrm{Opt}(\Phi)$ and $\#\mathrm{Opt}(\Phi)$ can be computed in time $\mathrm{poly}(m)$.

Proof. $\widetilde{L}(\Phi) = 0$ means that each ϕ_i is either '0', '1', 'x_j' or '$\overline{x}_j$' (for some j). For each x_i, define $S_i(0), S_i(1)$ as

$$S_i(0) := \sum_{j:\phi_j=x_i} w_j(0) + \sum_{j:\phi_j=\overline{x}_i} w_j(1),$$

$$S_i(1) := \sum_{j:\phi_j=x_i} w_j(1) + \sum_{j:\phi_j=\overline{x}_i} w_j(0).$$

If $S_i(0) = S_i(1)$, then let $x_i = *$ (don't care), if $S_i(0) > S_i(1)$, then let $x_i = 0$, otherwise let $x_i = 1$. This assignment achieves $\mathrm{Opt}(\Phi)$ where we assign arbitrary values to don't care variables. If the number of don't care variables is d, then $\#\mathrm{Opt}(\Phi) = 2^d$. □

Let $V' = \{x_{i_1}, x_{i_2}....\}$ be the set of Boolean variables that appear in constraints of size at least two. Note that $|V'| \leq \widetilde{L}(\Phi)$. For each assignment $a_1, a_2, \ldots \in \{0, 1\}$, we can compute

$$\mathrm{Opt}(\Phi[x_{i_1} = a_1, x_{i_2} = a_2, \ldots]), \#\mathrm{Opt}(\Phi[x_{i_1} = a_1, x_{i_2} = a_2, \ldots])$$

in time $\mathrm{poly}(m)$ since $\widetilde{L}(\Phi[x_{i_1} = a_1, x_{i_2} = a_2, \ldots]) = 0$. Let

$$K = \max_{a_1,a_2,\ldots\in\{0,1\}} \mathrm{Opt}(\Phi[x_{i_1} = a_1, x_{i_2} = a_2, \ldots]),$$

$$N = \sum \#\mathrm{Opt}(\Phi[x_{i_1} = a_1, x_{i_2} = a_2, \ldots]),$$

where the summation is over

$$\{a_1, a_2, \ldots \in \{0, 1\} \mid \mathrm{Opt}(\Phi[x_{i_1} = a_1, x_{i_2} = a_2, \ldots]) = K\}.$$

We have $K = \mathrm{Opt}(\Phi)$ and $N = \#\mathrm{Opt}(\Phi)$ and can compute them in time $\mathrm{poly}(m)2^{\widetilde{L}(\Phi)}$. □

Now we are ready to present our algorithm for Max formula SAT as shown in Figure 2.

This is a slight modification of Santhanam's greedy restriction algorithm for De Morgan formula SAT [27]. The difference is (1) we use a different definition of size and frequency, and (2) in the case of $\widetilde{L}(\Phi) < \frac{n}{2}$, we need an additional lemma. We have the following theorem.

Theorem 3. *Given an instance Φ of Max formula SAT with n variables and $\widetilde{L}(\Phi) = m = cn$, **EvalFormula** computes $\mathrm{Opt}(\Phi)$ and $\#\mathrm{Opt}(\Phi)$ in time $\mathrm{poly}(m)$ $\times 2^{(1-\frac{1}{32c^2})n}$.*

Since a disjunction of ℓ literals can be regarded as a De Morgan formula of size ℓ, the above theorem immediately implies the following.

Corollary 1. *Given an instance Φ of Max ℓ-SAT with n variables and $m = cn$ constraints, **EvalFormula** computes $\mathrm{Opt}(\Phi)$ and $\#\mathrm{Opt}(\Phi)$ in time $\mathrm{poly}(m)$ $\times 2^{(1-\frac{1}{32\ell^2c^2})n}$.*

EvalFormula($\Phi = \{(\phi_1, w_1), \ldots, (\phi_m, w_m)\}$: **instance**, n: **integer**)
01: **if** $\widetilde{L}(\Phi) = cn < n/2$,
02:     compute $\mathrm{Opt}(\Phi), \#\mathrm{Opt}(\Phi)$ by Lemma 1 and return $(\mathrm{Opt}(\Phi), \#\mathrm{Opt}(\Phi))$.
03: **else**
04:     $x = \arg\max_{x \in \mathrm{var}(\Phi)} \widetilde{\mathrm{freq}}_\Phi(x)$.
05:     $(K_0, N_0) \leftarrow$ **EvalFormula**($\Phi[x = 0], n - 1$).
06:     $(K_1, N_1) \leftarrow$ **EvalFormula**($\Phi[x = 1], n - 1$).
07:     **if** $K_0 = K_1$.
08:         return $(K_0, N_0 + N_1)$
09:     **else**
10:         $i = \arg\max_{j \in \{0,1\}} \{K_j\}$ and return (K_i, N_i).

Fig. 2. Max formula SAT algorithm

Proof (of Theorem 3). The proof of Theorem 3 is almost identical to the proof of Theorem 5.2 in Chen et al. [7], which gives a running time analysis of Santhanam's algorithm for De Morgan formula SAT based on a super-martingale approach.

For an instance Φ of Max formula SAT on n variables, define $\Phi_0 = \Phi$. For $1 \leq i \leq n$, we define Φ_i to be an instance of Max formula SAT obtained from Φ_{i-1} by uniformly at random assigning the most frequent variable of Φ_{i-1}. We need the following lemma.

Lemma 2 (Shrinkage Lemma with Respect to $\widetilde{L}(\cdot)$). *Let Φ be an instance of Max formula SAT on n variables. For any $k \geq 4$, we have*

$$\Pr\left[\widetilde{L}(\Phi_{n-k}) \geq 2 \cdot \widetilde{L}(\Phi) \cdot \left(\frac{k}{n}\right)^{3/2}\right] < 2^{-k}.$$

Let Φ be an instance of Max formula SAT with n variables and $\widetilde{L}(\Phi) = m = cn$ for some constant $c > 0$. Let $p = \left(\frac{1}{4c}\right)^2$ and $k = pn$. We construct the following computation tree.

The root is labeled Φ. At first, we pick up the most frequent variable x with respect to $\widetilde{L}(\Phi)$ and set the variable first to 0 then to 1. By simplification, Φ branches to the two Max formula SAT instances such as $\Phi[x = 0]$ and $\Phi[x = 1]$. The children of the current node are labeled by these instances. Repeating this procedure until we can make a full binary tree of depth exactly $n - k$. Note that this computation tree can be made in time $\mathrm{poly}(m)2^{n-k}$. Using Lemma 2, for all but at most 2^{-k} fraction of the leaves have the size $\widetilde{L}(\Phi_{n-k}) < 2\widetilde{L}(\Phi)\left(\frac{k}{n}\right)^{3/2} = 2cnp^{3/2} = 2cp^{1/2} \cdot pn = \frac{1}{2}pn = \frac{k}{2}$. For instances with $\widetilde{L}(\Phi_{n-k}) < k/2$, we can compute $\mathrm{Opt}(\Phi_{n-k}), \#\mathrm{Opt}(\Phi_{n-k})$ in time $\mathrm{poly}(m)2^{k/2}$ by Lemma 1. For instances with $\widetilde{L}(\Phi_{n-k}) \geq k/2$, we can compute $\mathrm{Opt}(\Phi_{n-k}), \#\mathrm{Opt}(\Phi_{n-k})$ in time $\mathrm{poly}(m)2^k$ by brute force search. The overall running time of **EvalFormula** is bounded by $\mathrm{poly}(m)2^{n-k}\{2^{k/2}(1 - 2^{-k}) + 2^k \cdot 2^{-k}\} = \mathrm{poly}(m)2^{(1-\frac{1}{32c^2})n}$. $\square$

4 A Deterministic Algorithm for Max SAT

In this section, we present our deterministic algorithm for sparse instances of Max SAT based on the combination of the algorithm for Max formula SAT and width reduction.

Before presenting our algorithm for Max SAT, let us recall Schuler's width reduction technique for SAT [28]. Let Φ be an instance of SAT, i.e., $\Phi = \{\phi_1, \phi_2, \ldots, \phi_m\}$ where each ϕ_i is a disjunction of literals. Assume for some i and $\ell' > \ell$, $\phi_i = (l_1 \vee \cdots \vee l_\ell \vee l_{\ell+1} \vee \cdots \vee l_{\ell'})$ and define $\Phi_L := \{\Phi \backslash \{\phi_i\}\} \cup \{(l_1 \vee \cdots \vee l_\ell)\}$ and $\Phi_R := \Phi[l_1 = \cdots l_\ell = 0]$. Then we can observe that Φ is satisfiable if and only if at least one of Φ_L, Φ_R is satisfiable. Note that compared with Φ, the number of clauses with width more than ℓ decreases in Φ_L, and the number of variables decreases in Φ_R. The step of producing two instances Φ_L, Φ_R from Φ is the one step of width reduction and by applying the reduction recursively, we obtain a set of instances where the width of each constraint is at most ℓ. After obtaining a set of ℓ-SAT instances, we can apply an algorithm for ℓ-SAT which runs in time $\text{poly}(m)2^{(1-\mu(\ell))n}$ for some $\mu(\ell) > 0$ and decide the satisfiability of Φ. Schuler makes use of this observation to design an algorithm for SAT.

We see that width reduction is possible for Max SAT instances. Our algorithm for Max SAT is shown in Figure 3.

MaxSAT$(\Phi = \{(\phi_1, w_1), \ldots, (\phi_m, w_m)\}$: **instance**, ℓ, n: **integer**)
01: **if** $\forall \phi_i \in \Phi, |\text{var}(\phi_i)| \leq \ell$,
02:     return **EvalFormula**(Φ, n).
03: **else**
04:     Pick arbitrary $\phi_i = (l_1 \vee \cdots \vee l_{\ell'})$ such that $\ell' > \ell$.
05:     $\Phi_L \leftarrow \{\Phi \backslash \{(\phi_i, w_i)\}\} \cup \{(l_1 \vee \cdots \vee l_\ell, \widetilde{w_i})\}$.
06:     $(K_L, N_L) \leftarrow$ **MaxSAT**(Φ_L, ℓ, n).
07:     $\Phi_R \leftarrow \Phi[l_1 = \cdots = l_\ell = 0]$.
08:     $(K_R, N_R) \leftarrow$ **MaxSAT**$(\Phi_R, \ell, n - \ell)$.
09:     **if** $K_L = K_R$,
10:         return $(K_L, N_L + N_R)$
11:     **else**
12:         $i = \arg\max_{j \in \{L,R\}} \{K_j\}$ and return (K_i, N_i).

Fig. 3. Max SAT algorithm

The correctness of **MaxSAT** is guaranteed by the following claim.

Claim. In **MaxSAT**, if **else** condition of line 3 holds,
then $\text{Opt}(\Phi) = \max\{K_L, K_R\}$.

Proof. First note that $\text{Opt}(\Phi) \geq \max\{K_L, K_R\}$. Let a be an assignment that maximizes the total weight of satisfied constraints, i.e., $a = \arg\max_a \text{Val}(\Phi, a)$. If $l_1 = \cdots = l_\ell = 0$ according to a, then $\text{Opt}(\Phi) = K_R$. Otherwise, a satisfies $(l_1 \vee \cdots \vee l_\ell)$ and $\text{Opt}(\Phi) = K_L$ holds. □

Now we describe our main theorem and its proof.

Theorem 4. *Given an instance Φ of Max SAT with n variables and $m = cn$ clauses, **MaxSAT** with appropriately chosen ℓ computes $\mathrm{Opt}(\Phi)$ and $\#\mathrm{Opt}(\Phi)$ in time $O(\mathrm{poly}(m)2^{(1-\mu(c))n})$, where $\mu(c)$ is $\Omega\left(\frac{1}{c^2 \log^2 c}\right)$ and $c > 4$.*

Proof. The overall structure of the proof is similar to the analysis of width reduction for SAT by Calabro et al. [4]. We think of the execution of **MaxSAT** as a rooted binary tree T, i.e., the root of T is labeled by an input instance Φ and for each node labeled with Ψ, its left (right) child is labeled with Ψ_L (Ψ_R, resp.). If Ψ is an instance of Max ℓ-SAT, i.e., every constraint ψ_i of Ψ satisfies $|\mathrm{var}(\psi_i)| \leq \ell$, then the node labeled with Ψ is a leaf.

Let us consider a path p from the root to a leaf v labeled with Ψ. We denote by L and R the number of left and right children p selects to reach v. It is easy to see that (1) $L \leq m$ since the number of constraints is m, (2) $R \leq n/\ell$ since a right branch eliminates ℓ variables at a time, and (3) Ψ is defined over at most $n - R\ell$ variables. Furthermore, the number of leaves which are reachable by exactly R times of right branches is at most $\binom{m+R}{R}$. Let $T(n, m, \ell)$ denote the running time of **EvalFormula** on instances of Max ℓ-SAT with n variables and $m = cn$ clauses, then we can upper bound the running time of **MaxSAT** as:

$$\mathrm{poly}(m)\left(\sum_{R=0}^{\frac{n}{\ell}} \binom{m+R}{R} T(n - R\ell, m, \ell)\right)$$

$$= \mathrm{poly}(m)\left(\sum_{R=0}^{\frac{n}{2\ell}-1} \binom{m+R}{R} T(n - R\ell, m, \ell) + \sum_{R=\frac{n}{2\ell}}^{\frac{n}{\ell}} \binom{m+R}{R} T(n - R\ell, m, \ell)\right).$$

We first upper bound the second summation above.

$$\mathrm{poly}(m)\left(\sum_{R=\frac{n}{2\ell}}^{\frac{n}{\ell}} \binom{m+R}{R} T(n - R\ell, m, \ell)\right)$$

$$\leq \mathrm{poly}(m)\left(\sum_{R=\frac{n}{2\ell}}^{\frac{n}{\ell}} \binom{m+R}{R} 2^{n-R\ell}\right)$$

$$\leq \mathrm{poly}(m)\binom{m+\frac{n}{2\ell}}{\frac{n}{2\ell}} 2^{n/2} \leq O(\mathrm{poly}(m)2^{(1-\mu(c))n}),$$

where the last two inequalities hold when $\ell = \alpha \log c$ for sufficiently large constant $\alpha > 0$.

We move on the analysis of the following summation.

$$\text{poly}(m) \left(\sum_{R=0}^{\frac{n}{2\ell}-1} \binom{m+R}{R} T(n - R\ell, m, \ell) \right)$$

$$= \text{poly}(m) \left(\sum_{R=0}^{\frac{n}{2\ell}-1} \binom{m+R}{R} 2^{\left(1 - \frac{1}{32\ell^2(\frac{cn}{n-R\ell})^2}\right)(n-R\ell)} \right)$$

$$\leq \text{poly}(m) \left(\sum_{R=0}^{\frac{n}{2\ell}-1} \binom{m+R}{R} 2^{\left(1 - \frac{1}{32\ell^2(2c)^2}\right)(n-R\ell)} \right)$$

$$\leq \text{poly}(m) \left(\sum_{R=0}^{m+\frac{n}{2\ell}} \binom{m+\frac{n}{2\ell}}{R} 2^{\left(1 - \frac{1}{32\ell^2(2c)^2}\right)(n-R\ell)} \right)$$

$$= \text{poly}(m) \left(2^{\left(1 - \frac{1}{32\ell^2(2c)^2}\right)n} \sum_{R=0}^{m+\frac{n}{2\ell}} \binom{m+\frac{n}{2\ell}}{R} 2^{-\left(1 - \frac{1}{32\ell^2(2c)^2}\right)R\ell} \right)$$

$$= \text{poly}(m) \left(2^{\left(1 - \frac{1}{32\ell^2(2c)^2}\right)n} \left(1 + 2^{-\left(1 - \frac{1}{32\ell^2(2c)^2}\right)\ell} \right)^{m+\frac{n}{2\ell}} \right)$$

$$\leq \text{poly}(m) \left(2^{\left(1 - \frac{1}{32\ell^2(2c)^2}\right)n} \left(e^{2^{-\left(1 - \frac{1}{32\ell^2(2c)^2}\right)\ell}} \right)^{m+\frac{n}{2\ell}} \right)$$

$$\leq \text{poly}(m) \left(2^{\left(1 - \frac{1}{32\ell^2(2c)^2}\right)n + \frac{2m}{2^{\left(1 - \frac{1}{32\ell^2(2c)^2}\right)\ell}}} \right).$$

Now we set $\ell = \beta \log c$ for sufficiently large constant $\beta > 0$, then the exponent is

$$\left(1 - \frac{1}{32\ell^2(2c)^2} \right) n + \frac{2m}{2^{\left(1 - \frac{1}{32\ell^2(2c)^2}\right)\ell}}$$

$$= \left(1 - \frac{1}{128\beta^2 c^2 \log^2 c} \right) n + \frac{2cn}{2^{\beta \log c - \frac{1}{128\beta c^2 \log c}}}$$

$$= \left(1 - \frac{1}{128\beta^2 c^2 \log^2 c} + \frac{2^{\frac{1}{128\beta c^2 \log c}+1}}{c^{\beta-1}} \right) n$$

$$\leq \left(1 - \frac{1}{256\beta^2 c^2 \log^2 c} \right) n,$$

where the last inequality is by the choice of β and $c > 4$. This completes the proof.

$\square$

Remark 1. Since we are interested in the running time of **MaxSAT** when c goes to infinity, we do not give a precise upper bound for small c. However, we can show that the running time is of the form $\mathrm{poly}(m)2^{(1-\mu(c))n}$ for some $\mu(c) > 0$ even when c is small.

5 A Deterministic Algorithm for Counting Problems

In this section, we show how to compute $\#\mathrm{Val}_{\geq K}(\Phi)$ in moderately exponential time and exponential space.

Theorem 5. *Given an instance Φ of Max SAT with n variables and $m = cn$ clauses, **MaxSAT2** with appropriately chosen ℓ computes $\#\mathrm{Val}_{\geq K}(\Phi)$ in time $O(\mathrm{poly}(m)2^{(1-\mu(c))n})$ and exponential space, where $\mu(c)$ is $\Omega\left(\frac{1}{c^2 \log^2 c}\right)$.*

We need the following lemma.

Lemma 3. *Given an instance Φ of Max formula SAT with n variables, m constraints and $\widetilde{L}(\Phi) \leq n$, $\#\mathrm{Val}_{\geq K}(\Phi)$ can be computed in time $\mathrm{poly}(m)2^{\widetilde{L}(\Phi)} \cdot 2^{(n-\widetilde{L}(\Phi))/2}$ and exponential space.*

Proof. We first show the following claim.

Claim. Given an instance Φ of Max formula SAT with n variables and $\widetilde{L}(\Phi) = 0$, $\#\mathrm{Val}_{\geq K}(\Phi)$ can be computed in time $\mathrm{poly}(m)2^{n/2}$ and exponential space.

Proof. $\widetilde{L}(\Phi) = 0$ means that each ϕ_i is either '0', '1', 'x_j' or '$\overline{x}_j$' (for some j). Without loss of generality, we can assume that ϕ_i is either 'x_j'or '$\overline{x}_j$' by removing each ϕ_i of the form '0' or '1' from Φ and replacing K by $K - w_i(\phi_i)$.

We use a slight modification of the algorithm for the subset-sum problem due to [16]. For brevity, let us assume that n is even and let $V_1 := \{x_1, \dots, x_{n/2}\}, V_2 := \{x_{n/2+1}, \dots, x_n\}$. Then Φ can be represented as a disjoint union of Φ_1, Φ_2 where Φ_i only depends on V_i. Define $W_i := \{\mathrm{Val}(\Phi_i, a) \mid a \in \{0, 1\}^{n/2}\}$. We do not think of W_i as a multiset and the size of W_i can be less than $2^{n/2}$.

We are to construct arrays $\{(s_j^i, t_j^i)\}_{j \in W_i}$ for $i = 1, 2$ such that (1) s_j^i is the jth largest value in W_i, (2) t_j^1 is the number of assignments $a \in \{0, 1\}^{n/2}$ such that $\mathrm{Val}(\Phi_1, a) = s_j^1$, and (3) t_j^2 is the number of assignments $a \in \{0, 1\}^{n/2}$ such that $\mathrm{Val}(\Phi_2, a) \geq s_j^2$.

First, we assume that we have constructed the arrays and show how to compute $\mathrm{Val}_{\geq K}(\Phi)$. Let s be an integer and set $s = 0$. For each s_j^1, we can find the smallest $s_{j'}^2$ such that $s_j^1 + s_{j'}^2 \geq K$ in time $\mathrm{poly}(n)$ using binary search. If such $s_{j'}^2$ can be found, then update s as $s + t_j^1 \cdot t_{j'}^2$. Repeat this for every $s_j^1 \in W_1$. Then, we have $s = \#\mathrm{Val}_{\geq K}(\Phi)$ in the end.

To construct $\{(s_j^i, t_j^i)\}_{j \in W_i}$, we first enumerate $\mathrm{Val}(\Phi_i, a)$ for every $a \in \{0, 1\}^{n/2}$ and sort them in the decreasing order in time $\mathrm{poly}(n)2^{n/2}$. Then, we can compute $\{(s_j^i, t_j^i)\}_{j \in W_i}$ from $j = 1$ inductively in time $\mathrm{poly}(n)2^{n/2}$. $\qquad\square$

Let $V' = \{x_{i_1}, x_{i_2}, \ldots\}$ be the set of Boolean variables that appear in constraints of size at least two. Note that $|V'| \leq \widetilde{L}(\varPhi)$. For each assignment $a_1, a_2, \ldots \in \{0,1\}$, we can compute

$$\#\mathrm{Val}_{\geq K}(\varPhi[x_{i_1} = a_1, x_{i_2} = a_2, \ldots])$$

in time $\mathrm{poly}(m)2^{n/2}$ since $\widetilde{L}(\varPhi[x_{i_1} = a_1, x_{i_2} = a_2, \ldots]) = 0$. Let

$$N = \sum_{a_1, a_2, \ldots \in \{0,1\}} \#\mathrm{Val}_{\geq K}(\varPhi[x_{i_1} = a_1, x_{i_2} = a_2, \ldots]).$$

We have $N = \#\mathrm{Val}_{\geq K}(\varPhi)$ and can compute it in time $\mathrm{poly}(m)2^{\widetilde{L}(\varPhi)} \cdot 2^{(n-\widetilde{L}(\varPhi))/2}$. $\square$

Now we are ready to present our algorithm for $\#\mathrm{Val}_{\geq K}(\varPhi)$ as shown in Figures 4,5. The running time analysis is almost the same as that for Theorem 4 and we omit it.

EvalFormula2$(\varPhi = \{(\phi_1, w_1), \ldots, (\phi_m, w_m)\}$: **instance**, n: **integer**)
01: **if** $\widetilde{L}(\varPhi) = cn < n/2$,
02:     compute $\#\mathrm{Val}_{\geq K}(\varPhi)$ by Lemma 3 and return $\#\mathrm{Val}_{\geq K}(\varPhi)$.
03: **else**
04:     $x = \arg\max_{x \in \mathrm{var}(\varPhi)} \widetilde{\mathrm{freq}}_\varPhi(x)$.
05:     $N_0 \leftarrow$ **EvalFormula2**$(\varPhi[x = 0], n - 1)$.
06:     $N_1 \leftarrow$ **EvalFormula2**$(\varPhi[x = 1], n - 1)$.
07:     **return** $N_0 + N_1$.

Fig. 4. Max formula SAT algorithm

MaxSAT2$(\varPhi = \{(\phi_1, w_1), \ldots, (\phi_m, w_m)\}$: **instance**, ℓ, n: **integer**)
01: **if** $\forall \phi_i \in \varPhi, |\mathrm{var}(\phi_i)| \leq \ell$,
02:     **return** **EvalFormula2**$(\varPhi, n)$.
03: **else**
04:     Pick arbitrary $\phi_i = (l_1 \vee \cdots \vee l_{\ell'})$ such that $\ell' > \ell$.
05:     $\varPhi_L \leftarrow \{\varPhi \setminus \{(\phi_i, w_i)\}\} \cup \{(l_1 \vee \cdots \vee l_\ell, \widetilde{w_i})\}$.
06:     $N_L \leftarrow$ **MaxSAT2**$(\varPhi_L, \ell, n)$.
07:     $\varPhi_R \leftarrow \varPhi[l_1 = \cdots = l_\ell = 0]$.
08:     $N_R \leftarrow$ **MaxSAT2**$(\varPhi_R, \ell, n - \ell)$.
09:     **return** $N_L + N_R$.

Fig. 5. Max SAT algorithm

6 A Randomized Algorithm for Max ℓ-SAT

In this section, we present our randomized algorithm for Max ℓ-SAT.

The basic idea of our algorithm is as follows: If $\widetilde{L}(\varPhi) < n/2$, then we use Lemma 1. Otherwise, we choose a subset U of the set of variables $V = \{x_1, x_2, \ldots, x_n\}$ uniformly at random with $|U| = (1 - p)n$, where p is chosen appropriately according to $\widetilde{L}(\varPhi)$. If the number of constraints which depend on at least two variables in $V \setminus U$ is 'small,' i.e., $|\{\phi_i \in \varPhi \mid |\mathrm{var}(\phi_i) \setminus U| \geq 2\}| < pn/(2\ell)$ holds, then we say U is *good*. Assume U is good and $U = \{x_{i_1}, x_{i_2}, \ldots, x_{i_{|U|}}\}$. Then, for every $a \in \{0, 1\}^{|U|}$, $\widetilde{L}(\varPhi[x_{i_1} = a_1, x_{i_2} = a_2, \ldots]) < \ell \cdot pn/(2\ell) = pn/2$ and $\mathrm{Opt}(\varPhi[x_{i_1} = a_1, x_{i_2} = a_2, \ldots])$ can be computed in time $\mathrm{poly}(m)2^{pn/2}$. Therefore, we can compute $\mathrm{Opt}(\varPhi)$ in time $\mathrm{poly}(m)2^{pn/2} \cdot 2^{(1-p)n} = \mathrm{poly}(m)2^{(1-p/2)n}$.

Our randomized algorithm for Max ℓ-SAT is shown in Figure 6.

MaxℓSAT($\varPhi = \{(\phi_1, w_1), \ldots, (\phi_m, w_m)\}$: **instance,** n: **integer**)
01: **if** $\widetilde{L}(\varPhi) = cn < n/2$,
02:     compute $\mathrm{Opt}(\varPhi)$ by Lemma 1 and return $\mathrm{Opt}(\varPhi)$.
03: **else**
04:     Pick $U \subseteq V, |U| = (1-p)n = \left(1 - \frac{1}{c\ell^3}\right)n$ uniformly at random.
05:     **if** $|\{\phi_i \in \varPhi \mid |\mathrm{var}(\phi_i) \setminus U| \geq 2\}| \geq pn/(2\ell)$,
06:         return $\bot$.
07:     **else** /* we assume $U = \{x_{i_1}, x_{i_2}, \ldots, x_{i_{|U|}}\}$ */
08:         **for each** $a \in \{0, 1\}^{|U|}$,
09:             $\varPhi_{U,a} \leftarrow \varPhi[x_{i_1} = a, x_{i_2} = a_2, \ldots]).$
10:             compute $\mathrm{Opt}(\varPhi_{U,a})$ by Lemma 1.
11:         return $\max_{a \in \{0,1\}^{|U|}} \mathrm{Opt}(\varPhi_{U,a})$.

Fig. 6. Max ℓ-SAT algorithm

Note that if **MaxℓSAT** does not return $\bot$, then it returns $\mathrm{Opt}(\varPhi)$. We show that U is good with constant probability. Similar calculation can be found in, e.g., [5].

Lemma 4. *In* **MaxℓSAT**, *if* **else** *condition of line 3 holds, then* **MaxℓSAT** *returns $\bot$ with probability at most $1/2$.*

Proof. Let X_i be a random variable such that $X_i = 1$ if $|\mathrm{var}(\phi_i) \setminus U| \geq 2$, otherwise $X_i = 0$.
Since $\mathbf{Pr}_U[X_i = 1] \leq \binom{\ell}{2}p^2$ by the union bound, we have

$$\mathbf{E}_U\left[\sum_{i \in [m]} X_i\right] \leq \binom{\ell}{2}p^2 m = \binom{\ell}{2}p^2 cn.$$

By Markov's inequality,

$$\Pr_{U}\left[\sum_{i\in[m]} X_i \geq \frac{pn}{2\ell}\right] \leq \frac{\binom{\ell}{2}p^2cn}{\frac{pn}{2\ell}} = c\ell^2(\ell-1)p.$$

Setting p yields the consequence of the lemma.   □

By combining the preceding argument and the above lemma, we have the following theorem.

Theorem 6. *Given an instance Φ of Max ℓ-SAT with n variables and $m=cn$ constraints, with the probability at least $\frac{1}{2}$,* **MaxℓSAT** *computes* $\mathrm{Opt}(\Phi)$ *in time* $\mathrm{poly}(m)2^{\left(1-\frac{1}{2c\ell^3}\right)n}$.

MaxℓSAT can be easily modified to compute $\#\mathrm{Opt}(\Phi)$ and $\#\mathrm{Val}_{\geq K}(\Phi)$ as our deterministic algorithms. If we use **MaxℓSAT** instead of **EvalFormula** in **MaxSAT** and **MaxSAT2**, we have the following theorems.

Theorem 7. *Given an instance Φ of Max SAT with n variables and $m = cn$ clauses,* $\mathrm{Opt}(\Phi)$ *and* $\#\mathrm{Opt}(\Phi)$ *can be computed in expected time* $O(\mathrm{poly}(m)$ $\times 2^{(1-\mu(c))n})$, *where $\mu(c)$ is* $\Omega\left(\frac{1}{c\log^3 c}\right)$.

Theorem 8. *Given an instance Φ of Max SAT with n variables and $m = cn$ clauses,* $\#\mathrm{Val}_{\geq K}(\Phi)$ *can be computed in expected time* $O(\mathrm{poly}(m)2^{(1-\mu(c))n})$ *and exponential space, where $\mu(c)$ is* $\Omega\left(\frac{1}{c\log^3 c}\right)$.

7   Concluding Remarks

In this paper, we present moderately exponential time polynomial space algorithms for sparse instances of Max SAT. There are several possible directions for future work. First, since our algorithm is partly inspired by the recent development in the study of exact algorithms for the circuit satisfiability problem [7, 17, 18, 27, 31, 33], we may hope that it is possible to make use of such algorithms and their analysis to improve existing algorithms or design new algorithms for Max CSPs or other problems related to SAT. Second, our algorithm can count the number of assignments that achieve total weights greater than a given objective value K in exponential space, however, it seems not obvious to obtain the same result by polynomial space algorithms. Finally, to improve $\mu(c)$ to subpolynomial in $1/c$, say $\mu(c) = \Omega(1/\mathrm{poly}(\log c))$ is a challenging goal since it implies non-trivial algorithms for arbitrary size of Max 3-SAT instances.

References

1. Bansal, N., Raman, V.: Upper bounds for MaxSAT: Further improved. In: Aggarwal, A.K., Pandu Rangan, C. (eds.) ISAAC 1999. LNCS, vol. 1741, pp. 247–258. Springer, Heidelberg (1999)
2. Binkele-Raible, D., Fernau, H.: A new upper bound for Max-2-SAT: A graph-theoretic approach. J. Discrete Algorithms 8(4), 388–401 (2010)

3. Bliznets, I., Golovnev, A.: A new algorithm for parameterized MAX-SAT. In: Thilikos, D.M., Woeginger, G.J. (eds.) IPEC 2012. LNCS, vol. 7535, pp. 37–48. Springer, Heidelberg (2012)

4. Calabro, C., Impagliazzo, R., Paturi, R.: A duality between clause width and clause density for SAT. In: Proceedings of the 21st Annual IEEE Conference on Computational Complexity (CCC), pp. 252–260 (2006)

5. Calabro, C., Impagliazzo, R., Paturi, R.: The complexity of satisfiability of small depth circuits. In: Chen, J., Fomin, F.V. (eds.) IWPEC 2009. LNCS, vol. 5917, pp. 75–85. Springer, Heidelberg (2009)

6. Chen, J., Kanj, I.A.: Improved exact algorithms for MAX-SAT. Discrete Applied Mathematics 142(1-3), 17–27 (2004)

7. Chen, R., Kabanets, V., Kolokolova, A., Shaltiel, R., Zuckerman, D.: Mining circuit lower bound proofs for meta-algorithms. Electronic Colloquium on Computational Complexity (ECCC) TR13-057 (2013)

8. Dantsin, E., Wolpert, A.: MAX-SAT for formulas with constant clause density can be solved faster than in $O(2^n)$ time. In: Biere, A., Gomes, C.P. (eds.) SAT 2006. LNCS, vol. 4121, pp. 266–276. Springer, Heidelberg (2006)

9. Gaspers, S., Sorkin, G.B.: A universally fastest algorithm for Max 2-SAT, Max 2-CSP, and everything in between. J. Comput. Syst. Sci. 78(1), 305–335 (2012)

10. Gramm, J., Hirsch, E.A., Niedermeier, R., Rossmanith, P.: Worst-case upper bounds for MAX-2-SAT with an application to MAX-CUT. Discrete Applied Mathematics 130(2), 139–155 (2003)

11. Gramm, J., Niedermeier, R.: Faster exact solutions for MAX2SAT. In: Bongiovanni, G., Petreschi, R., Gambosi, G. (eds.) CIAC 2000. LNCS, vol. 1767, pp. 174–186. Springer, Heidelberg (2000)

12. Gutin, G., Yeo, A.: Constraint satisfaction problems parameterized above or below tight bounds: A survey. In: Bodlaender, H.L., Downey, R., Fomin, F.V., Marx, D. (eds.) Fellows Festschrift 2012. LNCS, vol. 7370, pp. 257–286. Springer, Heidelberg (2012)

13. Håstad, J.: The shrinkage exponent of De Morgan formulas is 2. SIAM J. Comput. 27(1), 48–64 (1998)

14. Hirsch, E.A.: A new algorithm for MAX-2-SAT. In: Reichel, H., Tison, S. (eds.) STACS 2000. LNCS, vol. 1770, pp. 65–73. Springer, Heidelberg (2000)

15. Hirsch, E.A.: Worst-case study of local search for MAX-k-SAT. Discrete Applied Mathematics 130(2), 173–184 (2003)

16. Horowitz, E., Sahni, S.: Computing partitions with applications to the knapsack problem. J. ACM 21(2), 277–292 (1974)

17. Impagliazzo, R., Matthews, W., Paturi, R.: A satisfiability algorithm for AC^0. In: Proceedings of the 23rd Annual ACM-SIAM Symposium on Discrete Algorithms (SODA), pp. 961–972 (2012)

18. Impagliazzo, R., Paturi, R., Schneider, S.: A satisfiability algorithm for sparse depth two threshold circuits. In: Proceedings of the 54th Annual IEEE Symposium on Foundations of Computer Science (FOCS), pp. 479–488 (2013)

19. Kneis, J., Mölle, D., Richter, S., Rossmanith, P.: On the parameterized complexity of exact satisfiability problems. In: Jedrzejowicz, J., Szepietowski, A. (eds.) MFCS 2005. LNCS, vol. 3618, pp. 568–579. Springer, Heidelberg (2005)

20. Kneis, J., Mölle, D., Richter, S., Rossmanith, P.: A bound on the pathwidth of sparse graphs with applications to exact algorithms. SIAM J. Discrete Math. 23(1), 407–427 (2009)

21. Koivisto, M.: Optimal 2-constraint satisfaction via sum-product algorithms. Inf. Process. Lett. 98(1), 24–28 (2006)

22. Kojevnikov, A., Kulikov, A.S.: A new approach to proving upper bounds for MAX-2-SAT. In: Proceedings of the 17th Annual ACM-SIAM Symposium on Discrete Algorithms (SODA), pp. 11–17 (2006)
23. Kulikov, A.S.: Automated generation of simplification rules for SAT and MAXSAT. In: Bacchus, F., Walsh, T. (eds.) SAT 2005. LNCS, vol. 3569, pp. 430–436. Springer, Heidelberg (2005)
24. Kulikov, A.S., Kutzkov, K.: New bounds for MAX-SAT by clause learning. In: Diekert, V., Volkov, M.V., Voronkov, A. (eds.) CSR 2007. LNCS, vol. 4649, pp. 194–204. Springer, Heidelberg (2007)
25. Mahajan, M., Raman, V.: Parameterizing above guaranteed values: MaxSAT and MaxCUT. J. Algorithms 31(2), 335–354 (1999)
26. Niedermeier, R., Rossmanith, P.: New upper bounds for maximum satisfiability. J. Algorithms 36(1), 63–88 (2000)
27. Santhanam, R.: Fighting perebor: New and improved algorithms for formula and QBF satisfiability. In: Proceedings of the 51th Annual IEEE Symposium on Foundations of Computer Science (FOCS), pp. 183–192 (2010)
28. Schuler, R.: An algorithm for the satisfiability problem of formulas in conjunctive normal form. J. Algorithms 54(1), 40–44 (2005)
29. Scott, A.D., Sorkin, G.B.: Faster algorithms for MAX CUT and MAX CSP, with polynomial expected time for sparse instances. In: Arora, S., Jansen, K., Rolim, J.D.P., Sahai, A. (eds.) RANDOM 2003 and APPROX 2003. LNCS, vol. 2764, pp. 382–395. Springer, Heidelberg (2003)
30. Scott, A.D., Sorkin, G.B.: Linear-programming design and analysis of fast algorithms for Max 2-CSP. Discrete Optimization 4(3-4), 260–287 (2007)
31. Seto, K., Tamaki, S.: A satisfiability algorithm and average-case hardness for formulas over the full binary basis. Computational Complexity 22(2), 245–274 (2013)
32. Williams, R.: A new algorithm for optimal 2-constraint satisfaction and its implications. Theor. Comput. Sci. 348(2-3), 357–365 (2005)
33. Williams, R.: Non-uniform ACC circuit lower bounds. In: Proceedings of the 26th Annual IEEE Conference on Computational Complexity (CCC), pp. 115–125 (2011)

MUS Extraction Using Clausal Proofs

Anton Belov[1], Marijn J.H. Heule[2], and Joao Marques-Silva[1,3]

[1] Complex and Adaptive Systems Laboratory, University College Dublin
[2] The University of Texas at Austin
[3] IST/INESC-ID, Technical University of Lisbon, Portugal

Abstract. Recent work introduced an effective method for extraction of reduced unsatisfiable cores of CNF formulas as a by-product of validation of clausal proofs emitted by conflict-driven clause learning SAT solvers. In this paper, we demonstrate that this method for *trimming* CNF formulas can also benefit state-of-the-art tools for the computation of a Minimal Unsatisfiable Subformula (MUS). Furthermore, we propose a number of techniques that improve the quality of trimming, and demonstrate a significant positive impact on the performance of MUS extractors from the improved trimming.

1 Introduction

Recent years has seen a significant progress in efficient extraction of a Minimal Unsatisfiable Subformula (MUS) from a CNF formula [1,2,3,4,5]. However, most of the formulas that can be tackled relatively easily by SAT solvers are still too hard for today's MUS extraction tools. In the context of MUS extraction, the term *trimming* refers to a preprocessing step, whereby the input formula is replaced with a smaller unsatisfiable core. Trimming is typically performed by a repeated invocation of a SAT solver starting from the input formula (e.g. [2,6,7]), but the technique seldom pays off in practice. This is evidenced by the fact that none the state-of-the-art MUS extractors MUSer2 [8], TarmoMUS [9], and HaifaMUC [3], employ an explicit trimming step.

In this paper, we propose techniques to trim CNF formulas using clausal proofs [10]. A clausal proof is a sequence of clauses that includes the empty clause and that are entailed by the input formula. Clausal proofs can easily be emitted by SAT solvers and afterwards be used to extract an unsatisfiable core. Trimming based on clausal proofs can substantially reduce the size of CNF formulas, and a recent tool DRUPtrim [11] made this approach very efficient. Hence, as suggested in [11], it might be effective for the computation of MUSes.

We, first, confirm empirically the effectiveness of trimming using clausal proofs for MUS extraction. In addition, we present three techniques to strengthen MUS extraction tools via clausal proofs. These techniques are designed to both reduce the size of the trimmed formula further, and to compute a useful resolution graph — a crucial component of resolution-based MUS extractors. The first technique converts a clausal proof into a resolution graph. This conversion requires significantly less memory compared to computing the graph within a SAT solver.

C. Sinz and U. Egly (Eds.): SAT 2014, LNCS 8561, pp. 48–57, 2014.

The second technique, called *layered trimming*, was developed to reduce the size of cores and the resolution graphs. This is achieved by solving a formula multiple times while increasing the bound of variable elimination [12,13]. Our third technique adds interaction between the trimming procedure and an MUS extractor. If the extractor gets stuck, it provides the current over-approximation of an MUS to a trimmer to obtain a new resolution graph. The process is repeated until a MUS is found. Experimental results with the MUS extractors MUSer2 and HaifaMUC show that the proposed techniques can boost the performance of the extractors, particularly on hard benchmarks.

2   Preliminaries

We assume familiarity with propositional logic, its clausal fragment, and commonly used terminology of the area of SAT (cf. [14]). We focus on formulas in CNF (*formulas*, from hence on), which we treat as (finite) (multi-)sets of clauses. Given a formula F we denote the set of variables that occur in F by $Var(F)$. An *assignment* τ for F is a map $\tau : Var(F) \rightarrow \{0,1\}$. Assignments are extended to formulas according to the semantics of classical propositional logic. If $\tau(F) = 1$, then τ is a *model* of F. If a formula F has (resp. does not have) a model, then F is *satisfiable* (resp. *unsatisfiable*). A clause C is *redundant* in F if $F \setminus \{C\} \equiv F$. Given two clauses $C_1 = (x \vee A)$ and $C_2 = (\bar{x} \vee B)$, the *resolution rule* infers the clause $C = (A \vee B)$, called the *resolvent* of C_1 and C_2 on x. We write $C = C_1 \otimes_x C_2$, and refer to C_1 and C_2 as the *antecedents* of C.

Resolution Graph. A *resolution graph* is a directed acyclic graph in which the leaf nodes (no incoming edges) represent the input clauses of a given formula. The remaining nodes in the resolution graph are resolvents. The incoming edges represent the antecedents of a resolvent. In case a node has more than two antecedents, it can be constructed using a sequence of resolution steps. The *size* of a resolution graph is the number of its edges.

Clausal Proofs. For a CNF formula F, *unit propagation* simplifies F based on unit clauses; that is, it repeats the following until fixpoint: if there is a unit clause $(l) \in F$, remove all clauses containing l and remove $\bar{l}$ from all clauses. We refer as a *lemma* to a clause that is logically implied by a given formula. A clausal proof [7] is a sequence of lemmas. To validate whether a lemma L in a clausal proof is indeed logically implied, one can check if unit propagation of $\bar{L}$, the assignment that falsifies all literals in L, results in a conflict. This check is also known as *reverse unit propagation* (RUP) [15]. Clausal proofs are significantly smaller than resolution proofs, and only minor modifications of a SAT solver are required to emit clausal proofs [11]. Clausal proofs can be converted into resolution graphs by marking clauses involved in a conflict as the antecedents of a lemma. Fig. 1 shows a CNF formula, a clausal proof and a resolution graph that can be obtained from it.

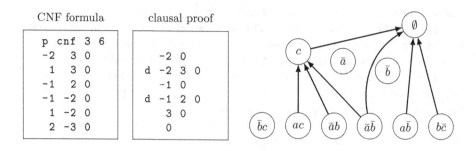

Fig. 1. An example CNF formula in DIMACS format and a clausal proof in DRUP format next to it. The files use three variables a, b, and c that are represented by 1, 2, and 3, respectively. Each line in the proof refers to a clause addition step (no prefix) or a clause deletion step (d prefix). The -2 3 0 represents the clause $(\bar{b} \vee c)$. On the right, a resolution graph that can be computed from the clausal proof. For all clauses in the graph only the literals are shown: $\bar{b}c$ means $(\bar{b} \vee c)$.

Minimal Unsatisfiability. A CNF formula F is *minimal unsatisfiable* if (i) F is unsatisfiable, and (ii) for any clause $C \in F$, the formula $F \setminus \{C\}$ is satisfiable. A CNF formula F' is a *minimal unsatisfiable subformula* (MUS) of a formula F if $F' \subseteq F$ and F' minimal unsatisfiable. While this paper focuses on MUS extraction on plain CNF formulas, the discussion can be extended to the computation of group-MUSes [16,17] without difficultly.

Practical MUS extraction algorithms are based on detection of *necessary* clauses. A clause C is *necessary* for F if F is unsatisfiable and $F \setminus \{C\}$ is satisfiable [18]. A basic *deletion*-based algorithm for MUS [6] extraction operates in the following manner. Starting from an unsatisfiable formula F, we repeat the following until fixpoint. Pick an unexamined clause $C \in F$ and solve $F \setminus \{C\}$. If the result is unsatisfiable, C is permanently removed from F, otherwise it is included in the computed MUS. The basic deletion algorithm with *clause-set refinement* and *model rotation* [19] is among the top performing algorithms for industrially relevant instances [20,2,3]. Additionally, reuse of lemmas and heuristic values between invocations of the SAT solver is crucial for performance [20]. Currently, the two prevalent approaches to achieving lemma reuse in the context of MUS extraction are the *assumption-based* and the *resolution-based* [17].

The assumption-based approach to MUS extraction relies on the incremental SAT solving paradigm [21]. Each clause C_i of the input CNF formula F is augmented with a fresh *assumption literal* a_i, and the modified formula is loaded once into an incremental SAT solver. To test a clause C_i for necessity the SAT solver is invoked under assumptions a_i and $\bar{a}_j$, for $j \neq i$. If the outcome is SAT, C_i is necessary for the given formula, and, as an optimization, can be *finalized* by adding a unit clause $(\bar{a}_i)$ to the SAT solver. If the outcome is UNSAT, C_i is "removed" by adding the a unit clause (a_i) to the SAT solver.

The resolution-based approach to MUS extraction relies on a modified SAT solver that constructs a resolution graph during search *explicitly*. The initial graph that represents a refutation of the input formula F is constructed when the formula is shown to be UNSAT. To test a clause $C \in F$ for necessity a resolution-based MUS extractor temporarily disables C and all of its descendant lemmas in the resolution graph, and invokes the SAT solver's search procedure on the remaining clauses. If the outcome is SAT, the lemmas are put back. In the case of the UNSAT outcome, the solver constructs a new resolution graph. A particularly important, in our context, feature of resolution-based MUS extractors is that the initial resolution graph can be provided as an input to the extractor, thus eliminating the cost of the first UNSAT call.

3   Trimming Strategies

Proofs of unsatisfiability can be used to *trim* a CNF formula F, i.e., remove the clauses from F that were not required to validate the proof. In this section we propose three new strategies to combine a trimming utility for clausal proofs with a MUS extraction tool. Spending a significant amount of time on trimming can improve the overall performance, particularly on hard formulas.

Trimming via Clausal Proofs. Resolution graphs are a crucial component of resolution-based MUS extraction tools. However, computing them while solving a given formula can be very costly, especially in terms of memory. Alternatively, one can compute a trimmed formula and a resolution graph by validating a clausal proof [11]. This approach requires significantly less memory.

We refer to a *trimmer* as a tool that, given a formula F and a clausal proof P, computes a trimmed formula F_{trim} and a resolution graph G_{res}. An example of such a tool is DRUPtrim [11]. Emitting clausal proofs is now supported by many state-of-the-art CDCL solvers. Consequently, we can compute trimmed formulas and resolution graphs efficiently using these solvers with a trimmer. The trimmed formulas are useful for assumption-based MUS extractors, while the resolution graphs are useful for resolution-based MUS extractors.

Fig. 2 shows the pseudo-code of the TrimExtract procedure that repeatedly trims a given formula F using a clausal proof that is returned by a SAT solver (the Solve procedure). The Trim procedure returns a trimmed formula and a resolution graph, which can be given to an extractor tool. The main heuristic in this loop deals with when to stop trimming and start extracting. Throughout our experiments we observed that it is best to switch to extraction if a trimmed formula is only a few percent smaller than the formula from the prior iteration.

TrimExtract (formula F)

TE1   **forever do**

TE2 $\langle F_{\text{trim}}, G_{\text{res}} \rangle := \text{Trim } (F, \text{ Solve } (F))$ // trim using a clausal proof

TE3   **if** $|F_{\text{trim}}| \approx |F|$ **then return** Extract $(F_{\text{core}}, G_{\text{res}})$ // switch to extractor

TE4 $F := F_{\text{trim}}$ // F_{core} becomes F for the next iteration

Fig. 2. Pseudo-code of TrimExtract that combines a trimming and MUS extractor tool

Trim (formula F, clausal proof)
  return ⟨trimmed F, resolution graph⟩

Solve (formula F)
  return clausal proof of solving F

Simplify (formula F)
  return ⟨simplified F, clausal proof⟩

LayeredTrim (formula F, iterations k)
1   $P := \emptyset$ // start with empty proof
2   $W := F$ // make a working copy
3   **for** $i \in \{1, ..., k\}$ **do**
4   ⟨$W_{\text{simp}}, P_{\text{simp}}$⟩ := Simplify($W$)
5   ⟨W, G⟩ := Trim(W_{simp}, Solve(W_{simp}))
6     $P := P \cup P_{\text{simp}}$
7   **return** Trim($F, P \cup$ Solve(W))

Fig. 3. The LayeredTrim procedure and the required subprocedures

Layered Trimming. Most lemmas in resolution proofs have hundreds of antecedents [11]. This has a number of disadvantages. First, storing the graph requires a lot of memory. Second, in a resolution-based MUS extractor, high connectivity causes large parts of the proof being disabled during each SAT call. Third, high connectivity causes additional clauses to be brought into the core.

Our next strategy, called *layered trimming*, aims at reducing the size of the resolutions graphs by adding lemmas with only two antecedents using variable elimination (VE) [12,13]. VE replaces some clauses by redundant and irredundant clauses. Trimming the VE preprocessed formula will remove some redundant clauses, allowing more applications of VE. In general, the smaller the formula, the faster the solver and hence the shorter the proof. After each iteration of layered trimming, the number of vertices in the proof is similar. Yet more and more vertices are produced by VE (two antecedents), while fewer vertices are produced by CDCL solving (many antecedents). This reduces the connectivity.

The pseudo-code of LayeredTrim is shown in Fig. 3. We initialize a working formula W with the input formula F. In each iteration, W first gets simplified using the variable elimination procedure, resulting in the formula W_{simp}. The simplification steps are emitted as a, possibly partial, clausal proof P_{simp}, and we start to build a proof P which accumulates all of the simplification steps. Next we solve and trim W_{simp}, and take this trimmed version as our new working formula W. At each iteration proof P is extended by the new simplification steps. After k iterations, we merge the layered simplification proof P with the final proof returned by Solve. Finally, Trim will use this merged proof and the original formula to compute a core and resolution graph.

Iterative Trimming. Our third strategy, called *iterative trimming*, adds interaction between the trimming and the resolution-based MUS extraction tools. Two observations inspired iterative trimming. First, despite the substantial trimming, and despite the availability of the resolution graph at the beginning of MUS extraction, the extractor can get stuck while solving hard instances. This indicates that the current resolution graph might no longer be useful. Thus, we terminate the extractor and pass the current over-approximation of an MUS from the extractor to the trimming tool in order to obtain a new core and a new resolution graph. The MUS extractor is then invoked again on the new graph, and this iterative process continues until an MUS is computed. Second, while layered trimming can be very effective on hard formulas, it is quite significantly more expensive than (plain) trimming, and so for instances that are already easy for MUS extraction after plain trimming, the effort does not pay off. If a good overall performance of an MUS extractor is more important than scalability, we can postpone layered trimming until the extractor indicates that the formula is hard.

4 Empirical Study

To evaluate the impact of our trimming strategies, we implemented a Python-based framework, called DMUSer, on top of the following tools: Glucose-3.0 [22] solver to emit clausal proofs in DRUP format and for simplification; DRUPtrim [11] to trim formulas and emit resolution graphs in TraceCheck format [23]; MUSer2 [8], an assumption-based MUS extractor with Glucose-3.0; and HaifaMUC [3], a resolution-based MUS extractor, modified to support TraceCheck input[1].

The benchmark set consists of 295 instances from the MUS Competition 2011 and 60 instances from SAT Competition 2009[2] which Glucose-3.0 could refute within 1 minute. We removed 31 instances from the MUS track — most of them were extremely easy — as some of the tools used by DMUSer produced errors.

All experiments were performed on 2 x Intel E5-2620 (2GHz) cluster nodes, with 1800 seconds CPU time and 4 GB memory limits per experiment. The reported CPU runtimes for trimming-based configurations include the runtime for both the trimming and the MUS extraction stages.

Table 1. The number of solved, timed- and memmed- out instances (out of 324), and the descriptive statistics of the CPU runtime (sec) of various configurations. The average is taken over the solved instances only, while the median is over all instances.

	MUSer2	HaifaMUC	Tr-M2	Tr-HM	LTr-M2	LTr-HM	ITr-HM-A	ITr-HM-B
Num. solved	250	258	257	266	273	277	**280**	276
Num.TO/MO	26/48	40/26	39/28	51/7	32/19	47/0	44/0	48/0
Med. CPU time	45.08	30.65	40.64	23.70	54.07	33.87	35.77	**23.16**
Avg. CPU time	97.58	110.09	102.95	102.03	162.62	108.52	117.07	112.12

The results of our study are presented in plots in Table 1 and Fig. 4 and 5. We use the following abbreviations. MUSer2 and HaifaMUC represent these MUS extractors running directly on the input instance, while the other configurations represent trimming followed (or interleaved) with MUS extraction: Tr- configurations use the original trimming of [11], LTr- configurations use the layered

---

[1] The source code of DMUSer and the benchmark set used for the evaluation are available from https://bitbucket.org/anton_belov/dmuser

[2] The benchmarks are available via http://satcompetition.org/

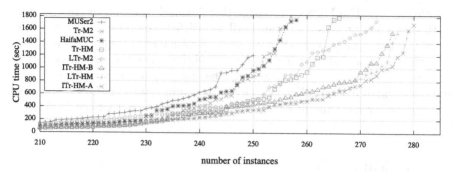

Fig. 4. Comparison of various trimming and MUS extraction techniques

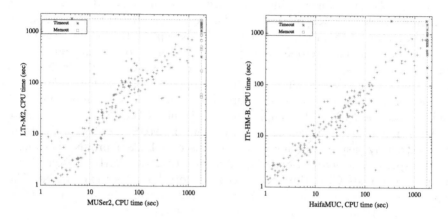

Fig. 5. Comparative performance of various trimming-based configurations against MUS extractors on non-trimmed formulas in terms of CPU runtime. Timeout of 1800 seconds is represented by the dashed (green) lines.

trimming, and `ITr-` use the iterative trimming; the `-M2` configurations perform MUS extraction on the trimmed CNF formula using `MUSer2`, while the `-HM` configurations compute MUSes using the TraceCheck version of `HaifaMUC` on the resolution graph of the trimmed formula.

Impact of Trimming. DRUP-based trimming for MUS extraction was suggested already in [11], but the impact has not been previously evaluated. Our results demonstrate that trimming is indeed an effective preprocessing technique for computing MUSes. The median reduction in the size of the input formula due to trimming is over 6x, with the average (resp. median) size of the trimmed formula being 1.5x (resp. 1.08x) of the size of the computed MUS. Comparing `MUSer2` vs. `Tr-M2` and `HaifaMUC` vs. `Tr-HM` on the cactus plot in Fig. 4 and in Table 1 we observe a notable performance improvement both with respect to `MUSer2` and to `HaifaMUC` (7 and 8 extra instances, respectively). Importantly, the MUS extractors run out of memory on fewer instances when executed on the trimmed formulas. Also, notice the decrease in median runtimes, indicating that trimming has an overall positive impact even on the relatively easy instances.

Impact of Layered Trimming. Our experimental data confirms the intuition that motivates the layered trimming technique: the lower connectivity in the resolution graph results in improved trimming and a lower memory consumption. We did not use a fixed k for the iterations, but repeated the loop until the solving time no longer decreased. On average (resp. median) the size of the cores produced by the layered trimming is 1.42x (resp. 1.06x) smaller than the size of the cores produced by (plain) trimming, and constitutes a mere 1.04x (resp. 1.01x) of the size of the computed MUS. Layered trimming resulted in solving 16 (resp. 11) extra instances using MUSer2 (resp. HaifaMUC) which is also clear from the cactus plot in Fig. 4. Notably, as seen in Table 1, LTr-HM had not ran out of memory on any instance – thanks to the smaller resolution graphs. The proposed technique, however, is not without drawbacks: observe the increase in the median runtime due to layered trimming, shown in Table 1. The scatter plot on the left of Fig. 5, that compares MUSer2 vs. LTr-M2, gives some clues: layered trimming is a too heavy-weight technique for many of the easy instances. This might be undesirable in some applications.

Impact of Iterative Trimming. We experimented with two configurations of the iterative trimming algorithm. The configuration ITr-HM-A starts HaifaMUC with the resolution graph obtained from running the layered trimming approach. The algorithm aborts MUS extraction and returns to (plain) trimming when any SAT call takes too much time — 10 seconds, increasing linearly with every iteration. ITr-HM-A solves extra 3 instances, and performs notably better on the difficult instances, but the performance slightly decreases on easier instances.

The second configuration of iterative trimming, ITr-HM-B, is designed to alleviate the weaker performance of the proposed algorithms on the easier instances. This configuration starts the with resolution graph of plain trimming and aborts after some time (100 seconds in ITr-HM-B) and switches to ITr-HM-A starting with the resolution graph obtained by layered trimming. Although ITr-HM-B solves one less instance than LTr-HM, Fig. 5 (right) demonstrates that on most easier instances the configuration outperforms HaifaMUC, while still maintaining the significantly improved performance on the difficult instances.

5   Conclusions

We presented three trimming strategies to improve the performance of MUS extractors. Clausal proof based trimming is particularly useful when dealing with hard instances, but it can be costly on easy instances. Our layered trimming strategy reduces the memory consumption of resolution-based MUS extractors and can be useful in other applications that prefer low connectivity in resolution graphs. By applying the iterative trimming strategy, the performance of our MUS extraction tool is improved on both the easy and the hard instances.

Acknowledgements. The first and third authors are supported by SFI PI grant BEACON (09/IN.1/I2618), and by FCT grants ATTEST (CMU-PT/ELE/0009/2009) and POLARIS (PTDC/EIA-CCO/123051/2010). The second author is supported by DARPA contract number N66001-10-2-4087 and by the NSF under grant CCF-1153558. We thank Vadim Ryvchin for adding TraceCheck support to `HaifaMUC`.

References

1. Marques-Silva, J., Janota, M., Belov, A.: Minimal sets over monotone predicates in boolean formulae. In: Sharygina, N., Veith, H. (eds.) CAV 2013. LNCS, vol. 8044, pp. 592–607. Springer, Heidelberg (2013)
2. Belov, A., Lynce, I., Marques-Silva, J.: Towards efficient MUS extraction. AI Communications 25(2), 97–116 (2012)
3. Nadel, A., Ryvchin, V., Strichman, O.: Efficient MUS extraction with resolution. In: [24], pp.197–200
4. Audemard, G., Lagniez, J.M., Simon, L.: Improving glucose for incremental SAT solving with assumptions: Application to MUS extraction, In: [25], pp. 309–317
5. Lagniez, J.M., Biere, A.: Factoring out assumptions to speed up MUS extraction. In: [25], pp. 276–292
6. Dershowitz, N., Hanna, Z., Nadel, A.: A scalable algorithm for minimal unsatisfiable core extraction. In: Biere, A., Gomes, C.P. (eds.) SAT 2006. LNCS, vol. 4121, pp. 36–41. Springer, Heidelberg (2006)
7. Zhang, L., Malik, S.: Validating SAT solvers using an independent resolution-based checker: Practical implementations and other applications. In: DATE, pp. 10880–10885. IEEE Computer Society (2003)
8. Belov, A., Marques-Silva, J.: MUSer2: An efficient MUS extractor. Journal of Satisfiability 8, 123–128 (2012)
9. Wieringa, S., Heljanko, K.: Asynchronous multi-core incremental SAT solving. In: Piterman, N., Smolka, S.A. (eds.) TACAS 2013. LNCS, vol. 7795, pp. 139–153. Springer, Heidelberg (2013)
10. Goldberg, E.I., Novikov, Y.: Verification of proofs of unsatisfiability for CNF formulas. In: DATE, pp. 10886–10891 (2003)
11. Heule, M.J.H., Hunt Jr., W.A., Wetzler, N.: Trimming while checking clausal proofs, In: [24], pp. 181–188
12. Davis, M., Putnam, H.: A computing procedure for quantification theory. J. ACM 7(3), 201–215 (1960)
13. Eén, N., Biere, A.: Effective preprocessing in SAT through variable and clause elimination. In: Bacchus, F., Walsh, T. (eds.) SAT 2005. LNCS, vol. 3569, pp. 61–75. Springer, Heidelberg (2005)
14. Biere, A., Heule, M.J.H., van Maaren, H., Walsh, T. (eds.): Handbook of Satisfiability. Frontiers in Artificial Intelligence and Applications, vol. 185. IOS Press (2009)
15. Van Gelder, A.: Verifying RUP proofs of propositional unsatisfiability. In: ISAIM (2008)
16. Liffiton, M.H., Sakallah, K.A.: Algorithms for computing minimal unsatisfiable subsets of constraints. J. Autom. Reasoning 40(1), 1–33 (2008)
17. Nadel, A.: Boosting minimal unsatisfiable core extraction. In: FMCAD, pp. 121–128 (October 2010)

18. Kullmann, O., Lynce, I., Marques-Silva, J.: Categorisation of clauses in conjunctive normal forms: Minimally unsatisfiable sub-clause-sets and the lean kernel. In: Biere, A., Gomes, C.P. (eds.) SAT 2006. LNCS, vol. 4121, pp. 22–35. Springer, Heidelberg (2006)
19. Marques-Silva, J., Lynce, I.: On improving MUS extraction algorithms. In: Sakallah, K.A., Simon, L. (eds.) SAT 2011. LNCS, vol. 6695, pp. 159–173. Springer, Heidelberg (2011)
20. Marques-Silva, J.: Minimal unsatisfiability: Models, algorithms and applications. In: ISMVL, pp. 9–14 (2010)
21. Eén, N., Sörensson, N.: Temporal induction by incremental SAT solving. Electr. Notes Theor. Comput. Sci. 89(4), 543–560 (2003)
22. Audemard, G., Simon, L.: Predicting learnt clauses quality in modern SAT solvers. In: Boutilier, C. (ed.) IJCAI, pp. 399–404 (2009)
23. Jussila, T., Sinz, C., Biere, A.: Extended resolution proofs for symbolic SAT solving with quantification. In: Biere, A., Gomes, C.P. (eds.) SAT 2006. LNCS, vol. 4121, pp. 54–60. Springer, Heidelberg (2006)
24. Formal Methods in Computer-Aided Design, FMCAD 2013, Portland, OR, USA, October 20-23. IEEE (2013)
25. Järvisalo, M., Van Gelder, A. (eds.): SAT 2013. LNCS, vol. 7962. Springer, Heidelberg (2013)

On Computing Preferred MUSes and MCSes*

Joao Marques-Silva[1,2] and Alessandro Previti[1]

[1] CASL, University College Dublin, Ireland
[2] IST/INESC-ID, Technical University of Lisbon, Portugal

Abstract. Minimal Unsatisfiable Subsets (MUSes) and Minimal Correction Subsets (MCSes) are essential tools for the analysis of unsatisfiable formulas. MUSes and MCSes find a growing number of applications, that include abstraction refinement in software verification, type debugging, software package management and software configuration, among many others. In some applications, there can exist preferences over which clauses to include in computed MUSes or MCSes, but also in computed Maximal Satisfiable Subsets (MSSes). Moreover, different definitions of preferred MUSes, MCSes and MSSes can be considered. This paper revisits existing definitions of preferred MUSes, MCSes and MSSes of unsatisfiable formulas, and develops a preliminary characterization of the computational complexity of computing preferred MUSes, MCSes and MSSes. Moreover, the paper investigates which of the existing algorithms and pruning techniques can be applied for computing preferred MUSes, MCSes and MSSes. Finally, the paper shows that the computation of preferred sets can have significant impact in practical performance.

1 Introduction

Reasoning about over-constrained systems finds a wide range of practical application, and has been the subject of extensive research over the last three decades (e.g. [35,19,29,22,38]). For the concrete case of propositional formulas, Minimal Unsatisfiable Subsets (MUSes), Minimal Correction Subsets (MCSes) and Maximal Satisfiable Subsets (MSSes) are essential tools for the analysis of unsatisfiable propositional formulas in Conjunctive Normal Form (CNF). MUSes, MCSes and MSSes find a growing number of applications, that include abstraction refinement in software verification (e.g. [23]), software (re-)configuration (e.g. [20]), software package management (e.g. [18]), type debugging (e.g. [3]), abstract argumentation frameworks (e.g. [39]), fault localization in C code (e.g. [21]), design debugging (e.g. [10]), post-silicon debugging (e.g. [40]), among many others. MSSes/MCSes are also tightly related with Maximum Satisfiability (MaxSAT) (e.g. [1,30]). In recent years, there has been extensive work on

---

* This work is partially supported by SFI grant BEACON (09/IN.1/I2618), by FCT grant POLARIS (PTDC/EIA-CCO/123051/2010), and INESC-IDs multiannual PIDDAC funding PEst-OE/EEI/LA0021/2013.

C. Sinz and U. Egly (Eds.): SAT 2014, LNCS 8561, pp. 58–74, 2014.

developing efficient algorithms for computing MUSes [16,28,5,27,24,31] and MC-Ses/MSSes [3,33,14,32,26]. Despite all the new algorithms and techniques proposed, the issue of computing preferred sets, namely MUSes, MCSes and MSSes, has seldom been investigated. The main exception is [22], that studies preferred explanations (MUSes in our context) and relaxations (MSSes in our context) for the case of generalized constraints. Nevertheless, the importance of preferences in a number of fields is well-known (e.g. [8,36] and references therein). Moreover, preferences in SAT have also been the subject of comprehensive research in recent years, which is summarized for example in [13]. To our best knowledge, the computation of preferred MUSes, MCSes or MSSes, their complexity, their algorithms and associated techniques, have not been studied before.

This paper represents a first step towards a systematic analysis of computing preferred MUSes, MCSes and MSSes. The first topic addressed by the paper are the possible definitions of what a preferred set (either MUS, MCS or MSS) represents. This paper focuses solely on the definitions of preference considered in [22]. Given these definitions, the paper then studies the computational complexity of computing preferred sets (either MUSes, MCSes or MSSes), focusing on hardness results. As the paper shows, computing MUSes or MCSes for the natural (lexicographic) definition of preference is hard for the second level of the polynomial hierarchy[1]. This in part explains the alternative definition of preference (i.e. anti-lexicographic) used in [22] for MUSes. For this alternative definition of preferred set, computing a preferred MUS (or MCS) is in FP^{NP}, and so can be computed with a polynomial number of calls to an NP oracle. Similarly, as the paper shows, computing a (lexicographically) preferred MSS is in FP^{NP}, and so it is *easier*[2] than computing a lexicographic preferred MUS (or MCS). Moreover, the paper investigates whether well-known algorithms for computing MUSes/MCSes/MSSes can be extended for computing preferred sets, and concludes that some algorithms cannot be extended for computing some preferred sets. In addition, the paper investigates whether well-known and widely used techniques for reducing the number of NP oracle calls can be extended for computing preferred sets and, again, concludes that several techniques cannot be used for computing preferred sets. Finally, the paper assesses the performance impact of computing preferred sets in state of the art tools for computing MUSes/MCSes/MSSes. Although the paper addresses propositional formulas in CNF, the results can be generalized to more expressive domains, including CSP and SMT.

The paper is organized as follows. Section 2 introduces the notation and definitions used in the rest of the paper. Section 3 develops preliminary complexity results regarding the computation of preferred MUSes/MCSes/MSSes. Section 4 surveys well-known algorithms and investigates which can be extended

---

[1] The complexity classes considered in this paper are briefly reviewed in Section 2. Similarly, the paper develops complexity characterizations assuming an NP oracle which, for the purposes of this paper, can be seen as equivalent to a SAT solver call.

[2] Under the assumption the polynomial hierarchy does not collapse. This is a generally accepted conjecture that is also (implicitly) assumed throughout the paper.

for computing preferred sets. Similarly, Section 5 surveys some of the best known techniques for reducing the number of NP oracle calls, and investigates which can be extended for computing preferred sets. Section 6 assesses the practical impact of some of the paper's results. The paper concludes in Section 7.

2   Preliminaries

This section introduces the notation and definitions used throughout the paper.

2.1   CNF Formulas, MUSes, MCSes and MSSes

Standard propositional logic definitions apply (e.g. [6]). CNF formulas are defined over a set of propositional variables. A CNF formula $\mathcal{F}$ is a conjunction of clauses, also interpreted as a set of clauses. A clause is a disjunction of literals, also interpreted as a set of literals. A literal is a variable or its complement. $m \triangleq |\mathcal{F}|$ represents the number of clauses. The set of variables associated with a CNF formula $\mathcal{F}$ is denoted by $X \triangleq \mathsf{var}(\mathcal{F})$. The clauses of $\mathcal{F}$ are *soft*, meaning that these clauses can be relaxed (i.e. not satisfied). In contrast, an optional set of *hard* clauses $\mathcal{H}$ may be specified. (For simplicity it is assumed that $\mathcal{F} \cap \mathcal{H} = \emptyset$. Moreover, $\mathcal{H}$ is implicit in some definitions.) MUSes, MCSes and MSSes are defined over $\mathcal{F}$ taking into account the hard clauses in $\mathcal{H}$ as follows:

Definition 1 (Minimal Unsatisfiable Subset (MUS)). *Let $\mathcal{F}$ denote the set of soft clauses and $\mathcal{H}$ denote the set of hard clauses, such that $\mathcal{H} \cup \mathcal{F}$ is unsatisfiable. $\mathcal{M} \subseteq \mathcal{F}$ is a Minimal Unsatisfiable Subset (MUS) iff $\mathcal{H} \cup \mathcal{M}$ is unsatisfiable and $\forall_{\mathcal{M}' \subsetneq \mathcal{M}}, \mathcal{H} \cup \mathcal{M}'$ is satisfiable.*

Definition 2 (Minimal Correction Subset (MCS)). *Let $\mathcal{F}$ denote the set of soft clauses and $\mathcal{H}$ denote the set of hard clauses, such that $\mathcal{H} \cup \mathcal{F}$ is unsatisfiable. $\mathcal{C} \subseteq \mathcal{F}$ is a Minimal Correction Subset (MCS) iff $\mathcal{H} \cup \mathcal{F} \setminus \mathcal{C}$ is satisfiable and $\forall_{\mathcal{C}' \subsetneq \mathcal{C}}, \mathcal{H} \cup \mathcal{F} \setminus \mathcal{C}'$ is unsatisfiable.*

Definition 3 (Maximal Satisfiable Subset (MSS)). *Let $\mathcal{F}$ denote the set of soft clauses and $\mathcal{H}$ denote the set of hard clauses, such that $\mathcal{H} \cup \mathcal{F}$ is unsatisfiable. $\mathcal{S} \subseteq \mathcal{F}$ is a Maximal Satisfiable Subset (MSS) iff $\mathcal{H} \cup \mathcal{S}$ is satisfiable and $\forall_{\mathcal{S}' \supsetneq \mathcal{S}}, \mathcal{H} \cup \mathcal{S}'$ is unsatisfiable.*

2.2   Set Membership Problems

To establish some of the complexity results in Section 3, the following set membership/non-membership decision problems are defined.

Definition 4 (MUS Membership). *Given $c \in \mathcal{F}$, the MUS membership problem is to decide whether there exists an MUS $\mathcal{M} \subseteq \mathcal{F}$ such that $c \in \mathcal{M}$.*

Definition 5 (MCS Membership). *Given $c \in \mathcal{F}$, the MCS membership problem is to decide whether there exists an MCS $\mathcal{C} \subseteq \mathcal{F}$ such that $c \in \mathcal{C}$.*

Definition 6 (MSS Non-membership). *Given $c \in \mathcal{F}$, the MSS non-membership problem is to decide whether there exists an MSS $\mathcal{S} \subseteq \mathcal{F}$ such that $c \notin \mathcal{S}$.*

Additional membership/non-membership problems could be defined, but are unnecessary for the purposes of the paper. Clearly, the above definitions can be extended to a set of target clauses $\mathcal{T} \subseteq \mathcal{F}$, such that membership (resp. non-membership) consists is deciding whether all clauses in $\mathcal{T}$ are (resp. are not) in some set.

2.3   Preferred MUSes, MCSes and MSSes

This section revisits the definitions from [22] adapted to CNF formulas.

Definition 7 (Precedence Operator, $\prec$). *Let $c_1, c_2 \in \mathcal{F}$. $c_1 \prec c_2$ denotes that clause c_1 is preferred to clause c_2.*

A strict total order $<$ is assumed between the clauses of $\mathcal{F}$. This can be viewed as a linearization of the partial order induced by the $\prec$. (See [22] for a detailed discussion of the linearization $<$). As a result, the clauses of $\mathcal{F}$ are listed as $\langle c_1, c_2, \ldots, c_m \rangle$, respecting the strict total order $<$. Throughout the paper, sequences are used to represent preferences among clauses in a set, e.g. $\langle c_1, c_2, \ldots, c_m \rangle$.

Definition 8 (Lexicographic preference, L-preference). *Given a strict total order $<$ on $\mathcal{F}$, a set $\mathcal{A} \subseteq \mathcal{F}$ is* lexicographically *preferred to another set $\mathcal{B} \subseteq \mathcal{F}$, denoted $\mathcal{A} <_{\text{lex}} \mathcal{B}$, iff $\exists_{1 \leq k \leq m}, c_k \in \mathcal{A} \setminus \mathcal{B}$ and $\mathcal{A} \cap \{c_1, \ldots, c_{k-1}\} = \mathcal{B} \cap \{c_1, \ldots, c_{k-1}\}$.*

Definition 9 (Anti-lexicographic preference, A-preference). *Given a strict total order $<$ on $\mathcal{F}$, a set $\mathcal{A} \subseteq \mathcal{F}$ is* anti-lexicographically *preferred to another set $\mathcal{B} \subseteq \mathcal{F}$, denoted $\mathcal{A} <_{\text{antilex}} \mathcal{B}$, iff $\exists_{1 \leq k \leq m}, c_k \in \mathcal{B} \setminus \mathcal{A}$ and $\mathcal{A} \cap \{c_{k+1}, \ldots, c_m\} = \mathcal{B} \cap \{c_{k+1}, \ldots, c_m\}$.*

Definition 10 (L-preferred/A-preferred MUSes/MCSes/MSSes). *An MUS/MCS/MSS $\mathcal{A}$ is L-preferred (resp. A-preferred) if for all MUS/MCS/MSS $\mathcal{B} \neq \mathcal{A}$, $\mathcal{A} <_{\text{lex}} \mathcal{B}$ (resp. $\mathcal{A} <_{\text{antilex}} \mathcal{B}$).*

For simplicity, throughout the paper we use the expression *preferred set* to denote L-preferred or A-preferred MUSes, MCSes or MSSes. Moreover, given a strict total order for a given set of clauses, there exists a unique preferred set.

Example 1. Let $\mathcal{H} = \emptyset$ and let $\mathcal{F}$ be given by the following (soft) clauses:

c_1	c_2	c_3	c_4	c_5	c_6	c_7
$(\bar{x}_1 \vee \bar{x}_2)$	(x_1)	$(x_5 \vee x_6)$	$(\bar{x}_1 \vee \bar{x}_3 \vee \bar{x}_4)$	(x_3)	(x_4)	(x_2)

where the strict total order is: $\langle c_1, c_2, \ldots, c_7 \rangle$. The MSSes, MCSes and MUSes of this CNF formula are shown in Table 1, which also shows which are L-preferred or A-preferred. Observe that an L-preferred set preferably includes the most preferred clauses, and so captures the more intuitive notion of preference. In contrast, an A-preferred set preferably discards the least preferred clauses.

Table 1. MSSes, MCSes & MUSes for Example 1

Set	Type	L or A -preferred?
$\{c_1, c_3, c_4, c_5, c_6, c_7\}$	MSS	✗
$\{c_1, c_2, c_3, c_4, c_5\}$	MSS	✓(L & A)
$\{c_1, c_2, c_3, c_4, c_6\}$	MSS	✗
$\{c_1, c_2, c_3, c_5, c_6\}$	MSS	✗
$\{c_2, c_3, c_4, c_5, c_7\}$	MSS	✗
$\{c_2, c_3, c_4, c_6, c_7\}$	MSS	✗
$\{c_2, c_3, c_5, c_6, c_7\}$	MSS	✗
$\{c_2\}$	MCS	✓(A)
$\{c_6, c_7\}$	MCS	✗
$\{c_5, c_7\}$	MCS	✗
$\{c_4, c_7\}$	MCS	✗
$\{c_1, c_6\}$	MCS	✗
$\{c_1, c_5\}$	MCS	✗
$\{c_1, c_4\}$	MCS	✓(L)
$\{c_1, c_2, c_7\}$	MUS	✓(L)
$\{c_2, c_4, c_5, c_6\}$	MUS	✓(A)

Example 2. To further highlight the differences between L-preferred and A-preferred sets, consider a set $\mathcal{F} = \{c_1, c_2, c_3, c_4\}$ with the strict total order $\langle c_1, c_2, c_3, c_4 \rangle$, and let the minimal sets of interest be $\mathcal{S}_1 = \{c_1, c_4\}$ and $\mathcal{S}_2 = \{c_2, c_3\}$. Thus, $\mathcal{S}_1$ is L-preferred, whereas $\mathcal{S}_2$ is A-preferred.

2.4 Computational Complexity

Standard computational complexity definitions are used throughout the paper [15,34]. The notation is adapted from [34] and more recent papers (e.g. [25]). Besides the well-known complexity classes P, NP, coNP and D^P, the following additional complexity classes of decision problems are used in the paper:

- P^{NP} = Δ_2^p: class of decision problems solvable in deterministic polynomial time by executing a polynomial number of calls to an NP oracle.
- NP^{NP} = Σ_2^p: class of decision problems solvable in non-deterministic polynomial time by executing a polynomial number of calls to an NP oracle.

Moreover, the following classes of function problems are used in the paper (see [34] for a definition of function problem):

- FP: class of function problems solvable in deterministic polynomial time.
- FP^{NP}: class of function problems solvable in deterministic polynomial time by executing a polynomial number of calls to an NP oracle.
- FP$^{NP}[\mathcal{O}(\log n)]$: class of function problems solvable in deterministic polynomial time by executing a logarithmic number of calls to an NP oracle.

- $FP^{\Sigma_2^p}$: class of function problems solvable in deterministic polynomial time by executing a polynomial number of calls to a Σ_2^p oracle.

The classes listed above can be extended to other levels of the polynomial hierarchies of decision and function problems (e.g. [34]).

2.5 Related Work

Preferences have been extensively studied in the recent past (e.g. [8,36,13]). The computation of preferred sets of explanations and relaxations in the area of constraints is studied in [22]. To our best knowledge, there is no work studying the computational complexity of computing preferred MUSes, MCSes and MSSes, nor investigating the restrictions/limitations of existing algorithms and techniques for computing preferred sets. There exists a large body of work on extracting MUSes, MCSes and MSSes. Recent work includes [28,14,5,32,27,26,24,31]. Well-known past work includes [35,12,11,4,22,3,33]. The use of MaxSAT for computing MCSes/MSSes is well-known. Recent work is surveyed in [1,30]. Complexity results on computing MSSes are available from [9].

3 Complexity Results

This section develops computational complexity results for computing L-preferred and A-preferred MUSes, MCSes and MSSes. This characterization is relevant, because it allows us to understand which preferred sets we can expect to be able to compute with at most a polynomial number of calls to an NP oracle. We start in Section 3.1 by developing some results which are used in later sections.

3.1 Membership Tests

We first characterize the computational complexity of testing whether a clause is in some MUS, in some MCS, not in some MSS, and in some MSS.

Proposition 1 (MUS Membership). *The MUS membership problem is Σ_2^p-complete [25].*

Proof. The proof in [25, Theorem 4, page 216], regarding deciding clause membership in an irredundant subformula, can be used, since the proof of hardness is actually for an unsatisfiable formula.   □

Proposition 2 (MCS Membership). *The MCS membership problem is Σ_2^p-complete.*

Proof. (Sketch) MCS membership is in Σ_2^p. Let c be the target clause. Guess $\mathcal{C} \subseteq \mathcal{F}$, with $c \in \mathcal{C}$, such that $\mathcal{F} \setminus \mathcal{C}$ is satisfiable, and $\mathcal{C}$ is irreducible. To prove Σ_2^p-hardness, we reduce MUS membership to MCS membership. By minimal hitting set duality between MUSes and MCSes (e.g. [7,3], but also [35]), a clause c is in some MUS iff c is in some MCS. Thus, MUS membership can be decided with MCS membership.   □

Proposition 3 (MSS Non-membership). *The MSS non-membership problem is Σ_2^p-complete.*

Proof. Similarly to MCS membership, MSS non-membership is clearly in Σ_2^p. To prove Σ_2^p-hardness, we reduce MSS non-membership to MCS membership. Let $c \in \mathcal{F}$ be a target clause. c is in some MCS of $\mathcal{F}$ iff c is not in some MSS of $\mathcal{F}$. But deciding whether c is not in some MSS of $\mathcal{F}$ is the MSS non-membership problem.   □

Proposition 4. *The MUS/MCS membership problems and the MSS non-membership problem extended to a target set $\mathcal{T} \subsetneq \mathcal{F}$ are Σ_2^p-complete.*

Proof. (Sketch]) MUS/MCS membership and MSS non-membership of a single clause (see Proposition 1, Proposition 2 and Proposition 3) are restrictions of the general case when $|\mathcal{T}| = 1$. The proof of membership generalizes the argument in the proof of Proposition 2.   □

In contrast, MSS membership is relatively easier to decide, as shown below.

Proposition 5 (MSS Membership). *MSS membership is in P if $\mathcal{H} = \emptyset$ and $\mathcal{F}$ has no empty clauses. If $\mathcal{H} \neq \emptyset$, and $\mathcal{H}$ is an arbitrary set of clauses, then MSS membership is NP-complete.*

Proof. Let $\mathcal{H} = \emptyset$. The proof is by contradiction. A clause $c \in \mathcal{F}$ is in some MSS of $\mathcal{F}$ iff it is not in every MCS of $\mathcal{F}$. Assume c was included in every MCS of $\mathcal{F}$. Then, by minimal hitting set duality [7], there would be an MUS with a single non-empty clause c; a contradiction. We now consider the case when $\mathcal{H} \neq \emptyset$. MSS membership is clearly in NP. Let c be the target clause. Guess a truth assignment. If clause c and the hard clauses are satisfied, then c is in some MSS. Completeness is straightforward, by reducing SAT to MSS membership. Given a CNF formula $\mathcal{G}$, just pick any clause $c_j \in \mathcal{G}$, make c_j soft and the other clauses of $\mathcal{G}$ hard, i.e. $\mathcal{F} = \{c_j\}$, $\mathcal{H} = \mathcal{G} \setminus \{c_j\}$, and check the membership of c_j in some MSS of $\mathcal{F}$ given $\mathcal{H}$. Note that $\mathcal{H}$ may be unsatisfiable. This also indicates there is no MSS for the constructed formula, and so c_j is not in some MSS of $\mathcal{F}$. □

3.2   L-Preferred and A-Preferred MUSes, MCSes and MSSes

Algorithms for computing A-preferred MUSes can be adapted from [22]. Moreover, the relaxation algorithm outlined in [22] shows that computing an L-preferred MSS is in FP^{NP}. This result is revisited later in this section.

Proposition 6. *Computing an L-preferred MUS (resp. MCS) is hard for Σ_2^p.*

Proof. We reduce the MUS (resp. MCS) membership problem to the problem of computing an L-preferred MUS (resp. MCS). Let $c \in \mathcal{F}$ be the target clause for MUS (resp. MCS) membership. Make c preferred over all other clauses in $\mathcal{F}$. c is in an MUS (resp. MCS) iff the L-preferred MUS (resp. MCS) contains c. □

Proposition 7. *Computing an L-preferred MUS (resp. MCS) is in* $FP^{\Sigma_2^p}$.

Proof. We describe a polynomial time algorithm that uses an oracle for Σ_2^p to construct a preferred MUS (resp. MCS). Let S denote the set to be constructed, initially set to $\emptyset$. Analyze all clauses of $\mathcal{F}$, by decreasing preference given a total order on the clauses of $\mathcal{F}$. For each clause c_i check whether $S \cup \{c_i\}$ is in some MUS (resp. MCS). By Proposition 4, this check can be done with a Σ_2^p oracle. If the outcome is true, then add c_i to S. The algorithm executes a linear number of iterations given by the number of clauses of $\mathcal{F}$, each time invoking a Σ_2^p oracle. Thus, computing an L-preferred MUS (resp. MCS) is in $FP^{\Sigma_2^p}$. $\square$

Proposition 8. *Computing an A-preferred MSS is hard for* Σ_2^p.

Proof. We reduce MSS non-membership to computing an A-preferred MSS. Let $c \in \mathcal{F}$ be the target clause for MSS non-membership. Make c the least preferred clause over all clauses in $\mathcal{F}$. c is not in an MSS iff the A-preferred MSS does not contains c. $\square$

Proposition 9. *Computing an A-preferred MSS is in* $FP^{\Sigma_2^p}$.

Proof. The proof mimics that of Proposition 7. $\square$

In contrast, computing an L-preferred MSS is (believed to be) easier.

Proposition 10. *Computing an L-preferred MSS is in* FP^{NP}.

Proof. One possible algorithm considers the total order of clauses, in decreasing order of preference, for possible inclusion in a set S. If the formula composed of S and target clause c is satisfiable, then c is added to S. The proof that the computed MSS is L-preferred is by induction on the total order of clauses. For the basis case, let c_0 be the most preferred clause. If c_0 can be satisfied, then it is included in S. Clearly, there can be no MSSes without c_0 that are preferred over an MSS that includes c_0. The inductive step is similar, by considering the clauses already in S. $\square$

It should be pointed out that the algorithms outlined in [22] can also be viewed as the sketch of a proof of FP^{NP} membership for the function problem of computing an L-preferred MSS. Moreover, the algorithm described in the proof above corresponds to the plain Grow procedure, as described in [3], and without the optimizations proposed in [32].

Proposition 11. *Computing an A-preferred MUS (resp. MCS) is in* FP^{NP}.

Proof. This is immediate from [22], since the QuickXplain algorithm is shown to compute (A-)preferred explanations (i.e. MUSes in our context). Moreover, the proof of Proposition 10 can easily be adapted for the case of A-preferred MCSes (but also as an alternative for A-preferred MUSes), by considering clauses from least to most preferred when deciding MSS membership. Clauses not included in the MSS are in an A-preferred MCS. $\square$

Table 2. Summary of complexity results

	Membership			Hardness		
	MUS	MCS	MSS	MUS	MCS	MSS
L-preferred	$FP^{\Sigma_2^P}$	$FP^{\Sigma_2^P}$	FP^{NP}	Σ_2^P	Σ_2^P	D^P
A-preferred	FP^{NP}	FP^{NP}	$FP^{\Sigma_2^P}$	D^P	D^P	Σ_2^P

A summary of the results in this section is shown in Table 2. Observe that it is simple, but less significant, to prove that computing an A-preferred MUS/MCS and an L-preferred MSS are all hard for D^P.

Remark 1. It seems unlikely that computing L-preferred MSSes and A-preferred MUSes/MCSes could be in $FP^{NP}[\mathcal{O}(\log n)]$, since each clause must be considered separately for inclusion in a preferred set, given the strict order of clauses imposed by the preference. This issue is left for future research.

The results of this section, as well as the insights provided by Example 1, motivate the following result.

Proposition 12. *Consider a set of soft clauses $\mathcal{F}$, with a strict total order $<$, $\langle c_1, c_2, \ldots, c_m \rangle$, and a inverse strict total order $<^{-1}$, $\langle c_m, c_{m-1}, \ldots, c_1 \rangle$. Then,*
1. *$\mathcal{C}$ is an A-preferred MCS, given $<$, iff $\mathcal{S} = \mathcal{F} \setminus \mathcal{C}$ is an L-preferred MSS, given $<^{-1}$.*
2. *$\mathcal{C}$ is an L-preferred MCS, given $<$, iff $\mathcal{S} = \mathcal{F} \setminus \mathcal{C}$ is an A-preferred MSS, given $<^{-1}$.*

Proof. (Sketch)
1. Consider an MSS algorithm that analyzes each clause according to the strict total order $<^{-1}$, starting from c_m and terminating in c_1, e.g. the Grow procedure from [3,32]. Selected clauses are included in the MSS and discarded clauses are included in the MCS. At step k, $m \geq k \geq 1$, if the algorithm selects the clause c_{m-k+1} to be in the MSS, then it is not included in the MCS. Any MCS that includes c_{m-k+1} (and shares the same clauses for smaller values of k is less preferred than the MCS selected by the algorithm. Moreover, the algorithm adds to the MSS the clauses most preferred given $<^{-1}$, and so the MSS is L-preferred for $<^{-1}$.
2. A similar argument can be used.   □

4   Algorithms for Computing Preferred Sets

This section samples some of the best known algorithms for computing preferred sets, namely MUSes, MCSes and MSSes, and investigates which can be used for computing preferred sets, namely L-preferred MSSes and A-preferred MUSes and MCSes. The analysis of preferred sets algorithms is mostly informal, due to

Table 3. Summary of algorithm analysis

	✗	✓, $\langle c_1, \ldots, c_m \rangle$	✓, $\langle c_m, \ldots, c_1 \rangle$
A-preferred MUS	–	INS, Dicho, QXP	DEL, PRG
A-preferred MCS	MxSAT, CLD	QXP	Grow, PRG
L-preferred MSS	MxSAT, CLD	Grow, PRG	QXP

lack of space and due to not including a formal description of the algorithms. The purpose of this section is thus to outline the key insights that explain why some algorithms cannot be used (at least as originally described) for computing preferred sets whereas others (possibly with additional constraints) can. The following algorithms are considered:

- For MUS extraction: (i) Deletion [4,11] (DEL); (ii) Insertion [12] (INS); (iii) Dichotomic [17] (Dicho); (iv) QuickXplain [22] (QXP); and (v) Progression [27] (PRG).
- For MCS & MSS extraction: (i) Grow procedure [3,32] (Grow); (ii) QuickXplain [22,33,14,27]; (iii) Clause D [26] (CLD); (iv) MaxSAT algorithms (e.g. [30,1]) (MxSAT); and (v) Progression [27] (PRG).

The main results of this section are summarized in Table 3. For each type of preferred set, Table 3 indicates which algorithms cannot be used (✗) for computing preferred sets and which can (✓). Another distinction is the order in which the clauses need to be considered by each algorithm. As shown in Table 3, for all the algorithms considered, the order is either from most to least preferred or from least to most preferred. This is an immediate consequence Proposition 12 and highlights the apparent duality between computing an L-preferred MSS and an A-preferred MCS. The remainder of this section analyzes each of the algorithms listed in Table 3.

It is well-known that MaxSAT algorithms (e.g. [1,30]) can be used for computing MCSes and MSSes. In general MaxSAT algorithms do not analyze clauses according to a strict total order, and so these algorithms cannot be readily applied to computing A-preferred MCSes (or L-preferred MSSes). An alternative would be to analyze the clauses according to a strict total order (e.g. [2]), but this would correspond to an algorithm that analyzes each clause separately, e.g. Grow or Deletion (see below).

The recently proposed Clause D (CLD) algorithm [26] starts from an initial assignment, that satisfies some of the clauses and falsifies the other clauses; clearly, falsified clauses do not have complemented literals and so the algorithm operates on the disjunction of the falsified clauses, attempting to satisfy additional clauses. As shown in Proposition 14 (see list item 3), the removal of satisfied clauses hinders the ability to compute L-preferred or A-preferred sets. As a result, the Clause D algorithm cannot be used for computing an A-preferred MCS (nor an L-preferred MSS).

The Insertion (INS) algorithm for MUS extraction [12] can be organized to compute an A-preferred MUS provided the clauses are analyzed from most to least preferred. The INS algorithm adds clauses to a working set until the working set becomes unsatisfiable. The last clause added to the set is deemed necessary for the MUS and added to a set representing an under-approximation of the MUS. If the clauses are analyzed from most to least preferred, the selected clause is the most preferred among the clauses that could be declared necessary. An MUS that picks a less preferred clause is guaranteed not to be A-preferred. Thus, the INS algorithm computes an A-preferred MUS.

The Dichotomic algorithm for MUS extraction [17] can be organized to compute A-preferred MUS provided the clauses are analyzed from most to least preferred. The rationale is the same as for the Insertion algorithm. The QuickX-plain (QXP) algorithm computes A-preferred explanations (MUSes in our context) [22], where clauses are sorted from most to least preferred. The rationale is as follows. The QXP algorithm splits a given target set in two halves and, given its organization, the algorithm is biased towards selecting clauses from the first set (i.e. the most preferred clauses). Moreover, this idea is applied recursively.

The Deletion (DEL) algorithm for MUS extraction [11,4] can be organized to compute an A-preferred MUS provided the clauses are analyzed from least to most preferred. The rationale is that the algorithm preferably drops clauses, and so it gives preference to drop the least preferred clauses. If a clause is deemed necessary for the MUS, then it is necessary for any other MUS given the set of selected clauses, and any other MUSes would have picked least preferred clauses.

The recently proposed Progression (PRG) algorithm for MUS extraction [27, Algorithm 3] can be organized to compute an A-preferred MUS provided the clauses are analyzed from least to most preferred. The operation of the PRG algorithm can be seen as similar to that of the DEL algorithm, with the exception that the set of clauses analyzed in each iteration grows exponentially, and binary search is used to find a necessary clause given the clauses to its *left* (i.e. more preferred clauses). Thus, the rationale is similar to the Deletion algorithm in that the PRG algorithm gives preference to drop the least preferred clauses.

The Grow procedure from [3] can be used for computing L-preferred MSSes provided the clauses are analyzed from most to least preferred. The computed MSS is L-preferred; no other MSS can include clauses that are preferred. By Proposition 12, the Grow procedure can also be used for computing A-preferred MCSes. The QXP algorithm can be adapted for computing A-preferred MCSes (and so also L-preferred MSSes) [33,14,27], being referred to as FastDiag in [14]. We assume the organization for computing minimal sets from [27, Algorithm 1]. The rationale is the same as for the QXP algorithm applied to computing A-preferred MUSes. The organization of the algorithm is biased towards selecting clauses from the first set, which contains the most preferred clauses, and this bias is applied recursively. By Proposition 12, the QXP algorithm can also be used for computing L-preferred MSSes.

The PRG algorithm can be used for computing minimal sets subject to a monotone predicate [27], and so can be used for computing MCSes. The rationale

Table 4. Computing an L-preferred MSS and an A-preferred MCS with Grow [3,32]

L-preferred MSS				A-preferred MCS			
$\mathcal{R}$	c_i	$\mathcal{C}$	Decision	$\mathcal{R}$	c_i	$\mathcal{C}$	Decision
$\langle c_1,\dots,c_7\rangle$	c_1	$\emptyset$	$\mathcal{C} \leftarrow \mathcal{C} \cup \{c_1\}$	$\langle c_7,\dots,c_1\rangle$	c_7	$\emptyset$	$\mathcal{C} \leftarrow \mathcal{C} \cup \{c_7\}$
$\langle c_2,\dots,c_7\rangle$	c_2	$\{c_1\}$	$\mathcal{C} \leftarrow \mathcal{C} \cup \{c_2\}$	$\langle c_6,\dots,c_1\rangle$	c_6	$\{c_7\}$	$\mathcal{C} \leftarrow \mathcal{C} \cup \{c_6\}$
$\langle c_3,\dots,c_7\rangle$	c_3	$\{c_1,c_2\}$	$\mathcal{C} \leftarrow \mathcal{C} \cup \{c_3\}$	$\langle c_5,\dots,c_1\rangle$	c_5	$\{c_6,c_7\}$	$\mathcal{C} \leftarrow \mathcal{C} \cup \{c_5\}$
$\langle c_4,\dots,c_7\rangle$	c_4	$\{c_1,c_2,c_3\}$	$\mathcal{C} \leftarrow \mathcal{C} \cup \{c_4\}$	$\langle c_4,\dots,c_1\rangle$	c_4	$\{c_5,c_6,c_7\}$	$\mathcal{C} \leftarrow \mathcal{C} \cup \{c_4\}$
$\langle c_5,c_6,c_7\rangle$	c_5	$\{c_1,\dots,c_4\}$	$\mathcal{C} \leftarrow \mathcal{C} \cup \{c_5\}$	$\langle c_3,c_2,c_1\rangle$	c_3	$\{c_4,\dots,c_7\}$	$\mathcal{C} \leftarrow \mathcal{C} \cup \{c_3\}$
$\langle c_6,c_7\rangle$	c_6	$\{c_1,\dots,c_5\}$	$\mathcal{C} \leftarrow \mathcal{C}$	$\langle c_2,c_1\rangle$	c_2	$\{c_3,\dots,c_7\}$	$\mathcal{C} \leftarrow \mathcal{C}$
$\langle c_7\rangle$	c_7	$\{c_1,\dots,c_5\}$	$\mathcal{C} \leftarrow \mathcal{C}$	$\langle c_1\rangle$	c_1	$\{c_3,\dots,c_7\}$	$\mathcal{C} \leftarrow \mathcal{C} \cup \{c_1\}$
Final		MSS: $\{c_1,\dots,c_5\}$, MCS: $\{c_6,c_7\}$		Final		MSS: $\{c_1,c_3,\dots,c_7\}$, MCS: $\{c_2\}$	

for why the PRG algorithm computes an A-preferred MCS is the same as for computing an A-preferred MUS. By Proposition 12, the PRG algorithm can also be used for computing L-preferred MSSes.

Example 3. The operation of the Grow procedure [3,32] for computing an L-preferred MSS and an A-preferred MCS is illustrated in Table 4, using the set of clauses and the precedence relation from Example 1. $\mathcal{R}$ represents the set of clauses that the algorithm has yet to analyze. $\mathcal{C}$ represents the set of clauses the algorithm has decided to include in a preferred set. When the Grow procedure terminates, $\mathcal{C}$ represents an MSS of $\mathcal{F}$. Observe that, for the A-preferred MCS, the algorithm exploits Proposition 12.

5   Pruning Techniques for Computing Preferred Sets

This section analyzes techniques commonly used for reducing (or pruning) the number of SAT oracle calls when computing MUSes, MCSes or MSSes. Such pruning techniques are widely regarded as essential in some settings, including MUS and (more recently) MCS extraction. Concrete examples for MUS extraction include clause set refinement [4,5], and model rotation [28,5]. Concrete examples for MCS/MSS extraction include filtering of satisfied clauses [32,26], exploiting backbone literals [26], and disjoint unsatisfiable core identification [26].

We start by establishing a general result about deciding whether to remove or include a set of clauses while computing L-preferred MSSes and A-preferred MUSes/MCSes.

Definition 11 (Deletion-safe & Addition-safe Sets). *Let $\mathcal{F}$ be the set of target soft clauses. Let $\mathcal{S} \subseteq \mathcal{F}$ be the preferred set. Let $\mathcal{D} \subseteq \mathcal{F}$ be a set of clauses. If $\mathcal{D} \cap \mathcal{S} = \emptyset$, then $\mathcal{D}$ is deletion-safe. If $\mathcal{D} \subseteq \mathcal{S}$, then $\mathcal{D}$ is addition-safe.*

Let $\mathcal{F}$ be the set of target soft clauses, and let $\mathcal{S} \subseteq \mathcal{F}$ be a preferred set. Consider the execution of a preferred set algorithm, where $\mathcal{C} \subseteq \mathcal{S}$ represents the clauses from $\mathcal{S}$ already identified by the algorithm, and $\mathcal{R} \subseteq \mathcal{F}$ represents the clauses not yet decided to be included or not in a preferred set. Now, if $\mathcal{D} \subseteq \mathcal{R}$ is deletion-safe, then $\mathcal{R}$ can be set to $\mathcal{R} \setminus \mathcal{D}$. Clearly, by definition of deletion-safe set, $\mathcal{D} \cap \mathcal{S} = \emptyset$, the clauses in $\mathcal{D}$ cannot be added to $\mathcal{S}$ and so can be dropped from $\mathcal{R}$. Similarly, if $\mathcal{D} \subseteq \mathcal{R}$ is addition-safe, then $\mathcal{R}$ can be refined to $\mathcal{R} \setminus \mathcal{D}$, and $\mathcal{C}$ be extended to $\mathcal{C} \cup \mathcal{D}$. By definition of addition-safe set, $\mathcal{D} \subseteq \mathcal{S}$ and so the clauses in $\mathcal{D}$ can be added to $\mathcal{C}$ and dropped from $\mathcal{R}$. Thus the following holds.

Proposition 13. *Given $\mathcal{F}$, $\mathcal{S}$, $\mathcal{C}$ and $\mathcal{R}$ as above, and $\mathcal{D} \subseteq \mathcal{R}$, then*
- *If $\mathcal{D}$ is deletion-safe, then $\mathcal{S} \subseteq \mathcal{C} \cup (\mathcal{R} \setminus \mathcal{D})$ and $\mathcal{R}$ can be updated to $\mathcal{R} \setminus \mathcal{D}$.*
- *If $\mathcal{D}$ is addition-safe, then $\mathcal{S} \subseteq (\mathcal{C} \cup \mathcal{D}) \cup (\mathcal{R} \setminus \mathcal{D})$, $\mathcal{R}$ can be updated to $\mathcal{R} \setminus \mathcal{D}$, and $\mathcal{C}$ updated to $\mathcal{C} \cup \mathcal{D}$.*

Depending on the pruning technique considered, the computed sets of clauses may or may not be deletion-safe or addition-safe. Some of the most well-known pruning techniques used for computing either MUSes, MCSes or MSSes are analyzed in the following proposition.

Proposition 14. *The following holds:*
1. *Clauses not included in an unsatisfiable core (i.e. the technique of clause set refinement [4,28]) are not guaranteed to be deletion-safe for A-preferred MUSes.*
2. *Clauses identified by model rotation [28,5] are addition-safe for A-preferred MUSes.*
3. *Clauses satisfied by some initial truth assignment (e.g. [32]) are not guaranteed to be addition-safe for L-preferred MSSes and for A-preferred MCSes.*
4. *Clauses satisfied (resp. falsified) by backbone literals [26] are addition-safe (resp. deletion-safe) for L-preferred MSSes and deletion-safe(resp. addition-safe) for A-preferred MCSes.*
5. *The identification of disjoint unsatisfiable cores [26] (and so the satisfied clauses that eventually result) are not guaranteed to be deletion-safe for A-preferred MCSes (and for L-preferred MSSes).*

Proof. (Sketch)
1. Given a computed unsatisfiable core, the clauses not in the core are not guaranteed to respect the strict total order of clauses, and so these clauses are not deletion-safe for A-preferred MUSes.
2. Model rotation identifies clauses that *must* be included in any MUS given that some other clauses are already included in an MUS. Since these clauses *must* be included in any MUS, they are addition-safe for A-preferred MUSes.
3. The clauses satisfied by an initial assignment need not respect the strict partial order of clauses, and so these clauses are not guaranteed to be addition-safe (for L-preferred MSSes) nor deletion-safe (for A-preferred MCSes).
4. Backbone literals represent literals that take a fixed value for some clauses in an MCS to be falsified. Any clause that is satisfied (resp. falsified) by these backbone literals cannot be included in any MCS (resp. MSS) and so must be included in some MSS (resp. MCS).
5. Unsatisfiable cores can be iteratively removed until a remaining set of clauses is satisfied. As above, these satisfied clauses need not respect the strict partial

order of clauses, and so these clauses are not guaranteed to be addition-safe (for L-preferred MSSes) nor deletion-safe (for A-preferred MCSes).   □

Example 4. Consider the clauses in Example 1, and the truth assignment $\{x_1 = 1, x_2 = 1, x_3 = 1, x_4 = 1, x_5 = 1\}$, which satisfies the clauses $\{c_2, c_3, c_5, c_6, c_7\}$. Clearly, removing these clauses from the set of clauses to consider for the MCS prevents finding the A-preferred MCS $\{c_2\}$. Similarly, if the computed unsatisfiable core is $\{c_1, c_2, c_7\}$, then clause set refinement would drop the other clauses, thus preventing the identification of the A-preferred MUS $\{c_2, c_4, c_5, c_6\}$.

6   Assessing Practical Impact

The negative results of the previous sections impose important restrictions on the type of preferred sets we can expect to be able to compute efficiently. Given the objective of using a polynomial number of calls to a SAT oracle (as current algorithms do), one can only consider L-preferred MSSes and A-preferred MUSes and MCSes, as was also proposed in [22]. Moreover, the paper also identifies restrictions on the type of algorithms and techniques that can be used. As a result, the computation of preferred sets is expected to be significantly less efficient in practice than what is currently achieved by state of the art algorithms.

To evaluate the effect of the limitations identified in the previous sections, a set of 79 (non-trimmed) instances from the MUS track of the 2011 SAT competition[3] were selected, such that the default run time of MUSer [5], i.e. with clause set refinement, did not exceed 50s. The experiments were run on a HPC cluster with 2xIntel E5-2620 @2GHz (6 core) nodes, each with 128 GB of memory, with a timeout of 3600 seconds. Figure 1 shows a scatter plot comparing the run times of MUSer with and without clause set refinement. (Observe that these represent plain MUS extraction instances, with no preferences specified.) As the plot shows, the performance difference is significant; for all instances the use of clause refinement improves the running time. In addition, without clause set refinement MUSer is unable to solve 46 instances (out of 79). For most of the instances solved, the run times increase by more than one order magnitude. As can be concluded, the limitation of not using clause set refinement has a significant impact in the ability of MUSer to solve problem instances within the given timeout. The main conclusion is that the computation of A-preferred MUSes (and similarly of A-preferred MCSes and L-preferred MSSes) should be restricted to small instances.

7   Discussion and Conclusions

In some settings, there exist preferences over which MUSes, MCSes or MSSes to compute. This paper revisits well-known definitions of preferred constraints and preferred explanations and relaxations for over-constrained sets of constraints [22], namely lexicographic (L-) and anti-lexicographic (A-) preferred

---

[3] http://www.satcompetition.org/2011/

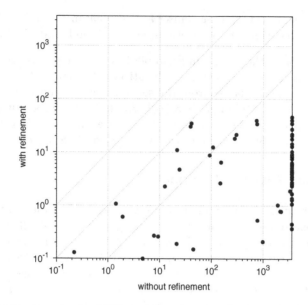

Fig. 1. Scatter plot for MUSer with and without clause set refinement

sets, and casts these definitions in the setting of unsatisfiable CNF formulas, namely MUSes, MCSes & MSSes. The paper shows that computing L-preferred MUSes/MCSes and A-preferred MSSes is hard for the second level of the polynomial hierarchy, whereas computing A-preferred MUSes/MCSes and L-preferred MSSes is in FP^{NP}, i.e. can be solved with a polynomial number of calls to an NP oracle. The practical significance of these results is that in practice, for efficiency reasons, a less natural definition of preferred set is used when computing MUSes or MCSes.

In addition, the paper investigates well-known algorithms for computing MUSes, MCSes and MSSes, and shows that some algorithms cannot be used for computing preferred sets. For example, the recently proposed CLD algorithm [26] cannot be used for computing preferred MCSes or MSSes. Moreover, the paper also investigates well-known, widely used, techniques for reducing the number of SAT solver calls, and shows that several of these techniques cannot be used when computing preferred sets. The consequences of these negative results are significant, since these prevent using well-known very effective pruning techniques when computing preferred sets. Experimental results on selected MUS problem instances illustrate the negative impact of this limitation.

A number of research directions can be envisioned. A natural line of research consists of analyzing other algorithms as well as other techniques for computing preferred sets. Moreover, a more comprehensive experimental evaluation will provide a better understanding of the practical consequences of the negative results proved in this paper. As a result, a main line of research is to develop practical solutions for computing preferred MUSes, MCSes and MSSes, that mitigate the negative results in this paper.

References

1. Ansótegui, C., Bonet, M.L., Levy, J.: SAT-based MaxSAT algorithms. Artif. Intell. 196, 77–105 (2013)
2. Argelich, J., Lynce, I., Marques-Silva, J.: On solving boolean multilevel optimization problems. In: Boutilier, C. (ed.) IJCAI, pp. 393–398 (2009)
3. Bailey, J., Stuckey, P.J.: Discovery of minimal unsatisfiable subsets of constraints using hitting set dualization. In: Hermenegildo, M.V., Cabeza, D. (eds.) PADL 2004. LNCS, vol. 3350, pp. 174–186. Springer, Heidelberg (2005)
4. Bakker, R.R., Dikker, F., Tempelman, F., Wognum, P.M.: Diagnosing and solving over-determined constraint satisfaction problems. In: Bajcsy, R. (ed.) IJCAI, pp. 276–281. Morgan Kaufmann (1993)
5. Belov, A., Lynce, I., Marques-Silva, J.: Towards efficient MUS extraction. AI Commun. 25(2), 97–116 (2012)
6. Biere, A., Heule, M., van Maaren, H., Walsh, T. (eds.): Handbook of Satisfiability. IOS Press (2009)
7. Birnbaum, E., Lozinskii, E.L.: Consistent subsets of inconsistent systems: structure and behaviour. J. Exp. Theor. Artif. Intell. 15(1), 25–46 (2003)
8. Boutilier, C., Brafman, R.I., Domshlak, C., Hoos, H.H., Poole, D.: CP-nets: A tool for representing and reasoning with conditional ceteris paribus preference statements. J. Artif. Intell. Res. (JAIR) 21, 135–191 (2004)
9. Cayrol, C., Lagasquie-Schiex, M.-C., Schiex, T.: Nonmonotonic reasoning: From complexity to algorithms. Ann. Math. Artif. Intell. 22(3-4), 207–236 (1998)
10. Chen, Y., Safarpour, S., Marques-Silva, J., Veneris, A.G.: Automated design debugging with maximum satisfiability. IEEE Trans. on CAD of Integrated Circuits and Systems 29(11), 1804–1817 (2010)
11. Chinneck, J.W., Dravnieks, E.W.: Locating minimal infeasible constraint sets in linear programs. INFORMS Journal on Computing 3(2), 157–168 (1991)
12. de Siqueira, J.L.,, N., Puget, J.-F.: Explanation-based generalisation of failures. In: ECAI, pp. 339–344 (1988)
13. Di Rosa, E., Giunchiglia, E.: Combining approaches for solving satisfiability problems with qualitative preferences. AI Commun. 26(4), 395–408 (2013)
14. Felfernig, A., Schubert, M., Zehentner, C.: An efficient diagnosis algorithm for inconsistent constraint sets. AI EDAM 26(1), 53–62 (2012)
15. Garey, M.R., Johnson, D.S.: Computers and Intractability: A Guide to the Theory of NP-Completeness. W.H. Freeman (1979)
16. Grégoire, É., Mazure, B., Piette, C.: On approaches to explaining infeasibility of sets of Boolean clauses. In: ICTAI (1), pp. 74–83. IEEE Press (2008)
17. Hemery, F., Lecoutre, C., Sais, L., Boussemart, F.: Extracting MUCs from constraint networks. In: Brewka, G., Coradeschi, S., Perini, A., Traverso, P. (eds.) ECAI. Frontiers in Artificial Intelligence and Applications, vol. 141, pp. 113–117. IOS Press (2006)
18. Ignatiev, A., Janota, M., Marques-Silva, J.: Towards efficient optimization in package management systems. In: ICSE (May 2014)
19. Jampel, M., Freuder, E.C., Maher, M.J.: CP-WS 1995. LNCS, vol. 1106. Springer (1996)
20. Janota, M., Botterweck, G., Marques-Silva, J.: On lazy and eager interactive reconfiguration. In: Collet, P., Wasowski, A., Weyer, T. (eds.) VaMoS. ACM (2014)
21. Jose, M., Majumdar, R.: Cause clue clauses: error localization using maximum satisfiability. In: Hall, M.W., Padua, D.A. (eds.) PLDI, pp. 437–446. ACM (2011)

22. Junker, U.: QUICKXPLAIN: Preferred explanations and relaxations for over-constrained problems. In: McGuinness, D.L., Ferguson, G. (eds.) AAAI, pp. 167–172. AAAI Press / The MIT Press (2004)
23. Komuravelli, A., Gurfinkel, A., Chaki, S., Clarke, E.M.: Automatic abstraction in SMT-based unbounded software model checking. In: Sharygina and Veith [37], pp. 846–862
24. Lagniez, J.-M., Biere, A.: Factoring out assumptions to speed up MUS extraction. In: Järvisalo, M., Van Gelder, A. (eds.) SAT 2013. LNCS, vol. 7962, pp. 276–292. Springer, Heidelberg (2013)
25. Liberatore, P.: Redundancy in logic I: CNF propositional formulae. Artif. Intell. 163(2), 203–232 (2005)
26. Marques-Silva, J., Heras, F., Janota, M., Previti, A., Belov, A.: On computing minimal correction subsets. In: Rossi, F. (ed.) IJCAI. IJCAI/AAAI (2013)
27. Marques-Silva, J., Janota, M., Belov, A.: Minimal sets over monotone predicates in boolean formulae. In: Sharygina and Veith [37], pp. 592–607
28. Marques-Silva, J., Lynce, I.: On improving MUS extraction algorithms. In: Sakallah, K.A., Simon, L. (eds.) SAT 2011. LNCS, vol. 6695, pp. 159–173. Springer, Heidelberg (2011)
29. Meseguer, P., Bouhmala, N., Bouzoubaa, T., Irgens, M., Sánchez, M.: Current approaches for solving over-constrained problems. Constraints 8(1), 9–39 (2003)
30. Morgado, A., Heras, F., Liffiton, M.H., Planes, J., Marques-Silva, J.: Iterative and core-guided MaxSAT solving: A survey and assessment. Constraints 18(4), 478–534 (2013)
31. Nadel, A., Ryvchin, V., Strichman, O.: Efficient MUS extraction with resolution. In: FMCAD, pp. 197–200. IEEE (2013)
32. Nöhrer, A., Biere, A., Egyed, A.: Managing SAT inconsistencies with HUMUS. In: Eisenecker, U.W., Apel, S., Gnesi, S. (eds.) VaMoS, pp. 83–91. ACM (2012)
33. O'Callaghan, B., O'Sullivan, B., Freuder, E.C.: Generating corrective explanations for interactive constraint satisfaction. In: van Beek, P. (ed.) CP 2005. LNCS, vol. 3709, pp. 445–459. Springer, Heidelberg (2005)
34. Papadimitriou, C.H.: Computational Complexity. Addison-Wesley (1993)
35. Reiter, R.: A theory of diagnosis from first principles. Artif. Intell. 32(1), 57–95 (1987)
36. Rossi, F., Venable, K.B., Walsh, T.: A Short Introduction to Preferences: Between Artificial Intelligence and Social Choice. Morgan & Claypool Publishers (2011)
37. Sharygina, N., Veith, H. (eds.): CAV 2013. LNCS, vol. 8044. Springer, Heidelberg (2013)
38. van Hoeve, W.-J.: Over-Constrained Problems. In: Hybrid Optimization: The 10 Years of CPAIOR, pp. 191–225. Springer (2011)
39. Wallner, J.P., Weissenbacher, G., Woltran, S.: Advanced SAT techniques for abstract argumentation. In: Leite, J., Son, T.C., Torroni, P., van der Torre, L., Woltran, S. (eds.) CLIMA XIV 2013. LNCS (LNAI), vol. 8143, pp. 138–154. Springer, Heidelberg (2013)
40. Zhu, C.S., Weissenbacher, G., Malik, S.: Post-silicon fault localisation using maximum satisfiability and backbones. In: Bjesse, P., Slobodová, A. (eds.) FMCAD, pp. 63–66. FMCAD Inc. (2011)

Conditional Lower Bounds
for Failed Literals and Related Techniques*

Matti Järvisalo and Janne H. Korhonen

HIIT & Department of Computer Science, University of Helsinki, Finland

Abstract. We prove time-complexity lower bounds for various practically relevant probing-based CNF simplification techniques, namely failed literal detection and related techniques. Specifically, we show that improved algorithms for these simplification techniques would give a $2^{\delta n}$ time algorithm for CNF-SAT for some $\delta < 1$, violating the Strong Exponential Time Hypothesis.

1 Introduction

Automated formula simplification at the conjunctive normal form (CNF) level is today an integral part of the SAT solving workflow, often notably speeding up Boolean satisfiability (SAT) solving of real-world application instances. Indeed, various polynomial-time techniques have been proposed for simplifying CNF formulas before (i.e., in preprocessing) and during (i.e., in inprocessing [26]) search for satisfiability; see e.g. [1–3, 6, 11, 14, 15, 20, 21, 25, 29, 30, 33]. However, formula simplification tends to come with a price. While stronger simplification might be achieved by using more computational effort, in practice time used for simplification should not outweigh the benefits of the simplifications in terms of the overall solving time (i.e., the combined time used for simplification and search). While SAT solver developers keep on searching for more efficient ways of implementing simplification techniques, our formal understanding of the time complexity of different simplification techniques is rather limited at present. This paper take steps towards a more in-depth understanding of the price of simplification: we prove lower bounds for different *probing-based* CNF simplification techniques.

Unit propagation is a common basis for many different simplification techniques [3, 12, 13, 17, 18, 20, 21, 31, 32, 35, 38]. A key example is *failed literal elimination* [13, 31, 34], which aims at deducing unit clauses via checking whether assuming a truth value for a single variable results in a conflict by unit propagation. Failed literals is a key technique used during search within lookahead DPLL solvers [23] for both search tree pruning as well as a basis of branching heuristics [22, 27, 28, 31, 34]. Furthermore, in combination with conflict-driven

---

* This work was supported by Academy of Finland Finnish Centre of Excellence in Computational Inference Research COIN (grant #251170; M.J.) and Helsinki Doctoral Programme in Computer Science – Advanced Computing and Intelligent Systems (J.K.).

C. Sinz and U. Egly (Eds.): SAT 2014, LNCS 8561, pp. 75–84, 2014.

clause learning (CDCL) solvers, failed literals can be detected during preprocessing as well as during search, e.g., by inprocessing CDCL SAT solvers such as
Lingeling [4]. Various clause elimination and clause strengthening techniques are
essentially generalisations of failed literals, probing for either conflicts or specific
literal dependencies using unit propagation by assuming one or more literals at
a time.

1.1   Contributions

Our main result is a conditional lower bound for the *failed literal existence* problem, i.e., that of deciding whether a given CNF formula contains a failed literal.
Since the fixpoint of unit propagation can be computed in time $O(n + m)$ on
CNF formulas with n clauses and m variables, failed literal existence has a simple algorithm with running time $O(n(n + m))$: for each literal $\ell \in F$, run unit
propagation on $F \wedge (\ell)$ and see if a conflict is derived. An iterative application of
this simple algorithm gives a $O(n^2(n + m))$ algorithm for applying *failed literal
elimination* until fixpoint.

In practice, the quadratic running time of the simple algorithm for failed literal
existence can be too time consuming. However, as our main result, formalized as
Theorem 1, we show that non-negligible improvements over the simple algorithm
would give an improved algorithm for CNF-SAT.

Theorem 1. *Let $\varepsilon > 0$. If there is a $O((N + M)^{2-\varepsilon})$ algorithm for failed literal
existence on Horn-3-CNF formulas with N variables and M clauses, then there is
a $2^{(1-\varepsilon/2)n}$ poly(n, m) time algorithm for CNF-SAT on formulas with n variables
and m clauses*

In other words, any such improvement, even in the restricted setting of Horn-
3-CNF formulas, would give us a exponential speedup over brute force for CNF-
SAT, improving upon the state of the art. Indeed, this would violate the *strong
exponential time hypothesis (SETH)* [8, 24] stating that

$$\lim_{k\to\infty} \inf\{\delta \colon k\text{-CNF can be solved in time } O(2^{\delta n})\} = 1 \,.$$

In particular, SETH would imply that CNF-SAT with unrestricted clause length
cannot be solved in time $2^{(1-\varepsilon)n}$ poly(n, m) for any $\varepsilon > 0$. Thus Theorem 1 gives
a conditional lower bound against faster algorithms for failed literal existence.

Corollary 1. *Failed literal existence cannot be solved on Horn-3-CNF formulas
with N variables and M clauses in time $O((N + M)^{2-\varepsilon})$ for any $\varepsilon > 0$ unless
SETH fails.*

This result falls in line with other recent work investigating lower bounds
based on SETH [7, 9, 37]. While SETH itself is an extremely strong complexity
assumption, our result can be interpreted as showing that any attempt to improve upon the simple $O(n(n + m))$ algorithm for finding a failed literal faces
barriers equivalent to improving the worst-case performance of CNF-SAT algorithms.

A detailed proof of Theorem 1 is presented in Section 3. As outlined in Section 4, minor variations of the proof also give the same quadratic lower bound for several related problems: checking the existence of *asymmetric tautologies* and *asymmetric literals*, as well as for checking whether a binary CSP restricted to domain-size 3 is singleton arc consistent.

2   Preliminaries

We assume that the reader is familiar with standard definitions on propositional satisfiability. When convenient, a clause is seen as a set of literals and a CNF formula as a set of clauses. Recall that the subclass k-CNF consists of CNF formulas consisting of clauses of length $\leq k$; Horn consists of CNF formulas in which each clause has at most one positive literal; and that Horn-k-CNF is the intersection of k-CNF and Horn.

Given a CNF formula F with clauses $(\neg l_1), \ldots, (\neg l_k)$, and $(l_1 \vee \cdots l_k \vee l_{k+1})$, the *unit resolution rule* allows to extend F by letting $F := F \wedge (l_{k+1})$, i.e., allows the derivation of the unit clause (l_{k+1}) from F in one step. *Unit propagation* on F applies the unit resolution rule until fixpoint, and we write $F \vdash_{\mathrm{up}} (l)$ if unit propagation on F derives the unit clause (l).

A literal $l \in F$ is a *failed literal* in F if unit propagation derives a conflict on $F \wedge (l)$, that is, we have $F \vdash_{\mathrm{up}} (\ell'), (\neg \ell')$ for some literal ℓ'. In particular, this implies that F is logically equivalent to $F \wedge (\neg l)$, which can be used to simplify F by letting $F := F \wedge (\neg l)$ if $l \in F$ fails in F; this is called the *failed literal rule*. *Failed literal elimination* refers to applying the failed literal rule until fixpoint.

3   Proof of the Failed Literal Existence Lower Bound

On a high level, our strategy for proving Theorem 1 follows that of Pătraşcu and Williams [37]. In particular, assume that the following conditions hold.

(i) For some $\varepsilon > 0$, there is a $O\big((N + M)^{2-\varepsilon}\big)$ algorithm for failed literal existence on Horn-3-CNF formulas with N variables and M clauses.

(ii) There is a reduction that maps a CNF formula F with n variables and m clauses to a Horn-3-CNF F_{fl} with N variables and M clauses such that

    ii.a F_{fl} has a failed literal if and only if F is satisfiable,

    ii.b $N, M \leq 2^{n/2} \operatorname{poly}(n, m)$, and

    ii.c F_{fl} can be constructed in time $2^{n/2} \operatorname{poly}(n, m)$.

Lemma 1. *Conditions (i) and (ii) imply that CNF-SAT has an algorithm with running time $2^{(1-\varepsilon/2)n} \operatorname{poly}(n, m)$.*

Proof. On input CNF formula F, we (1) construct F_{fl}, and (2) use the algorithm for failed literal existence to decide whether F_{fl} has a failed literal, and thus whether F is satisfiable. The first step takes $2^{n/2} \operatorname{poly}(n, m)$ time and the second step takes $O\big((N + M)^{2-\varepsilon}\big)$ time, that is, $2^{(1-\varepsilon/2)n} \operatorname{poly}(n, m)$ time. $\qquad\square$

Thus in order to prove Theorem 1, it suffices to construct a reduction satisfying Condition (ii) above. In what follows, we first present a *reduction template* (in Section 3.1) which can be used as a base construction for obtaining reductions from CNF-SAT to failed literal existence as well as other related existence problems. We then instantiate the reduction for failed literal existence and show how to obtain Horn-3-CNF formulas from the reduction (in Section 3.2). Instantiations for related problems, namely, the existence problem for asymmetric tautologies and literals, and checking singleton arc consistency of binary CSPs of domain size three, are presented in Section 4.

3.1   A Reduction Template

Let $F = C_1 \wedge C_2 \wedge \cdots \wedge C_m$ be a CNF formula over variables $x_1, x_2, \ldots, x_n$. Without loss of generality, we assume that n is even and $n \geq 4$. We split the variable set into *high* variables $x_1, x_2 \ldots, x_{n/2}$ and the *low* variables $x_{n/2+1}, x_{n/2+2} \ldots, x_n$. Let $P = \{p_1, p_2, \ldots, p_{2^{n/2}}\}$ be the set of all truth assignments into the high variables, and similarly let $Q = \{q_1, q_2, \ldots, q_{2^{n/2}}\}$ be the set of all truth assignments into the low variables. For $p \in P$ and $q \in Q$, denote by pq the assignment into variables $x_1, x_2, \ldots, x_n$ obtained by combining p and q.

We now construct a new formula F' from F as follows. The variable set of F' is

$$\{y_r : r \in P \cup Q\} \cup \{c_i : i = 1, 2, \ldots, m\} \cup \{w\},$$

and the clauses of F' are given by the following three rules.

(1) For each partial assignment $p \in P$ and clause $C_i \in F$, if p satisfies C_i we include the clause $(\neg y_p \vee c_i)$, or equivalently, $(y_p \rightarrow c_i)$.

(2) For each partial assignment $q \in Q$, we include the clause

$$\left(y_q \vee \bigvee_{i:\, q(C_i) \neq 1} \neg c_i \right), \qquad \text{or equivalently,} \qquad \left(\left(\bigwedge_{i:\, q(C_i) \neq 1} c_i \right) \rightarrow y_q \right).$$

In words, this clause states that having $c_i = 1$ for all clauses $C_i \in F$ that are not satisfied by q implies $y_q = 1$. Without loss of generality, we will assume that the original formula F contains the tautological clauses $(x_1 \vee \neg x_1)$ and $(x_2 \vee \neg x_2)$. This ensures that clauses generated by this rule have length at least 3; in particular, they are not units.

(3) For each partial assignment $q \in Q$, we include the clause $(\neg y_q \vee w)$, or equivalently, $(y_q \rightarrow w)$.

Intuitively, the important feature of F' is how unit propagation behaves on the formula. The variables of F' can be seen to be arranged in layers, as illustrated in Figure 1. These layers are (1) the high variables $\{y_p : p \in P\}$, (2) the clause variables $\{c_1, c_2, \ldots, c_m\}$, (3) the low variables $\{y_q : q \in Q\}$, and (4) the terminal variable $\{w\}$. Clauses are implications from variables on layer i to a single variable on layer $i+1$. As all the clauses are Horn, positive units will propagate

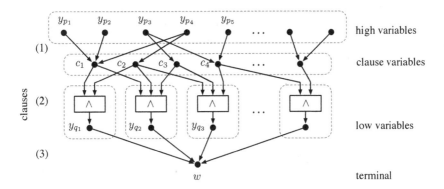

Fig. 1. Illustration of the reduction

only downwards towards the terminal w, and negative units will propagate only upwards toward the variables y_p.

Furthermore, the "satisfiability gadgets" consisting of the clauses (2) control the flow of unit propagation between variables y_p and w. That is, the only way for unit propagation to pass through these gadgets is that we start from a literal y_p for an assignment p that can be extended to a satisfying assignment for F. More formally, we make the following observations about F'.

Lemma 2. *Given a CNF formula F with n variables and m clauses, F' is a CNF formula with N clauses and M variables such that*

(a) *F' is a Horn-CNF formula,*
(b) *F' is satisfied by assigning all variables to 0,*
(c) *$N = 2^{n/2} + m + 1$,*
(d) *$M = 2^{n/2} \operatorname{poly}(m)$, and*
(e) *F' can be constructed in time $2^{n/2} \operatorname{poly}(n, m)$.*

Lemma 3. *Let $p \in P$ be fixed.*

(a) *If there is a $q \in Q$ such that pq satisfies F, then $F' \wedge (y_p) \vdash_{up} (w)$.*
(b) *if there is no $q \in Q$ such that pq satisfies F, then unit propagation on $F' \wedge (y_p)$ derives exactly the units (c_i) for i such that $p(C_i) = 1$.*
(c) *Unit propagation on $F' \wedge (\neg w)$ derives exactly the units $(\neg y_q)$ for $q \in Q$.*

Proof. Fix $p \in P$ and $q \in Q$. We start by making the following simple observation: the assignment pq satisfies F if and only if $\{i\colon p(C_i) \neq 1\} \cap \{i\colon q(C_i) \neq 1\} = \emptyset$, which in turn is equivalent to $\{i\colon q(C_i) \neq 1\} \subseteq \{i\colon p(C_i) = 1\}$.

For (a), assume that there is $q \in Q$ such that pq satisfies F. For any i such that $p(C_i) = 1$, we have $(y_p \to c_i)$, and thus $F' \wedge (y_p) \vdash_{up} (c_i)$. Since pq satisfies F, we have by the earlier observation that $F' \wedge (y_p) \vdash_{up} (c_i)$ for all i such that $q(C_i) \neq 1$. Thus unit propagation derives (y_q) using the clause $\left(\bigwedge_{q(C_i) \neq 1} c_i \right) \to y_q$, and, further, (w) using the clause $(y_q \to w)$.

For (b), note that unit resolution on $F' \wedge (y_p)$ immediately derives (c_i) for i such that $p(C_i) = 1$, and no other units are immediately derived or already present in F'. Since pq does not satisfy F for any $q \in Q$, we have $\{i \colon p(C_i) \neq 1\} \cap \{i \colon q(C_i) \neq 1\} \neq \emptyset$. Thus, unit propagation does not derive (y_q) from the clause $\left(\bigwedge_{q(C_i) \neq 1} c_i \right) \to y_q$, nor does it derive (w).

For (c), note that $F' \wedge (\neg w) \vdash_{\mathrm{up}} (\neg y_q)$ for all $q \in Q$, using the clause $(y_q \to w)$. Since each clause $\left(\bigwedge_{q(C_i) \neq 1} c_i \right) \to y_q$ has length at least 3 and variables w and y_q do not appear in any other clauses, no further units are derived. □

3.2 The Lower Bound

To complete the reduction to failed literal existence, we construct the formula F_{fl} by adding the clause $(\neg w \vee \neg y_p)$ or equivalently, $(w \to \neg y_p)$ to F' for each $p \in P$. Lemma 2 also holds for F_{fl}. Furthermore, we have the following.

Lemma 4. *Let ℓ be a literal in F_{fl}. We have that*

(a) *ℓ is a failed literal in F_{fl} if $\ell = y_p$ for some $p \in P$ and there is a $q \in Q$ such that pq satisfies F, and*
(b) *ℓ is not a failed literal otherwise.*

Proof. First, we note that (a) follows from Lemma 3(a), as we have $F_{\mathrm{fl}} \wedge (y_p) \vdash_{\mathrm{up}}$ (w) and $(w \to \neg y_p)$ if p satisfies the conditions of (a). Thus, it suffices to show that there are no other failed literals.

Now let ℓ be literal that does not satisfy the requirements of (a). We show that $F_{\mathrm{fl}} \wedge (\ell)$ is satisfiable, so ℓ cannot be a failed literal. There are three cases to consider.

1. If ℓ is a negative literal, assigning all variables to 0 satisfies $F_{\mathrm{fl}} \wedge (\ell)$, as all clauses are non-unit Horn clauses.
2. If ℓ is a positive literal and ℓ is not of form y_p, then assigning variables in the set $\{\ell, w\} \cup \{y_q \colon q \in Q\}$ to 1 and other variables to 0 satisfies $F_{\mathrm{fl}} \wedge (\ell)$.
3. If $\ell = y_p$ for some $p \in P$ and pq does not satisfy F for any $q \in Q$, then by Lemma 3(b) assigning y_p and all variables c_i such that $p(C_i) = 1$ to 1 and other variables to 0 satisfies $F_{\mathrm{fl}} \wedge (y_p)$.

□

The formula F_{fl} is still not necessarily 3-CNF. However, standard rewriting of a long clause as clauses of length 3 preserves the Horn property: rewriting a Horn clause $(x_1 \wedge \cdots \wedge x_{k-1}) \to l)$, where l is either x_k or $\neg x_k$, using fresh variables $a_3, \ldots, a_{k-1}$ gives the Horn clauses

$$((x_1 \wedge x_2) \to a_3) \wedge ((a_3 \wedge x_3) \to a_4) \wedge \cdots \wedge ((a_{k-1} \wedge x_{k-1}) \to l).$$

This rewriting preserves unit propagations over the original clause, and hence does not affect the existence of failed literals.

Theorem 1 follows now from Lemmas 1–4 and the rewriting from Horn-CNF to Horn-3-CNF.

4   Extensions

The reduction template described in Section 3.1 can be used to prove similar lower bounds for existence problems over other notions related to failed literals. Here we consider the existence problem for asymmetric tautologies and literals, and checking singleton arc consistency of binary CSPs of domain size three.

For a CNF formula F, a clause $C = (l_1 \vee \cdots \vee l_k) \in F$ is an *asymmetric tautology* [19] if unit propagation derives a conflict on $(F \setminus C) \wedge (\neg l_1) \wedge \cdots \wedge (\neg l_k)$. If C is an asymmetric tautology, then F can be simplified by letting $F := F \setminus C$. A clause $C \in F$, a literal ℓ is an *asymmetric literal in* C if unit propagation on $F \vee (\ell)$ derives (ℓ') for some $\ell' \in C \setminus \{\ell\}$. If ℓ is an asymmetric tautology in C, then F can be simplified by letting $F := (F \setminus C) \wedge (C \setminus l)$. The notion of asymmetric literals is a generalization of hidden literals [20]. Both asymmetric tautologies and asymmetric literals are detected e.g. by vivification and during inprocessing within the Lingeling SAT solver [4].

Furthermore, we consider the problem of *checking singleton arc consistency* [10, 36] of constraint satisfaction problems (CSPs) with constraints over pairs of variable (i.e., binary CSPs) and with variable domains restricted to cardinality three. On CNF formulas, checking singleton arc consistency is equivalent to checking for the existence of failed literals, and extends to naturally to CSPs[1]. Here we consider the *restricted setting* of $(3, 2)$-CSPs, i.e., binary CSPs with variable domains restricted to cardinality three.

Theorem 2. *Let $\varepsilon > 0$. There is a $2^{(1-\varepsilon/2)n} \operatorname{poly}(n, m)$ time algorithm for CNF-SAT on formulas with n variables and m clauses if one of the following holds.*

(a) *There is a $O\big((N + M)^{2-\varepsilon}\big)$ algorithm for asymmetric tautology existence over Horn-3-CNF formulas with N variables and M clauses.*

(b) *There is a $O\big((N + M)^{2-\varepsilon}\big)$ algorithm for asymmetric literal existence over Horn-3-CNF formulas with N variables and M clauses.*

(c) *There is a $O\big((N + M)^{2-\varepsilon}\big)$ algorithm for checking singleton arc consistency over $(3, 2)$-CSPs with N variables and M constraints.*

Proof sketch. For (a) and (b), we start from the reduction template given in Section 3.1 and extend it in a similar manner as in Section 3.2 with following differences. For (a), we construct formula F_{at} by adding the clause $(y_p \to w)$ to F' for each $p \in P$. This new clause is an asymmetric tautology if and only if p can be extended to a satisfying assignment for F, and there are no other asymmetric tautologies in F_{at} (compare with Lemma 4).

---

[1] Enforcing arc consistency refers to reducing the variable domains of a CSP until fixpoint using the following rule: if there is a variable x and a value v in the domain $D(x)$ of x that is not supported by a constraint in the CSP, then let $D(x) := D(x) \setminus \{v\}$. A CSP is singleton arc consistent if for all x and $v \in D(x)$, enforcing singleton arc consistency on the CSP after assigning $D(x) := \{v\}$ does not result in an empty domain for some variable.

For (b), we construct the formula F_{al} by adding the clause $(w \vee y_p)$ to F' for each $p \in P$; again, only possible asymmetric literals arise from these clauses and correspond to satisfying assignments of F as before. Replacing w with $\neg w$ in all clauses gives a Horn formula.

For (c), we transform the Horn-3-CNF formula F_{fl} obtained from the failed literal reduction into $(3, 2)$-CSP instance, by emulating each length 3 Horn clause with two binary constraints over domain of size 3. That is, we replace each clause $((a \wedge b) \to c)$ by two constraints C and C' as follows. For C, we set $\text{var}(C) = \{a, c\}$ and allow all assignments with $a = 0$ or $a = 2$, and assignments $a = 1, c = 0$ and $a = 1, c = 2$. For C', we set $\text{var}(C') = \{b, c\}$ and allow all assignments with $b = 0$ or $b = 2$, and assignments $b = 1, c = 0$ and $b = 1, c = 1$. The resulting CSP instance is singleton arc consistent if and only if F_{fl} has no failed literals. □

5   Concluding Remarks

We established a connection between the strong exponential time hypothesis and the existence of subquadratic algorithms for checking whether a given CNF formula contains at least one failed literal, as well as several other related simplification techniques. Any improvement over the obvious algorithm for failed literal existence, even on Horn-3-CNF formulas, would require a major algorithmic breakthrough.

However, several questions related to our results remain open. So far, we were unable to establish a similar lower bound for failed literal existence for 2-CNF formulas, and as such cannot rule out the possibility of a subquadratic algorithm for the 2-CNF case. In contrast, our proofs require clauses of width 3, or domain size 3 in the case of binary CSPs. Similarly, it is open whether there exist $O((n + m)^{3-\epsilon})$ time algorithms for computing the *fixpoint* of failed literal elimination. Proving a conditional lower bound for the fixpoint computation would be a stronger result than our lower bound for failed literal existence. Again, it is not clear whether such a lower bound can be established, especially in the limited setting of Horn-CNFs; note that for 2-CNFs, computing the fixpoint can be done in time $O(n(n + m))$, as all failed literals in the fixpoint fail without eliminating other failed literals first. Another possible extension of our results would be to prove a more general conditional lower bound for *k-step lookahead*, i.e., the extension of failed literal existence / failed literal elimination (i.e., 1-step lookahead) to testing sets of $k > 1$ literals instead of a single literal at a time [16].

Finally, a converse result—that is, showing that faster CNF-SAT algorithms would imply faster algorithms for failed literal existence—would give an even tighter connection between the complexity of the two problems.

References

1. Bacchus, F.: Enhancing Davis Putnam with extended binary clause reasoning. In: Proc. AAAI, pp. 613–619. AAAI Press / The MIT Press (2002)
2. Bacchus, F., Winter, J.: Effective preprocessing with hyper-resolution and equality reduction. In: Giunchiglia, E., Tacchella, A. (eds.) SAT 2003. LNCS, vol. 2919, pp. 341–355. Springer, Heidelberg (2004)
3. Berre, D.L.: Exploiting the real power of unit propagation lookahead. Electronic Notes in Discrete Mathematics 9, 59–80 (2001)
4. Biere, A.: Lingeling, Plingeling and Treengeling entering the SAT Competition 2013. In: Proceedings of SAT Competition 2013. Department of Computer Science Series of Publications B, vol. B-2013-1, pp. 51–52. University of Helsinki (2013)
5. Biere, A., Heule, M., van Maaren, H., Walsh, T. (eds.): Handbook of Satisfiability. Frontiers in Artificial Intelligence and Applications, vol. 85. IOS Press (2009)
6. Brafman, R.I.: A simplifier for propositional formulas with many binary clauses. IEEE Transactions on Systems, Man, and Cybernetics, Part B 34(1), 52–59 (2004)
7. Calabro, C., Impagliazzo, R., Kabanets, V., Paturi, R.: The complexity of unique k-SAT: An isolation lemma for k-CNFs. Journal of Computer and System Sciences 74(3), 386–393 (2008)
8. Calabro, C., Impagliazzo, R., Paturi, R.: The complexity of satisfiability of small depth circuits. In: Chen, J., Fomin, F.V. (eds.) IWPEC 2009. LNCS, vol. 5917, pp. 75–85. Springer, Heidelberg (2009)
9. Cygan, M., Dell, H., Lokshtanov, D., Marx, D., Nederlof, J., Okamoto, Y., Paturi, R., Saurabh, S., Wahlstrom, M.: On problems as hard as CNF-SAT. In: Proc. CCC, pp. 74–84. IEEE (2012)
10. Debruyne, R., Bessière, C.: Some practicable filtering techniques for the constraint satisfaction problem. In: Proc. IJCAI, pp. 412–417. Morgan Kaufmann (1997)
11. Eén, N., Biere, A.: Effective preprocessing in SAT through variable and clause elimination. In: Bacchus, F., Walsh, T. (eds.) SAT 2005. LNCS, vol. 3569, pp. 61–75. Springer, Heidelberg (2005)
12. Fourdrinoy, O., Grégoire, É., Mazure, B., Sais, L.: Reducing hard SAT instances to polynomial ones. In: Proc. IRI, pp. 18–23, IEEE Systems, Man, and Cybernetics Society (2007)
13. Freeman, J.: Improvements to propositional satisfiability search algorithms. PhD thesis. University of Pennsylvania, USA (1995)
14. Gelder, A.V.: Toward leaner binary-clause reasoning in a satisfiability solver. Ann. Math. Artif. Intell. 43(1), 239–253 (2005)
15. Gershman, R., Strichman, O.: Cost-effective hyper-resolution for preprocessing CNF formulas. In: Bacchus, F., Walsh, T. (eds.) SAT 2005. LNCS, vol. 3569, pp. 423–429. Springer, Heidelberg (2005)
16. Gwynne, M., Kullmann, O.: Generalising unit-refutation completeness and SLUR via nested input resolution. J. Autom. Reasoning 52(1), 31–65 (2014)
17. Han, H., Somenzi, F.: Alembic: An efficient algorithm for CNF preprocessing. In: Proc. DAC, pp. 582–587. IEEE (2007)
18. Heule, M., Dufour, M., van Zwieten, J., van Maaren, H.: March_eq: Implementing additional reasoning into an efficient look-ahead SAT solver. In: Hoos, H.H., Mitchell, D.G. (eds.) SAT 2004. LNCS, vol. 3542, pp. 345–359. Springer, Heidelberg (2005)
19. Heule, M., Järvisalo, M., Biere, A.: Clause elimination procedures for CNF formulas. In: Fermüller, C.G., Voronkov, A. (eds.) LPAR-17. LNCS, vol. 6397, pp. 357–371. Springer, Heidelberg (2010)

20. Heule, M.J.H., Järvisalo, M., Biere, A.: Efficient CNF simplification based on binary implication graphs. In: Sakallah, K.A., Simon, L. (eds.) SAT 2011. LNCS, vol. 6695, pp. 201–215. Springer, Heidelberg (2011)
21. Heule, M.J.H., Järvisalo, M., Biere, A.: Revisiting hyper binary resolution. In: Gomes, C., Sellmann, M. (eds.) CPAIOR 2013. LNCS, vol. 7874, pp. 77–93. Springer, Heidelberg (2013)
22. Heule, M., van Maaren, H.: Aligning CNF- and equivalence-reasoning. In: Hoos, H.H., Mitchell, D.G. (eds.) SAT 2004. LNCS, vol. 3542, pp. 145–156. Springer, Heidelberg (2005)
23. Heule, M., van Maaren, H.: Look-ahead based SAT solvers. In: Biere, et al. (eds.) [5], pp. 155–184
24. Impagliazzo, R., Paturi, R., Zane, F.: Which problems have strongly exponential complexity? Journal of Computer and System Sciences 63(4), 512–530 (2001)
25. Järvisalo, M., Biere, A., Heule, M.: Simulating circuit-level simplifications on CNF. J. Autom. Reasoning 49(4), 583–619 (2012)
26. Järvisalo, M., Heule, M.J.H., Biere, A.: Inprocessing rules. In: Gramlich, B., Miller, D., Sattler, U. (eds.) IJCAR 2012. LNCS (LNAI), vol. 7364, pp. 355–370. Springer, Heidelberg (2012)
27. Kullmann, O.: Investigating the behaviour of a SAT solver on random formulas (2002)
28. Kullmann, O.: Fundaments of branching heuristics. In: Biere, et al. (eds.) [5], pp. 205–244
29. Li, C.M.: Equivalency reasoning to solve a class of hard SAT problems. Inf. Process. Lett. 76(1-2), 75–81 (2000)
30. Li, C.M.: Integrating equivalency reasoning into Davis-Putnam procedure. In: Proc. AAAI, pp. 291–296. AAAI Press / The MIT Press (2000)
31. Li, C.M.: Anbulagan: Heuristics based on unit propagation for satisfiability problems. In: Proc. IJCAI, pp. 366–371. Morgan Kaufmann (1997)
32. Lynce, I., Marques-Silva, J.P.: Probing-based preprocessing techniques for propositional satisfiability. In: Proc. ICTAI, p. 105. IEEE Computer Society (2003)
33. Manthey, N., Heule, M.J.H., Biere, A.: Automated reencoding of boolean formulas. In: Biere, A., Nahir, A., Vos, T. (eds.) HVC. LNCS, vol. 7857, pp. 102–117. Springer, Heidelberg (2013)
34. Niemelä, I., Simons, P.: Smodels - an implementation of the stable model and well-founded semantics for normal LP. In: Fuhrbach, U., Dix, J., Nerode, A. (eds.) LPNMR 1997. LNCS, vol. 1265, pp. 421–430. Springer, Heidelberg (1997)
35. Piette, C., Hamadi, Y., Sais, L.: Vivifying propositional clausal formulae. In: Proc. ECAI. FAIA, pp. 525–529. IOS Press (2008)
36. Prosser, P., Stergiou, K., Walsh, T.: Singleton consistencies. In: Dechter, R. (ed.) CP 2000. LNCS, vol. 1894, pp. 353–368. Springer, Heidelberg (2000)
37. Ptracu, M., Williams, R.: On the possibility of faster SAT algorithms. In: Proc. SODA, pp. 1065–1075 (2010)
38. Sheeran, M., Stålmarck, G.: A tutorial on Stålmarck's proof procedure for propositional logic. Formal Methods in System Design 16(1), 23–58 (2000)

Fixed-Parameter Tractable Reductions to SAT

Ronald de Haan* and Stefan Szeider*

Institute of Information Systems,
Vienna University of Technology,
Vienna, Austria

Abstract. Today's SAT solvers have an enormous importance and impact in many practical settings. They are used as efficient back-end to solve many NP-complete problems. However, many computational problems are located at the second level of the Polynomial Hierarchy or even higher, and hence polynomial-time transformations to SAT are not possible, unless the hierarchy collapses. In certain cases one can break through these complexity barriers by fixed-parameter tractable (fpt) reductions which exploit structural aspects of problem instances in terms of problem parameters. Recent research established a general theoretical framework that supports the classification of parameterized problems on whether they admit such an fpt-reduction to SAT or not. We use this framework to analyze some problems that are related to Boolean satisfiability. We consider several natural parameterizations of these problems, and we identify for which of these an fpt-reduction to SAT is possible. The problems that we look at are related to minimizing an implicant of a DNF formula, minimizing a DNF formula, and satisfiability of quantified Boolean formulas.

1 Introduction

Modern SAT solvers have an enormous importance and impact in many practical settings that require solutions to NP-complete problems. In fact, due to the success of SAT, NP-complete problems have lost their scariness, as in many cases one can efficiently encode NP-complete problems to SAT and solve them by means of a SAT solver [8,20,31,38]. However, many important computational problems are located above the first level of the Polynomial Hierarchy (PH) and thus considered significantly "harder" than SAT. Hence we cannot hope for polynomial-time reductions from these problems to SAT, as such transformations would cause the (unexpected) collapse of the PH.

Realistic problem instances are not random and often contain some kind of structure. Recent research succeeded to exploit such structure to break the complexity barriers between levels of the PH [15,36]. The idea is to exploit problem structure in terms of a problem *parameter*, and to develop reductions to SAT

---

* Supported by the European Research Council (ERC), project 239962 (COMPLEX REASON), and the Austrian Science Fund (FWF), project P26200 (Parameterized Compilation).

C. Sinz and U. Egly (Eds.): SAT 2014, LNCS 8561, pp. 85–102, 2014.

that can be computed efficiently as long as the problem parameter is reasonably small. The theory of *parameterized complexity* [13,17,33] provides exactly the right type of reduction suitable for this purpose, called *fixed-parameter tractable reductions*, or *fpt-reductions* for short. Now, for a suitable choice of the parameter, one can aim at developing fpt-reductions from the hard problem under consideration to SAT.

Such positive results go significantly beyond the state-of-the-art of current research in parameterized complexity. By shifting the scope from fixed-parameter tractability to fpt-reducibility (to SAT), parameters can be less restrictive and hence larger classes of inputs can be processed efficiently. Therefore, the potential for positive tractability results is greatly enlarged. In fact, there are some known reductions that, in retrospect, can be seen as fpt-reductions to SAT. A prominent example is Bounded Model Checking [6,7], which can be seen as an fpt-reduction from the model checking problem for linear temporal logic (LTL), which is PSPACE-complete, to SAT, where the parameter is an upper bound on the size of a counterexample.

Recently, extending the work of Flum and Grohe [16], we initiated the development of a general theoretical framework to support the classification of hard problems on whether they admit an fpt-reduction to SAT or not [24]. This framework provides a hardness theory that can be used to provide evidence that certain problems do not admit an fpt-reduction to SAT, similar to NP-hardness which provides evidence against polynomial-time tractability [18] and W[1]-hardness which provides evidence against fixed-parameter tractability [13]. For an overview of the parameterized complexity classes in this framework and the relation between them, see Figure 1.

New Contributions. We use this new framework to analyze problems related to Boolean satisfiability. We focus on problems that are located at the second level of the PH, i.e., problems complete for Σ_2^P. This initiates a structured investigation of fpt-reducibility to SAT of problems related to Boolean satisfiability that are "beyond NP." Concretely, we look at the following problems, consider several parameterizations of these problems, and identify for which of these parameterized problems an fpt-reduction to SAT is possible and for which this is not possible:

- minimizing an implicant of a formula in disjunctive normal form (DNF) (parameterizations: the size of the minimized implicant, and the difference in size between the original and the minimized implicant);
- minimizing a DNF formula (parameterizations: the size of the minimized formula, and the difference in size between the original and the minimized formula); and
- the satisfiability problem of quantified Boolean formulas (QBFs) (parameterizations: the treewidth of the incidence graph of the formula restricted to several subsets of variables).

In particular, we show that minimizing an implicant of a DNF formula does not become significantly easier when the minimized implicant is small (Proposition 1), nor when the difference in size between the original and the minimized implicant is small (Proposition 2). The problem of reducing a DNF formula in size (while preserving logical equivalence) also does not become significantly easier when the difference in size is small (Proposition 3). However, the problem of reducing a DNF formula to an equivalent DNF formula that is small can be done with a small number of SAT calls (Theorem 1). Moreover, we show that deciding satisfiability of a quantified Boolean formula with one quantifier alternation can be reduced to a single SAT instance if the variables in the second quantifier block interact with each other in a tree-like fashion (Theorem 2), whereas a similar restriction on the variables in the first quantifier block does not make the problem any easier (Proposition 5).

Related Work. Many of the decision problems analyzed in this paper have been studied before in a classical complexity setting [19,40,41]. The logic minimization problems that we consider in this paper have been studied since the 1950s (cf. [41]). The problem of minimizing an implicant of a DNF formula plays a central role in the analysis of logic minimization problems [41]. Variants of the minimization problems that we consider, where a subset-minimal solution is sought, are often solved by calling SAT solvers as subroutines. One example of such work is related to identifying minimal unsatisfiable subsets (MUSes) of a CNF formula [3,23,32]. Recent work on MUS extraction indicates that reducing the number of SAT calls made in these algorithms is beneficial for the practical performance of these algorithms [32]. Decision procedures using SAT solvers as a subroutine have also been used to solve problems that lie at the second level of the PH, e.g., problems related to abstract argumentation [14].

2 Preliminaries

Propositional Logic and Quantified Boolean Formulas. A *literal* is a propositional variable x or a negated variable $\neg x$. The *complement* $\overline{x}$ of a positive literal x is $\neg x$, and the complement $\overline{\neg x}$ of a negative literal $\neg x$ is x. For literals $l \in \{x, \neg x\}$, we let $\mathrm{Var}(l) = x$ denote the variable occurring in l. A *clause* is a finite set of literals, not containing a complementary pair x, $\neg x$, and is interpreted as the disjunction of these literals. A *term* is a finite set of literals, not containing a complementary pair x, $\neg x$, and is interpreted as the conjunction of these literals. We let $\top$ denote the empty clause. A formula in *conjunctive normal form (CNF)* is a finite set of clauses, interpreted as the conjunction of these clauses. A formula in *disjunctive normal form (DNF)* is a finite set of terms, interpreted as the disjunction of these terms. We say that a DNF formula φ is a *term-wise subformula* of another DNF formula φ' if for all terms $t \in \varphi$ there exists a term $t' \in \varphi'$ such that $t \subseteq t'$. We define the *size* $\|\varphi\|$ of a DNF formula φ to be $\sum_{t \in \varphi} |t|$; the number of terms of φ is denoted by $|\varphi|$. For a DNF formula φ, the set $\mathrm{Var}(\varphi)$ denotes the set of all variables x such that some term of φ contains x or $\neg x$. We use the standard notion of *(truth) assignments* $\alpha : \mathrm{Var}(\varphi) \to \{0, 1\}$ for Boolean formulas and *truth* of a formula under such an assignment. We denote the problem of deciding whether a propositional formula φ is satisfiable by SAT, and the problem of deciding whether φ is not satisfiable by UNSAT. Let φ be a DNF formula. We say that a set C of literals is an *implicant of* φ if all assignments that satisfy $\bigwedge_{l \in C} l$ also satisfy φ. Let $\gamma = \{x_1 \mapsto d_1, \ldots, x_n \mapsto d_n\}$ be a function that maps some variables of a formula φ to other variables or to truth values. We let $\varphi[\gamma]$ denote the application of such a substitution γ to the formula φ. We also write $\varphi[x_1 \mapsto d_1, \ldots, x_n \mapsto d_n]$ to denote $\varphi[\gamma]$.

A *(prenex) quantified Boolean formula (QBF)* is a formula of the form $Q_1 X_1 Q_2 X_2 \ldots Q_m X_m \psi$, where each Q_i is either $\forall$ or $\exists$, the X_i are disjoint sets of propositional variables, and ψ is a Boolean formula over the variables in $\bigcup_{i=1}^m X_i$. We call ψ the *matrix* of the formula. Truth of such formulas is defined in the usual way. We say that a QBF is *in QDNF* if the matrix is in DNF. For the remainder of this paper, we will restrict our attention to QDNF formulas. Consider the following decision problem.

$\exists\forall$-SAT
Instance: A QDNF $\varphi = \exists X.\forall Y.\psi$, where ψ is quantifier-free.
Question: Is φ satisfiable?

The complexity class consisting of all problems that are polynomial-time reducible to $\exists\forall$-SAT is denoted by Σ_2^P, and its co-class is denoted by Π_2^P. These classes form the second level of the PH [35].

Parameterized Complexity. We introduce some core notions from parameterized complexity theory. For an in-depth treatment we refer to other sources [13,17,33]. A *parameterized problem* L is a subset of $\Sigma^* \times \mathbb{N}$ for some finite alphabet Σ. For an instance $(I, k) \in \Sigma^* \times \mathbb{N}$, we call I the *main part* and k the *parameter*. The following generalization of polynomial time computability is commonly regarded as the tractability notion of parameterized complexity theory. A parameterized problem L is *fixed-parameter tractable* if there exists a computable function f and a constant c such that there exists an algorithm that decides whether $(I, k) \in L$ in time $O(f(k)\|I\|^c)$, where $\|I\|$ denotes the size of I. Such an algorithm is called an *fpt-algorithm*, and this amount of time is called *fpt-time*. FPT is the class of all fixed-parameter tractable decision problems. If the parameter is constant, then fpt-algorithms run in polynomial time where the order of the polynomial is independent of the parameter. This provides a good scalability in the parameter in contrast to running times of the form $\|I\|^k$, which are also polynomial for fixed k, but are already impractical for, say, $k > 3$. By XP we denote the class of all problems L for which it can be decided whether $(I, k) \in L$ in time $O(\|I\|^{f(k)})$, for some fixed computable function f.

Parameterized complexity also generalizes the notion of polynomial-time reductions. Let $L \subseteq \Sigma^* \times \mathbb{N}$ and $L' \subseteq (\Sigma')^* \times \mathbb{N}$ be two parameterized problems. A *(many-one) fpt-reduction* from L to L' is a mapping $R : \Sigma^* \times \mathbb{N} \to (\Sigma')^* \times \mathbb{N}$ from instances of L to instances of L' such that there exist some computable function $g : \mathbb{N} \to \mathbb{N}$ such that for all $(I, k) \in \Sigma^* \times \mathbb{N}$: (i) (I, k) is a yes-instance of L if and only if $(I', k') = R(I, k)$ is a yes-instance of L', (ii) $k' \leq g(k)$, and (iii) R is computable in fpt-time. Similarly, we call reductions that satisfy properties (i) and (ii) but that are computable in time $O(\|I\|^{f(k)})$, for some fixed computable function f, *xp-reductions*.

Let C be a classical complexity class, e.g., NP. The parameterized complexity class para-C is then defined as the class of all parameterized problems $L \subseteq \Sigma^* \times \mathbb{N}$, for some finite alphabet Σ, for which there exist an alphabet Π, a computable function $f : \mathbb{N} \to \Pi^*$, and a problem $P \subseteq \Sigma^* \times \Pi^*$ such that $P \in C$ and for all instances $(x, k) \in \Sigma^* \times \mathbb{N}$ of L we have that $(x, k) \in L$ if and only if $(x, f(k)) \in P$. Intuitively, the class para-C consists of all problems that are in C after a precomputation that only involves the parameter [16].

3 Fpt-Reductions to SAT

Problems in NP and co-NP can be encoded into SAT in such a way that the time required to produce the encoding and consequently also the size of the resulting SAT instance are polynomial in the input (the encoding is a polynomial-time many-one reduction). Typically, the SAT encodings of problems proposed for practical use are of this kind (cf. [37]). For problems that are "beyond NP," say for problems on the second level of the PH, such polynomial SAT encodings do not exist, unless the PH collapses. However, for such problems, there still could exist SAT encodings which can be produced in fpt-time in terms of some parameter associated with the problem. In fact, such fpt-time SAT encodings have been obtained for various problems on the second level of the PH [15,24,36]. The classes para-NP and para-co-NP contain exactly those parameterized problems that admit such a many-one fpt-reduction to SAT and UNSAT, respectively. Thus, with fpt-time encodings, one can go significantly beyond what is possible by conventional polynomial-time SAT encodings.

Consider the following example. The problem of deciding satisfiability of a QBF does not allow a polynomial-time SAT encoding. However, if the number of universal variables is small, one can use known methods in QBF solving to get an fpt-time encoding into SAT.

QBF-SAT(# ∀-vars)
Instance: A QBF φ.
Parameter: The number of universally quantified variables of φ.
Question: Is φ true?

The idea behind this encoding is to repeatedly use universal quantifier expansion [2,5]. Eliminating k many universally quantified variables in this manner leads to an existentially quantified formula that is at most a factor of 2^k larger than the original formula.

Fpt-time encodings to SAT also have their limits. Clearly, para-Σ_2^P-hard and para-Π_2^P-hard parameterized problems do not admit fpt-time encodings to SAT, even when the parameter is a constant, unless the PH collapses. There are problems that apparently do not admit fpt-time encodings to SAT, but are neither para-Σ_2^P-hard nor para-Π_2^P-hard. In recent work [24] we have introduced several complexity classes for such intermediate problems, including the following.

The parameterized complexity class $\exists^k \forall^*$ consists of all parameterized problems that can be many-one fpt-reduced to the following variant of quantified Boolean satisfiability that is based on truth assignments of restricted weight.

$\exists^k \forall^*$-WSAT

Instance: A quantified Boolean formula $\varphi = \exists X.\forall Y.\psi$, and an integer k.
Parameter: k.
Question: Does there exist a truth assignment α to X with weight k such that for all truth assignments β to Y the assignment $\alpha \cup \beta$ satisfies ψ?

For each problem in $\exists^k \forall^*$ there exists an xp-reduction to UNSAT. However, there is evidence that problems that are hard for $\exists^k \forall^*$ do not allow a many-one fpt-reduction to SAT [24]. Many natural parameterized problems from various domains are complete for the class $\exists^k \forall^*$, and for none of them an fpt-reduction to SAT or UNSAT has been found. If there exists an fpt-reduction to SAT for any $\exists^k \forall^*$-complete problem then this is the case for all $\exists^k \forall^*$-complete problems. The dual complexity class of $\exists^k \forall^*$ is denoted by $\forall^k \exists^*$, and has similar (yet dual) properties. Note that the notion of reducibility underlying hardness for all parameterized complexity classes mentioned above is that of many-one fpt-reductions. For a more detailed discussion on the complexity classes $\exists^k \forall^*$ and $\forall^k \exists^*$, we refer to previous work [24].

One can also enhance the power of polynomial-time SAT encodings by considering polynomial-time algorithms that can query a SAT solver multiple times. Such an approach has been shown to be quite effective in practice (see, e.g., [3,14,32]) and extends the scope of SAT solvers to problems in the class Δ_2^P, but not to problems that are Σ_2^P-hard or Π_2^P-hard. Also here, switching from polynomial-time to fpt-time provides a significant increase in power. The class para-Δ_2^P contains all parameterized problems that can be solved by an fpt-algorithm that can query a SAT solver multiple times (i.e., by an fpt-time Turing reduction to SAT).

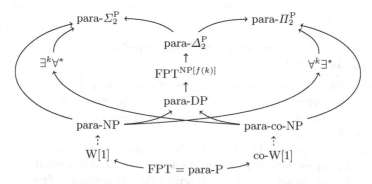

Fig. 1. Parameterized complexity classes up to the second level of the polynomial hierarchy. Arrows indicate inclusion relations. (We omit the full definition of some of the parameterized complexity classes depicted in the figure. For a detailed definition of these, we refer to other sources [13,17,35].)

An overview of all relevant parameterized complexity classes can be found in Figure 1. Locating problems in the complexity landscape as laid out in this figure can provide a guideline for practitioners, to indicate what algorithmic approaches are possible and where complexity theoretic obstacles lie.

There are two fundamental aspects that are relevant for an algorithm that makes queries to a SAT solver: (i) the running time of the algorithm (which does not take into account the time needed by the SAT solver to answer the queries) and (ii) the number of SAT calls. Results from classical *bounded query complexity* [25,42] suggests that if the running time is polynomial, then increasing the numbers of SAT calls increases the computing power. Several such separation results are known [12,30]. From a practical point of view, the number of SAT calls may seem to be relatively insignificant, assuming that the queries are easy for the solver, and the solver can reuse information from previous calls [4,26,43]. For a theoretical worst-case model, however, one must assume that all queries involve hard SAT instances, and that no information from previous calls can be reused. Therefore, in a theoretical analysis, it makes sense to study the number of SAT calls made by fpt-time algorithms. In the parameterized setting, it is natural to bound the number of SAT calls by a function of the parameter. This yields the class $\mathrm{FPT}^{\mathrm{NP}[f(k)]}$, which lies between para-DP (two calls) and para-Δ_2^{P} (unrestricted number of calls).

4   Minimization Problems for DNF Formulas

We consider several problems related to minimizing implicants of DNF formulas and minimizing DNF formulas. We consider several parameterizations, and we show that some of these allow an fpt-reduction to SAT, whereas others apparently do not.

The following decision problem, that is concerned with reducing a given implicant of a DNF formula in size, is Σ_2^{P}-complete [41].

SHORTEST-IMPLICANT-CORE
Instance: A DNF formula φ, an implicant C of φ of size n, and an integer m.
Question: Does there exist an implicant $C' \subseteq C$ of φ of size m?

We consider two parameterizations of this problem: (1) SHORTEST-IMPLICANT-CORE(core size), where the parameter $k = m$ is the size of the minimized implicant, and (2) SHORTEST-IMPLICANT-CORE(reduction size), where the parameter $k = n - m$ is the difference in size between the original implicant and the minimized implicant. We show that neither of these restrictions is enough to admit an fpt-reduction to SAT. All results can straightforwardly be extended to the variant of the problem where implicants of size at most m are accepted.

Next, consider the following decision problem, that is concerned with deciding whether a given DNF formula φ is logically equivalent to a DNF formula φ' of size m, and that is Σ_2^{P}-complete [41].

DNF-MINIMIZATION

Instance: A DNF formula φ of size n, and an integer m.
Question: Does there exist a term-wise subformula φ' of φ of size m such that $\varphi \equiv \varphi'$?

We consider the following two parameterizations of this problem: (1) DNF-MINIMIZATION(reduction size), where the parameter $k = n - m$ is the difference in size between the original formula φ and the minimized formula φ', and (2) DNF-MINIMIZATION(core size), where the parameter $k = m$ is the size of the minimized formula φ'. We show that the former parameterization is not enough to allow an fpt-reduction to SAT, but that for the latter parameterization, the problem can be solved with an fpt-algorithm that uses at most $\lceil \log_2 k \rceil + 1$ many SAT calls. Moreover, this algorithm works even for the case where equivalent DNF formulas that are not term-wise subformulas of φ are also accepted.

We will now set out to prove the complexity results mentioned in the discussion above. In order to prove $\exists^k \forall^*$-hardness of SHORTEST-IMPLICANT-CORE(core size), we need the following technical lemma (we omit its straightforward proof).

Lemma 1. *Let (φ, k) be an instance of $\exists^k \forall^*$-WSAT. In polynomial time, we can construct an equivalent instance (φ', k) of $\exists^k \forall^*$-WSAT with $\varphi' = \exists X. \forall Y. \psi$, such that for every assignment $\alpha : X \to \{0, 1\}$ that has weight $m \neq k$, it holds that $\forall Y. \psi[\alpha]$ is true.*

Proposition 1. SHORTEST-IMPLICANT-CORE(core size) *is $\exists^k \forall^*$-complete.*

Proof (sketch). To show hardness, we give a many-one fpt-reduction from $\exists^k \forall^*$-WSAT(DNF) to SHORTEST-IMPLICANT-CORE(core size). Intuitively, the choice for some $C' \subseteq C$ with $|C'| = k$ corresponds directly to the choice of some assignment $\alpha : X \to \{0, 1\}$ of weight k. Both involve a choice between $\binom{n}{k}$ many candidates, and in both cases verifying whether the chosen candidate witnesses that the instance is a yes-instance involves solving a co-NP-complete problem. Any implicant C' forces those variables x that are included in C' to be set to true (and the other variables are not forced to take any truth value). However, by Lemma 1, any assignment that sets more than k variables x to true will trivially satisfy ψ. Therefore, the only relevant assignment is the assignment that sets only those x to true that are forced to be true by some C' of length k, and hence the choice for such a C' corresponds exactly to the choice for some assignment α of weight k. To verify whether some C' of length k is an implicant of the formula φ is equivalent to checking whether the formula $\bigwedge_{c \in C'} c \wedge \varphi$ is valid, which in turn is equivalent to checking whether a formula $\forall Y. \psi[\alpha]$ is true, for some assignment α.

Let (φ, k) be an instance of $\exists^k \forall^*$-WSAT(DNF), with $\varphi = \exists X. \forall Y. \psi$. By Lemma 1, we may assume without loss of generality that for any assignment $\alpha : X \to \{0, 1\}$ of weight $m \neq k$, $\forall Y. \psi[\alpha]$ is true. We may also assume without loss of generality that $|X| > k$; if this were not the case, (φ, k) would trivially be a no-instance. We construct an instance (φ', C, k) of SHORTEST-IMPLICANT-CORE(core size) by letting $\text{Var}(\varphi') = X \cup Y$, $C = \bigwedge_{x \in X} x$, and $\varphi' = \psi$. Clearly,

φ' is a Boolean formula in DNF. Also, consider the assignment $\alpha : X \to \{0,1\}$ where $\alpha(x) = 1$ for all $x \in X$. We know that $\forall Y.\psi[\alpha]$ is true, since α has weight more than k. Therefore C is an implicant of φ'. We omit a detailed proof of correctness for this reduction.

To show membership in $\exists^k\forall^*$, we give a many-one fpt-reduction from SHORTEST-IMPLICANT-CORE(core size) to $\exists^k\forall^*$-WSAT. This reduction uses exactly the same similarity between the two problems, i.e., the fact that assignments of weight k correspond exactly to implicants of length k, and that verifying whether this choice witnesses that the instance is a yes-instance in both cases involves checking validity of a propositional formula. We describe the reduction, and omit a detailed proof of correctness. Let (φ, C, k) be an instance of SHORTEST-IMPLICANT-CORE(core size), where $C = \{c_1, \ldots, c_n\}$. We construct an instance (φ', k) of $\exists^k\forall^*$-WSAT, where $\varphi' = \exists X.\forall Y.\psi$, by defining $X = \{x_1, \ldots, x_n\}$, $Y = \mathrm{Var}(\varphi)$, $\psi = \psi^{X,Y}_{\mathrm{corr}} \to \varphi$, and $\psi^{X,Y}_{\mathrm{corr}} = \bigwedge_{1 \le i \le n}(x_i \to c_i)$. □

Proposition 2. SHORTEST-IMPLICANT-CORE(reduction size)   is $\exists^k\forall^*$-complete.

Proof (sketch). As an auxiliary problem, we consider the parameterized problem $\exists^{n-k}\forall^*$-WSAT, which is a variant of $\exists^k\forall^*$-WSAT. Given an input consisting of a QDNF $\varphi = \exists X.\forall Y.\psi$ with $|X| = n$ and an integer k, the problem is to decide whether there exists an assignment α to X with weight $n - k$ such that $\forall Y.\psi[\alpha]$ is true. The parameter for this problem is k. We claim that this problem has the following properties. We omit a proof of these claims.

Claim 1. $\exists^{n-k}\forall^*$-WSAT is $\exists^k\forall^*$-complete.

Claim 2. Let (φ, k) be an instance of $\exists^{n-k}\forall^*$-WSAT. In polynomial time, we can construct an equivalent instance (φ', k) of $\exists^{n-k}\forall^*$-WSAT with $\varphi' = \exists X.\forall Y.\psi$, such that for any assignment $\alpha : X \to \{0,1\}$ that has weight $m \neq (|X| - k)$, it holds that $\forall Y.\psi[\alpha]$ is true.

Using these claims, both membership and hardness for $\exists^k\forall^*$ follow straightforwardly using arguments similar to the $\exists^k\forall^*$-completeness proof of SHORTEST-IMPLICANT-CORE(core size). The fpt-reductions in the proof of Proposition 1 show that SHORTEST-IMPLICANT-CORE(reduction size) fpt-reduces to and from $\exists^{n-k}\forall^*$-WSAT. □

We can now turn to proving complexity results for the problems of minimizing DNF formulas.

Proposition 3. DNF-MINIMIZATION(reduction size) *is* $\exists^k\forall^*$-complete.

Proof (sketch). To show $\exists^k\forall^*$-hardness, we use the reduction from the literature that is used to show Σ^P_2-hardness of the unparameterized version of DNF-MINIMIZATION(reduction size). The polynomial-time reduction from the unparameterized version of SHORTEST-IMPLICANT-CORE(reduction size) to the unparameterized version of DNF-MINIMIZATION(reduction size) given by Umans [41, Theorem 2.2] is an fpt-reduction from SHORTEST-IMPLICANT-CORE(reduction size) to DNF-MINIMIZATION(reduction size).

To show membership in $\exists^k\forall^*$, one can give an fpt-reduction to $\exists^k\forall^*$-WSAT. We describe the main idea behind this reduction, and we omit a detailed proof. Given an instance (φ, k) of DNF-MINIMIZATION(reduction size) we construct an instance (φ', k) of $\exists^k\forall^*$-WSAT where the assignment to the existentially quantified variables of φ' represents the k many literal occurrences that are to be removed, and where universally quantified part of φ' is used to verify the equivalence of φ and the formula obtained from φ by removing the k literals chosen by the assignment to the existential variables. □

The following result, which we give without proof, gives some first upper and lower bounds on the complexity of DNF-MINIMIZATION(core size).

Proposition 4. DNF-MINIMIZATION(core size) *is* para-co-NP-*hard and is in* $\exists^k\forall^*$.

Next, we turn our attention to an fpt-algorithm that solves DNF-MINIMIZATION(core size) by using $f(k)$ many SAT calls, for some computable function f. In order to so, we will define the notion of relevant variables, and establish several lemmas that help us to describe and analyze the algorithm (the first of which we state without proof).

Let φ be a DNF formula and let $x \in \mathrm{Var}(\varphi)$ be a variable occurring in φ. We call x *relevant in* φ if there exists some assignment $\alpha : \mathrm{Var}(\varphi)\backslash\{x\} \to \{0,1\}$ such that $\varphi[\alpha \cup \{x \mapsto 0\}] \neq \varphi[\alpha \cup \{x \mapsto 1\}]$.

Lemma 2. *Let φ be a DNF formula and let φ' be a DNF formula of minimal size that it is equivalent to φ. Then for every variable $x \in \mathrm{Var}(\varphi)$ it holds that $x \in \mathrm{Var}(\varphi')$ if and only if x is relevant in φ.*

Lemma 3. *Given a DNF formula φ and a positive integer m (given in unary), deciding whether there are at least m variables that are relevant in φ is in NP.*

Proof. We describe a guess-and-check algorithm that decides the problem. The algorithm first guesses m distinct variables occurring in φ, and for each guessed variable x the algorithm guesses an assignment α_x to the remaining variables $\mathrm{Var}(\varphi)\backslash\{x\}$. Then, the algorithm verifies whether the guessed variables are really relevant by checking that, under α_x, assigning different values to x changes the outcome of the Boolean function represented by φ, i.e., $\varphi[\alpha_x \cup \{x \mapsto 0\}] \neq \varphi[\alpha_x \cup \{x \mapsto 1\}]$. It is straightforward to construct a SAT instance ψ that implements this guess-and-check procedure. Moreover, from any assignment that satisfies ψ it is easy to extract the relevant variables. □

Lemma 4. *Let $x_1, \ldots, x_k$ be propositional variables. There are $2^{O(k \log k)}$ many different DNF formulas ψ over the variables $x_1, \ldots, x_k$ that are of size k.*

Proof. Each suitable DNF formula $\psi = t_1 \vee \cdots \vee t_\ell$ can be formed by writing down a sequence $\sigma = (l_1, \ldots, l_k)$ of literals l_i over $x_1, \ldots, x_k$, and splitting this sequence into terms, i.e., choosing integers $1 = d_1 < \cdots < d_{\ell+1} = k + 1$ such that $t_i = \{l_{d_i}, \ldots, l_{d_{i+1}-1}\}$ for each $1 \leq i \leq \ell$. To see that there are $2^{O(k \log k)}$ many formulas ψ, it suffices to see that there are $O(k^k)$ many sequences σ, and $O(2^k)$ many choices for the integers d_i. □

Algorithm 1. Solving DNF-MINIMIZATION(core size) in fpt-time using ($\lceil \log_2 k \rceil + 1$) many SAT calls.

---

input : an instance (φ, k) of DNF-MINIMIZATION(core size)
output: YES iff $(\varphi, k) \in$ DNF-MINIMIZATION(core size)

rvars $\leftarrow \emptyset$;   // variables relevant in φ
$i \leftarrow 0; j \leftarrow k + 2$;   // bounds on # of rvars
while $i + 1 < j$ **do**   // logarithmic search for the # of rvars
$\quad | \quad \ell \leftarrow \lceil (i+j)/2 \rceil$;
$\quad | \quad$ query the SAT solver whether there exist at least ℓ variables
$\quad | \quad \quad \quad$ that are relevant in φ ; // for idea behind encoding, see Lemma 3
$\quad | \quad$ **if** *the SAT solver returns a model M* **then**
$\quad | \quad \quad | \quad$ rvars $\leftarrow$ the ℓ many relevant variables encoded by the model M ;
$\quad | \quad$ **else break**
if $|$rvars$| > k$ **then**
$\quad | \quad$ **return** NO ;   // too many rvars for any DNF of size $\leq k$
else
$\quad | \quad$ **foreach** *DNF formula ψ of size k over var's in* rvars **do** // $2^{O(k \log k)}$ many
$\quad | \quad \quad | \quad$ construct a formula φ_ψ that is unsatisfiable iff $\psi \equiv \varphi$;
$\quad | \quad \quad \quad \quad \quad$ // the formulas φ_ψ must be variable disjoint
$\quad | \quad$ query the SAT solver whether $\bigwedge_\psi \varphi_\psi$ is satisfiable ;
$\quad | \quad$ **if** *the SAT solver returns YES* **then**
$\quad | \quad \quad | \quad$ **return** NO ;   // no candidate ψ is equivalent to φ
$\quad | \quad$ **else**
$\quad | \quad \quad | \quad$ **return** YES ;   // some candidate ψ is equivalent to φ

Theorem 1. DNF-MINIMIZATION(core size) *can be solved by an fpt-algorithm that uses* ($\lceil \log_2 k \rceil + 1$) *many SAT calls, where the SAT solver returns a model for satisfiable formulas. Moreover, the first $\lceil \log_2 k \rceil$ many calls to the solver use SAT instances of size $O(k^2 n^2)$, whereas the last call uses a SAT instance of size $2^{O(k \log k)} \cdot n$, where n is the input size.*

Proof. The algorithm given in pseudo-code in Algorithm 1 solves the problem DNF-MINIMIZATION(core size) in the required time bounds. To obtain the required running time, we assume that each call to a SAT solver takes only a single time step. By Lemma 2, we know that any minimal equivalent formula of φ must contain all and only the variables that are relevant in φ. The algorithm firstly determines how many variables are relevant in φ. By Lemma 3, we know that this can be done with a binary search using $\lceil \log_2 k \rceil$ SAT calls. If there are more than k relevant variables, the algorithm rejects. Otherwise, the algorithm will have computed the set rvars of relevant variables. Next, with a single SAT call, it checks whether there exists some DNF formula ψ of size k over the variables in rvars. By Lemma 4, we know that there are $2^{O(k \log k)}$ many different DNF formulas ψ of size k over the variables in rvars. Verifying whether a particular DNF formula ψ is equivalent to the original formula φ can be done by checking whether the formula $\varphi_\psi = (\psi \wedge \neg\varphi) \vee (\neg\psi \wedge \varphi)$ is unsatisfiable. Verifying whether there exists some suitable DNF formula ψ that is equivalent to φ can be done by making variable-disjoint copies of all φ_ψ and checking whether the conjunction of these copies is unsatisfiable.   □

Note that the algorithm requires that the SAT solver returns a model if the query is satisfiable. Also, the algorithm can be modified straightforwardly to return a DNF formula ψ of size at most k that is equivalent to an input φ if such a formula ψ exists. It would need to search for this ψ that is equivalent to φ, for which it would need an additional $O(k \log k)$ many SAT calls (with instances of size $2^{O(k \log k)} \cdot \|\varphi\|$).

An interesting topic for further research is to investigate how many SAT calls are needed for an fpt-algorithm to produce an equivalent DNF when the SAT solver only returns whether or not the input is satisfiable, and does not return a satisfying assignment in case the input is satisfiable. For decision problems, the difference between these two interfaces to SAT solvers is (theoretically) not relevant, when allowing more than a constant number of calls to the solver [29, Lemma 6.3.4]. For instance, when allowing a logarithmic number of calls, using witnessed and non-witnessed SAT calls yields the same computational power [29, Corollary 6.3.5]. For function problems, on the other hand, the difference does seem to be relevant, in cases where the number of calls is bounded to logarithmically many in the input size (cf. [21, Theorem 5.4]) or bounded by a function of the parameter.

From a practical point of view, the algorithm given in Algorithm 1 might not be the best approach to solve the problem. The (single) instance produced for the last SAT call in the algorithm is rather large (exponential in k). However, this instance is equivalent to the conjunction of $2^{O(k \log k)}$ many instances of linear size, and these instances can be solved in parallel. Such a parallel approach involves more (yet easier) SAT calls, but might be more efficient in practice.

5   QBF Satisfiability and Treewidth

The graph parameter *treewidth* measures in a certain sense the tree-likeness of a graph (for a definition of treewidth, see, e.g., [10,11]). Many hard problems are fixed-parameter tractable when parameterized by the treewidth of a graph associated with the input [11,22]. By associating the following graph with a QDNF formulas one can apply the parameter treewidth also to QDNF formulas (for QCNF formulas the graph can be defined analogously, taking clauses instead of terms).

The *incidence graph* of a QDNF formula φ is the bipartite graph where one side of the partition consists of the variables and the other side consists of the terms; a variable and a term are adjacent if the variable appears positively or negatively in the term. The *incidence treewidth* of φ, in symbols incid.tw(φ), is the treewidth of the incidence graph of φ. It is well known that checking the truth of a QDNF formula whose number of quantifier alternations is bounded by a constant is fixed-parameter tractable when parameterized by incid.tw (this can be easily shown using Courcelle's Theorem [22]).

Bounding the treewidth of the entire incidence graph is very restrictive. In this section we investigate whether bounding the treewidth of certain subgraphs of the incidence graph is sufficient to reduce the complexity. To this aim we define the *existential incidence treewidth* of a QDNF formula φ, in symbols $\exists$-incid.tw(φ), as the treewidth of the incidence graph of φ after deletion of all universal variables. The *universal incidence treewidth*, in symbols $\forall$-incid.tw(φ), is the treewidth of the incidence graph of φ after deletion of all existential variables.

The existential and universal treewidth can be small for formulas whose incidence treewidth is arbitrarily large. Take for instance a QDNF formula φ whose incidence graph is an $n \times n$ square grid, as in Figure 2. In this example, $\exists$-incid.tw$(\varphi) = \forall$-incid.tw$(\varphi) = 2$ (since after the deletion of the universal or the existential variables the incidence graph becomes a collection of trivial path-like graphs), but incid.tw$(\varphi) = n$ [10]. Hence, a tractability result in terms of existential or universal incidence treewidth would apply to a significantly larger class of instances than a tractability result in terms of incidence treewidth. In the

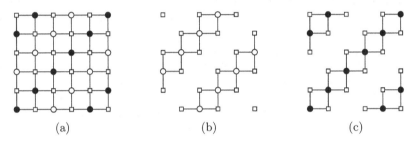

      (a)   (b)   (c)

Fig. 2. Incidence graph of a QDNF formula (a). Universal variables are drawn with black round shapes, existential variables with white round shapes, and terms are drawn with square shapes. Both deleting the universal variables (b) and the existential variables (c) significantly decreases the treewidth of the incidence graph.

following we pinpoint the exact complexity of checking the satisfiability of $\exists\forall$-QDNF formulas parameterized by $\exists$-incid.tw and $\forall$-incid.tw. We let $\exists\forall$-SAT($\forall$-incid.tw) denote the problem $\exists\forall$-SAT parameterized by $\forall$-incid.tw, and similarly we let $\exists\forall$-SAT($\exists$-incid.tw) denote the problem $\exists\forall$-SAT parameterized by $\exists$-incid.tw. We show that parameterizing by $\exists$-incid.tw does not decrease the complexity and leaves the problem on the second level of the PH, but for $\exists\forall$-SAT($\forall$-incid.tw) we get an fpt-reduction to SAT.

Proposition 5. $\exists\forall$-SAT($\exists$-incid.tw) *is para-Σ_2^P-complete.*

Proof. Membership in para-Σ_2^P is obvious. To show para-Σ_2^P-hardness, it suffices to show that the problem is already Σ_2^P-hard when the parameter value is restricted to 1 [16]. We show this by means of a reduction from $\exists\forall$-SAT. The idea of this reduction is to introduce for each existentially quantified variable x a corresponding universally quantified variable z_x that is used to represent the truth value assigned to x. Each of the existentially quantified variables then only directly interacts with universally quantified variables.

Take an arbitrary instance of $\exists\forall$-SAT, specified by $\varphi = \exists X.\forall Y.\psi(X,Y)$, where $\psi(X,Y)$ is in DNF. We introduce a new set $Z = \{\, z_x : x \in X \,\}$ of variables. It is straightforward to verify that $\varphi = \exists X.\forall Y.\psi(X,Y)$ is equivalent to the formula $\exists Z.\forall X.\forall Y.\chi$, where $\chi = \bigvee_{x \in X}[(x \wedge \neg z_x) \vee (\neg x \wedge z_x)] \vee \psi(X,Y)$. Clearly, if we now delete all universally quantified variables, the incidence graph of χ consists only of isolated paths of length 2, and therefore the treewidth is 1. This proves that $\exists\forall$-SAT($\exists$-incid.tw) is para-Σ_2^P-hard.   □

Theorem 2. $\exists\forall$-SAT($\forall$-incid.tw) *is para-NP-complete.*

Proof. Hardness for para-NP can be proven by showing that the problem is already NP-hard when restricted to instances where the parameter value is 1 [16]. In order to do this, one can reduce an instance of SAT to an instance of $\exists\forall$-SAT whose matrix is in DNF by using the standard Tseitin transformation, resulting in tree-like interactions between the universally quantified variables. Therefore, the resulting formula has universal incidence treewidth 1.

We now show para-NP-membership of $\exists\forall$-SAT($\forall$-incid.tw). We construct a CNF formula φ' that is satisfiable if and only if φ is true. The idea is to construct a formula that encodes the following guess-and-check algorithm. Firstly, the algorithm guesses an assignment γ to the existential variables. Note that the incidence graph of the formula instantiated with γ has a small treewidth, because instantiating with γ removes all existentially quantified variables. Then, the algorithm employs dynamic programming to exploit the fact that the incidence graph of the remaining formula has small treewidth to decide validity of the remaining formula. This dynamic programming approach is widely used to solve problems for instances where some graph representing the structure of the instance has small treewidth (cf. [9]).

Next, we show how to encode this guess-and-check algorithm into a formula φ' that is satisfiable if and only if the algorithm accepts. In order to do so, we formally define treewidth and tree decompositions of graphs. A tree decomposition of a graph $G = (V,E)$ is a pair $(\mathcal{T}, (B_t)_{t \in T})$ where $\mathcal{T} = (T,F)$ is a rooted tree and $(B_t)_{t \in T}$ is a family of subsets of V such that: (i) for every $v \in V$, the set $B^{-1}(v) = \{\, t \in T : v \in B_t \,\}$ induces a nonempty subtree of $\mathcal{T}$; and (ii) for every edge $\{v,w\} \in E$, there is a $t \in T$ such that $v, w \in B_t$. In order to simplify the proof, we will consider the following normal form of tree decompositions. We call a tree decomposition $(\mathcal{T}, (B_t)_{t \in T})$ *nice* if every node $t \in T$ is of one of the following four types: (*leaf node*) t has no children and $|B_t| = 1$; (*introduce node*) t has one child t' and $B_t = B_{t'} \cup \{v\}$ for some vertex $v \notin B_{t'}$; (*forget node*) t has one child t' and $B_t = B_{t'} \setminus \{v\}$ for some vertex $v \in B_{t'}$; or (*join node*) t has two children t_1, t_2 and $B_t = B_{t_1} = B_{t_2}$. Given any graph G and a tree decomposition of G of width k, a nice tree decomposition of G of width k can be computed in polynomial time [11,27].

Let $\varphi = \exists X.\forall Y.\psi$ be a quantified Boolean formula where $\psi = \delta_1 \vee \cdots \vee \delta_u$, and let $(\mathcal{T}, (B_t)_{t \in T})$ be a tree decomposition of width k of the incidence graph of φ after deletion of the existentially quantified variables. We may assume without loss of generality that $(\mathcal{T}, (B_t)_{t \in T})$ is a nice tree decomposition. We may also assume without loss of generality that for each $t \in T$, B_t contains some $y \in Y$.

We let $\mathrm{Var}(\varphi') = X \cup Z$ where $Z = \{ z_{t,\alpha,i} : t \in T, \alpha : \mathrm{Var}(t) \to \{0,1\}, 1 \leq i \leq u \}$. Intuitively, the variables $z_{t,\alpha,i}$ represent whether at least one assignment extending α (to the variables occurring in nodes t' below t) violates the term δ_i of ψ. We then construct φ' as follows by using the structure of the tree decomposition. For all $t \in T$, all $\alpha : \mathrm{Var}(t) \to \{0,1\}$, all $1 \leq i \leq u$, and each literal $l \in \delta_i$ such that $\mathrm{Var}(l) \in X$, we introduce the clause (I): $(\bar{l} \to z_{t,\alpha,i})$. Then, for all $t \in T$, all $\alpha : \mathrm{Var}(t) \to \{0,1\}$, and all $1 \leq i \leq u$ such that for some $l \in \delta_i$ it holds that $\mathrm{Var}(l) \in Y$ and $\alpha(l) = 0$, we introduce the clause (II): $(z_{t,\alpha,i})$. Next, let $t \in T$ be any introduction node with child t', and let $\alpha : \mathrm{Var}(t') \to \{0,1\}$ be an arbitrary assignment. For any assignment $\alpha' : \mathrm{Var}(t) \to \{0,1\}$ that extends α, and for each $1 \leq i \leq u$, we introduce the clause (III): $(z_{t',\alpha,i} \to z_{t,\alpha',i})$. Then, let $t \in T$ be any forget node with child t', and let $\alpha : \mathrm{Var}(t) \to \{0,1\}$ be an arbitrary assignment. For any assignment $\alpha' : \mathrm{Var}(t') \to \{0,1\}$ that extends α, and for each $1 \leq i \leq u$, we introduce the clause (IV): $(z_{t',\alpha',i} \to z_{t,\alpha,i})$. Next, let $t \in T$ be any join node with children t_1, t_2, and let $\alpha : \mathrm{Var}(t) \to \{0,1\}$ be an arbitrary assignment. For each $1 \leq i \leq u$, we introduce the clauses (V): $(z_{t_1,\alpha,i} \to z_{t,\alpha,i})$ and $(z_{t_2,\alpha,i} \to z_{t,\alpha,i})$. Finally, for the root node $t_{\mathrm{root}} \in T$ and for each $\alpha : \mathrm{Var}(t_{\mathrm{root}}) \to \{0,1\}$ we introduce the clause (VI): $\bigvee_{1 \leq i \leq u} \neg z_{t_{\mathrm{root}},\alpha,i}$. It is straightforward to verify that φ' contains $O(2^k|T|)$ many clauses. We claim that this reduction is correct, i.e., that φ is true if and only if φ' is satisfiable. We omit a detailed proof of this. $\qquad\square$

Instead of incidence graphs one can also use *primal graphs* to model the structure of QBF formulas (see, e.g., [1,34]). One can define corresponding parameters primal treewidth, universal primal treewidth, and existential primal treewidth. The proof of Proposition 5 shows that $\exists\forall$-SAT is para-Σ_2^P-hard when parameterized by existential primal treewidth. The parameter incidence treewidth is more general than primal treewidth in the sense that small primal treewidth implies small incidence treewidth [28,39], but the converse does not hold in general. Hence, Theorem 2 also holds for the parameter universal primal treewidth.

6 Conclusion

We studied the fpt-reducibility to SAT for several problems beyond NP under natural parameters. Our positive results show that in some cases it is possible to utilize structure in terms of parameters to break through the barriers between classical complexity classes. Parameters that admit an fpt-reduction to SAT can be less restrictive than parameters that provide fixed-parameter tractability of the problem itself, hence our approach extends the scope of fixed-parameter tractability. Additionally, we show that fpt-time algorithms that can query a SAT solver (i.e., fpt-time Turing reductions to SAT) exist for some problems that cannot be solved in polynomial-time with the help of queries to a SAT solver (unless the PH collapses), hence our approach also extends the scope of algorithms using SAT queries. Our negative results point out the limits of the approach, showing that some problems do, most likely, not admit fpt-reductions to SAT under certain natural parameters.

References

1. Atserias, A., Oliva, S.: Bounded-width QBF is PSPACE-complete. In: Portier, N., Wilke, T. (eds.) 30th International Symposium on Theoretical Aspects of Computer Science, STACS 2013, Kiel, Germany, February 27 - March 2. LIPIcs, vol. 20, pp. 44–54. Schloss Dagstuhl - Leibniz-Zentrum fuer Informatik (2013)
2. Ayari, A., Basin, D.: Qubos: Deciding quantified Boolean logic using propositional satisfiability solvers. In: Aagaard, M.D., O'Leary, J.W. (eds.) FMCAD 2002. LNCS, vol. 2517, pp. 187–201. Springer, Heidelberg (2002)
3. Belov, A., Lynce, I., Marques-Silva, J.: Towards efficient MUS extraction. AI Commun. 25(2), 97–116 (2012)
4. Benedetti, M., Bernardini, S.: Incremental compilation-to-SAT procedures. In: Hoos, H.H., Mitchell, D.G. (eds.) SAT 2004. LNCS, vol. 3542, pp. 46–58. Springer, Heidelberg (2005)
5. Biere, A.: Resolve and expand. In: Hoos, H.H., Mitchell, D.G. (eds.) SAT 2004. LNCS, vol. 3542, pp. 59–70. Springer, Heidelberg (2005)
6. Biere, A.: Bounded model checking. In: Biere, A., Heule, M., van Maaren, H., Walsh, T. (eds.) Handbook of Satisfiability, Frontiers in Artificial Intelligence and Applications, vol. 185, pp. 457–481. IOS Press (2009)
7. Biere, A., Cimatti, A., Clarke, E., Zhu, Y.: Symbolic model checking without BDDs. In: Cleaveland, W.R. (ed.) TACAS/ETAPS 1999. LNCS, vol. 1579, pp. 193–207. Springer, Heidelberg (1999)
8. Biere, A., Heule, M., van Maaren, H., Walsh, T. (eds.): Handbook of Satisfiability. Frontiers in Artificial Intelligence and Applications, vol. 185. IOS Press (2009)
9. Bodlaender, H.L.: Dynamic programming on graphs with bounded treewidth. In: Lepistö, T., Salomaa, A. (eds.) ICALP 1988. LNCS, vol. 317, pp. 105–118. Springer, Heidelberg (1988)
10. Bodlaender, H.L.: A partial k-arboretum of graphs with bounded treewidth. Theoretical Computer Science 209(1-2), 1–45 (1998)
11. Bodlaender, H.L.: Fixed-parameter tractability of treewidth and pathwidth. In: Bodlaender, H.L., Downey, R., Fomin, F.V., Marx, D. (eds.) Fellows Festschrift 2012. LNCS, vol. 7370, pp. 196–227. Springer, Heidelberg (2012)
12. Chang, R., Kadin, J.: The Boolean Hierarchy and the Polynomial Hierarchy: A closer connection. SIAM J. Comput. 25(2), 340–354 (1996)
13. Downey, R.G., Fellows, M.R.: Parameterized Complexity. Monographs in Computer Science. Springer, New York (1999)
14. Dvořák, W., Järvisalo, M., Wallner, J.P., Woltran, S.: Complexity-sensitive decision procedures for abstract argumentation. Artificial Intelligence 206(0), 53–78 (2014)
15. Fichte, J.K., Szeider, S.: Backdoors to normality for disjunctive logic programs. In: Proceedings of the Twenty-Seventh AAAI Conference on Artificial Intelligence, AAAI 2013, pp. 320–327. AAAI Press (2013)
16. Flum, J., Grohe, M.: Describing parameterized complexity classes. Information and Computation 187(2), 291–319 (2003)
17. Flum, J., Grohe, M.: Parameterized Complexity Theory, Texts in Theoretical Computer Science. An EATCS Series, vol. XIV. Springer, Berlin (2006)
18. Garey, M.R., Johnson, D.R.: Computers and Intractability. W. H. Freeman and Company, New York (1979)
19. Goldsmith, J., Hagen, M., Mundhenk, M.: Complexity of DNF minimization and isomorphism testing for monotone formulas. Information and Computation 206(6), 760–775 (2008)

20. Gomes, C.P., Kautz, H., Sabharwal, A., Selman, B.: Satisfiability solvers. In: Handbook of Knowledge Representation. Foundations of Artificial Intelligence, vol. 3, pp. 89–134. Elsevier (2008)
21. Gottlob, G., Fermüller, C.G.: Removing redundancy from a clause. Artificial Intelligence 61(2), 263–289 (1993)
22. Gottlob, G., Pichler, R., Wei, F.: Bounded treewidth as a key to tractability of knowledge representation and reasoning. Artificial Intelligence 174(1), 105–132 (2010)
23. Grégoire, E., Mazure, B., Piette, C.: On approaches to explaining infeasibility of sets of Boolean clauses. In: 20th IEEE International Conference on Tools with Artificial Intelligence (ICTAI 2008), Daytion, Ohio, USA, November 3-5 , pp. 74–83. IEEE Computer Society (2008)
24. de Haan, R., Szeider, S.: The parameterized complexity of reasoning problems beyond NP. In: Baral, C., De Giacomo, G., Eiter, T. (eds.) Principles of Knowledge Representation and Reasoning: Proceedings of the Fourteenth International Conference, Vienna, Austria, July 20-24. AAAI Press (2014)
25. Hartmanis, J.: New developments in structural complexity theory. Theoretical Computer Science 71(1), 79–93 (1990)
26. Hooker, J.N.: Solving the incremental satisfiability problem. J. Logic Programming 15(1&2), 177–186 (1993)
27. Kloks, T.: Treewidth: Computations and Approximations. Springer, Berlin (1994)
28. Kolaitis, P.G., Vardi, M.Y.: Conjunctive-query containment and constraint satisfaction. J. of Computer and System Sciences 61(2), 302–332 (2000), special issue on the Seventeenth ACM SIGACT-SIGMOD-SIGART Symposium on Principles of Database Systems (Seattle, WA, 1998)
29. Krajicek, J.: Bounded arithmetic, propositional logic and complexity theory. Cambridge University Press (1995)
30. Krentel, M.W.: The complexity of optimization problems. J. of Computer and System Sciences 36(3), 490–509 (1988)
31. Malik, S., Zhang, L.: Boolean satisfiability from theoretical hardness to practical success. Communications of the ACM 52(8), 76–82 (2009)
32. Marques-Silva, J., Janota, M., Belov, A.: Minimal sets over monotone predicates in Boolean formulae. In: Sharygina, N., Veith, H. (eds.) CAV 2013. LNCS, vol. 8044, pp. 592–607. Springer, Heidelberg (2013)
33. Niedermeier, R.: Invitation to Fixed-Parameter Algorithms. Oxford Lecture Series in Mathematics and its Applications. Oxford University Press, Oxford (2006)
34. Pan, G., Vardi, M.Y.: Fixed-parameter hierarchies inside PSPACE. In: Proceedings of 21th IEEE Symposium on Logic in Computer Science (LICS 2006), Seattle, WA, USA, August 12-15, pp. 27–36. IEEE Computer Society (2006)
35. Papadimitriou, C.H.: Computational Complexity. Addison-Wesley (1994)
36. Pfandler, A., Rümmele, S., Szeider, S.: Backdoors to abduction. In: Rossi, F. (ed.) Proceedings of the 23rd International Joint Conference on Artificial Intelligence, IJCAI 2013. AAAI Press/IJCAI (2013)
37. Prestwich, S.D.: CNF encodings. In: Biere, A., Heule, M., van Maaren, H., Walsh, T. (eds.) Handbook of Satisfiability, pp. 75–97. IOS Press (2009)
38. Sakallah, K.A., Marques-Silva, J.: Anatomy and empirical evaluation of modern SAT solvers. Bulletin of the European Association for Theoretical Computer Science 103, 96–121 (2011)
39. Samer, M., Szeider, S.: Constraint satisfaction with bounded treewidth revisited. J. of Computer and System Sciences 76(2), 103–114 (2010)

40. Stockmeyer, L.J.: The polynomial-time hierarchy. Theoretical Computer Science 3(1), 1–22 (1976)
41. Umans, C.: Approximability and Completeness in the Polynomial Hierarchy. Ph.D. thesis. University of California, Berkeley (2000)
42. Wagner, K.W.: Bounded query classes. SIAM J. Comput. 19(5), 833–846 (1990)
43. Whittemore, J., Kim, J., Sakallah, K.A.: SATIRE: A new incremental satisfiability engine. In: Proceedings of the 38th Design Automation Conference, DAC 2001, Las Vegas, NV, USA, June 18-22, pp. 542–545. ACM (2001)

On Reducing Maximum Independent Set
to Minimum Satisfiability

Alexey Ignatiev[1,3], Antonio Morgado[1], and Joao Marques-Silva[1,2]

[1] IST/INESC-ID, Lisbon, Portugal
[2] University College Dublin, Ireland
[3] ISDCT SB RAS, Irkutsk, Russia
{aign,ajrm}@sat.inesc-id.pt, jpms@ucd.ie

Abstract. Maximum Independent Set (MIS) is a well-known NP-hard graph problem, tightly related with other well known NP-hard graph problems, namely Minimum Vertex Cover (MVC) and Maximum Clique (MaxClq). This paper introduces a novel reduction of MIS into Minimum Satisfiability (MinSAT), thus, providing an alternative approach for solving MIS. The reduction naturally maps the vertices of a graph into clauses, without requiring the inclusion of hard clauses. Moreover, it is shown that the proposed reduction uses fewer variables and clauses than the existing alternative of mapping MIS into Maximum Satisfiability (MaxSAT). The paper develops a number of optimizations to the basic reduction, which significantly reduce the total number of variables used. The experimental evaluation considered the reductions described in the paper as well as existing state-of-the-art approaches. The results show that the proposed approaches based on MinSAT are competitive with existing approaches.

1 Introduction

Maximum Independent Set (MIS) is a well-known NP-hard graph problem, tightly related with other well known NP-hard graph problems, namely Minimum Vertex Cover (MVC) and Maximum Clique (MaxClq) [13]. These NP-hard graph problems find a wide range of practical applications, having been extensively studied in a number of settings over the last few decades (e.g. see [1, 4, 6–10, 12, 18, 23–25, 30, 35, 36, 42–46, 48, 49, 51, 56] and references therein). A large number of solutions have been developed for these NP-hard graph problems including complete algorithms, e.g. branch-and-bound search, but also incomplete algorithms, e.g. local search, genetic algorithms. These works also include recent algorithms for the MaxClq problem [35], as well as reductions of MaxClq to Maximum Satisfiability (MaxSAT) [36].

In contrast, and although work in Minimum Satisfiability (MinSAT) [27, 40] can be traced to the mid 90s, to our best knowledge, no *natural* applications have been described in the literature for MinSAT. For example, earlier work on MinSAT mainly targeted the development of inapproximability results for other combinatorial optimization problems. Admittedly, it is well-known how to reduce the different variants of MaxSAT to MinSAT (e.g. [19, 37]). For example, by flipping the polarity of literals when soft clauses are all unit, one obtains a MinSAT instance instead of a MaxSAT instance [37]. However, since algorithms for MaxSAT and MinSAT share many insights, these mappings are not expected to provide significant breakthroughs. Nevertheless, there has been significant recent research activity on algorithms for the MinSAT problem, and

C. Sinz and U. Egly (Eds.): SAT 2014, LNCS 8561, pp. 103–120, 2014.
© Springer International Publishing Switzerland 2014

its variants [19, 34, 37, 38], and so it is of interest to find combinatorial optimization problems that can be modeled as variants of MinSAT. A preliminary example towards achieving this goal is the encoding of WMaxCSP into weighted partial MinSAT, recently proposed in [3].

This paper represents another step towards identifying combinatorial problems that can be modeled as variants of MinSAT, in our case of (weighted) MinSAT. More concretely, this paper establishes a relationship between MIS (and other related NP-hard graph problems) and MinSAT, by showing how to reduce MIS to MinSAT. The reduction of MIS (and so of related graph problems) to MinSAT is significant due to the fact that it provides many concrete practical applications of MinSAT, something that to our best knowledge was not known. Besides the basic reduction (and associated proof), the paper also develops a number of optimizations which are shown to be crucially effective when solving MIS problem instances in practice.

The paper is organized as follows. Section 2 introduces the notation and definitions used throughout the paper. Section 3 develops the basic reduction of MIS to MinSAT. Section 4 develops optimizations to the basic reduction which in practice yield significant reductions in the number of used variables. Preliminary experimental results are analyzed in Section 5. Finally, Section 6 concludes the paper.

2   Preliminaries

This section briefly introduces the definitions used throughout the paper. Additional standard definitions can be found elsewhere (e.g. [5]). Boolean formulas are represented in calligraphic font: $\mathcal{F}, \mathcal{M}, \mathcal{S}, \mathcal{T}, \mathcal{U}, \mathcal{W}, \mathcal{F}'$, etc. A Boolean formula in conjunctive normal form (CNF) is defined as a finite set of finite sets of literals. Where appropriate, a CNF formula will also be understood as a conjunction of disjunctions of literals, where each disjunction represents a *clause* and a *literal* is a Boolean variable or its complement. Boolean variables are represented by $\{x, x_1, x_2, \ldots\}$, and literals by $\{l, l_1, l_2, \ldots\}$. The set of all variables of formula $\mathcal{F}$ is denoted by $\text{var}(\mathcal{F})$. The clauses of a formula are represented by $\{c, c_1, c_2, \ldots\}$. Two literals are said to be complementary, if one of the literals corresponds to a variable x, while the other corresponds to the negation of the variable, that is $\neg x$. An assignment is a mapping $\mathcal{A} : \text{var}(\mathcal{F}) \mapsto \{0, 1\}$. A clause is satisfied by an assignment if one of its literals is assigned value 1. A model of $\mathcal{F}$ is an assignment that satisfies all clauses in $\mathcal{F}$.

The standard definitions of MinSAT and MaxSAT are assumed (e.g. [32, 38]). In the context of MinSAT and MaxSAT, a formula $\mathcal{F}$ is viewed as a 2-tuple $(\mathcal{H}, \mathcal{R})$, where $\mathcal{H}$ denotes the *hard* clauses, which must be satisfied, and $\mathcal{R}$ denotes the *soft* (or *relaxable*) clauses. A weight can be associated with each clause, such that hard clauses have a special weight $\top$. Hence, the weight function is a mapping $w : \mathcal{H} \cup \mathcal{R} \to \{\top\} \cup \mathbb{N}$, such that $\forall_{c \in \mathcal{H}} w(c) = \top$ and $\sum_{c \in \mathcal{R}} w(c) < \top$. If no weight function is specified, it is assumed that $\forall_{c \in \mathcal{R}} w(c) = 1$.

The MinSAT and MaxSAT problems are defined as follows. The *Minimum/Maximum Satisfiability* (MinSAT/MaxSAT) problem consists in computing a subset of soft clauses $\mathcal{S} \subseteq \mathcal{R}$, that minimizes/maximizes the sum of the weights of the clauses in $\mathcal{S}$, such that $\mathcal{S} \cup \mathcal{H}$ is satisfiable while falsifying $\mathcal{R} \setminus \mathcal{S}$. Closely related to MinSAT, is the *Maximum Falsifiability* (MaxFalse) problem, which corresponds to computing the complement of a solution of the MinSAT problem (see [22]).

The paper also considers a number of optimization problems in graphs. Consider an undirected graph $\mathcal{G} = (V, E)$, where $r = |V|$ and $s = |E|$. An *Independent Set* (IS) is a set $I \subseteq V$ such that $\forall_{u,v \in I}$, $(u, v) \notin E$. A *Vertex Cover* (VC) is a set $C \subseteq V$ such that $\forall_{(u,v) \in E}$, $u \in C \vee v \in C$. Finally, a *Clique* (or complete subgraph) is a set $L \subseteq V$ such that $\forall_{u,v \in L}$, $u \neq v \Rightarrow (u, v) \in E$. Given an independent set $I \subseteq V$, a well-known result is that $V \setminus I$ is a vertex cover of $\mathcal{G}$ and I is a clique of the complemented graph $\mathcal{G}^C$ (e.g. [13]). The *Maximum Independent Set* (MIS) problem consists in computing an IS of largest size. The problem can be generalized to the case when a weight is associated with each vertex. More importantly, given the above relationships between ISes, VCes and cliques, a solution of the MIS problem also represents a solution for *Minimum Vertex Cover* (MVC) of a graph $\mathcal{G}$, as well as a *Maximum Clique* (MaxClq) of the complemented graph $\mathcal{G}^C$.

2.1 Related Work

As mentioned before, a solution for the MIS problem can be used to compute a solution for the MVC problem or the MaxClq problem (and vice-versa). Approaches to the considered problems can be divided in two main categories, either exact algorithms or heuristic methods. Heuristic algorithms try to obtain a solution quickly but do not guarantee the optimality of the solution returned. Local search has been extensively used as a way to obtain a heuristic solution to the problems (e.g. see [1, 4, 7–10, 44, 45] and references therein).

An approximation to the MIS problem can also be obtained by a greedy algorithm [14] that at each step selects a vertex to belong to the MIS and removes all other vertices that share an edge with the selected vertex. In [14, 39, 57], the greedy heuristic approach was explored as a way to obtain lower bounds in a branch-and-bound procedure for binate covering problems.

In contrast to the heuristic methods, exact algorithms guarantee the optimality of the solution returned. Many exact algorithms can be found in the literature, most of which are based on the branch-and-bound technique (e.g. [11, 15, 28, 35, 36, 42, 47, 50, 52–54]). An additional approach to solve MaxClq is characterized by encoding the MaxClq problem into MaxSAT, and use an off-the-shelf MaxSAT solver on the encoded instance. Nevertheless, such approach is not competitive with current state-of-the-art exact MaxClq solvers [36] (also confirmed by our experimental results, see Section 5).

Recent years have seen a growing interest in the development of algorithms and techniques for the MinSAT problem. Existing works can be categorized in two main areas of research: either by reducing the MinSAT problem into a MaxSAT problem (e.g. [20, 29, 34, 58]); or by proposing a dedicated MinSAT solver (e.g. [2, 3, 20, 37, 38]).

3 Reducing MIS to MinSAT

Reductions of MaxClq to MaxSAT are well-known (e.g. [35]). Since a solution to a MIS problem can be obtained by solving the MaxClq problem of the complemented graph, then MaxSAT algorithms can be used for computing solutions to MIS. Additionally, MaxSAT can be reduced into MinSAT using auxiliary variables [55], thus making it possible to solve MIS through MinSAT.

This section proposes a natural reduction of the MIS problem directly into MinSAT that does not require the addition of hard clauses. The proposed reduction, referred to

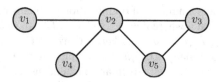

Fig. 1. Example graph

as *basic*, was recently mentioned in the context of Maximum Falsifiability [22]. Maximum Falsifiability (MaxFalse) was introduced in [22] as the problem of computing the complement of a MinSAT solution.

Consider an undirected graph $\mathcal{G} = (V, E)$, and a CNF formula $\mathcal{F}$. The idea of the basic reduction into MinSAT/MaxFalse is to associate with each vertex $v_i \in V$, a clause $c_i \in \mathcal{F}$. If c_i is falsified in the MinSAT/MaxFalse solution of $\mathcal{F}$, then the corresponding vertex v_i is included in the solution of the MIS of $\mathcal{G}$.

Each pair of clauses c_i, $c_j \in \mathcal{F}$, whose associated vertices have an edge between them in the graph $((v_i, v_j) \in E)$, cannot be allowed to be falsified simultaneously (due to the independence of vertices). As such, for each edge $(v_i, v_j) \in E$, a new variable x is created, and the literal x is added to c_i, while $\neg x$ is added to c_j. Any assignment to the variable x will force at least one of the clauses c_i or c_j to be satisfied.

Example 1. Consider the graph $\mathcal{G} = (V, E)$, with $V = \{v_1, v_2, v_3, v_4, v_5\}$ and $E = \{(v_1, v_2), (v_2, v_3), (v_2, v_4), (v_2, v_5), (v_3, v_5)\}$, as shown in Figure 1.

The basic reduction creates a new CNF formula $\mathcal{F}$. Each vertex v_i is represented by a clause c_i, and each edge of the graph (v_i, v_j) introduces a new variable $x_{i,j}$. Formula $\mathcal{F}$ is formed by the clauses in Equation 1.

$$
\begin{aligned}
c_1 &= x_{1,2} \\
c_2 &= \neg x_{1,2} \vee x_{2,3} \vee x_{2,4} \vee x_{2,5} \\
c_3 &= \neg x_{2,3} \vee x_{3,5} \\
c_4 &= \neg x_{2,4} \\
c_5 &= \neg x_{2,5} \vee \neg x_{3,5}
\end{aligned}
\tag{1}
$$

The following proposition proves the correctness of the basic reduction.

Proposition 1. *Given a graph $\mathcal{G} = (V, E)$. Let $\mathcal{F}$ be a CNF formula obtained from $\mathcal{G}$ by the above basic reduction.*
(1) Any MinSAT/MaxFalse solution of $\mathcal{F}$ represents an MIS of $\mathcal{G}$.
(2) Any MIS solution of $\mathcal{G}$ represents a MinSAT/MaxFalse solution of $\mathcal{F}$.

Proof. Here we consider just MaxFalse (for MinSAT the complement can be used).

(1) Consider a MaxFalse solution of $\mathcal{F}$, and let $\mathcal{F}' \subseteq \mathcal{F}$ be the set of clauses that are falsified by the MaxFalse solution. Then there is an assignment $\mathcal{A}$ that falsifies all clauses in $\mathcal{F}'$. Let $V' \subseteq V$ be a set of vertices that are associated to the clauses in $\mathcal{F}'$.

First we prove that the vertices in V' are independent. Assume by contradiction that the vertices in V' are not independent. Then there is (at least) one edge between two of its vertices, which means there is a variable x such that x belongs to one of the associated clauses in $\mathcal{F}'$ and $\neg x$ to another. Those two clauses cannot be simultaneously falsified, which is a contradiction since $\mathcal{A}$ falsifies all clauses in $\mathcal{F}'$.

Now we prove that V' is an MIS of $\mathcal{G}$. Assume by contradiction that V' is not maximum. Then there is a set $V'' \subseteq V$ such that V'' is an MIS and $|V''| > |V'|$. Let $\mathcal{F}'' \subseteq \mathcal{F}$ be the set of clauses associated to V''. Since the vertices in V'' are independent (no edges between them), the clauses in $\mathcal{F}''$ share no variables, and can be simultaneously falsified (consider an assignment that falsifies all literals in $\mathcal{F}''$). But $|\mathcal{F}''| = |V''| > |V'| = |\mathcal{F}'|$, which is a contradiction since $\mathcal{F}'$ is a MaxFalse/MinSAT solution.

(2) Consider $V' \subseteq V$ an MIS of $\mathcal{G}$ and let $\mathcal{F}' \subseteq \mathcal{F}$ be the set of clauses associated to the vertices in V'. The clauses in $\mathcal{F}'$ are simultaneously falsifiable, because the vertices in V' are independent, which means the clauses in $\mathcal{F}'$ share no variables, so the assignment that falsifies all literals in the $\mathcal{F}'$ is able to falsify all clauses in $\mathcal{F}'$.

Finally, we prove that $\mathcal{F}'$ is a MaxFalse solution. Consider by contradiction that $\mathcal{F}'$ is not a MaxFalse solution. Since the clauses in $\mathcal{F}'$ are simultaneously falsifiable, then there is a set $\mathcal{F}'' \subseteq \mathcal{F}$ such that $\mathcal{F}''$ is a MaxFalse solution and $|\mathcal{F}''| > |\mathcal{F}'|$. Let $V'' \subseteq V$ be the set of vertices associated to the clause in $\mathcal{F}''$. Since the clauses in $\mathcal{F}''$ are simultaneously falsifiable then the vertices in V'' are independent (otherwise there would be two different clauses in $\mathcal{F}''$ containing complementary literals, and the clauses in $\mathcal{F}''$ could not be simultaneously falsifiable). But $|V''| = |\mathcal{F}''| > |\mathcal{F}'| = |V'|$, which is a contradiction since V' is a MIS of $\mathcal{G}$. $\square$

Observe that, given a graph with r vertices and s edges, the basic reduction of MIS into MinSAT/MaxFalse always introduces s variables and r clauses[1]. As such, the basic reduction represents a polynomial-time reduction from MIS to MinSAT/MaxFalse[2].

Although the proposed reduction is not efficient in terms of the number of used variables, it is still more compact than the known reductions from MaxClq to MaxSAT (e.g see [35, 36]). To the best of our knowledge, there are two MaxSAT encodings of MaxClq described in the literature, both have to deal with not only soft clauses but also with hard clauses (partial MaxSAT instances are constructed). Given a graph $\mathcal{G} = (V, E)$, $|V| = r$, $|E| = s$, the first encoding produces r variables and r soft clauses, as well as $\frac{r \cdot (r-1)}{2} - s$ hard clauses (one hard clause for each edge of the complemented graph $\mathcal{G}^C$). The improved version of this encoding reduces the number of soft clauses to k, where $k \leq r$ is the number of disjoint independent sets of $\mathcal{G}$ computed heuristically. The reader is referred to [35, 36] for details.

Additionally observe that, the basic reduction does not always produce a formula with the minimal number of variables. That is, there are graphs for which a CNF formula with r clauses and less than s variables can be obtained, whose MinSAT/MaxFalse solution represents an MIS solution of the original graph. Section 4 introduces several techniques that reduce the number of variables.

---

[1] Note that the number of edges s is usually much higher than the number of vertices r and can be potentially close to $\frac{r \cdot (r-1)}{2}$.

[2] Observe that since MIS is an NP-hard problem, then the basic reduction provides an (alternative) natural proof of NP-hardness of both MinSAT and MaxFalse. The original proof of NP-hardness of MinSAT was demonstrated by reducing MaxSAT into MinSAT (see [27]). However, note that 2-MIS — the maximum independent set problem for a graph with a vertex degree bounded by 2 — can be trivially solved in polynomial time [17]. Therefore, the basic reduction does not cover the case of NP-hardness of 2-MinSAT/2-MaxFalse, even though they are also known to be NP-hard (see [27]).

4 Improvements to the Basic Reduction

Consider a graph $\mathcal{G} = (V, E)$, $|V| = r$, $|E| = s$. As it was shown in Section 3, the number of variables introduced by the basic reduction of the MIS problem for $\mathcal{G}$ to MinSAT/MaxFalse is equal to the number of edges of the graph $\mathcal{G}$, which is bounded by $\frac{r \cdot (r-1)}{2}$. This means that in the case of dense graphs with a large number of vertices the basic reduction generates CNF formulas with a large number of variables, which does not allow one to efficiently use this reduction in practice. This section describes 3 techniques for producing CNF formulas with a number of variables smaller by orders of magnitude compared to the basic reduction. In some cases decreasing the number of variables also leads to formula simplification by decreasing the number of clauses. Additionally, we also make the conjecture that given a graph $\mathcal{G}$, finding a formula $\mathcal{F}$ with the minimum number of variables cannot be done in polynomial time.

4.1 Greedy Approach

In contrast to introducing a new variable for each edge of graph $\mathcal{G}$, the greedy approach is able to use one variable for several edges. The greedy approach hinges on the idea that all edges incident to a vertex of $\mathcal{G}$ can be represented by one Boolean variable. As a result, the number of variables used to encode a graph into a formula is bounded by the number r of vertices of $\mathcal{G}$.

The pseudocode of the greedy algorithm is shown in Algorithm 1. For any graph $\mathcal{G} = (V, E)$ it constructs a set of clauses $\mathcal{F}$, each clause of which corresponds to a vertex of $\mathcal{G}$ and several edges of $\mathcal{G}$ can be encoded by the same pair of literals. Initially each clause of formula $\mathcal{F}$ is an empty set of literals (see lines 2–5). Since the graphs we consider are not directed, we assume that both (v_i, v_j) and (v_j, v_i) denote the same edge between vertices v_i and v_j of the graph. The idea of the algorithm is that each edge incident to a vertex v_i can be encoded by the same pair of literals x_i and $\neg x_i$. Therefore, at each iteration of the main loop, Algorithm 1 picks a vertex v_i of $\mathcal{G}$ with the maximum degree (line 7) and introduces a positive literal x_i to clause c_i (line 8). After that for each vertex v_j that has a connection to v_i, clause c_j gets a literal $\neg x_i$ (see line 10). All the considered edges (all the ones incident to v_i) are removed from graph $\mathcal{G}$. The loop continues until there are no edges in the graph that are not yet encoded.

Example 2. Consider the graph represented in Figure 1. Figure 2 illustrates how the greedy reduction works. First, it picks v_2 as a vertex with the maximum degree and adds a literal x_2 into c_2, while literal $\neg x_2$ is added into clauses c_1, c_3, c_4, and c_5. Then the corresponding edges are removed from the graph. The only edge in the graph that is not yet considered is the edge (v_3, v_5). The next vertex with the maximum degree is v_3. The algorithm adds literal x_3 into clause c_3, while $\neg x_3$ is added into c_5. The resulting set of clauses is shown in (2) below.

$$
\begin{aligned}
c_1 &= \neg x_2 \\
c_2 &= x_2 \\
c_3 &= \neg x_2 \vee x_3 \\
c_4 &= \neg x_2 \\
c_5 &= \neg x_2 \vee \neg x_3
\end{aligned}
\tag{2}
$$

Algorithm 1. Greedy reduction algorithm

---

1 **Function** Greedy($\mathcal{G} = (V, E)$)
2 $\quad \mathcal{F} \leftarrow \emptyset$
3 $\quad$ **foreach** $v_i \in V$ **do**
4 $\quad\quad c_i \leftarrow \emptyset$
5 $\quad\quad \mathcal{F} \leftarrow \mathcal{F} \cup c_i$

6 $\quad$ **while** $E \neq \emptyset$ **do**
7 $\quad\quad v_i \leftarrow$ vertex in V with maximum degree
8 $\quad\quad c_i \leftarrow c_i \cup \{x_i\}$
9 $\quad\quad$ **foreach** $v_j \in V$ *s.t.* $(v_i, v_j) \in E$ **do**
10 $\quad\quad\quad c_j \leftarrow c_j \cup \{\neg x_i\}$
11 $\quad\quad\quad E \leftarrow E \setminus \{(v_i, v_j)\}$

12 $\quad$ **return** $\mathcal{F}$

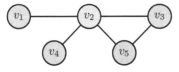

(a) Pick v_2 and encode its connections

(b) Pick v_3 and encode its connections

Fig. 2. Example on how the greedy approach works

Proposition 2. *Given a graph $\mathcal{G}$ with r vertices, the complexity of the greedy reduction algorithm for graph $\mathcal{G}$ is $\mathcal{O}(r^2)$.*

Proof. Observe that the algorithm has to traverse all the edges of graph $\mathcal{G}$, and each is traversed once. The trivial worst case scenario is when graph $\mathcal{G}$ is a clique — in this case graph $\mathcal{G}$ has r^2 edges. $\qquad\square$

Proposition 3. *Given a graph $\mathcal{G} = (V, E)$, let $\mathcal{F} = Greedy(\mathcal{G})$. $V' \subseteq V$ is an MIS of $\mathcal{G}$ iff the set $\mathcal{F}' \subseteq \mathcal{F}$ of clauses associated to the vertices in V' is a MaxFalse solution (the complement of a MinSAT solution) of $\mathcal{F}$.*

For the sake of succinctness and due to lack of space, we do not provide a proof of Proposition 3. However, the correctness of the greedy approach can be shown using an argumentation analogous to the proof of Proposition 1.

Also note that the proposed greedy algorithm is known to be not optimal in terms of the number of used variables. A simple example of a graph, for which the greedy approach is not optimal is shown in Figure 3. For this graph it is enough to introduce 4 variables, while the greedy algorithm introduces 5 variables. The resulting set of clauses produced by the greedy algorithm and the *optimal encoding* are shown in Figure 4.

Nevertheless, the greedy algorithm can be seen as a significant improvement over the basic approach. Given a graph $\mathcal{G} = (V, E)$, where $|V| = r$, the basic reduction can

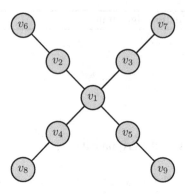

Fig. 3. Counterexample showing non-optimality of the greedy algorithm

$$c_1 = x_1$$
$$c_2 = \neg x_1 \vee x_2$$
$$c_3 = \neg x_1 \vee x_3$$
$$c_4 = \neg x_1 \vee x_4$$
$$c_5 = \neg x_1 \vee x_5$$
$$c_6 = \neg x_2$$
$$c_7 = \neg x_3$$
$$c_8 = \neg x_4$$
$$c_9 = \neg x_5$$

(a) Greedy encoding

$$c_1 = \neg x_2 \vee \neg x_3 \vee \neg x_4 \vee \neg x_5$$
$$c_2 = x_2$$
$$c_3 = x_3$$
$$c_4 = x_4$$
$$c_5 = x_5$$
$$c_6 = \neg x_2$$
$$c_7 = \neg x_3$$
$$c_8 = \neg x_4$$
$$c_9 = \neg x_5$$

(b) Optimal encoding

Fig. 4. Sets of clauses produced for the graph shown in Figure 3

potentially introduce $\mathcal{O}(r^2)$ variables while the number of variables used by the greedy algorithm is bounded by r.

4.2   Optimizations

This section provides a description of two additional heuristic techniques for minimizing the number of variables. Both techniques can be applied to the formulas produced by the considered basic and greedy reduction algorithms.

Variable Compatibility. Although CNF formulas produced by the greedy reduction are much more compact (in terms of the number of used variables) than the ones produced by the basic reduction, in some cases it might be still possible to reduce the number of variables even more. The following technique is referred to as *variable compatibility*. The idea of the variable compatibility method originates from the approaches used in the automata theory for the finite-state machine minimization, which makes use of the so-called compatibility graphs and merger tables [21, 26].

While the original method of simplifying finite-state machines (e.g. see [26]) operates with the so-called *compatible states*, here we use a notion of *compatible variables*. Variables decided to be compatible can replace each other and, thus, reduce the total number of variables. Assume that for a given graph $G = (V, E)$, where $|V| = r$, $|E| = s$, an MIS to MaxFalse reduction (either basic or greedy) produces a CNF formula $\mathcal{F}$ over a set X, $|X| = k$, variables. Note that for the case of the basic reduction $k \leq s$, for the greedy reduction, $k \leq r$.

The variable compatibility method consists in constructing and filling a $k \times k$ table, rows and columns of which are labeled by variables of X. Filling a cell of the table with coordinates (x_i, x_j) means that variables x_i and x_j are not compatible. Initially, all the cells of the table are empty (all variables are possibly compatible). The following rules can be used in order to conclude that two variables are not compatible.

1. A clause of $\mathcal{F}$ cannot be a *tautology*, i.e. it cannot contain literals x_i and $\neg x_i$ simultaneously. For example, given a clause $\neg x_1 \vee x_2$, variables x_1 and x_2 cannot replace each other. Otherwise, the clause is a tautology.
2. The structure of the clauses must enforce *no new connection* between vertices of the original graph $\mathcal{G}$. For example, given two clauses $\neg x_1 \vee x_2$ and $\neg x_3$, variables x_2 and x_3 cannot replace each other. Otherwise, there is an edge between the corresponding vertices in the original graph.

Example 3. Consider the graph represented in Figure 1. Although the clauses produced by the basic reduction for this graph are already shown in (1), for the sake of simplicity we represent it again with each variable having exactly one index (instead of two).

$$
\begin{aligned}
c_1 &= x_1 \\
c_2 &= \neg x_1 \vee x_2 \vee x_3 \vee x_4 \\
c_3 &= \neg x_2 \vee x_5 \\
c_4 &= \neg x_3 \\
c_5 &= \neg x_4 \vee \neg x_5
\end{aligned}
\tag{3}
$$

In order to determine the classes of *compatible variables*, the following compatibility table can be constructed:

	x_1	x_2	x_3	x_4	x_5
x_1	—				
x_2	$*_2 *_{1,3}$	—			
x_3	$*_2 *_{1,4}$		—		
x_4	$*_2 *_{1,5}$			—	
x_5	$*_{1,5}$	$*_3$	$*_{3,4}$		—

If two variables are not compatible because of violating rule 1, then the corresponding cell of the table is marked by *. If they are not compatible because of rule 2, the cell is marked by symbol *. The subscripts denote the clauses involved.

Note that rule 1 can be applied to each clause that contains both positive and negative literals. The second rule can be applied to a pair of clauses if there is *no edge* between the corresponding nodes in the original graph $\mathcal{G}$. For checking that, one can construct a graph $\mathcal{G}^C$, which is complement to $\mathcal{G}$ and use edges of $\mathcal{G}^C$ to check the validity of rule 2. In our example the edges of $\mathcal{G}^C$ are represented by the pairs of clauses: (c_1, c_3), (c_1, c_4), (c_1, c_5), (c_3, c_4), and (c_4, c_5).

After applying both rules and filling the table one can see that there are 3 compatibility classes: (x_2, x_3, x_4), (x_3, x_4), (x_4, x_5). Here, one can heuristically choose a compatibility class to use. Using the first compatibility class (x_2, x_3, x_4) enables us to use 3 variables instead of the original 5: x_1, x_2, and x_5 (variables x_3 and x_4 are replaced by x_2). Thus, as a result, the clauses after their modification are shown in (4).

$$
\begin{aligned}
c_1 &= x_1 \\
c_2 &= \neg x_1 \vee x_2 \\
c_3 &= \neg x_2 \vee x_5 \\
c_4 &= \neg x_2 \\
c_5 &= \neg x_2 \vee \neg x_5
\end{aligned}
\tag{4}
$$

An immediate observation is that time complexity of the variable compatibility heuristic is formed by the time required to check both rules 1 and 2. In order to check rule 1, in the worst case one has to traverse pairs of literals in each clause of $\mathcal{F}$. For checking rule 2, in the worst case one has to traverse pairs of literals in all pairs of clauses. Thus, the complexity of variable compatibility is $\mathcal{O}(k^2 \cdot r^2)$, where k and r are numbers of variables and clauses in the original formula $\mathcal{F}$, respectively.

Also observe that the example above illustrates the non-optimality of the variable compatibility technique. It was shown in Section 4.1 that it is enough to use only 2 variables for encoding graph $\mathcal{G}$ from Figure 1, while variable compatibility leaves 3 variables in formula $\mathcal{F}$ after doing the simplification.

Literal Compatibility. An immediate observation is that a possible improvement of variable compatibility, which was shown above to be not optimal, can be *literal compatibility*. The idea is that instead of computing compatibility classes for variables, one can try to determine classes of *compatible literals*. The two rules to apply are almost the same and can be seen as a generalization of the ones presented in Section 4.2:

1. *(no tautology)*, e.g. given a clause $\neg x_1 \vee x_2$, literals $\neg x_1$ and $\neg x_2$ are not compatible. Note that if literals x_1 and x_j are not compatible, then literals $\neg x_1$ and $\neg x_2$ are not compatible as well.
2. *(no new connection)*, e.g. given a pair of clauses: $\neg x_1 \vee x_2$ and $\neg x_3$, literals $\neg \neg x_1 = x_1$ and $\neg x_3$, $\neg x_2$ and $\neg x_3$ are not compatible.

Example 4. Consider the graph shown in Figure 1. In order to improve the basic reduction, one can construct the following compatibility table.

	x_1	$\neg x_1$	x_2	$\neg x_2$	x_3	$\neg x_3$	x_4	$\neg x_4$	x_5	$\neg x_5$
x_1	−	−								
$\neg x_1$	−	−								
x_2	$*_2*_{1,3}$		−	−						
$\neg x_2$		$*_2*_{1,3}$	−	−						
x_3	$*_2*_{1,4}$			$*_2*_{3,4}$	−	−				
$\neg x_3$		$*_2*_{1,4}$	$*_2*_{3,4}$		−	−				
x_4	$*_2*_{1,5}$			$*_2$		$*_2*_{4,5}$	−	−		
$\neg x_4$		$*_2*_{1,5}$	$*_2$		$*_2*_{4,5}$		−	−		
x_5	$*_{1,5}$	$*_{1,3}$	$*_3$			$*_{3,4}$	$*_{4,5}$	$*_5$	−	−
$\neg x_5$	$*_{1,3}$	$*_{1,5}$		$*_3$	$*_{4,5}$	$*_{3,4}$	$*_5$		−	−

The compatibility classes for literals x and $\neg x$ should be symmetric (e.g. see compatibility classes $(x_1, \neg x_2, \neg x_3, \neg x_4)$ and $(\neg x_1, x_2, x_3, x_4)$). Thus, we consider only classes for positive literals. So, according to the result table there are the following classes: $(x_1, \neg x_2, \neg x_3, \neg x_4)$, $(x_2, x_3, x_4, \neg x_5)$, (x_3, x_4), and (x_4, x_5). Using the first compatibility class, one can get 2 variables instead of the original 5. As a result, the clauses after their modification (using the first compatibility class) are:

$$
\begin{aligned}
c_1 &= x_1 \\
c_2 &= \neg x_1 \\
c_3 &= x_1 \vee x_5 \\
c_4 &= x_1 \\
c_5 &= x_1 \vee \neg x_5
\end{aligned}
\tag{5}
$$

Observe that time complexity of the literal compatibility heuristic is asymptotically the same ($\mathcal{O}(k^2 \cdot r^2)$) as the time complexity of variable compatibility. The only difference is that instead of k variables, for the case of literal compatibility $2k$ literals are considered.

Note that although literal compatibility is supposed to be more compact than variable compatibility, it is still not optimal in terms of the number of used variables. As an example, one can consider a graph shown in Figure 3 and use the literal compatibility technique to reduce the set of clauses produced by the greedy reduction (see Figure 4a). In this case, literal compatibility is not able to remove any variable and leaves the formula as it is, containing 5 variables. Recall that the optimal encoding for the graph presented in Figure 3 is shown in Figure 4b and contains 4 variables.

4.3 Further CNF Formula Simplification

All the techniques described in the previous sections can reduce the number of variables used when reducing MIS to MinSAT/MaxFalse by orders of magnitude (see Section 4.1). However, one can simplify the resulting formula even more. Here we give a brief explanation of how we can make the formula simpler after we have finished *removing* variables.

Recall that each clause of the formula being produced represents a vertex of the original graph. In many practical cases the number of removed variables is so large that some of the clauses that originally represent different vertices of the graph and, thus, have different literals, start duplicating each other. With a view to simplify the formula, one can keep just one version of each clause while removing all the duplications and making the formula weighted. Although it is not always the case, there are situations when formulas being produced get simplified by orders of magnitude. As an example, Table 1 shows the number of variables and clauses in the formulas produced for the *c-fat* family of DIMACS MaxClq[3] instances by the basic reduction, and the greedy reduction with and without variable compatibility optimization. Additionally it also reports the running time of the MinSatz solver for all the considered instances. Note that MinSatz could not solve 4 instances produced by the basic reduction while it was able to immediately report the answer for all the instances encoded by the greedy approach using variable compatibility.

---

[3] ftp://dimacs.rutgers.edu/pub/challenge/graph/benchmarks/clique/

Table 1. Example of formula sizes for the basic reduction with and without variable compatibility

<table>
<tr><th></th><th></th><th colspan="3">basic</th><th colspan="3">greedy</th><th colspan="3">greedy+vc</th></tr>
<tr><th></th><th></th><th># of vars</th><th># of clauses</th><th>r.t.</th><th># of vars</th><th># of clauses</th><th>r.t.</th><th># of vars</th><th># of clauses</th><th>r.t.</th></tr>
<tr><td rowspan="8">Instance</td><td>c-fat200-1</td><td>18366</td><td>200</td><td>0.4</td><td>188</td><td>200</td><td>0.05</td><td>35</td><td>37</td><td>0</td></tr>
<tr><td>c-fat200-2</td><td>16665</td><td>200</td><td>0.75</td><td>176</td><td>200</td><td>0.07</td><td>16</td><td>18</td><td>0</td></tr>
<tr><td>c-fat200-5</td><td>11427</td><td>200</td><td>0.96</td><td>142</td><td>200</td><td>0.07</td><td>5</td><td>7</td><td>0</td></tr>
<tr><td>c-fat500-1</td><td>120291</td><td>500</td><td>—</td><td>486</td><td>500</td><td>0.63</td><td>78</td><td>80</td><td>0</td></tr>
<tr><td>c-fat500-10</td><td>78123</td><td>500</td><td>—</td><td>374</td><td>500</td><td>0.53</td><td>6</td><td>8</td><td>0</td></tr>
<tr><td>c-fat500-2</td><td>115611</td><td>500</td><td>—</td><td>474</td><td>500</td><td>0.51</td><td>38</td><td>40</td><td>0</td></tr>
<tr><td>c-fat500-5</td><td>101559</td><td>500</td><td>—</td><td>436</td><td>500</td><td>0.37</td><td>14</td><td>16</td><td>0</td></tr>
</table>

4.4 Complexity of Reducing the Number of Variables

Previous sections 4.1 and 4.2 investigated the ways to simplify CNF formulas produced by the MIS reduction to MinSAT/MaxFalse in terms of the number of used variables. Note that the number of clauses of the result CNF formulas were considered to be fixed. We also showed that in this sense all the considered improvements to the reduction are not optimal in general, i.e. the number of used variables in general is greater or equal to the minimum number of variables introduced by a *potentially optimal* encoding.

Let us formulate a problem of finding an optimal encoding as follows: given a graph $\mathcal{G}$ on r vertices construct a CNF formula $\mathcal{F}$ with exactly r clauses and a *smallest possible* number of variables. Let us refer to this problem as *MinRed*. An important question is whether there is polynomial-time algorithm for solving MinRed. Although we do not know the answer to this question, our conjecture is that there is no polynomial time algorithm for solving MinRed. A support of this conjecture is that the MinRed problem seemingly is at least as hard as the Minimum Vertex Cover problem (MVC). The basic idea of the conjecture is both MinRed and MVC being minimization problems, MVC approximates MinRed. This means that given a graph $\mathcal{G} = (V, E)$, $|V| = r$, $|E| = s$, the size μ' of its MinRed solution is *always* lower or equal to the size μ'' of any vertex cover of $\mathcal{G}$ (including its MVC).

Indeed, any vertex cover of size μ'' of the graph $\mathcal{G}$ can be used to construct a CNF formula containing exactly r clauses and μ'' variables. To do so, one can apply an algorithm similar to the greedy reduction algorithm described in Section 4.1. Such an algorithm introduces a new variable x_i for each vertex v_i in the MVC of $\mathcal{G}$ and adds literal x_i to the corresponding clause c_i of $\mathcal{F}$ while adding its complementary literal $\neg x_i$ into all clauses c_j corresponding to the vertices v_j connected to v_i. Therefore, size μ'' of the MVC can be seen as an upper bound on the minimum possible number of variables in $\mathcal{F}$, i.e. $\mu' \leq \mu''$. Moreover, there are cases where the minimum number of variables in $\mathcal{F}$ is strictly less than the minimum vertex cover of $\mathcal{G}$. An example of such a situation is shown in Figure 5. For this graph the MVC solution has size 2 (it includes vertices v_1 and v_2). However, the minimum number of variables in formula $\mathcal{F}$ corresponding to $\mathcal{G}$ (the MinRed solution for $\mathcal{G}$) is 1. It should be noted again that although MVC's upper-bounding MinRed gives an intuition that it must be at least as hard to find a solution for MinRed as to find its upper bound (a solution for MVC), we acknowledge that we do not know whether this is indeed true and there is no polytime algorithm for MinRed nor we have a proof of this fact.

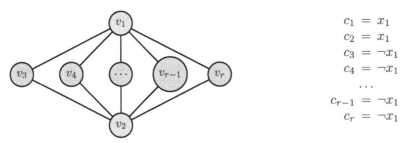

(a) Graph $\mathcal{G}$ and its MVC   (b) The optimal formula $\mathcal{F}$ for $\mathcal{G}$

Fig. 5. Example of a graph $\mathcal{G}$, for which the size of the MVC of $\mathcal{G}$ is greater than the minimum number of variables in the corresponding CNF formula $\mathcal{F}$

5   Experimental Results

This section presents the results obtained in the experimental evaluation with the proposed approaches. The experiments were performed on an Intel Xeon 5160 3GHz, with 4GB of memory, and running Fedora Linux operating system.

Although the MIS problem (and the other known NP-hard graph problems) is well studied, to the best of our knowledge there are not many native MIS instances available. Therefore, the classes of benchmarks used in the evaluation are described below.

1. We considered several known sets of native *crafted* MaxClq benchmarks, namely DIMACS MaxClq, FRB, and additional MaxClq instances studied in [35, 36, 38]. All the mentioned benchmark sets comprise the *Crafted MaxClq* set of benchmarks used in our experimental evaluation. The total number of instances in the Crafted MaxClq benchmark set is 117.
2. Another considered benchmark set includes native MIS instances. These instances are obtained from the Binate Covering Problem benchmarks (BCP) since it is known that solving MIS can be seen as an approximation for BCP (e.g. see [14,57]).

The total number of instances considered is 233.

The experimental evaluation is aimed at showing that the proposed reduction from MIS to MinSAT/MaxFalse can be seen as an efficient way to solve the MIS problem. In order to do so and since there are tight relationships between MIS and MaxClq, MaxSAT and MinSAT/MaxFalse, we used several approaches to MIS/MaxClq and, thus, the corresponding classes of dedicated solvers.

1. For both benchmark sets a native MaxClq solver called *MaxCLQ* was used. It is known to be one of the best native tools for MaxClq[4] (for a comprehensive comparison of different state-of-the-art tools for MaxClq see [31, 35, 36, 38]). In order to enable MaxCLQ to deal with MIS instances, they were trivially transformed into MaxClq by complementing the graphs.
2. MinSAT/MaxFalse instances were solved by *MinSatz*, which is a known branch-and-bound MinSAT solver (see [38]). In order to produce MinSAT/MaxFalse instances, we used the greedy reduction with variable compatibility and clause duplicates removal.

---

[4] Although it was reported in [31] that IncMaxCLQ was the best native solver for MaxClq, we were not able to run it in our experimental evaluation.

Table 2. Number of solved instances for different approaches

—	CLQ	MinSz	MaxSz-d	MaxSz-mf	MiFuMaX-d	MiFuMaX-mf	Direct MaxSAT VBS	MaxFalse VBS	VBS
Crafted Clq	66	59	43	36	19	30	45	74	76
BCP	63	65	61	55	56	53	66	72	76
Total	129	124	104	91	75	83	111	146	152

3. Alternatively, we also transformed the MinSAT instances into MaxSAT with the use of Tseitin variables (see [55]). The following MaxSAT solvers were applied to the obtained MaxSAT instances: MaxSatz [33] and MiFuMaX [41]. MiFuMaX was chosen since it is based on the widely used Fu&Malik's core-guided algorithm for MaxSAT (e.g. see [16]). The corresponding solvers in the evaluation are called *MaxSatz-mf* and *MiFuMaX-mf*, respectively.
4. Finally, we considered MaxSAT instances that encode the MaxClq problem *directly* (without doing the MIS-to-MaxFalse transformations). The algorithm is an improved MaxClq-to-MaxSAT encoding (see [36]) and uses enumeration of disjoint independent sets. For these instances MaxSatz and MiFuMaX solvers were also used (*MaxSatz-dir* and *MiFuMaX-dir* in the evaluation, respectively).

Figure 6 shows a cactus plot illustrating the performance of the considered solvers on the total set of all instances in both Crafted MaxClq and BCP benchmark sets. The best performance overall is shown by MaxCLQ, which is able to solve 129 instances out of 233. MinSatz comes second with 124 instances solved, which is 4% less than MaxCLQ's result. MaxSAT solvers dealing with both direct MaxClq-to-MaxSAT encodings and MaxSAT instances obtained from the MIS-to-MaxFalse encodings perform significantly worse (see Figure 6).

Note that the *virtual best solver* (VBS) among the MaxSAT solvers dealing with the direct MaxClq-to-MaxSAT encoding can solve 111 instances while the VBS incorporating all the approaches based on the MinSAT/MaxFalse encodings of MIS can solve 146 instances. The VBS among all the considered approaches is able to solve 152 instances, which is only 6 instances more than the MinSAT/MaxFalse approach. Moreover, the VBS of all the approaches based on the MinSAT/MaxFalse encodings is able solve 15 instances (6.4% out of all 233 instances) that none of the other considered approaches can solve. Table 2 shows a detailed information about the number of instances solved by different approaches to MIS/MaxClq.

The experimental results indicate that the direct MaxSAT approach to the MIS and MaxClq problems has the worst performance among the considered approaches. A possible reason of the MinSAT approach being so much better than MaxSAT is that the MinSAT encoding is more compact in terms of the number of introduced variables and clauses, which is a result of the techniques proposed in the paper. Furthermore, the advantage of the MinSAT encoding is also explained by the fact that it does not contain any hard clauses while the known MaxSAT encodings do use a large number of hard clauses. Although the best performance is shown by MaxCLQ, which is a native MaxClq algorithm, MinSatz is very close to MaxCLQ solving 4% fewer instances. It is also interesting that the MinSAT/MaxFalse approach can solve 6.4% instances that cannot be solved by other approaches (i.e. native MaxCLQ and direct MaxSAT). Moreover, although there are not many papers on MinSAT solving (especially if compared to MaxSAT), considering the virtual best solvers shows that the proposed MinSAT/MaxFalse approach to the MIS problem is a promising way to deal with the MIS/MaxClq problems.

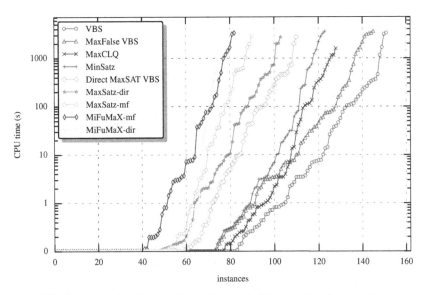

Fig. 6. Cactus plot showing the performance of the considered approaches

6   Conclusions

A number of new algorithms for the MinSAT problem have been proposed in recent years [19,34,37,38]. Nevertheless, besides being used in problem reductions, MinSAT has seldom been used for modeling and solving combinatorial optimization problems, the exception being [3]. This paper represents another step towards identifying combinatorial optimization problems which can be solved with MinSAT. The paper proposes a reduction from MIS (and so from related NP-hard graph problems) to MinSAT. The paper also develops a number of heuristics to reduce the obtained MinSAT formulas. In practice, the proposed techniques are very effective, and allow compacting the original MinSAT formulas to a fraction of their original size. The experimental results show that the obtained MinSAT formulas, solved with a standard MinSAT solver, allow obtaining results that are comparable to native solvers for MaxClq instances, and outperform those solvers on actual MIS instances. Moreover, the use of MinSAT comprehensively outperforms approaches based on using a reduction to MaxSAT. In addition, the results of the VBS solvers suggest that portfolios of solvers could significantly outperform a standalone solver. Overall, the experimental results are promising, and motivate the development of more efficient MinSAT solvers.

Future research work will focus on further improvements to the proposed MinSAT models. Another area of research is to implement portfolios of solvers for NP-hard graph problems, by exploiting some of the reductions proposed in recent years (including the ones in this paper). Finally, another area of research is to develop more efficient MinSAT solvers, e.g. similar to what has been done in the MaxSAT area in recent years.

Acknowledgments. This work is partially supported by SFI PI grant BEACON (09-/IN.1/I2618), FCT grant POLARIS (PTDC/EIA-CCO/123051/2010), and INESC-ID's multiannual PIDDAC funding PEst-OE/EEI/LA0021/2013.

References

1. Andrade, D.V., Resende, M.G.C., Werneck, R.F.F.: Fast local search for the maximum independent set problem. J. Heuristics 18(4), 525–547 (2012)
2. Ansotegui, C., Li, C.M., Manya, F., Zhu, Z.: A SAT-based approach to MinSAT. In: Escrig, M.T., Toledo, F.J., Golobardes, E. (eds.) CCIA 2002. LNCS (LNAI), vol. 2504, pp. 185–189. Springer, Heidelberg (2002)
3. Argelich, J., Li, C.-M., Manyà, F., Zhu, Z.: MinSAT versus MaxSAT for optimization problems. In: Schulte, C. (ed.) CP 2013. LNCS, vol. 8124, pp. 133–142. Springer, Heidelberg (2013)
4. Battiti, R., Protasi, M.: Reactive local search for the maximum clique problem. Algorithmica 29(4), 610–637 (2001)
5. Biere, A., Heule, M., van Maaren, H., Walsh, T. (eds.): Handbook of Satisfiability. Frontiers in Artificial Intelligence and Applications, vol. 185. IOS Press (2009)
6. Bomze, I.M., Budinich, M., Pardalos, P.M., Pelillo, M.: The maximum clique problem. In: Handbook of Combinatorial Optimization, pp. 1–74. Springer (1999)
7. Cai, S., Su, K., Chen, Q.: EWLS: A new local search for minimum vertex cover. In: Fox, M., Poole, D. (eds.) AAAI. AAAI Press (2010)
8. Cai, S., Su, K., Luo, C., Sattar, A.: NuMVC: An efficient local search algorithm for minimum vertex cover. J. Artif. Intell. Res. (JAIR) 46, 687–716 (2013)
9. Cai, S., Su, K., Sattar, A.: Local search with edge weighting and configuration checking heuristics for minimum vertex cover. Artif. Intell. 175(9-10), 1672–1696 (2011)
10. Cai, S., Su, K., Sattar, A.: Two new local search strategies for minimum vertex cover. In: Hoffmann, J., Selman, B. (eds.) AAAI. AAAI Press (2012)
11. Carraghan, R., Pardalos, P.M.: An exact algorithm for the maximum clique problem. Operations Research Letters 9(6), 375–382 (1990)
12. Chamaret, B., Josselin, S., Kuonen, P., Pizarroso, M., Salas-Manzanedo, B., Ubeda, S., Wagner, D.: Radio network optimization with maximum independent set search. In: IEEE 47th Vehicular Technology Conference, 1997, vol. 2, pp. 770–774 (May 1997)
13. Cormen, T.H., Leiserson, C.E., Rivest, R.L., Stein, C.: Introduction to algorithms. The MIT press (2009)
14. Coudert, O.: On solving covering problems. In: DAC, pp. 197–202 (1996)
15. Fahle, T.: Simple and fast: Improving a branch-and-bound algorithm for maximum clique. In: Möhring, R., Raman, R. (eds.) ESA 2002. LNCS, vol. 2461, pp. 485–498. Springer, Heidelberg (2002)
16. Fu, Z., Malik, S.: On solving the partial MAX-SAT problem. In: Biere, A., Gomes, C.P. (eds.) SAT 2006. LNCS, vol. 4121, pp. 252–265. Springer, Heidelberg (2006)
17. Garey, M.R., Johnson, D.S.: Computers and Intractability: A Guide to the Theory of NP-Completeness. W. H. Freeman (1979)
18. Gavril, F.: Algorithms for minimum coloring, maximum clique, minimum covering by cliques, and maximum independent set of a chordal graph. SIAM J. Comput. 1(2), 180–187 (1972)
19. Heras, F., Morgado, A., Planes, J., Marques-Silva, J.: Iterative SAT solving for minimum satisfiability. In: ICTAI, pp. 922–927 (2012)
20. Heras, F., Morgado, A., Planes, J., Marques-Silva, J.: Iterative sat solving for minimum satisfiability. In: ICTAI, vol. 1, pp. 922–927. IEEE (2012)
21. Hopcroft, J.E., Motwani, R., Ullman, J.D.: Introduction to automata theory, languages, and computation - international edition, 2nd edn. Addison-Wesley (2003)
22. Ignatiev, A., Morgado, A., Planes, J., Marques-Silva, J.: Maximal falsifiability: Definitions, algorithms, and applications. In: LPAR, pp. 439–456 (2013)
23. Jain, K., Padhye, J., Padmanabhan, V.N., Qiu, L.: Impact of interference on multi-hop wireless network performance. Wireless Networks 11(4), 471–487 (2005)

24. Johnson, D.S., Papadimitriou, C.H., Yannakakis, M.: On generating all maximal independent sets. Inf. Process. Lett. 27(3), 119–123 (1988)
25. Joseph, D., Meidanis, J., Tiwari, P.: Determining dna sequence similarity using maximum independent set algorithms for interval graphs. In: Nurmi, O., Ukkonen, E. (eds.) SWAT 1992. LNCS, vol. 621, pp. 326–337. Springer, Heidelberg (1992)
26. Kohavi, Z.: Switching and Finite Automata Theory. Tata McGraw-Hill (1978)
27. Kohli, R., Krishnamurti, R., Mirchandani, P.: The minimum satisfiability problem. SIAM J. Discrete Math. 7(2), 275–283 (1994)
28. Konc, J., Janezic, D.: An improved branch and bound algorithm for the maximum clique problem. In: MATCH, vol. 58, pp. 560–590 (2007)
29. Kügel, A.: Natural Max-SAT encoding of Min-SAT. In: Hamadi, Y., Schoenauer, M. (eds.) LION 2012. LNCS, vol. 7219, pp. 431–436. Springer, Heidelberg (2012)
30. Lawler, E.L., Lenstra, J.K., Kan, A.H.G.R.: Generating all maximal independent sets: NP-hardness and polynomial-time algorithms. SIAM J. Comput. 9(3), 558–565 (1980)
31. Li, C.M., Fang, Z., Xu, K.: Combining maxsat reasoning and incremental upper bound for the maximum clique problem. In: ICTAI, pp. 939–946 (2013)
32. Li, C.M., Manya, F.: MaxSAT, hard and soft constraints. In: Biere, et al. (eds.) [5], pp. 613–631
33. Li, C.M., Manyà, F., Planes, J.: New inference rules for max-sat. J. Artif. Intell. Res.(JAIR) 30, 321–359 (2007)
34. Li, C.M., Manyà, F., Quan, Z., Zhu, Z.: Exact MinSAT solving. In: Strichman, O., Szeider, S. (eds.) SAT 2010. LNCS, vol. 6175, pp. 363–368. Springer, Heidelberg (2010)
35. Li, C.M., Quan, Z.: Combining graph structure exploitation and propositional reasoning for the maximum clique problem. In: ICTAI, pp. 344–351 (2010)
36. Li, C.M., Quan, Z.: An efficient branch-and-bound algorithm based on maxsat for the maximum clique problem. In: AAAI, vol. 10, pp. 128–133 (2010)
37. Li, C.M., Zhu, Z., Manya, F., Simon, L.: Minimum satisfiability and its applications. In: IJCAI, pp. 605–610 (2011)
38. Li, C.M., Zhu, Z., Manya, F., Simon, L.: Optimizing with minimum satisfiability. Artif. Intell. 190, 32–44 (2012)
39. Manquinho, V.M., Silva, J.P.M.: Satisfiability-based algorithms for boolean optimization. Ann. Math. Artif. Intell. 40(3-4), 353–372 (2004)
40. Marathe, M.V., Ravi, S.S.: On approximation algorithms for the minimum satisfiability problem. Inf. Process. Lett. 58(1), 23–29 (1996)
41. MiFuMaX — a Literate MaxSAT Solver, http://sat.inesc-id.pt/~mikolas/sw/mifumax/book.pdf (accessed: January 31, 2014)
42. Östergård, P.R.J.: A fast algorithm for the maximum clique problem. Discrete Applied Mathematics 120(1-3), 197–207 (2002)
43. Pardalos, P.M., Xue, J.: The maximum clique problem. Journal of Global Optimization 4(3), 301–328 (1994)
44. Pullan, W.J.: Approximating the maximum vertex/edge weighted clique using local search. J. Heuristics 14(2), 117–134 (2008)
45. Pullan, W.J., Hoos, H.H.: Dynamic local search for the maximum clique problem. J. Artif. Intell. Res. (JAIR) 25, 159–185 (2006)
46. Ramaswami, R., Sivarajan, K.N.: Routing and wavelength assignment in all-optical networks. IEEE/ACM Trans. Netw. 3(5), 489–500 (1995)
47. Régin, J.-C.: Using constraint programming to solve the maximum clique problem. In: Rossi, F. (ed.) CP 2003. LNCS, vol. 2833, pp. 634–648. Springer, Heidelberg (2003)
48. Resende, M.G.C., Feo, T.A., Smith, S.H.: Algorithm 787: Fortran subroutines for approximate solution of maximum independent set problems using GRASP. ACM Trans. Math. Softw. 24(4), 386–394 (1998)

49. Robson, J.M.: Algorithms for maximum independent sets. J. Algorithms 7(3), 425–440 (1986)
50. San Segundo, P., Rodríguez-Losada, D., Jiménez, A.: An exact bit-parallel algorithm for the maximum clique problem. Computers & Operations Research 38(2), 571–581 (2011)
51. Tarjan, R.E., Trojanowski, A.E.: Finding a maximum independent set. SIAM J. Comput. 6(3), 537–546 (1977)
52. Tomita, E., Kameda, T.: An efficient branch-and-bound algorithm for finding a maximum clique with computational experiments. Journal of Global Optimization 37(1), 95–111 (2007)
53. Tomita, E., Seki, T.: An efficient branch-and-bound algorithm for finding a maximum clique. In: Calude, C.S., Dinneen, M.J., Vajnovszki, V. (eds.) DMTCS 2003. LNCS, vol. 2731, pp. 278–289. Springer, Heidelberg (2003)
54. Tomita, E., Sutani, Y., Higashi, T., Takahashi, S., Wakatsuki, M.: A simple and faster branch-and-bound algorithm for finding a maximum clique. In: Rahman, M. S., Fujita, S. (eds.) WALCOM 2010. LNCS, vol. 5942, pp. 102–112. Springer, Heidelberg (2010)
55. Tseitin, G.S.: On the complexity of derivation in propositional calculus. Studies in Constructive Mathematics and Mathematical Logic 2(115-125), 10–13 (1968)
56. Tsukiyama, S., Ide, M., Ariyoshi, H., Shirakawa, I.: A new algorithm for generating all the maximal independent sets. SIAM J. Comput. 6(3), 505–517 (1977)
57. Villa, T., Kam, T., Brayton, R.K., Sangiovanni-Vincentelli, A.L.: Explicit and implicit algorithms for binate covering problems. IEEE Trans. on CAD of Integrated Circuits and Systems 16(7), 677–691 (1997)
58. Zhu, Z., Li, C.-M., Manyà, F., Argelich, J.: A new encoding from MinSAT into MaxSAT. In: Milano, M. (ed.) CP 2012. LNCS, vol. 7514, pp. 455–463. Springer, Heidelberg (2012)

Long Proofs of (Seemingly) Simple Formulas

Mladen Mikša and Jakob Nordström

School of Computer Science and Communication
KTH Royal Institute of Technology
SE-100 44 Stockholm, Sweden

Abstract. In 2010, Spence and Van Gelder presented a family of CNF formulas based on combinatorial block designs. They showed empirically that this construction yielded small instances that were orders of magnitude harder for state-of-the-art SAT solvers than other benchmarks of comparable size, but left open the problem of proving theoretical lower bounds. We establish that these formulas are exponentially hard for resolution and even for polynomial calculus, which extends resolution with algebraic reasoning. We also present updated experimental data showing that these formulas are indeed still hard for current CDCL solvers, provided that these solvers do not also reason in terms of cardinality constraints (in which case the formulas can become very easy). Somewhat intriguingly, however, the very hardest instances in practice seem to arise from so-called fixed bandwidth matrices, which are provably easy for resolution and are also simple in practice if the solver is given a hint about the right branching order to use. This would seem to suggest that CDCL with current heuristics does not always search efficiently for short resolution proofs, despite the theoretical results of [Pipatsrisawat and Darwiche 2011] and [Atserias, Fichte, and Thurley 2011].

1 Introduction

Modern applied SAT solving is a true success story, with current state-of the art solvers based on *conflict-driven clause learning (CDCL)* [4,21,23] having delivered performance improvements of orders of magnitude larger than seemed possible just 15–20 years ago. From a theoretical perspective, however, the dominance of the CDCL paradigm is somewhat surprising in that it is ultimately based on the fairly weak *resolution* proof system [9]. Since it is possible in principle to extract a resolution refutation of an unsatisfiable formula from the execution trace of a CDCL solver running on it, lower bounds on resolution refutation length/size yield lower bounds on the running time of any CDCL solver trying to decide this formula.[1] By now, there is a fairly extensive literature on SAT instances for which exponential lower bounds are known, imposing firm restrictions on what kind of formulas the basic CDCL approach can hope to solve.

---

[1] Provided that the solver does not reason in terms of cardinality constraints or systems of linear equations and does not introduce new variables to apply extended resolution, in which case the theoretical lower bound guarantees no longer apply.

C. Sinz and U. Egly (Eds.): SAT 2014, LNCS 8561, pp. 121–137, 2014.
© Springer International Publishing Switzerland 2014

This suggests that an interesting question might be to turn the tables and ask for *maximally hard* instances. What are the smallest CNF formulas that are beyond reach of the currently best solvers? Pigeonhole principle (PHP) formulas were the first to be proven hard for resolution in the breakthrough result by Haken [15], but in terms of formula size N their hardness scales only as $\exp(\Omega(\sqrt[3]{N}))$. Two formula families with refutation length $\exp(\Omega(N))$ are Tseitin formulas[2] over so-called expander graphs and random k-CNF formulas, as shown by Urquhart [27] and Chvátal and Szemerédi [11], respectively. The strongest lower bounds to date in terms of the explicit constant in the exponent were established recently by Beck and Impagliazzo [5].

Spence [26] instead focused on empirical hardness and exhibited a family of 3-CNF formulas that seem practically infeasible even for very small instances (around 100 variables). These formulas can be briefly described as follows. Fix a set of $4n + 1$ variables. Randomly partition the variables into groups of 4 plus one group of 5. For each 4-group, write down the natural 3-CNF formula encoding the *positive cardinality constraint* that at least 2 variables must be true, and for the 5-group encode that a strict majority of 3 variables must be true. Do a second random variable partition into 4-groups plus one 5-group, but now encode *negative cardinality constraints* that the number of false variables is at least 2 and 3, respectively. By a counting argument, the CNF formula consisting of the conjunction of all these clauses must be unsatisfiable. Although [26] does not present any theoretical analysis, these formulas have a somewhat pigeonhole principle-like flavour and one can intuitively argue that they would seem likely to be hard provided that every moderately large set of positive cardinality constraints involves variables from many different negative constraints.

This construction was further developed by Van Gelder and Spence in [28], where the variable partitioning is done in terms of an $n \times n$ matrix with 4 non-zero entries in each row and column except that one extra non-zero entry is added to some empty cell. The variables in the formula correspond to the non-zero entries, each row is a positive cardinality constraint on its non-zero entries just as before, and each column provides a negative cardinality constraint. Equivalently, this formula can be constructed on a bipartite graph which is 4-regular on both sides except that one extra edge is added. In addition, there is a "no quadrangles" requirement in [28] that says that the graph contains no cycles of length 4. Just as above, it seems reasonable to believe that such formulas should be hard for resolution if the graph is a good expander. One such instance on 105 variables was issued by [28] as a "challenge formula" to be solved by any SAT solver in less than 24 hours, and in the concluding remarks the authors ask whether lower bounds on resolution length can be proven for formulas generated in this way.

1.1   Our Theoretical Results

We show that the formulas in [26,28] are exponentially hard for resolution if the collection of constraints have a certain expansion property, and that random

---

[2] Encoding that the sum of the vertex indegrees in an undirected graph is even.

instances of these formulas are expanding with overwhelming probability. Let U denote the set of positive cardinality constraints and V the set of negative constraints. Then we can represent the formulas in [28] (and [26]) as bipartite (multi-)graphs $G = (U \mathbin{\dot\cup} V, E)$, where edges are identified with variables and $x = (u, v) \in E$ if x occurs in both $u \in U$ and $v \in V$. Informally, we obtain the following lower bound for resolution (see Theorem 6 for the formal statement).

Theorem 1 (Informal). *If a 4-regular bipartite (multi-)graph G with one extra edge added is a sufficiently good expander, then the formula in [28] generated from G (or in [26] if G is a multigraph) requires resolution refutations of length $\exp(\Omega(n))$. In particular, random instances of these formulas require resolution length $\exp(\Omega(n))$ asymptotically almost surely.*

As a side note, we remark that the "no quadrangles" condition discussed above is not necessary (nor sufficient) for this theorem to hold—the more general notion of expansion is enough to guarantee that the formulas will be hard.

In one sentence, the proof works by reducing the formula to the pigeonhole principle on a 3-regular bipartite graph, which is then shown to be hard by a slight tweak of the techniques developed by Ben-Sasson and Wigderson [6]. A more detailed (if still incomplete) proof sketch is as follows. Start by fixing any complete matching in G (which can be shown to exist) and set all the matched edges plus the added extra edge to true. Also, set all remaining edges incident to the unique degree-5 vertex v^* on the right to false (this satisfies the negative constraint for v^*, which means that the corresponding clauses vanish). After this restriction, we are left with n constraints on the left which require that at least 1 out of the remaining 3 variables should be true, whereas on the right we have $n - 1$ constraints which all require that at most 1 remaining variable is true. But this is just a restricted PHP formula where each pigeon can go into one of three holes. Since we had a bipartite expander graph before restricting edges, and since not too many edges were removed, the restricted graph is still an expander. Now we can argue along the lines of [6] to obtain a linear lower bound on the refutation width from which an exponential length lower bound follows (and since restrictions can only make formulas easier, this lower bound must also hold for the original formula).

In fact, using tools from [2] one can show that the formulas are hard not only for resolution but also for *polynomial calculus resolution* [1,12], which adds the power of Gröbner basis computations to resolution.

Theorem 2 (Informal). *For 4-regular bipartite (multi-)graphs with one extra edge that are sufficiently good expanders the formulas in [26,28] require refutations of size $\exp(\Omega(n))$ in polynomial calculus resolution (PCR). In particular, randomly sampled instances of these formulas require PCR refutation size $\exp(\Omega(n))$ asymptotically almost surely.*

The technical details of this argument get substantially more involved, however. Thus, although Theorem 1 is strictly subsumed by Theorem 2, we focus mostly on the former theorem since it has a much cleaner and simpler proof that can be presented in full within the page limits of this extended abstract.

1.2 Our Empirical Results

We report results from running some current state-of-the-art SAT solvers on random instances of the formulas constructed by Spence [26] and Van Gelder and Spence [28], as well as on so-called *fixed bandwidth* versions of these formulas. The latter are formulas for which the non-zero entries on each row in the matrix appear on the diagonal and at some fixed (and small) horizontal offsets from the diagonal. Such matrices yield highly structured formulas, and as pointed out in [28] it is not hard to show that these formulas have polynomial-size refutations.

Our findings are that random instances of the formulas in [26,28] are very hard, and become infeasible for slightly above 100 variables. As could be expected, the formulas in [28] are somewhat harder than the original formulas in [26], since the former are guaranteed not to have any multi-edges in the bipartite graph representing the constraints and thus "spread out" variables better among different constraints. However, to our surprise the formulas that are hardest in practice are actually the ones generated from fixed bandwidth matrices. A priori, one possible explanation could be that although the formulas are theoretically easy, the constants hidden in the asymptotic notation are so bad that the instances are hard for all practical purposes. This appears not to be the case, however—when the SAT solver is explicitly given a good variable branching order the fixed bandwidth formulas are solved much more quickly. Thus, this raises the question whether this could perhaps be an example of formulas for which CDCL with current state-of-the-art heuristics fails to search effectively for resolution proofs. This stands in intriguing contrast to the theoretical results in [3,25], which are usually interpreted as saying that CDCL essentially harnesses the full power of resolution.

We have also done limited experiments with feeding the formulas in [26,28] to Sat4j [7], the latest version of which can detect (small) cardinality constraints [8]. It is not hard to see that if the SAT solver is given the power to count, then it could potentially figure out quickly that it cannot possibly be the case that a strict majority of the variables is both true and false simultaneously. Indeed, this is also what happens, and in particular Sat4j solves the challenge formula from [28] in less than a second.

1.3 Organization of This Paper

After reviewing some preliminaries in Section 2, we state and prove formal versions of our proof complexity lower bounds in Section 3. In Section 4, we report our experimental results. Section 5 contains some concluding remarks.

2 Proof Complexity Preliminaries

In what follows, we give a brief overview of the relevant proof complexity background. This material is standard and we refer to, e.g., the survey [24] for more details. All formulas in this paper are in conjunctive normal form (CNF), i.e., consist of conjunctions of clauses, where a clause is a disjunction of positive literals (unnegated variables) and negative literals (negated variables, denoted by overline). It is convenient to view clauses as sets, so that there is no repetition of literals and order is irrelevant. A k-CNF formula has all clauses of size at most k, which is always assumed to be some fixed (and, in this paper, small) constant.

A *resolution refutation* $\pi : F \vdash \bot$ of a formula F (sometimes also referred to as a *resolution proof* for F) is an ordered sequence of clauses $\pi = (D_1, \ldots, D_\tau)$ such that $D_\tau = \bot$ is the empty clause without literals, and each line D_i, $1 \leq i \leq \tau$, is either one of the clauses in F (an *axiom* clause) or is derived from clauses D_j, D_k in π with $j, k < i$ by the *resolution rule* $\frac{B \vee x \quad C \vee \overline{x}}{B \vee C}$ (where $B \vee C$ is the *resolvent* of $B \vee x$ and $C \vee \overline{x}$ on x). It is also sometimes technically convenient to add a *weakening rule* $\frac{B}{B \vee C}$ that allows to add literals to a previously derived clause. The *length* (or *size*) $L(\pi)$ of a resolution refutation π is the number of clauses in π. The *width* $W(C)$ of a clause C is the number of literals $|C|$, and the width $W(\pi)$ of a refutation π is the width of a largest clause in π. Taking the minimum over all refutations of F, we obtain the length $L_\mathcal{R}(F \vdash \bot)$ and width $W_\mathcal{R}(F \vdash \bot)$ of refuting F, respectively. It is not hard to show that all use of weakening can be eliminated from refutations without increasing these measures.

Resolution can be extended with algebraic reasoning to yield the proof system *polynomial calculus resolution (PCR)*, or more briefly just *polynomial calculus*.[3] For some fixed field $\mathbb{F}$ (which would be GF(2) in practical applications but can be any field in theory) we consider the polynomial ring $\mathbb{F}[x, \overline{x}, y, \overline{y}, \ldots]$ with x and $\overline{x}$ as distinct formal variables, and translate clauses $\bigvee_{x \in L^+} x \vee \bigvee_{y \in L^-} \overline{y}$ to monomials $\prod_{x \in L^+} x \cdot \prod_{y \in L^-} \overline{y}$. A *PCR refutation* π of F is then an ordered sequence of polynomials $\pi = (P_1, \ldots, P_\tau)$, expanded out as linear combinations of monomials, such that $P_\tau = 1$ and each line P_i, $1 \leq i \leq \tau$, is one of the following:

- a monomial encoding a clause in F;
- a *Boolean axiom* $x^2 - x$ or *complementarity axiom* $x + \overline{x} - 1$ for any variable x;
- a polynomial obtained from one or two previous polynomials by *linear combination* $\frac{Q \quad R}{\alpha Q + \beta R}$ or *multiplication* $\frac{Q}{xQ}$ for any $\alpha, \beta \in \mathbb{F}$ and any variable x.

The *size* $S(\pi)$ of a PCR refutation π is the number of monomials in π (counted with repetitions) and the *degree* $Deg(\pi)$ is the maximal degree of any monomial appearing in π. Taking the minimum over all PCR refutations, we define the size $S_{\mathcal{PCR}}(F \vdash \bot)$ and degree $Deg_{\mathcal{PCR}}(F \vdash \bot)$ of refuting F in PCR. When the proof system is clear from context, we will drop the subindices denoting resolution or PCR, respectively. It is straightforward to show that PCR can simulate resolution efficiently by simply mimicking the resolution steps in a refutation, and this simulation can be done without any noticeable blow up in size/length or degree/width. There are formulas, however, for which PCR can be exponentially stronger than resolution with respect to size/length.

---

[3] Strictly speaking, PCR as defined in [1] is a slight generalization of polynomial calculus [12], but here we will not be too careful in distinguishing between the two and the term "polynomial calculus" will refer to PCR unless specified otherwise.

A *restriction* ρ on F is a partial assignment to the variables of F. In a restricted formula $F\!\restriction_\rho$ (or refutation $\pi\!\restriction_\rho$) all clauses satisfied by ρ are removed and all other clauses have falsified literals removed. It is a well-known fact that restrictions preserve resolution refutations, so that if π is a resolution refutation of F, then $\pi\!\restriction_\rho$ is a refutation of $F\!\restriction_\rho$ (possibly using weakening) in at most the same length and width. For polynomials, we think of 0 as true and 1 as false. Thus, if a restriction satisfies a literal in a monomial that monomial vanishes, and all falsified literals in a monomial get replaced by 1 and vanish. Again it holds that if π is a PCR refutation of F, then $\pi\!\restriction_\rho$ is a PCR refutation of $F\!\restriction_\rho$ (after a simple postprocessing step to take care of cancelling monomials and to adjust for that multiplication can only be done one variable at a time). This restricted refutation will have at most the same size and degree (except possibly for a constant factor in size due to postprocessing multiplications).

3   Theoretical Hardness Results

In this section, we present our proof complexity lower bounds. We will focus below on the formulas in [28]. The proof for the formulas in [26] is very similar in spirit but contains some further technical complications, and we defer the discussion of this to the end of this section. Let us start by giving an explicit, formal definition of these formulas, which we will refer to as *subset cardinality formulas*.

Definition 3 (Subset cardinality formula). *Suppose that* $G = (U \,\dot\cup\, V, E)$ *is a 4-regular bipartite (multi-)graph except that one extra edge has been added. Then the* subset cardinality formula $SC(G)$ *over* G *has variables* $x_e, e \in E$, *and clauses:*

- $x_{e_1} \vee x_{e_2} \vee x_{e_3}$ *for every triple* e_1, e_2, e_3 *of edges incident to* $u \in U$,
- $\overline{x}_{e_1} \vee \overline{x}_{e_2} \vee \overline{x}_{e_3}$ *for every triple* e_1, e_2, e_3 *of edges incident to* $v \in V$.

As noted before, an easy counting argument shows that these formulas are unsatisfiable. Intuitively, the hardness of proving this unsatisfiability should depend on the structure of the underlying graph G. We remind the reader that compared to [28], the "no quadrangles" condition mentioned in Section 1 is missing in Definition 3. This is because this condition is neither necessary nor sufficient to obtain lower bounds. Expressed in terms of the graph G, what quadrangle-freeness means is that there are no 4-cycles, which is essentially saying that no constraints in G have a very "localized structure." However, the fixed bandwidth formulas already discussed in Section 1 can be constructed to be quadrangle-free, but are still guaranteed to be easy for resolution. Therefore, in order for our lower bound proof to go through we need the more general condition that the graph G should be an expander as defined next.

Definition 4 (Expander). *A bipartite graph* $G = (U \,\dot\cup\, V, E)$ *is an* (s, δ)-expander *if for each vertex set* $U' \subseteq U, |U'| \leq s$, *it holds that* $|N(U')| \geq \delta|U'|$, *where* $N(U') = \{v \in V \mid \exists (u, v) \in E \text{ for } u \in U'\}$ *is the set of neighbours of* U'.

The key idea in our lower bound proof is to apply a suitably chosen restriction to reduce subset cardinality formulas to so-called *graph pigeonhole principle formulas PHP(G)*. These formulas are also defined in terms of bipartite graphs $G = (U \,\dot\cup\, V, E)$ and encode that every "pigeon" vertex on the left, i.e., in U, needs to have at least one of the edges incident to it set to true, while every "hole" vertex on the right, i.e., in V, must have at most one edge incident to it set to true. Ben-Sasson and Wigderson [6] showed that random instances of such formulas are hard for resolution if the left degree is at least 5, and modifying their techniques slightly we prove that left degree 3 is sufficient provided that the graphs have good enough expansion. The proof is by showing a resolution width lower bound and then applying the lower bound on length in terms of width in [6]. An analogous result can be proven also for polynomial calculus by using techniques from Alekhnovich and Razborov [2] to obtain a degree lower bound and then applying the lower bound on size in terms of degree in Impagliazzo et al. [17], which yields the following lemma.

Lemma 5. *Suppose that $G = (U \,\dot\cup\, V, E)$ is a 3-regular $\left(\epsilon n, \frac{3}{2} + \delta\right)$-expander for some constant $\epsilon, \delta > 0$ and $|U| = |V| = n$, and let G' be the graph obtained by removing any vertex from V in G and its incident edges. Then the resolution refutation length of the graph pigeonhole principle $PHP(G')$ is $\exp(\Omega(n))$, and the same bound holds for PCR size.*

Let us first show how Lemma 5 can be used to establish the lower bound for subset cardinality formulas and then present a proof of the lemma for the case of resolution. The argument for polynomial calculus is more involved and we will only be able to sketch it due to space constraints.

Theorem 6. *Suppose that $G = (U \,\dot\cup\, V, E)$ is a 4-regular $\left(\epsilon n, \frac{5}{2} + \delta\right)$-expander for $|U| = |V| = n$ and some constants $\epsilon, \delta > 0$, and let G' be obtained from G by adding an arbitrary edge from U to V. Then any polynomial calculus refutation of $SC(G')$ must have size $\exp\left(\Omega(n)\right)$ (and hence the same lower bound holds for resolution length).*

Proof. We want to restrict the subset cardinality formula $SC(G')$ to get a graph pigeonhole principle formula. By a standard argument, which can be found for instance in [10], we know that any 4-regular bipartite graph G has a perfect matching. Fix such a matching M and let $M' = M \cup \{(u', v')\}$, where (u', v') denotes the edge added to G. We apply the following restriction ρ to $SC(G')$:

$$\rho\left(x_{(u,v)}\right) = \begin{cases} \top & \text{if } (u, v) \in M', \\ \bot & \text{if } v = v' \text{ and } (u, v') \notin M', \\ * & \text{otherwise (i.e., the variable is unassigned).} \end{cases} \tag{1}$$

This reduces the original formula $SC(G')$ to $PHP(G'')$ on the graph G'' obtained by removing the matching M and also the vertex v' with incident edges from G. To see this, consider what happens with the clauses encoding the constraints.

For every vertex $u \in U \setminus \{u'\}$, which has four edges $e_i, 1 \leq i \leq 4$, incident to it, we have the clauses $\{x_{e_1} \vee x_{e_2} \vee x_{e_3}, x_{e_1} \vee x_{e_2} \vee x_{e_4}, x_{e_1} \vee x_{e_3} \vee x_{e_4}, x_{e_2} \vee x_{e_3} \vee x_{e_4}\}$ in $SC(G')$. After applying ρ, the one edge that is in the matching M will be set to true, satisfying all of these clauses but one. If in addition u is one of the vertices neighbouring v', the remaining constraint will shrink to a 2-clause. The constraint corresponding to u' is similarly reduced. In this case, we have five incident edges $e_i, 1 \leq i \leq 5$, and two of them are set to true. If, for instance, we have $e_4 \in M$ and $e_5 = (u', v')$, then the only clause that is not satisfied is $x_{e_1} \vee x_{e_2} \vee x_{e_3}$, which corresponds to the pigeon axiom for the vertex u' in G''.

For a constraint $v \in V \setminus \{v'\}$ with neighbours $e_i, 1 \leq i \leq 4$, the clause set is the same as for $U \setminus \{u'\}$ except that every variable is negated. If $e_4 \in M$, then after the restriction we are left with the set of clauses $\{\overline{x}_{e_1} \vee \overline{x}_{e_2} \vee \overline{x}_{e_3}, \overline{x}_{e_1} \vee \overline{x}_{e_2}, \overline{x}_{e_1} \vee \overline{x}_{e_3}, \overline{x}_{e_2} \vee \overline{x}_{e_3}\}$, where the last three clauses are the hole axioms for the vertex v in G'' and the first clause can be ignored since it follows by weakening of any of the other clauses. Since ρ satisfies the constraint v' the clauses encoding this constraint vanish. This shows that $SC(G')\restriction_\rho$ is indeed equal to $PHP(G'')$.

Now all that remains is to observe that G'' can be obtained by removing a right vertex from a 3-regular bipartite $(\epsilon n, \frac{3}{2} + \delta)$-expander. This is so since deleting the matching M from G decreases all vertex degrees from 4 to 3 and lowers the expansion factor by at most an additive 1. Applying Lemma 5 we conclude that $PHP(G'')$ requires polynomial calculus size (and hence resolution length) $\exp(\Omega(n))$. As restrictions do not increase the length/size of refutations, the same lower bound must hold also for $SC(G')$.

It remains to prove Lemma 5. We give a full proof of the lemma for resolution below, but due to space constraints we can only outline the argument for polynomial calculus. For both resolution and polynomial calculus we need a stronger notion of expansion as defined next.

Definition 7 (Boundary expander). *A bipartite graph $G = (U \dot\cup V, E)$ is an (s, δ)-boundary expander if for every set of vertices $U' \subseteq U, |U'| \leq s$, it holds that $|\partial(U')| \geq \delta|U'|$, where $v \in \partial(U')$ if there is exactly one vertex $u \in U'$ that is a neighbour of v.*

Using the following connection between usual expansion and boundary expansion (which is straightforward to show and is stated here without proof) we can prove Lemma 5.

Proposition 8. *Every d-regular (s, δ)-expander is also an $(s, 2\delta - d)$-boundary expander.*

Proof (of Lemma 5 for resolution). Since G is an $(\epsilon n, 2\delta)$-boundary expander by Proposition 8, even after removing a vertex in V it must hold for G' that every set of vertices $U'' \subseteq U, |U''| \leq \epsilon n$ satisfies $|\partial_{G'}(U'')| \geq 2\delta|U'| - 1$.

Let us also observe that the connected component $G^c = (U^c \dot\cup V^c, E^c)$ of G to which the vertex v' belongs must be a 3-regular graph with $|U^c| > \epsilon n$. This is so since if $|U^c| \leq \epsilon n$, it would follow from the expansion of G that $|V^c| =$

$|N_{G^c}(U^c)| \geq \left(\frac{3}{2} + \delta\right)|U^c| > |U^c|$. But $|U^c| \neq |V^c|$ implies that G^c cannot be a 3-regular bipartite graph, which is a contradiction. Furthermore, for every proper subset $U'' \subsetneq U^c$ it must hold that $|N(U'')| > |U''|$, since otherwise U'' and its neighbours $N(U'')$ would form a disconnected component in G^c. Hence, when we remove the vertex v' from G^c we have $|N(U'')| \geq |U''|$ for every proper subset $U'' \subsetneq U^c$. By Hall's theorem, this implies that every proper subset $U'' \subsetneq U^c$ has a matching in G^c. This shows that any refutation of $PHP(G')$ must use all the pigeons in G^c, i.e., at least ϵn pigeon axiom clauses, to show that $PHP(G')$ is unsatisfiable, since the formula becomes satisfiable if just one of these pigeon axioms is removed.

Now we can employ the progress measure on refutations developed in [6] to show that the width of refuting $PHP(G')$ is lower bounded by $\epsilon \delta n - 1$. This follows by a straightforward adaptation of the argument in Sections 5 and 6.2 of [6], which yields a width lower bound analogous to that in Theorem 4.15. By appealing to the lower bound on length in terms of width in Corollary 3.6 in [6] we obtain a lower bound on the resolution refutation length of $\exp(\Omega(n))$.

The proof of Lemma 5 for polynomial calculus is similar in that we prove a degree lower bound and then use the lower bound on size in terms of degree in [17] (which is an exact analogue of the result in [6] for resolution). We closely follow the proof of Theorem 4.14 in [2], from which one can derive that any refutation of the graph pigeonhole principle on an (s, δ)-boundary expander requires degree $\delta s/2$. This is almost what we need, except that we lose an additive 1 when we remove a vertex from the right. Nevertheless, the proof still goes through if we subtract 1 everywhere, yielding a degree lower bound of $\delta s/2 - 1$. The only point where we argue a bit differently than [2] is when we need to show that at least s pigeons have to be used to prove unsatisfiability. But we have already shown this claim in the proof of the resolution width lower bound and we can reuse the same argument in the proof of Lemma 5 for polynomial calculus.

This proves that the formulas in [28] are hard for polynomial calculus (and hence also for resolution) if the underlying graph is an expander. In order to establish that randomly sampled instances of such formulas are hard, we just need the fact that randomly sampled graphs are likely to be expanders. The following theorem tells us what we need to know.

Theorem 9 ([16]). *Let $d \geq 3$ be a fixed integer. Then for every $\delta, 0 < \delta < \frac{1}{2}$, there exists an $\epsilon > 0$ such that almost all d-regular bipartite graphs G with n vertices on each side are $\left(\epsilon n, d - \frac{3}{2} + \delta\right)$-expanders.*

Corollary 10. *The formula $SC(G)$ for a random 4-regular bipartite graph G with an arbitrary extra edge added requires polynomial calculus refutations (and hence also resolution refutations) of exponential size asymptotically almost surely.*

Proof. Use Theorem 9 with $d = 4$ together with Theorem 6.

Let us conclude this section by discussing how the lower bound proof above for the formulas in [28], i.e., subset cardinality formulas $SC(G)$ for ordinary graphs G, can be made to work also for the formulas in [26].

Following the description in Section 1, the formulas in [26] can be defined in terms of permutations of $[4n + 1]$. We can construct a bipartite multigraph $G(\sigma)$ from a permutation σ of $[4n + 1]$ as follows. We first partition $[4n + 1]$ into subsets $\{1, 2, 3, 4\}, \{5, 6, 7, 8\} \ldots, \{4n - 3, 4n - 2, 4n - 1, 4n, 4n + 1\}$ and identify the vertices in V with these subsets. Second, by using the partition into subsets $\{\sigma(1), \sigma(2), \sigma(3), \sigma(4)\}, \ldots, \{\sigma(4n - 3), \sigma(4n - 2), \sigma(4n - 1), \sigma(4n), \sigma(4n + 1)\}$ we obtain the vertices in U. Then, for every number that is in both in $u \in U$ and $v \in V$ we add an edge between u and v. In this way, we obtain a subset cardinality formula on the multigraph $G(\sigma)$, which we will denote by $SC^*(G(\sigma))$ (or more briefly just $SC^*(\sigma)$) to highlight that the formula is generated from a multigraph obtained from a permutation.

In order to show that the formula $SC^*(\sigma)$ requires polynomial calculus refutations of exponential size if the multigraph $G(\sigma)$ is an expander, there are three issues we need to address in the proof for standard graphs above:

- Firstly, our graph theoretic claims should now be made for multigraphs instead of ordinary graphs. This is not a problem, however, since all the claims we need are still true in this setting and since multiple copies of an edge can be eliminated by setting the corresponding variables to false.
- Secondly, the degree-5 vertices are not necessarily connected in $G(\sigma)$. Because of this, we need to modify the restriction used to reduce the formula to a graph pigeonhole principle formula. We still find a matching and set all the edges in it to true, but now we choose two special edges incident to the degree-5 vertices u' and v' and set their values to true and false, respectively. In this way the graph we get has a vertex on the right with two of the edges incident to it set to true, one from the matching and one from the special edge incident to u'. This forces the values of the remaining two edges to be false, which satisfies the constraint v' and gives us the graph pigeonhole principle formula required by Lemma 5.
- Thirdly, a slightly subtle point is that we do not require $G(\sigma)$ to be expanding, but rather a slightly modified multigraph $G^M(\sigma)$. To form $G^M(\sigma)$, we start by removing from $G(\sigma)$ one of the edges incident to the degree-5 vertex in U and one of the edges incident to the degree-5 vertex in V. The resulting multigraph has two vertices of degree 3, which we connect with an edge in order to form the 4-regular multigraph $G^M(\sigma)$.

In order to show that randomly sampled instances of the formulas $SC^*(\sigma))$ are hard, we note that the model of a random graph used to prove Theorem 9 is actually based on random permutations. Hence, the claim that random 4-regular graphs are good expanders holds for random permutations as well, which implies that almost all instances of the formulas in [26] are hard for polynomial calculus.

Theorem 11. *Let σ be a permutation of $[4n + 1]$ and let $SC^*(\sigma)$ be the corresponding subset cardinality formula. If the multigraph $G^M(\sigma)$ is an $(\epsilon n, \frac{5}{2} + \delta)$-expander for some constants $\epsilon, \delta > 0$, then any polynomial calculus refutation (and resolution refutation) of $SC^*(\sigma)$ has size $\exp(\Omega(n))$.*

In particular, for a random permutation σ the formula $SC^*(\sigma)$ requires polynomial calculus refutation of exponential size asymptotically almost surely.

4   Empirical Results on SAT Solver Performance

For our experiments we used the SAT solvers Glucose 2.2 [14], March-rw [19], and Lingeling-ala [18]. The experiments were run under Linux on a computer with two quad-core AMD Opteron 2.2 GHz CPUs (2374 HE) and 16 GB of memory, where only one solver was running on the computer at any given time. We limited the solver running time to 1 hour per instance. For the experiments with fixed variable ordering we used a version of MiniSat 2.2.0 [22] modified so that the solver always branches on unset variable in fixed order.

The CNF formula instances were obtained as follows:

1. The formulas $SC^*(\sigma)$ from [26] were generated by taking one fixed partition of $[4n+1]$ into $\{1, 2, 3, 4\}$, $\{5, 6, 7, 8\}$, ..., $\{4n-3, 4n-2, 4n-1, 4n, 4n+1\}$ and one random partition into 4-groups plus one 5-group, and then encoding positive and negative cardinality constraints, respectively, on these two partitions.

2. For the formulas $SC(G)$ from [28] we started with a random (non-bipartite) 4-regular graph, took the bipartite double cover (with two copies v_L, v_R of each vertex v and edges (u_L, v_R) for all edges (u, v) in the original graph), and finally added a random edge.[4]

3. The fixed bandwidth formulas were constructed from an $n \times n$ matrix with ones in the first row on positions $1, 2, 4, 8$ and zeroes everywhere else, and with every subsequent row being a cyclic shift one step to the right of the preceding row. Finally, an extra one was added to the top right cell of the matrix if this was a zero, and otherwise to the nearest cell containing a zero.[5]

For each CNF formula we ran each SAT solver three times (with different random seeds), and for randomly generated formulas we ran on three different CNF formulas for each parameter value. The values in the plots are the medians of these values. We also performed exactly the same set of experiments on randomly shuffled version of the formulas (with randomly permuted clauses, variables, and polarities), but this random shuffling did not affect the results in any significant way and so we do not display these plots.

We present the results of our experiments in Figure 1 with one subplot per solver.[6] As can be seen from these plots, all three versions of the formulas become infeasible for around 100–120 variables. Comparing to our experiments on random 3-CNF formulas and Tseitin formulas on random 3-regular graphs in Figure 2, it should be clear that all three flavours of the formulas from [26,28] that we investigated were substantially harder than random formulas. Notice

---

[4] We remark that, strictly speaking, this does not yield uniformly random instances but we just wanted to obtain some instances with "good enough" randomness (and hence expansion) on which we could run experiments.

[5] Note that this construction yields quadrangle-free instances for large enough n, except possibly for quadrangles involving the added extra top-right entry.

[6] The code for generating the CNF instances and complete data for the experiments can be found at http://www.csc.kth.se/~jakobn/publications/sat14/

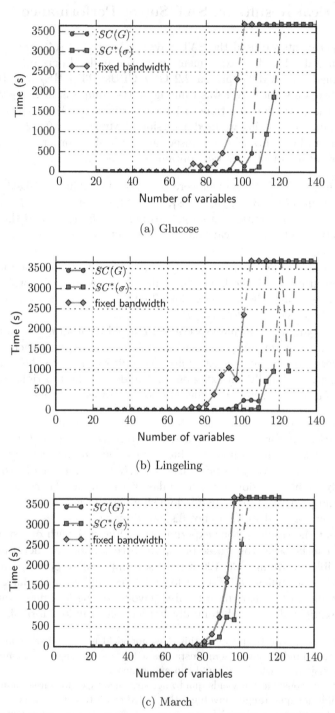

(a) Glucose

(b) Lingeling

(c) March

Fig. 1. SAT solver performance on variants of the formulas in [26,28]

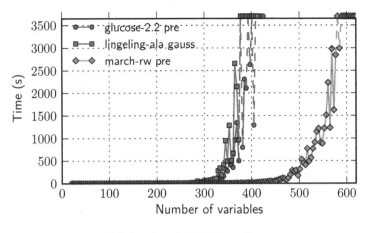

(a) Random 3-CNF formulas

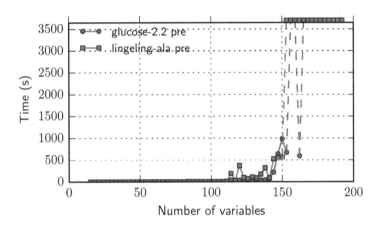

(b) Tseitin formulas on random graphs

Fig. 2. SAT solver performance on two well-known hard formula families

that for Tseitin formulas we do not present results for March and that Lingeling was run without Gaussian elimination. The reason is that March and Lingeling with Gaussian elimination solve Tseitin formulas in less than a second for even the largest instances we have tried.

Comparing random instances of formulas $SC^*(\sigma)$ and $SC(G)$ with fixed bandwidth instances, we can see that the easiest ones are $SC^*(\sigma)$ while $SC(G)$ are somewhat harder. This is as expected—by construction, for $SC(G)$ we are guaranteed that no pair of positive and negative constraints share more than one variable, whereas for $SC^*(\sigma)$ it could happen in principle that a positive and a negative constraint act on two, three, or even four common variables.

Somewhat counter-intuitively, however, the instances that are hardest in our practical experiments are the theoretically easy fixed bandwidth formulas.

In order to investigate whether the hardness of fixed bandwidth formulas could be attributed to hidden constants in the asymptotics—i.e., that the polynomial upper bounds on resolution length are so large in practice that the fixed bandwidth formulas are infeasible for all practical purposes—we ran a modified version of MiniSat on these formulas which always branched on variables row by row and in every row column by column. Intuitively, this seems to be the appropriate variable ordering if one is to recover the polynomial-length resolution refutation presented in [28]. And indeed, MiniSat run on fixed bandwidth formulas with fixed variable ordering performed much better than any of the other solvers on random instances of $SC^*(\sigma)$ and $SC(G)$ formulas. (We also verified that fixed variable ordering is not a good idea in general—as expected, MiniSat with fixed variable ordering performs poorly on random instances of $SC^*(\sigma)$ and $SC(G)$ formulas.)

Given the latest advances in SAT solving technology, with solvers going beyond resolution by incorporating elements of algebraic reasoning (Gröbner bases) and geometric reasoning (pseudo-Boolean solvers), a natural question is whether the formulas in [26,28] remain hard for such solvers.

Regarding algebraic solvers, we are not aware of any general-purpose solvers that can compete with CDCL solvers, but as mentioned the theoretical lower bounds that we prove for resolution hold also for polynomial calculus, which is a proof system for formalizing the reasoning in solvers based on Gröbner basis computations. Also, one can note that the algebraic reasoning in terms of Gaussian elimination in Lingeling does not seem to help.

For pseudo-Boolean solvers, which can be seen to search for proofs in (more or less restricted version of) the cutting planes proof system [13], the story could potentially be very different. As noted multiple times already, the formulas $SC^*(\sigma)$ and $SC(G)$ are just encodings of a fairly simple counting principle, and in contrast to resolution and polynomial calculus the cutting planes proof system knows how to count. Thus, pseudo-Boolean solvers with enough well-developed methods of cardinality constraints reasoning should have the potential to solve these formulas quickly. This indeed appears to be the case as reported in [8], and our own (albeit limited) experiments also show this.

5   Concluding Remarks

In this work, we establish that the formulas constructed by Spence [26] and Van Gelder and Spence [28] are exponentially hard for resolution and also for polynomial calculus resolution (PCR), which extends resolution with Gröbner basis computations. Formally, we prove that if the bipartite (multi-)graph describing the constraints encoded by the formula is expanding, then this implies exponential lower bounds on proof size in resolution and PCR. Furthermore, we show that random instances of these formulas are almost surely expanding, meaning that the exponential lower bound applies with high probability.

We also investigate the performance of some current state-of-the-art SAT solvers on these formulas, and find that small instances are indeed much harder than, e.g., random 3-CNF formulas with the same number of variables. Somewhat surprisingly, however, the very hardest formulas in our experiments are versions of the formulas in [26,28] generated from fixed bandwidth matrices. This is intriguing, since such formulas are easy for resolution, and since the current conventional wisdom (based on [3,25]) seems to be that CDCL solvers can search efficiently for short resolution proofs. In view of this, an interesting (albeit very speculative) question is whether perhaps these fixed bandwidth matrix formulas could be used to show formally that CDCL with VSIDS, 1UIP, and phase saving, say, does *not* polynomially simulate resolution.

Since the formulas in [26,28] encode what is in essence a fairly simple counting argument, SAT solvers that can reason efficiently with cardinality constraints could potentially solve these formulas fast. This indeed turns out to be the case for the latest version of Sat4j [8]. It would be interesting to investigate whether the formulas in [26,28] could be slightly obfuscated to make them hard also for solvers with cardinality constraints. If so, this could yield small benchmark formulas that are hard not only for standard CDCL solvers but also for solvers extended with algebraic and/or geometric reasoning.

Another candidate construction of small but very hard CNF formulas is the one presented by Markström [20]. It would be interesting to investigate what theoretical hardness results can be established for these formulas (for resolution and proof systems stronger than resolution) and how the practical hardness scales compared to the constructions by Spence and Van Gelder [26,28]. In particular, an interesting question is whether these formulas, too, become easy for CDCL solvers enhanced with cardinality constraints reasoning.

Acknowledgements. The authors are very grateful to Massimo Lauria and Marc Vinyals for stimulating discussions and for invaluable practical help with setting up and evaluating the experiments. We wish to thank Niklas Sörensson for explaining how to fix the variable decision order in MiniSat, and Daniel Le Berre for sharing data about the performance of the latest version of Sat4j on our benchmark formulas. We are also grateful to Allen Van Gelder for comments on a preliminary write-up of some of the results in this paper, as well as for introducing us to this problem in the first place. Finally, we thank several participants of the workshop *Theoretical Foundations of Applied SAT Solving (14w5101)* at the Banff International Research Station in January 2014 for interesting conversations on themes related to this work.

The authors were funded by the European Research Council under the European Union's Seventh Framework Programme (FP7/2007–2013) / ERC grant agreement no. 279611. The second author was also supported by the Swedish Research Council grants 621-2010-4797 and 621-2012-5645.

References

1. Alekhnovich, M., Ben-Sasson, E., Razborov, A.A., Wigderson, A.: Space complexity in propositional calculus. SIAM Journal on Computing 31(4), 1184–1211 (2002), preliminary version appeared in STOC 2000
2. Alekhnovich, M., Razborov, A.A.: Lower bounds for polynomial calculus: Nonbinomial case. Proceedings of the Steklov Institute of Mathematics 242, 18–35 (2003), http://people.cs.uchicago.edu/razborov/files/misha.pdf, Preliminary version appeared in FOCS 2001
3. Atserias, A., Fichte, J.K., Thurley, M.: Clause-learning algorithms with many restarts and bounded-width resolution. Journal of Artificial Intelligence Research 40, 353–373 (2011), preliminary version appeared in SAT 2009
4. Bayardo Jr., R.J., Schrag, R.: Using CSP look-back techniques to solve real-world SAT instances. In: Proceedings of the 14th National Conference on Artificial Intelligence (AAAI 1997), pp. 203–208 (July 1997)
5. Beck, C., Impagliazzo, R.: Strong ETH holds for regular resolution. In: Proceedings of the 45th Annual ACM Symposium on Theory of Computing (STOC 2013), pp. 487–494 (May 2013)
6. Ben-Sasson, E., Wigderson, A.: Short proofs are narrow—resolution made simple. Journal of the ACM 48 48(2), 149–169 (2001), preliminary version appeared in STOC 1999
7. Le Berre, D., Parrain, A.: The Sat4j library, release 2.2. Journal on Satisfiability, Boolean Modeling and Computation 7, 59–64 (2010), system description
8. Biere, A., Le Berre, D., Lonca, E., Manthey, N.: Detecting cardinality constraints in CNF. In: Sinz, C., Egly, U. (eds.) SAT 2014. LNCS, vol. 8561, pp. 285–301. Springer, Heidelberg (2014)
9. Blake, A.: Canonical Expressions in Boolean Algebra. Ph.D. thesis, University of Chicago (1937)
10. Bondy, J.A., Murty, U.S.R.: Graph Theory. Springer (2008)
11. Chvátal, V., Szemerédi, E.: Many hard examples for resolution. Journal of the ACM 35(4), 759–768 (1988)
12. Clegg, M., Edmonds, J., Impagliazzo, R.: Using the Groebner basis algorithm to find proofs of unsatisfiability. In: Proceedings of the 28th Annual ACM Symposium on Theory of Computing (STOC 1996), pp. 174–183 (May 1996)
13. Cook, W., Coullard, C.R., Turán, G.: On the complexity of cutting-plane proofs. Discrete Applied Mathematics 18(1), 25–38 (1987)
14. The Glucose SAT solver, http://www.labri.fr/perso/lsimon/glucose/.
15. Haken, A.: The intractability of resolution. Theoretical Computer Science 39(2-3), 297–308 (1985)
16. Hoory, S., Linial, N., Wigderson, A.: Expander graphs and their applications. Bulletin of the American Mathematical Society 43(4), 439–561 (2006)
17. Impagliazzo, R., Pudlák, P., Sgall, J.: Lower bounds for the polynomial calculus and the Gröbner basis algorithm. Computational Complexity 8(2), 127–144 (1999)
18. Lingeling and Plingeling, http://fmv.jku.at/lingeling/
19. March, http://www.st.ewi.tudelft.nl/~marijn/sat/march_dl.php
20. Markström, K.: Locality and hard SAT-instances. Journal on Satisfiability, Boolean Modeling and Computation 2(1-4), 221–227 (2006)
21. Marques-Silva, J.P., Sakallah, K.A.: GRASP: A search algorithm for propositional satisfiability. IEEE Transactions on Computers 48(5), 506–521 (1999), preliminary version appeared in ICCAD 1996

22. The MiniSat page, http://minisat.se/
23. Moskewicz, M.W., Madigan, C.F., Zhao, Y., Zhang, L., Malik, S.: Chaff: Engineering an efficient SAT solver. In: Proceedings of the 38th Design Automation Conference (DAC 2001), pp. 530–535 (June 2001)
24. Nordström, J.: Pebble games, proof complexity and time-space trade-offs. Logical Methods in Computer Science 9, 15:1–15:63 (2013)
25. Pipatsrisawat, K., Darwiche, A.: On the power of clause-learning SAT solvers as resolution engines. Artificial Intelligence 175, 512–525 (2011), preliminary version appeared in CP 2009
26. Spence, I.: sgen1: A generator of small but difficult satisfiability benchmarks. Journal of Experimental Algorithmics 15, 1.2:1.1–1.2:1.15 (2010)
27. Urquhart, A.: Hard examples for resolution. Journal of the ACM 34(1), 209–219 (1987)
28. Van Gelder, A., Spence, I.: Zero-one designs produce small hard SAT instances. In: Strichman, O., Szeider, S. (eds.) SAT 2010. LNCS, vol. 6175, pp. 388–397. Springer, Heidelberg (2010)

Proof Complexity and the Kneser-Lovász Theorem

Gabriel Istrate[1,2,*] and Adrian Crăciun[1,2]

[1] Dept. of Computer Science, West University of Timişoara,
Timişoara, RO-300223, Romania
[2] e-Austria Research Institute, Bd. V. Pârvan 4, cam. 045 B,
Timişoara, RO-300223, Romania
gabrielistrate@acm.org

Abstract. We investigate the proof complexity of a class of propositional formulas expressing a combinatorial principle known as *the Kneser-Lovász Theorem*. This is a family of propositional tautologies, indexed by an nonnegative integer parameter k that generalizes the Pigeonhole Principle (obtained for $k = 1$).

We show, for all fixed k, $2^{\Omega(n)}$ lower bounds on resolution complexity and exponential lower bounds for bounded depth Frege proofs. These results hold even for the more restricted class of formulas encoding Schrijver's strenghtening of the Kneser-Lovász Theorem. On the other hand for the cases $k = 2, 3$ (for which combinatorial proofs of the Kneser-Lovász Theorem are known) we give polynomial size Frege ($k = 2$), respectively extended Frege ($k = 3$) proofs. The paper concludes with a brief announcement of the results (presented in subsequent work) on the complexity of the general case of the Kneser-Lovász theorem.

1 Introduction

One of the most interesting approaches in discrete mathematics is the use of topological methods to prove results having a purely combinatorial nature. The approach started with Lovász's proof [18] of a combinatorial statement raised as an open problem by Kneser in 1955 (see [16] for a historical account). A significant amount of work has resulted from this conjecture (to get a feel for the advances consult [20,11]).

Methods from topological combinatorics raise interesting challenges from a complexity-theoretic point of view: they are non-constructive, often based on principles that appear to lack polynomial time algorithms (e.g. Sperner's Lemma and the Borsuk-Ulam Theorem [22]). The concepts involved (simplicial complexes, chains, chain maps) seem to require intrinsically exponential-size representations.

In this paper we raise the possibility of using statements from topological combinatorics as a source of interesting candidates for proof complexity. In particular we view the Kneser-Lovász theorem as a statement on the unsatisfiability of a certain class of propositional formulas, and investigate the complexity of proving their unsatisfiability.

We were initially motivated by the problem of separating the Frege and extended Frege proof systems. Various candidate formulas have been proposed (see [2] for a discussion). The question whether the non-elementary nature of mathematical proofs of Kneser's theorem translates into hardness and separation results in propositional complexity is a natural one to ask.

---

* Corresponding author.

C. Sinz and U. Egly (Eds.): SAT 2014, LNCS 8561, pp. 138–153, 2014.

A slightly different perspective on this problem is the following: Matoušek obtained [19] a "purely combinatorial" proof of the Kneser-Lovász theorem, a proof that does not explicitly mention any topological concept. While combinatorial, Matoušek's proof is nonconstructive: the approach in [19] "hides" in purely combinatorial terms the application of the so-called *Octahedral Tucker Lemma*, a discrete variant of the Borsuk-Ulam theorem. Searching for the object guaranteed to exist by this principle, though "constructive" in theory [7] is likely to be intractable, as the associated search problem is likely to be complete for the class PPAD [21][1].

Thus another perspective on the main question we are interested in is **under what circumstances do cases of the Kneser-Lovász theorem have combinatorial proofs of polynomial size**. This depends, of course, on the proof system considered, making the question fit the "bounded reverse mathematics" program of Cook and Nguyen [5]. A natural boundary seems to be the class of Frege proofs: for $k = 1$ the Kneser-Lovász theorem is equivalent to the pigeonhole principle (PHP) that has polynomial size TC^0-Frege proofs, but exponential lower bounds in resolution [3] and bounded depth Frege. On the other hand **obtaining a similar upper bound** for the general case would be quite significant, as **it would seem to require completely bypassing the techniques from Algebraic Topology** starting instead from radically different principles.

Our contributions (and the outline of the paper) can be summarized as follows: In Section 3 we give a reduction between $Kneser_{k,n}$ and $Kneser_{k+1,n}$ for arbitrary $k \geq 1$. As an application we infer that existing lower bounds for PHP apply to formulas $Kneser_{k,n}$ for any fixed value of k. In Section 4 we investigate cases $k = 2, 3$ (when the Kneser-Lovász theorem has combinatorial proofs). We give Frege proofs (for $k = 2$) and extended Frege proofs (for $k = 3$), both having polynomial size.

As usual in the case of bounded reverse mathematics, our positive results could have been made uniform by stating them (more carefully) as expressibility results in certain logics: for instance our result for the case $k = 2$ of the Kneser-Lovász theorem could be strengthened to an expressibility result in logical theory VNC^1 [5]. We will not pursue this approach in the paper, deferring it to the journal version.

2   Preliminaries

Throughout this paper k will be a fixed constant greater or equal to 1. Given a set of integers A, we will denote by $\binom{A}{k}$ the set of cardinality k subsets of set A. We will write $\binom{n}{k}$ in the previous definition in case $A = \{1, 2, \ldots, n\}$, for some $n \geq 1$. $A \subseteq [n]$ will be called *stable* if for no $1 \leq i \leq n$ both i and $i + 1$ (mod n) are in A. Also denote by $A_{\leq k}$ (called "firsts of A") the set of smallest (at most) k elements of A.

The Kneser-Lovász theorem informally states that if we color the size k subsets of set $\{1, 2, \ldots, n\}$ with $n - 2k + 1$ colors, then two disjoint sets must have the same color. Formally:

Proposition 1. *Given $n \geq 2k \geq 1$ and a function $c : \binom{n}{k} \to [n - 2k + 1]$, there exist two disjoint sets A, B and a color $1 \leq l \leq n - 2k + 1$ with $c(A) = c(B) = l$.*

---

[1] We thank Prof. Pálvölgyi for noting that while the Regular Tucker Lemma is indeed complete, the PPAD-completeness of the *Octahedral* Tucker Lemma is open.

An even stronger form was proved by Schrijver [24]: Proposition 1 is true if we limit the domain of c to all stable subsets of $[n]$ of cardinality k (we will denote this collection of such sets by $\binom{n}{k}_{stab}$):

Proposition 2. *Given $n \geq 2k \geq 1$ and a function $c : \binom{n}{k}_{stab} \to [n - 2k + 1]$, there exist two disjoint sets A, B and a color $1 \leq l \leq n - 2k + 1$ with $c(A) = c(B) = l$.*

The Kneser-Lovász Theorem can be seen as a statement about the chromatic number of a particular graph: define the graph $KG_{n,k}$ to consist of the subsets of cardinality k of $[n]$, connected by an edge when the corresponding sets are disjoint (Figure 2). Then the Kneser-Lovász Theorem is equivalent to $\chi(KG_{n,k}) \geq n - 2k + 2$ (in fact $\chi(KG_{n,k}) = n - 2k + 2$, since the upper bound is easy [20]).

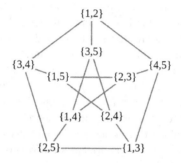

Fig. 1. Kneser graph $KG_{5,2}$ a.k.a. the Petersen graph. The Kneser-Lovász Theorem states that the chromatic number of this graph is 3. Schrijver's Theorem claims that a similar result holds for the interior star only.

We assume familiarity with the basics of proof complexity, as presented for instance in [12], in particular with resolution complexity (the size measure will be denoted by res). We will also need concepts of Frege and extended Frege (EF) proof systems together with methods and results in [3]. We will state our positive results using the sequent calculus system LK [12], a system p-equivalent to Frege proofs.

Definition 1. *Let PHP_n^m be the formula $\bigwedge_{i=1}^{m} (\bigvee_{l=1}^{n} X_{i,l}) \vdash \bigvee_{i \neq j} [\bigvee_{l=1}^{n} (X_{i,l} \wedge X_{j,l})].$*

PHP_n^{n+1} has polynomial size Frege proofs [3]. An important ingredient of the proof is the representation of natural numbers as sequences of bits, with every bit being expressed as the truth value of a certain formula. We will use a similar strategy. In particular quantities such as $\binom{n}{2}$ will refer to the logical encoding of the binary expansion of integer $\frac{n \cdot (n-1)}{2}$. We will further identify statements such as "$A = B$" or "$A \leq B$" with the logical formulas expressing them. The approach of Buss uses *counting*, defining a set of families of formulas $Count_n$, such that $Count_n(Y_1, \ldots, Y_n)$ yields the binary encoding of the number of variables $Y_1, \ldots, Y_n$ that are *TRUE*. Such formulas can be computed by polynomial size Frege proofs [3]. We will often drop the index n from notation if its value is self-evident. We will further need several simple intentional properties of function $Count$ with respect to combinatorics. Formal arguments are deferred to the journal version.

Lemma 1. *Let $n \leq m$. and let $X_1, \ldots X_n, Y_1, \ldots Y_m$ be boolean variables . In LK one can give polynomial-size proofs of the following facts:*

1. $X_1 \wedge X_2 \wedge \ldots X_n \vdash Count_n(X_1, X_2, \ldots, X_n) = n$,

2. $\vdash Count_{\binom{n}{2}}(X_1 \wedge X_2, \ldots, X_i \wedge X_j, \ldots, X_{n-1} \wedge X_n) = \binom{Count_n(X_1, X_2, \ldots, X_n)}{2}$,

3. $\vdash Count_{n^2}((X_i \wedge \{i \neq j\})_{i,j=1}^n) = Count_n(X_1, X_2, \ldots, X_n) \cdot (n-1)$,

4. $(X_1 \leq Y_1), \ldots, (X_n \leq Y_n) \vdash (Count_n((X_i)_{i=1}^n) \leq Count_m((Y_j)_{j=1}^m)$.

Finally a *variable substitution* in a formula will refer in this paper to substituting every variable by some other variable (not necessarily in a 1-1 manner).

2.1 Propositional Formulation of the Kneser-Lovász Theorem

We define a variable $X_{A,l}$ for every set $A \in \binom{n}{k}$ of cardinality k, and partition class $P_l := c^{-1}(\{l\})$. $X_{A,l}$ is intended to be $TRUE$ iff $A \in P_l$ and zero otherwise.

Definition 2. *Denote by*

- $Ant_{k,n}$ *the formula* $\bigwedge_{A \in \binom{n}{k}} (\bigvee_{l=1}^{n-2k+1} X_{A,l})$,

- $Cons_{k,n}$ *the formula* $\bigvee_{\substack{A,B \in \binom{n}{k} \\ A \cap B = \emptyset}} (\bigvee_{l=1}^{n-2k+1} (X_{A,l} \wedge X_{B,l}))$,

- $Onto_{k,n}$ *the formula* $\bigvee_{A \in \binom{n}{k}} (\bigvee_{\substack{l,s=1 \\ l \neq s}}^{n-2k+1} (\overline{X_{A,l}} \vee \overline{X_{A,s}}))$,

- *Finally, denote by $Kneser_{k,n}$ the formula $[Ant_{k,n} \vdash Cons_{k,n}]$. $Kneser_{k,n}$ is (by [18]) a tautology with $(n - 2k + 1) \cdot \binom{n}{k}$ variable,*
- *We will also encode the* onto *version of the Kneser-Lovász Theorem. Indeed, denote by $Kneser_{k,n}^{onto}$ the formula $[Ant_{k,n} \wedge Onto_{k,n} \vdash Cons_{k,n}]$.*

Note that formula $Kneser_{1,n}$ is essentially the Pigeonhole principle PHP_{n-1}^n.

3   Lower Bounds: Resolution Complexity and Bounded-Depth Frege Proofs

The following result provides a mapping between two different Kneser graphs showing that many lower bounds on the complexity of the pigeonhole principle apply directly to any family $(Kneser_{k,n})_n$:

Theorem 1. *For all $k \geq 1, n \geq 3$, there exists a variable substitution Φ_k,*
$\Phi_k : Var(Kneser_{k+1,n}) \longrightarrow Var(Kneser_{k,n-2})$ *such that $\Phi_k(Kneser_{k+1,n})$ is a formula consisting precisely of the clauses of $Kneser_{k,n-2}$ (perhaps repeated and in a different order).*

Proof. For simplicity we will use different notations for the variables of the two formulas: we assume that $Var(Kneser_{k+1,n}) = \{X_{A,i}\}$ and $Var(Kneser_{k,n-2}) = \{Y_{A,i}\}$, with obvious (different) ranges for i and A.

Let $A \in \binom{n}{k+1}$. For $1 \leq i \leq n - 2(k+1) + 1 = n - 2k - 1$ define $\Phi_k(X_{A,i})$ by:

- **Case 1:** $A_{\leq k} \subseteq [n-2]$: Define $\Phi_k(X_{A,i}) = Y_{A_{\leq k},i}$.
- **Case 2:** $A_{\leq k} \not\subseteq [n-2]$: In this case necessarily both $n-1$ and n are members of A. Let $A = P \cup \{n-1,n\}$, $|P| = k - 1$. Let $\lambda = max\{j : j \leq n-2, j \notin P\}$. Define $\Phi_k(X_{A,i}) = Y_{P \cup \{\lambda\},i}$.

Formula $Kneser_{k,n-2}$ has clauses of two types

- (a). Clauses of type $Y_{A,1} \vee Y_{A,2} \vee \ldots Y_{A,n-2k-1}$, with $A \in \binom{n-2}{k}$.
- (b). Clauses of type $\overline{Y_{A,i}} \vee \overline{Y_{B,i}}$ with $1 \leq i \leq n-2k-1$, $A, B \subseteq \binom{n-2}{k}$, $A \cap B = \emptyset$.

As Φ_k preserves the second index, every clause of type (a) of $Kneser_{k+1,n}$ maps via Φ_k to a clause of type (a) of $Kneser_{k,n-2}$. On the other hand every clause of type (a) is the image through Φ_k of some clause of $Kneser_{k,n+1}$, for instance of clause $X_{C,1} \vee X_{C,2} \vee \ldots \vee X_{C,n-2k-1}$, where $C = A \cup \{n-1\}$.

As for clause $\overline{X_{A,i}} \vee \overline{X_{B,i}}$ of type (b), again we use the fact that Φ_k preserves the second index, and prove that the substituted variables correspond to disjoint subsets:

- **Case I:** A, B both fall in Case 1. of the definition of Φ_k.
  Denote for simplicity $C = A_{\leq k}, D = B_{\leq k}$, hence $\Phi_k(\overline{X_{A,i}} \vee \overline{X_{B,i}}) = \overline{Y_{C,i}} \vee \overline{Y_{D,i}}$). It follows that C, D are disjoint (as $A \cap B = \emptyset$ and $C \subseteq A, D \subseteq B$). Note that the converse is also true: every clause $\overline{Y_{C,i}} \vee \overline{Y_{D,i}}$ is the image of clause $\overline{X_{A,i}} \vee \overline{X_{B,i}}$, with $A = C \cup \{n-1\}, B = D \cup \{n\}$.
- **Case II:** One of the sets, say A, falls under Case 2, the other one, B, falls under Case 1 (note that A and B cannot both fall under Case 2, as they would both contain $n-1, n$ and they would no longer be disjoint). In this case $C = P \cup \{\lambda\}, D = B_{\leq k}$. As $\{\lambda+1,\ldots,n\} \subset A$ and $A \cap B = \emptyset$, $\lambda+1,\ldots,n \notin B$. Therefore, even though it might be possible that $\lambda \in B$, certainly $\lambda \notin B_{\leq k}$ (since there are no elements in B larger than λ). Thus $C \cap D = (P \cup \{\lambda\}) \cap B_{\leq k} \subseteq A \cap B = \emptyset$.

□

The previous result can be applied k times to show the following two lower bounds:

Theorem 2. *For any fixed $k \geq 1$ we have $res(Kneser_{n,k}) = 2^{\Omega(n)}$ (where the constant might depend on k).*

Proof. The result follows from the following claim, similar to more powerful (less trivial) results of this type explicitly stated, e.g. in [1]:

Lemma 2. *Let Φ be a propositional formula let $X \xrightarrow{\phi} Y$ be a variable substitution and let $\Xi = \Phi[X \xrightarrow{\phi} Y]$ be the resulting formula. Assume that $P = C_1, C_2, \ldots, C_r$ is a resolution refutation of Φ and let $\phi(P) = \phi(C_1), \phi(C_2), \ldots, \phi(C_r)$. Then $\phi(P)$ is a resolution refutation of Ξ. Consequently $res(\Xi) \leq res(\Phi)$.*

$\square$

Similarly

Theorem 3. *For any fixed $k \geq 1$ and arbitrary $d \geq 1$ there exists $\epsilon_d > 0$ such that the family $(Kneser_{n,k})$ has $\Omega(2^{n^{\epsilon_d}})$ depth-d Frege proofs*

Proof. We employ the the corresponding bound for $PHP^n_{n-1}(= Kneser_{1,n})$ [13], [23].

$\square$

3.1 Extension: Lower Bounds on the Proof Complexity of Schrijver's Theorem

We can prove (stronger) bounds similar to those of Theorems 2 and 3 for Schrijver's formulas by noting that the following variant of Theorem 1 holds:

Theorem 4. *For every $k \geq 1, n \geq 3$ there exists a variable substitution Φ_k, $\Phi_k : Var(Sch_{k+1,n}) \longrightarrow Var(Sch_{k,n-2})$ such that $\Phi_k(Sch_{k+1,n})$ is a formula consisting precisely of the clauses of $Sch_{k,n-2}$ (perhaps repeated and in a different order).*

Proof. Substitution Φ_k is exactly the same as the one in the proof of Theorem 1. In this case we need to further argue three things:

(1) If Φ_k maps $X_{A,i}$ onto $Y_{C,i}$ and A is stable then so is C.
(2) Every clause $Y_{C,1} \vee Y_{C,2} \vee \ldots \vee Y_{C,n-2k-1}$ of $Sch_{k,n-2}$ is the image of a clause $X_{A,1} \vee X_{A,2} \vee \ldots \vee X_{A,n-2k-1}$ with A stable.
(3) Every clause $\overline{Y_{C,i}} \vee \overline{Y_{D,i}}$ of $Sch_{k,n-2}$ is the image of a clause $\overline{X_{A,i}} \vee \overline{X_{B,i}}$ with A, B disjoint and *stable*.

(1) If $A_{\leq k} \subseteq [n-2]$ then $C = A_{\leq k}$ satisfies the stability condition everywhere except perhaps at elements 1 and n-2. But if $1 \in C \subseteq A$ then $n \notin A$ (as A is stable). Similarly $n - 1 \notin A$. This contradicts the fact that A must contain one of $n - 1, n$. On the other hand it is not possible that A falls under Case 2, as it would have to contain successive elements $n - 1, n$.
(2) Since C is stable, one of $1, n - 2$ is **not** in C. Define A to consist of C together with the unique element in $n - 1, n$ not forbidden by stability.
(3) Similarly to (2): given disjoint stable sets C, D in $[n - 2]$ obtain A and B by adding the elements $n - 1, n$ to C, D, one to each set, respecting the stability condition. This is possible as C and D are disjoint. For instance, if $n - 2 \in C$ then $1 \notin C$, and we distribute n in C and $n - 1$ in D.

$\square$

4   The Cases $k = 2$ and $k = 3$ of the Kneser-Lovász Theorem

Unlike the general case, for $k \in \{2, 3\}$ Kneser's conjecture has combinatorial proofs [25], [8]. This facts motivates the following theorem, similar to the one proved in [3] for the Pigeonhole Principle:

Theorem 5. *The following are true:*

- *(a) The class of formulas $Kneser_{2,n}^{onto}$ has polynomial size Frege proofs.*
- *(b) The class of formulas $Kneser_{3,n}^{onto}$ has polynomial size extended Frege proofs.*

Proof. Informally, the basis for the combinatorial proofs in [8], [25] of cases $k \in \{2,3\}$ is the following claim, only valid for these values of k: any partition of $\binom{n}{k}$ into classes $P_1, P_2, \ldots, P_{n-2k+1}$ contains at least one class P_j such that either $\bigcap_{A \in P_j} A \neq \emptyset$. or $A \cap B = \emptyset$ for some $A, B \in P_j$.

This claim could be used as the basis for the propositional simulation of the proofs from [25] and [8], respectively. This strategy only leads to *extended Frege*, rather than Frege proofs for $Kneser_{k,n}$. The reason is that we eliminate one element from $\{1, \ldots, n\}$ and one class from the partition. Similar to the case of PHP in [3], doing so involves renaming, leading to extended Frege proofs.

For $k = 2$ we will bypass the problem above by giving a stronger, counting-based proof of $Kneser_{2,n}$. We will then explain why a similar strategy apparently does not work for $k = 3$ as well. In both situations, $k \in \{2,3\}$ below we first present the mathematical argument, then discuss how to formalize it in (extended) Frege.

4.1 Case $k = 2$

Intuitively, the proof will employ the following strategy: we will partition the color classes of a $n - 2k + 1$ coloring into two parts: *(a)* classes whose member sets have one element in common, and there are at least four subsets, *(b)* all the other classes (we will show, in fact, that each such class has at most three elements).

The largest class of the first type has at most $n - 1$ elements. More generally, the first k largest such classes should have at most $n - 1 + n - 2 + \ldots + n - k$ elements. The resulting upper bound is largest when $k = n - 3$ (i.e. there are no classes of the second type), but even then the upper bound is less than $\binom{n}{2}$, the number of sets to color with $n - 3$ colors.

Semantic Proof. The result follows from the following sequence of claims:

Lemma 3. *Given any $(n - 3)$-coloring c of $\binom{n}{2}$ and color $1 \leq l \leq n - 3$, at least one of the following alternatives is true:*

1. *there exist two disjoint sets $D, E \in c^{-1}(l)$.*
2. *$|c^{-1}(l)| \leq 3$.*
3. *there exists $x \in [n]$, $x \in \bigcap_{A \in c^{-1}(l)} A$.*

Proof. Assume that $D = \{a, b\} \in c^{-1}(l)$ and there is a set $E \in c^{-1}(l)$, $a \notin E$, then either $D \cap E = \emptyset$ or $E = \{b, c\}$, for some c. If $\bigcap_{A \in c^{-1}(l)} A = \emptyset$ then there exists another set F with $b \notin F$. F has to intersect both D and E, thus $F = \{a, c\}$. Hence $|c^{-1}(l)| \leq 3$.   □

Define, for $r \geq 0$

$$p_r = |\{1 \leq \lambda \leq r : |c^{-1}(\lambda)| \geq 4 \text{ and } \bigcap_{A \in c^{-1}(\lambda)} A \neq \emptyset\}|,$$

$$s_r = |\{i \in [n] : \bigcap_{A \in c^{-1}(\lambda)} A = \{i\} \text{ for some } 1 \leq \lambda \leq r \text{ with } |c^{-1}(\lambda)| \geq 4\}|,$$

(call such an i counted by s_r *special*)

$$M_r = \sum_{i=1}^{r} |c^{-1}(i)|, \ N_r = p_r(n-1) - \frac{p_r(p_r-1)}{2} + 3(r - p_r).$$

Lemma 4. *Sequences M_r, N_r are monotonically increasing.*

Proof. First $p_{r+1} - p_r \in \{0, 1\}$. Next $M_{r+1} - M_r = |c^{-1}(r+1)| \geq 0$. Finally, $N_{r+1} - N_r = 3$ if $p_{r+1} = p_r$, $N_{r+1} - N_r = (n-1) - p_r$ if $p_{r+1} - p_r = 1$. In this latter case $p_r = p_{r+1} - 1 \leq (n-3) - 1 = n - 4$ hence $N_{r+1} - N_r \geq 3$. □

We now prove the following result:

Lemma 5. *For $1 \leq r \leq n - 3$, $M_r \leq N_r$.*

Proof. First $s_r(n-1) - \frac{s_r(s_r-1)}{2} \leq p_r(n-1) - \frac{p_r(p_r-1)}{2}$. Indeed, the left hand side is

$$s_r(n-1) - (0 + 1 + \ldots s_r - 1) =$$
$$= (n-1) + (n-1-1) + (n-1-2) + \ldots + (n-1-s_r+1)$$
$$= (n-1) + (n-2) + \ldots + (n - s_r)$$

and similarly for the right-hand side. The desired inequality follows from the fact that $s_r \leq p_r$, valid since a special i may be counted for two different λ.

We prove the lemma by showing the stronger inequality

$$M_r \leq s_r(n-1) - \frac{s_r(s_r-1)}{2} + 3(r - p_r). \tag{1}$$

The first two terms of the right-hand side of (1) count sets $\{p, q\} \in \binom{n}{2}$ with at least one special element. Indeed $s_r(n-1)$ is the number of *pairs* (i, j) with $i \neq j$ and i special. This formula overcounts *sets* with at least one special element when j is special too (and set $\{i, j\}$ is counted for both pairs (i, j) and (j, i)). The number of such pairs is precisely $\frac{s_r(s_r-1)}{2}$.

Now M_r sums up cardinalities of color classes 1 to r. For those λ's in $[r]$ such that $|c^{-1}(\lambda)| \geq 4$ and all sets in the color class intersects at a special i, all these sets contain a special value, hence they are also counted by the right-hand side of (1). The difference is made by the remaining λ's (there are $r - p_r$ of them). By Lemma 3 they add at most $3(r - p_r)$ sets to M_r, establishing the desired result. □

Lemma 6. *We have* $N_{n-3} \leq \binom{n}{2} - 3$.

Proof. $N_{n-3} = (n-1) + (n-2) + \ldots + (n - p_{n-3}) + 3(n - 3 - p_{n-3})$. *But* $3(n - 3 - p_{n-3}) \leq 3 + 4 + \ldots + (n - p_{n-3} - 1)$ *hence*

$$N_{n-3} \leq 3 + 4 + \ldots + (n-1) = n(n-1)/2 - 1 - 2 = \binom{n}{2} - 3.$$

□

Now Theorem (5) (a) follows by setting $r = n - 3$. The right-hand side is $\binom{n}{2} - 3$. But there are $M_{n-3} = \binom{n}{2}$ sets to cover.

□

Propositional Simulation. Now we start translating the above proof into sequent calculus LK. We will sketch the nontrivial steps of the translation. Tedious but straightforward computations shows that all these steps amount to polynomial length proofs.

Lemma 3 can be polynomially simulated as follows:

Lemma 7. *For $n \geq 5$ and $1 \leq l \leq n - 3$ define the propositional formula* $Int_{n,l}[(X_{S,l})_{S \in \binom{n}{2}}]$ *to be*

$$\bigvee_{\substack{D,E \in \binom{n}{2} \\ D \cap E = \emptyset}} (X_{D,l} \wedge X_{E,l}) \vee [Count((X_{S,l})_{S \in \binom{n}{2}}) \leq 3] \vee \bigvee_{i \in [n]} (\bigwedge_{i \notin S} \overline{X_{S,l}}).$$

Here Count are Buss's counting formulas. Then for every $1 \leq l \leq n - 3$ formula $Ant_{n,2} \vdash Int_{n,l}$ *has proofs of polynomial length in sequent calculus LK.*

Proof. We will apply the following trivial

Lemma 8. *Let A, B, C, D be four distinct subsets of cardinality 2 of $[n]$. Then at least one of the following alternatives holds:*

– *At least two sets among A, B, C, D are disjoint.*
– $|A \cup B \cup C \cup D| = 5$ *and* $|A \cap B \cap C \cap D| = 1$.

The lemma will be used "at the meta level", that is it will not be codified propositionally, but simply used to argue for the correctness of the proof.

Define (only for notational convenience, not as part of the Frege proof) shorthand

$$Z_{A,B,C,D}^l := X_{A,l} \wedge X_{B,l} \wedge X_{C,l} \wedge X_{D,l}.$$

Now for any $1 \leq l \leq n - 3$,

$$Ant_{2,n}, \neg [Count((X_{S,l})_{S \in \binom{n}{2}}) \leq 3] \vdash \bigvee_{\substack{A,\ldots,D \\ distinct}} (Z_{A,B,C,D}^l).$$

On the other hand, when $|A \cap B \cap C \cap D| = \emptyset$ two of these sets must be disjoint,

hence for such sets $Z_{A,B,C,D}^l \vdash \bigvee_{\substack{E,F \in \{A,\ldots,D\} \\ E \cap F = \emptyset}} (X_{E,l} \wedge X_{F,l}).$

As for any $n \geq 5$ any two disjoint sets in $\binom{n}{2}$ are part of a 4-tuple of sets in $\binom{n}{2}$

$$\bigvee_{\substack{E,F \in \{A,B,C,D\} \\ E \cap F = \emptyset}} (X_{E,l} \wedge X_{F,l}) \vdash \bigvee_{\substack{E,F \in \binom{n}{2} \\ E \cap F = \emptyset}} (X_{E,l} \wedge X_{F,l}), \text{ hence}$$

$$Ant_{n,2}, \neg[Count(X_{A,l}) \leq 3] \vdash \bigvee_{\substack{E,F \in \{A...D\} \\ E \cap F = \emptyset}} (X_{E,l} \wedge X_{F,l}) \vee \bigvee_{\substack{A,B,C,D \subseteq [n] \\ |A \cap B \cap C \cap D| = 1}} Z_{A,B,C,D}^{l}.$$

$$(2)$$

Now we rewrite

$$\bigvee_{\substack{A,B,C,D \subseteq [n] \\ |A \cap B \cap C \cap D| = 1}} Z_{A,B,C,D}^{l} = \bigvee_{i \in [n]} \left(\bigvee_{A \cap B \cap C \cap D = \{i\}} Z_{A,B,C,D}^{l} \right).$$

Fix an arbitrary 4-tuple (A, B, C, D), $A \cap B \cap C \cap D = \{i\}$. For any $H \in \binom{n}{2}$, $H \not\ni i$ one of the sets A, B, C, D is disjoint from H. Hence by modus ponens (cut) with $E = H$ and $F \in \{A, B, C, D\}$ with $H \cap F = \emptyset$

$$Ant_{2,n}, Z_{A,B,C,D}^{l}, \bigwedge_{\substack{E,F \in \binom{n}{2} \\ E \cap F = \emptyset}} (\overline{X_{E,l}} \vee \overline{X_{F,l}}) \vdash \overline{X_{H,l}}$$

By repeatedly introducing ANDs in the conclusion, then OR in the antecedent

$$Ant_{2,n}, \bigvee_{\substack{A,B,C,D \subseteq [n] \\ A \cap B \cap C \cap D = \{i\}}} Z_{A,B,C,D}^{l}, \bigwedge_{\substack{E,F \in \binom{n}{2} \\ E \cap F = \emptyset}} (\overline{X_{E,l}} \vee \overline{X_{F,l}}) \vdash \bigwedge_{\substack{H \in \binom{n}{2} \\ H \not\ni i}} \overline{X_{H,l}}.$$

By repeated introduction of ORs in both the antecedent and the conclusion

$$Ant_{2,n}, \bigvee_{i \in [n]} \left(\bigvee_{\substack{A,B,C,D \subseteq [n] \\ A \cap B \cap C \cap D = \{i\}}} Z_{A,B,C,D}^{l} \right), \bigwedge_{\substack{E,F \in \binom{n}{2} \\ E \cap F = \emptyset}} (\overline{X_{E,l}} \vee \overline{X_{F,l}}) \vdash \bigvee_{i \in [n]} \left(\bigwedge_{\substack{H \in \binom{n}{2} \\ H \not\ni i}} \overline{X_{H,l}} \right).$$

Taking into account (2) and moving the third antecedent on the right-hand side we get the proof of Lemma 7.

$\square$

Definition 3. *Define for $i \in [n], l \in [n-3]$ formula*

$$Special_{i,l}((X_{S,l})_{S \in \binom{n}{2}}) \equiv \left[(Count((X_{S,l})_{S \in \binom{n}{2}}) \geq 4) \right] \wedge \left[\left(\bigwedge_{\substack{B \in \binom{n}{2} \\ B \not\ni i}} \overline{X_{B,l}} \right) \right].$$

For $r \in [n-3]$ let q_r be the number of indices $i \in [n]$ such that there is a color l, $1 \leq l \leq r$ with $Special_{i,l}((X_{S,l})_{S \in \binom{n}{2}}) = TRUE$.

Remark 1. Semantically we have $q_r = s_r$ (in q_r we do not require that the intersection of all sets $B \in \binom{n}{2} \cap c^{-1}(l)$ have cardinality *exactly one*, but that is true if $Count((X_{S,l})_{S \in \binom{n}{2}}) \geq 4)$

Given $r \leq n - 3$ we can compute, using a Frege proof, the binary representation of q_r. as $q_r = Count(\{i \in [n] \mid \bigvee_{l=1}^{r} Special_{i,l}\})$. Now define for $0 \leq r \leq n - 3$

$$M_r = |\{A \in \binom{n}{2}: \bigvee_{1 \leq l \leq r} X_{A,l}\}| \text{ (semantically } = \sum_{i=1}^{r} |c^{-1}(i)|),$$

$$M_r^{(1)} = |\{A \in \binom{n}{2}: \bigvee_{1 \leq l \leq r} (X_{A,l} \wedge [Count(X_{S,l}) \leq 3]) \}|,$$

$$M_r^{(2)} = |\{A \in \binom{n}{2}: \bigvee_{1 \leq l \leq r} (X_{A,l} \wedge [Count((X_{S,l})_{S \in \binom{n}{2}}) \geq 4]) \}|,$$

$$Q_r^{(1)} = |\{l \mid (1 \leq l \leq r) \wedge [Count(X_{S,l}) \leq 3] \}|.$$

One can easily prove in LK the following

Lemma 9. $Ant_{2,n} \wedge Onto_{2,n} \vdash [M_r = M_r^{(1)} + M_r^{(2)}]$.

Lemma 10. *One can compute in LK the binary expansions of $M_r^{(1)}$, $Q_r^{(1)}$ and prove that $Ant_{2,n} \wedge Onto_{2,n} \vdash [M_r^{(1)} \leq 3 \cdot Q_r^{(1)}]$.*

Proof. For the first part we use Buss's counting approach. For the second, define

$$W_l = \begin{cases} 1 \text{ if } Count(X_{S,l}) \leq 3, \\ 0 \text{ otherwise.} \end{cases} \text{ and } Y_l = \begin{cases} Count(X_{S,l}) \text{ if } Count(X_{S,l}) \leq 3, \\ 0 \qquad\qquad\qquad \text{ otherwise.} \end{cases}$$

Then (one can readily prove in LK that) $Y_l \leq 3W_l$. Summing up we get $M_r^{(1)} \leq 3Q_r^{(1)}$. The proof (using the fact that the cardinal of a union of disjoint sets is the sum of cardinals of individual subsets) can easily be simulated in LK.

$\square$

Definition 4. *Let*

$$P_r^{(2)} = |\{A \in \binom{n}{2}: \bigvee_{1 \leq l \leq r} X_{A,l} \wedge [(\bigwedge_{B \not\ni First(A)} \overline{X_{B,l}}) \oplus (\bigwedge_{B \not\ni Second(A)} \overline{X_{B,l}})]\}|,$$

where

- *$First(A)$ is the smallest element in A, $Second(A)$ is the largest.*
- *$P \oplus Q$ in the above expression is a shorthand for $(P \wedge \overline{Q}) \vee (Q \wedge \overline{P})$. Since there are $O(n^2)$ sets B to consider, the size of the formula after expanding to CNF is $O(n^4)$.*

Lemma 11. *One can prove in LK that*

$$Ant_{2,n} \wedge Onto_{2,n} \wedge \neg Cons_{2,n} \vdash [M_r^{(2)} \leq P_r^{(2)}].$$

Proof. The inequality follows in the following way: From Lemma 7

$$Ant_{2,n} \vdash Int_{n,l}, \text{ hence}$$

$$Ant_{2,n} \wedge \neg Cons_{2,n} \wedge X_{A,l} \wedge [Count((X_{S,l})_{S \in \binom{n}{2}}) \geq 4] \vdash \bigvee_{i \in [n]} Special_{i,l}.$$

Now assume $X_{A,l} \wedge [Count((X_{S,l})_{S \in \binom{n}{2}}) \geq 4]$. For $i \neq First(A), Second(A)$ set A is among the B's in the conjunction defining $Special_{i,l}$, so all these formulas evaluate to $FALSE$. Furthermore, if $X_{A,l}$ and $Count((X_{S,l})_{S \in \binom{n}{2}}) \geq 4$ then exactly one of the two remaining terms, $\bigwedge_{B \neq First(A)} \overline{X_{B,l}}$ and $\bigwedge_{B \neq Second(A)} \overline{X_{B,l}}$ also simplifies to $FALSE$. Indeed, there is a set $B \neq A$ with $X_{B,l}$. B does not contain one of $First(A), Second(A)$, hence $\overline{X_{B,l}}$ appears in exactly one of the corresponding conjunctions, making it $FALSE$.

Hence every set A counted by $M_r^{(2)}$ is among those counted by $P_r^{(2)}$ and, by $Onto_{2,n}$, only in one such set.

□

Define $U_r = |\{A \in \binom{n}{2} : (\bigvee_{\lambda=1}^{r} Special_{First(A),\lambda}) \wedge (\bigvee_{\nu=1}^{r} Special_{Second(A),\nu})\}|.$

Lemma 12. *We have (and can prove in polynomial size in LK)*

$$Ant_{n,2} \vdash [U_r = |\{(i,j) : i < j \in [n] \text{ and } (\bigvee_{\lambda=1}^{r} Special_{i,\lambda}) \wedge$$

$$\wedge (\bigvee_{\nu=1}^{r} Special_{j,\nu})\}| = \binom{q_r}{2}].$$

Proof. The first equality amounts to no more than semantic reinterpretation. The last equality follows from Lemma 1 (2).

□

Lemma 13. $Ant_{2,n} \wedge Onto_{2,n} \vdash [U_r + P_r^{(2)} \leq q_r \cdot (n-1)]$ *has poly-size LK proofs.*

Proof. See the technical report version [10].

□

Corollary 1. *LK can efficiently prove formulas:*

(1). $Ant_{2,n} \wedge Onto_{2,n} \wedge \neg Cons_{2,n} \vdash [M_r + U_r \leq q_r(n-1) + 3 \cdot Q_r^1]$,

(2). $Ant_{2,n} \wedge Onto_{2,n} \vdash [M_{n-3} = \binom{n}{2}]$,

(3). $Ant_{2,n} \wedge Onto_{2,n} \vdash [q_{n-3} \leq n-3]$,

(4). $Ant_{2,n} \wedge Onto_{2,n} \vdash [q_{n-3}(n-1) + 3 \cdot Q_{n-3}^1 + \binom{n-3}{2}] \leq$

$\leq (n-3) \cdot (n-1) + U_{n-3}]$.

Proof. See [10].

□

Now we can put everything together to prove Theorem 5 (a): by (1)

$$Ant_{2,n} \wedge Onto_{2,n} \wedge \neg Cons_{2,n} \vdash [M_{n-3} + U_{n-3} \leq q_{n-3}(n-1) + 3 \cdot Q_{n-3}^1].$$

Adding relation (4), taking into account (2) and simplifying by $U_{n-3} + 3Q_{n-3}^1$ we get

$$Ant_{n,2} \wedge Onto_{n,2} \wedge \neg Cons_{2,n} \vdash \binom{n}{2} + \binom{n-3}{2} \leq (n-1)(n-3),$$

or, equivalently

$$Ant_{n,2} \wedge Onto_{n,2} \wedge \neg Cons_{2,n} \vdash [2n^2 - 8n + 12 \leq 2n^2 - 8n + 6] \vdash \square$$

Moving $\neg Cons_{2,n}$ to the other side we get the desired result.

□

4.2   Case $k = 3$

An analog of Lemma 3 holds for $k = 3$ (see [10] for a proof that can be efficiently simulated in EF):

Lemma 14. *[8] For any $1 \leq \lambda \leq n - 5$ at least one of the following is true:*

- *$c^{-1}(\lambda)$ contains two disjoint sets*
- *$|c^{-1}(\lambda)| \leq 3n - 8$, or*
- *there exists $x \in \bigcap_{A \in c^{-1}(\lambda)} A$.*

Assuming this claim we settle the case $k = 3$. The argument we give is simpler than the argument in [8]. Full details are deferred to the journal version.

Lemma 15. *[8] Kneser's conjecture is true for $k = 3$.*

Proof. By induction. The base case $n = 7$ can be verified directly. Assume that one could give a coloring c of the Kneser graph $KG_{n,3}$ with $n - 5$ colors. If there is a color λ with $x \in \bigcap_{A \in c^{-1}(l)} A$ then one could eliminate both element x and color λ, obtaining a $n - 6$ coloring of graph $KG_{n-1,3}$, thus contradicting the inductive hypothesis.

If no color class contains two disjoint sets then all of them satisfy $|c^{-1}(l)| \leq 3n - 8$. But then we would have $\binom{n}{3} \leq (n-5)(3n-8)$. This is false for $n \geq 7$.

□

We could try to give a Frege proof of the case $k = 3$ based on counting principles, using the following strategy, similar to the one used in case $k = 2$:

1. Count, using a Frege proof, the number p_r of sets $c^{-1}(l)$, $1 \leq l \leq t$ such that $|c^{-1}(\lambda)| \geq 3n - 7$ (implicitly $\cap_{A \in c^{-1}(l)} A \neq \emptyset$).
2. Define $M_r^{(3)} = \sum_{i=1}^r |c^{-1}(i)|$ and

$$N_r^{(3)} = \binom{n-1}{2} + \binom{n-2}{2} + \ldots + \binom{n-p_r}{2} + (n - 5 - p_r)(3n - 7).$$

3. Show inductively that $M_r^{(3)} \leq N_r^{(3)}$.
4. Obtain a contradiction from $M_{n-5}^{(3)} = \binom{n}{3}$ and $N_{n-3}^{(3)} < \binom{n}{3}$.

Although some of this program can be carried through, **this approach does not seem to work.** The inequality that critically fails is the last one: when $k = 2$ we showed that $N_{n-3}^{(2)} < \binom{n}{2}$ as the maximum of the upper bound was obtained for $p_r = n - 3$. For $k = 3$, though, such a statement is not true. Indeed, since *(a)* we need to give upper bound estimates on the size of $n - 5$ color classes, and *(b)* $3n - 8$, the bound on the size of independent sets is growing with n, one cannot guarantee that $N_{n-3}^{(3)} < \binom{n}{3}$ for all possible values of p_r. For instance, if $p_r = n - 6$ (an event we cannot exclude) the resulting upper bound, $\binom{n-1}{2} + \binom{n-2}{2} + \ldots + \binom{6}{2} + (3n - 8) = \binom{n}{3} - 10 - 6 - 3 - 1 + (3n - 7) = \binom{n}{3} + (3n - 27)$ is not smaller than $\binom{n}{3}$ for $n \geq 10$. For this reason when $k = 3$ we will have to do with the *extended Frege* proof described above.

Lemma 14 can be efficiently simulated in EF (actually in Frege) via a straightforward but tedious adaptation of the argument in [8] (see [10]).

On the other hand it may still be possible (and we conjecture that this can be done) to obtain a Frege proof by a more refined version of the above counting approach.

□

5   Heads Up: The General Case of the Kneser-Lovász Theorem

In this section we briefly announce the other results on the proof complexity of the Kneser-Lovász Theorem presented in a companion paper [9]. Unlike the cases $k = 2, 3$, cases $k \geq 4$ apparently require proof systems more powerful than EF. Indeed, the general case of the Kneser-Lovász theorem follows by a combinatorial result known as the *octahedral Tucker lemma* [20]. The propositional counterpart of this implication is the existence of a variable substitution that transforms the propositional encoding of the octahedral Tucker lemma into the Kneser-Lovász formulae.

Though the formalization of the octahedral Tucker lemma yields a formula of exponential size, the octahedral Tucker lemma admits [9] a (nonstandard) version leading to polynomial-size formulas that is sufficient to prove the Kneser-Lovász theorem. However, even this version seems to require exponentially long EF proofs. The reason is that we prove the Octahedral Tucker Lemma by reduction to a Tseitin-type statement, crucially, though, to one on a complete graph K_m of **exponential size** ($m = O(n! \cdot 2^n)$).

The (exponentially long) proofs of such statements can be generated implicitly [15]. However, not only the proof steps but the very *formulas* involved in the proof may have exponential size and need to be generated implicitly. Implicit proofs with implicitly generated formulas have been previously considered in the literature [14]. We postpone the discussion of further technical details to [9].

6   Combinatorial Topology and Benchmark Generation

Our results naturally suggest using the family of tautologies $Kneser_{k,n}$ and, more generally, families of propositional formulas translating results in combinatorial topology [11], as benchmark generators. Indeed, we believe it is interesting to investigate solvability of such formulas in practice.

However, such a task runs into the issue that naively generated formulas of this type are quickly becoming large: for example, formula $Kneser_{4,9}$ has $\binom{9}{4} \cdot (9 - 2 \cdot 4 + 1) =$ 252 variables. One can alleviate this problem to some extent by using the propositional translation of Schrijver's theorem instead. An even more promising approach (that however requires more investigations) is to use the generalization of the Kneser-Lovász theorem known as Dolnikov's theorem [20]. The interest of this combinatorial result is that it provides lower bounds on the chromatic number of so-called generalized Kneser graphs; moreover, *every* graph is a generalized Kneser graph. However, the lower bounds provided by Dolnikov's theorem (the 2-colorability defect of the associated hypergraph) are nonconstructive and not always tight, as it is the case for Kneser graphs. Expansion is one property that could limit the growth of the propositional translation of colorability statements for such graphs. Thus finding a family of expander graphs for which the bounds are effective and tight is an interesting research question.

7   Conclusions, Open Problems and Acknowledgments

Our work raises several open questions:

1. Does $Kneser_{2,n}$ have polynomial size cutting plane proofs/OBDD with projection, as PHP does [6,4]?
2. Does family $Kneser_{3,n}$ have polynomial size Frege proofs? We believe so.
3. Are $(Kneser_{k,n})$, $k \geq 4$, hard for Frege/EF proofs? Though proving such a result seems beyond current methods, we would not be surprised if they were.
4. There is a reasonably sophisticated literature dealing with extensions of the Kneser-Lovász Theorem (see e.g. [11]) or other results in combinatorial topology [20,17]. Do they yield interesting results in bounded reverse mathematics?

Acknowledgement. This work has been supported by CNCS IDEI Grant PN-II-ID-PCE-2011-3-0981 "Structure and computational difficulty in combinatorial optimization: an interdisciplinary approach".

References

1. Ben-Sasson, E., Nordström, J.: Understanding space in proof complexity: Separations and trade-offs via substitutions. In: Proceedings of the Second Symposium on Innovations in Computer Science, pp. 401–416 (2011)
2. Bonet, M., Buss, S., Pitassi, T.: Are there hard examples for Frege Systems? In: Clote, P., Remmel, J. (eds.) Feasible Mathematics II, pp. 30–56 (1995)
3. Buss, S.: Polynomial size proofs of the propositional pigeonhole principle. Journal of Symbolic Logic 52(4), 916–927 (1987)
4. Chén, W., Zhang, W.: A direct construction of polynomial-size OBDD proof of pigeon hole problem. Information Processing Letters 109(10), 472–477 (2009)
5. Cook, S., Nguyen, P.: Logical foundations of proof complexity. Cambridge University Press (2010)
6. Cook, W., Coullard, C., Turán, G.: On the complexity of cutting-plane proofs. Discrete Applied Mathematics 18(1), 25–38 (1987)
7. Freund, R., Todd, M.: A constructive proof of Tucker's combinatorial lemma. Journal of Combinatorial Theory, Series A 30(3), 321–325 (1981)
8. Garey, M., Johnson, D.: The complexity of near-optimal graph coloring. Journal of the ACM 23(1), 43–49 (1976)
9. Istrate, G., Crăciun, A.: Proof complexity and the kneser-lovasz theorem. In: SAT 2014. LNCS, vol. 8561, pp. 139–154. Springer, Heidelberg (2014)
10. Istrate, G., Crăciun, A.: Proof complexity and the kneser-lovász theorem. Tech. rep., arXiv.org Report 1402.4338 (2014)
11. Kozlov, D.: Combinatorial Algebraic Topologya. Springer (2008)
12. Krajicek, J.: Bounded Arithmetic, Propositional Logic and Complexity Theory. Cambridge University Press (1995)
13. Krajicek, J., Pudlák, P., Woods, A.: Exponential lower bound to the size of bounded depth Frege proofs of the pigeonhole principle. Random Structures and Algorithms 7(1), 15–39 (1995)
14. Krajicek, J.: Diagonalization in proof complexity. Fundamenta Mathematicae 182, 181–192 (2004)
15. Krajíček, J.: Implicit proofs. Journal of Symbolic Logic 69(2), 387–397 (2004)
16. de Longueville, M.: 25 years proof of the Kneser conjecture: The advent of topological combinatorics. EMS Newsletter 53, 16–19 (2004)
17. de Longueville, M.: A Course in Topological Combinatorics. Springer (2012)
18. Lovász, L.: Kneser's conjecture, chromatic number, and homotopy. Journal of Combinatorial Theory, Series A 25, 319–324 (1978)
19. Matoušek, J.: A combinatorial proof of Kneser's conjecture. Combinatorica 24(1), 163–170 (2004)
20. Matoušek, J.: Using the Borsuk-Ulam Theorem, 2nd edn. Springer (2008)
21. Pálvölgyi, D.: 2D-TUCKER is PPAD-complete. In: Leonardi, S. (ed.) WINE 2009. LNCS, vol. 5929, pp. 569–574. Springer, Heidelberg (2009)
22. Papadimitriou, C.H.: On the complexity of the parity argument and other inefficient proofs of existence. Journal of Computer and System Sciences 48(3), 498–532 (1994)
23. Pitassi, T., Beame, P., Impagliazzo, R.: Exponential lower bounds for the pigeonhole principle. Computational Complexity 3(2), 97–140 (1993)
24. Schrijver, A.: Vertex-critical subgraphs of Kneser graphs. Nieuw Arch. Wiskd., III. Ser. 26, 454–461 (1978)
25. Stahl, S.: n-tuple colorings and associated graphs. Journal of Combinatorial Theory B 20(3), 185–203 (1976)

QBF Resolution Systems
and Their Proof Complexities[*]

Valeriy Balabanov[1], Magdalena Widl[2], and Jie-Hong R. Jiang[1]

[1] National Taiwan University
[2] Vienna University of Technology
balabasik@gmail.com, widl@kr.tuwien.ac.at, jhjiang@ntu.edu.tw

Abstract. Quantified Boolean formula (QBF) evaluation has a broad range of applications in computer science and is gaining increasing attention. Recent progress has shown that for a certain family of formulas, Q-resolution, which forms the foundation of learning in modern search-based QBF solvers, is exponentially inferior in proof size to two of its extensions: Q-resolution with resolution over universal literals (QU-resolution) and long-distance Q-resolution (LQ-resolution). The relative proof power between LQ-resolution and QU-resolution, however, remains unknown. In this paper, we show their incomparability by exponential separations on two families of QBFs, and further propose a combination of the two resolution methods to achieve an even more powerful proof system. These results may shed light on solver development with enhanced learning mechanisms. In addition, we show how QBF Skolem/Herbrand certificate extraction can benefit from polynomial LQ-resolution proofs in contrast to their exponential Q-resolution counterparts.

1 Introduction

Quantified Boolean formulas (QBFs) can naturally express many decision problems encountered in verification [4,16], planning [15], two-player games [8], electronic design automation [10,12], and other fields in computer science. QBFs extend formulas of propositional logic by adding quantifiers over the (Boolean) variables, which makes them more expressive and allows a more compact representation of logical constraints. Their efficient evaluation has significant practical impacts and is gaining more and more research attention. State of the art evaluation methods for QBF have been considerably influenced by the advancement of satisfiability (SAT) solving of propositional logic [14] and contain methods based on SAT techniques like conflict-driven clause learning (CDCL) [14]. However, possibly to its higher complexity, QBF evaluation remains premature for robust industrial applications and awaits new insights for a breakthrough.

Resolution is a fundamental technique in automated reasoning, in particular for SAT [20]. CDCL, the key technique for efficiency in modern SAT solvers,

---

[*] This work was supported in part by the National Science Council (NSC) of Taiwan under grants 101-2923-E-002-015-MY2 and 102-2221-E-002-232, by the Vienna Science and Technology Fund (WWTF) through project ICT10-018, and by the Austrian Science Fund (FWF) under grant S11409-N23.

C. Sinz and U. Egly (Eds.): SAT 2014, LNCS 8561, pp. 154–169, 2014.
© Springer International Publishing Switzerland 2014

can be considered as a guided resolution process. Not surprisingly, resolution also plays an essential role in the learning mechanism (QCDCL) of modern QBF solvers [2,7,9,13] In QBF, the existence of more than one sound resolution rule enables different proof systems. In particular, Q-resolution [11], which allows resolution only over existential variables and uses universal reduction to remove universal variables, and its extensions by allowing resolution over universal variables (QU-resolution) [18] and by allowing tautological long-distance derivations (LQ-resolution) [1,19], have been proposed.

Recent studies have shown that members of a certain family of QBFs [11] have proofs in QU-resolution or LQ-resolution of polynomial size in the formula size but any Q-resolution proof is claimed to be of exponential size [6,11,18]. On the practical side, an embedding of LQ-resolution in the QCDCL-based solver DepQBF [13] has resulted in significant performance gains [6]. This gives rise to the question whether other resolution systems can have similar impacts. Also, the relative proof complexity between QU-resolution and LQ-resolution remains unknown.

In addition to its contribution to learning, resolution can produce a syntactic proof of the truth or falsity of a QBF. However, besides a validation of the decision result, many applications require a semantic certificate to represent a concrete solution. Such certificates are typically represented in terms of Skolem functions for true QBFs and Herbrand functions for false QBFs. They can be extracted from Q-resolution proofs in time linear in the proof size [1] and their size is usually related to the proof size. Thus, the study of certificate extraction from the potentially smaller QU-resolution and LQ-resolution proofs is very important.

The quests for efficient QBF evaluation and for the extraction of compact QBF (counter)models motivate the investigation of more powerful resolution systems. In this work, we present the following related results. First, we show the incomparability of QU-resolution and LQ-resolution with respect to their proof complexities. To this end, we construct two families of QBFs for which either of the two calculi has only proofs of exponential size, but the other can produce proofs of polynomial size. Second, we define two stronger proof systems and show an exponential separation to QU- and LQ-resolution for one of them. Third, we propose a new procedure for (counter)model extraction from resolution proofs in all the discussed proof systems. Finally, we present an experimental evaluation of the new certificate extraction method.

2   Preliminaries

A Boolean variable over the domain $\{\top \text{ (true)}, \bot \text{ (false)}\}$ appears in a propositional formula ϕ as a *positive literal* or a *negative literal*. We refer to the opposite polarity of a (positive or negative) literal l by $\bar{l}$ and to the variable of a literal l by $\text{var}(l)$. We use $\text{lit}(v) \in \{v, \bar{v}\}$ to refer to either literal of a variable v. A propositional conjunctive normal form (CNF) formula is a conjunction of *clauses*, each of which is a disjunction of literals. We denote a CNF formula by a set of clauses, a clause by a set of literals, and the empty clause by $\square$. We use the Boolean connectives $\neg, \wedge, \vee, \rightarrow, \leftrightarrow$ with their standard interpretation.

Given a set V of Boolean variables, a set $L = V \cup \{\bar{v} \mid v \in V\}$ of positive and negative literals over V, and the existential ($\exists$) and universal ($\forall$) quantifiers, a *quantified Boolean formula* (QBF) $\mathcal{P}.\phi$ in *prenex conjunctive normal form* (PCNF) consists of the prefix $\mathcal{P} = Q_1 v_1 \ldots Q_k v_k$ with $Q_i \in \{\exists, \forall\}$, $v_i \in V$, and $v_i \neq v_j$ if $i \neq j$, and the CNF matrix $\phi \subset 2^L$. All QBFs in this work are assumed to be in PCNF, to be *closed*, i.e., all literals in the matrix are quantified in the prefix, and to be free of tautological clauses. For each variable $v_i \in V$, its *quantifier level* $\mathsf{lev}(v_i)$ is the number of alternations between $\exists$ and $\forall$ quantifiers from Q_1 to Q_i. We apply this definition also to literals, i.e., $\mathsf{lev}(l) = \mathsf{lev}(\mathsf{var}(l))$. The *quantifier index* of v_i is $\mathsf{idx}(v_i) = i$. Similarly, for literal l, $\mathsf{idx}(l) = \mathsf{idx}(\mathsf{var}(l))$. The set V of variables is partitioned into the set $V_\exists = \{v_i \in V \mid Q_i = \exists\}$ of *existential variables* and the set $V_\forall = \{v_i \in V \mid Q_i = \forall\}$ of *universal variables*. We use letters from the beginning of the Latin alphabet for existential variables/literals, letters from the end for universal variables/literals, and v for either.

A (partial) *assignment* to a QBF $\Phi = \mathcal{P}.\phi$ is a set $\sigma \subset L$ where it holds that if $l \in \sigma$ then $\bar{l} \notin \sigma$. The *assignment condition* $\mathsf{cond}(\sigma)$ is the conjunction $(\bigwedge_{l \in \sigma} l)$ of literals in σ. A clause C is evaluated under an assignment σ to $C_{\lceil \sigma}$ such that $C_{\lceil \sigma} = \top$ if $C \cap \sigma \neq \emptyset$, $C_{\lceil \sigma} = \bot$ if $C \setminus \{v \mid \bar{v} \in \sigma\} = \emptyset$, and $C_{\lceil \sigma} = C \setminus \{v \mid \bar{v} \in \sigma\}$ otherwise. A QBF Φ is evaluated under an assignment σ to $\Phi_{\lceil \sigma}$ by replacing each $C \in \phi$ by $C_{\lceil \sigma}$. The QBF $\forall x \mathcal{P}.\phi$ is true if and only if $\mathcal{P}.\phi_{\lceil \{x\}}$ and $\mathcal{P}.\phi_{\lceil \{\bar{x}\}}$ are true. The QBF $\exists e \mathcal{P}.\phi$ is true if and only if $\mathcal{P}.\phi_{\lceil \{e\}}$ or $\mathcal{P}.\phi_{\lceil \{\bar{e}\}}$ is true.

A clause containing a variable in both polarities is tautological. In QBF reasoning the derivation of such clauses can be useful under certain conditions. A universal variable x contained in a clause C as both x and $\bar{x}$ is called a *merged variable*. A *merged literal* l^* is used to replace both literals l and $\bar{l}$ in C. We define $\mathsf{var}(l^*) = \mathsf{var}(l)$, $\mathsf{lev}(l^*) = \mathsf{lev}(l)$, and $\mathsf{idx}(l^*) = \mathsf{idx}(l)$.

The QBF proof systems considered in this work are based on the two derivation rules *resolution* and *universal reduction*. Given two clauses C_1 and C_2, and a *pivot variable* p with $p \in C_1, \bar{p} \in C_2$, resolution produces the clause $\mathsf{resolve}(C_1, p, C_2) = C_1 \setminus \{p\} \cup C_2 \setminus \{\bar{p}\}$. We call this rule an *ordinary* resolution if the following condition holds: For all (merged or regular) literals $l_1 \in C_1 \setminus \{p\}$ and $l_2 \in C_2 \setminus \{\bar{p}\}$ it holds that if $\mathsf{var}(l_1) = \mathsf{var}(l_2)$ then $l_1 = l_2$ and l_1 is not merged. Otherwise we refer to it as *long-distance* resolution. We further distinguish ordinary resolution into $\mathsf{resolve}_\exists$ if $p \in V_\exists$ and $\mathsf{resolve}_\forall$ if $p \in V_\forall$. We call long-distance resolution over pivot $p \in V_\exists$ *proper* and denote it by $\mathsf{resolve}_{\exists L}$ if the following *index restriction* holds: For all (merged or regular) literals $l_1 \in C_1 \setminus \{p\}$ and $l_2 \in C_2 \setminus \{\bar{p}\}$ it holds that if $\mathsf{var}(l_1) = \mathsf{var}(l_2)$ and either $l_1 \neq l_2$ or l_1 is merged, then $\mathsf{var}(l_1) \in V_\forall$ and $\mathsf{idx}(l_1) = \mathsf{idx}(l_2) > \mathsf{idx}(p)$. Note that since $p \in V_\exists$ and $l_1, l_2 \in V_\forall$, lev can be used instead of idx. Given a clause C, universal reduction produces the clause $\mathsf{reduce}(C) = C \setminus \{l \mid \mathsf{var}(l) \in V_\forall \text{ and } \mathsf{lev}(l) > \mathsf{lev}(l') \text{ for all } l' \in C \text{ with } \mathsf{var}(l') \in V_\exists\}$, i.e., it removes from C all universal variables whose quantifier levels are greater than the largest level of any existential variable in C. Note that reduce applies to merged literals from C in the same way as it applies to regular literals.

The following three QBF resolution proof systems are sound and complete: Q-resolution [11] contains the derivation rules reduce and $\mathsf{resolve}_\exists$. QU-resolution [17] and LQ-resolution [1] extend Q-resolution by the rules $\mathsf{resolve}_\forall$ and $\mathsf{resolve}_{\exists L}$, respectively. A {Q,QU,LQ}-resolution proof Π of the falsity of a QBF $\Phi = \mathcal{P}.\phi$ is a directed acyclic graph (DAG) representing clauses derived from ϕ by repeated applications of the respective rules in process of deriving $\square$. The operation reduce is applied to any clause in Π from which it can remove a literal. (Note that the definition of a QU-resolution proof in [17] does not include the mandatory application of reduce. We discuss the influence of arbitrarily *postponing* the reduce operation in Section 3.1.) We call application of a derivation rule a *step*. The *size* of Π is the number of clauses in Π that are derived by resolution (not by reduction). By *topological order* we refer to any order following the derivation steps in Π from the clauses in ϕ to $\square$.

To witness the falsity (truth) of a QBF, a countermodel (model) can be built in terms of *Herbrand* (*Skolem*) functions. A false (true) QBF $\Phi = \mathcal{P}.\phi$ warrants the existence of a Herbrand (Skolem) function h_v (s_v) for each $v \in V_\forall$ ($v \in V_\exists$) referring only to the variables $\{e \in V_\exists \mid \text{lev}(e) < \text{lev}(v)\}$ ($\{x \in V_\forall \mid \text{lev}(x) < \text{lev}(v)\}$) such that substituting each appearance of a variable v in ϕ by its function h_v (s_v) makes the resultant formula, denoted $\Phi[\mathcal{H}]$ for $\mathcal{H} = \{h_v \mid v \in V_\forall\}$ ($\Phi[\mathcal{S}]$ for $\mathcal{S} = \{s_v \mid v \in V_\exists\}$), unsatisfiable (tautological).

3 Resolution Proof Systems and Their Complexities

In this section, we first show an exponential gap between the proof complexities of LQ-resolution and QU-resolution with respect to two families of QBFs obtained by modifications of a family of QBFs introduced in [11] (in the sequel called "KBKF family"). Then we introduce two new resolution proof systems, both of which are extensions of Q-resolution, and show an exponential separation between QU-resolution, LQ-resolution, and one of the new resolution systems.

3.1 Incomparability of LQ- and QU-resolutions

We first give an intuition of how to engineer a false QBF that inhibits $\text{resolve}_\forall$ and $\text{resolve}_{\exists L}$ steps in any of its resolution proofs. Ex. 1 shows a false QBF for which any resolution proof cannot contain $\text{resolve}_\forall$ or $\text{resolve}_{\exists L}$ steps.

Ex. 1. Consider the false QBF $\Phi = \exists a \forall x \forall y \exists b.(a, x, y, b)(\overline{a}, \overline{x}, \overline{y}, b)(x, y, \overline{b})(\overline{x}, \overline{y}, \overline{b})$. The falsity of Φ is shown by the Herbrand functions $h_y = h_x = a$. Let Π be a QU-resolution proof of Φ. Since $\text{lev}(x) = \text{lev}(y)$, the universal reduction reduce always removes both x and y at once. Thus, any clause in Π either contains both x and y in the same polarity, or neither x nor y in any polarity. It follows that Π cannot contain any clause derived by $\text{resolve}_\forall$. Alternatively, let Π be an LQ-resolution proof of Φ. Due to the level restriction, any $\text{resolve}_{\exists L}$ step must have a as pivot variable, so $\text{resolve}_{\exists L}((a, x, y, b), a, (\overline{a}, \overline{x}, \overline{y}, b)) = (x^*, y^*, b)$ is the only possible such step. However, this resolvent can never be used in a derivation of $\square$, because the necessary pivot literal $\overline{b}$ always occurs in clauses together with literals of x and y, which forbids any further resolution.

Definition 1 reproduces the definition of the KBKF family [11]. Theorem 3.2 in [11] claims that any Q-resolution proof for members of this family is of size exponential in t [11], but its proof is not completely given. It has further been shown that there exist a QU-resolution proof [17] and an LQ-resolution proof [6] of size polynomial in t. For the remainder of this section it is important to keep in mind that for all $i \in [1..t]$, $\mathsf{lev}(e_i) = \mathsf{lev}(d_i) < \mathsf{lev}(x_i)$ and $\mathsf{lev}(x_t) < \mathsf{lev}(f_i)$.

Definition 1 (KBKF family[11]). *For $t > 1$, the tth member KBKF[t] of the KBKF family consists of the following prefix and clauses:*

$$\exists d_1 e_1 \, \forall x_1 \, \exists d_2 e_2 \, \forall x_2 \, .. \, \exists d_t e_t \, \forall x_t \, \exists f_1 .. f_t$$
$$\begin{aligned}
B &= (\bar{d}_1, \bar{e}_1) \\
D_i &= (d_i, x_i, \bar{d}_{i+1}, \bar{e}_{i+1}) & E_i &= (e_i, \bar{x}_i, \bar{d}_{i+1}, \bar{e}_{i+1}) & \text{for } i \in [1..t-1] \\
D_t &= (d_t, x_t, \bar{f}_1, .., \bar{f}_t) & E_t &= (e_t, \bar{x}_t, \bar{f}_1, .., \bar{f}_t) \\
F_i &= (x_i, f_i) & F_i' &= (\bar{x}_i, f_i) & \text{for } i \in [1..t]
\end{aligned}$$

We now apply ideas from Ex. 1 to transform the KBKF family into the family KBKF-qu, for which, based on Theorem 3.2 in [11], the smallest QU-refutations are of exponential size but there exist LQ-refutations of size polynomial in t. It follows from the existence of these proofs that the members of this family are false. For $t > 1$, KBKF-qu[t] is obtained from KBKF[t] by adding fresh universal variables y_i to some clauses.

Definition 2 (KBKF-qu family). *For $t > 1$, the tth member KBKF-qu[t] of the KBKF-qu family consists of the following prefix and clauses:*

$$\exists d_1 e_1 \, \forall x_1 y_1 \, \exists d_2 e_2 \, \forall x_2 y_2 \, .. \, \exists d_t e_t \, \forall x_t y_t \, \exists f_1 .. f_t$$
$$\begin{aligned}
B &= (\bar{d}_1, \bar{e}_1) \\
D_i &= (d_i, x_i, y_i, \bar{d}_{i+1}, \bar{e}_{i+1}) & E_i &= (e_i, \bar{x}_i, \bar{y}_i, \bar{d}_{i+1}, \bar{e}_{i+1}) & \text{for } i \in [1..t-1] \\
D_t &= (d_t, x_t, y_t, \bar{f}_1, .., \bar{f}_t) & E_t &= (e_t, \bar{x}_t, \bar{y}_t, \bar{f}_1, .., \bar{f}_t) \\
F_1 &= (x_i, y_i, f_i) & F_i' &= (\bar{x}_i, \bar{y}_i, f_i) & \text{for } i \in [1..t]
\end{aligned}$$

The following proposition shows that the shortest Q-refutation for KBKF-qu[t] is at least as long as the shortest Q-refutation for KBKF[t].

Proposition 1. *Given a false QBF $\Phi = Q_1 v_1 \, .. \, Q_k v_k. \, C_1 \wedge C_2 \wedge .. \wedge C_n$ over the set V of variables, it holds that for any variable $v \in V$, if $\Phi^* = Q_1 v_1 \, .. \, Q_k v_k. \, C_1 \wedge .. \wedge (C_j \cup \{\mathsf{lit}(v)\}) \wedge .. \wedge C_n$ is false, then the smallest $\{Q, QU, LQ\}$-resolution proof for Φ^* is at least as large as that for Φ.*

The validity of Proposition 1 can be understood by the fact that removing the literal $\mathsf{lit}(v)$ from the clause $(C_j \cup \{\mathsf{lit}(v)\})$ can only decrease the proof size of Φ^*. Note that adding a fresh variable $v \notin V$ to $\mathcal{P}$ influences neither the satisfiability of Φ, nor the validity of any of its Q-resolution proofs. Thus Proposition 1 can be extended for addition of fresh variables to $\mathcal{P}$ and their literals to ϕ.

Theorem 1. *For $t > 1$ there exists an LQ-refutation of polynomial size for KBKF-qu[t], but any QU-refutation for KBKF-qu[t] is of exponential size in t (based on Theorem 3.2 in [11]).*

Proof. Except for the clause B, each clause of KBKF-qu[t] contains two universal variables x, y with the same level and the same polarity. For any QU-refutation, in order to have a resolve$_\forall$ step over two clauses C_1 and C_2 with x (respectively y)

as pivot, y (x) must be removed from one of the clauses, which can only be done by reduce. Whenever y (x) is reduced, so is x (y). Therefore, any QU-refutation will be a Q-refutation , and by Theorem 3.2 in [11] and Proposition 1, the shortest Q-refutation for KBKF-qu is exponential. On the other hand, by following the method proposed in Proposition 1 of [6], a polynomial LQ-refutation can be obtained.   □

We continue with the following modification of the KBKF family that inhibits resolve$_{\exists L}$ steps but allows polynomial QU-refutations. For $t > 1$, KBKF-lq[t] is retrieved from KBKF[t] by adding literals $\overline{f}_1, .., \overline{f}_t$ to clauses B, D_i and E_i, and literals $\overline{f}_{i+1}, .., \overline{f}_t$ to clauses F_i and F_i', for all $i \in [1..t-1]$.

Definition 3 (KBKF-lq family). *For $t > 1$, the tth member KBKF-lq[t] of the KBKF-lq family consists of the following prefix and clauses:*

$$\exists d_1 e_1 \ \forall x_1 \ \exists d_2 e_2 \ \forall x_2 \ .. \ \exists d_t e_t \ \forall x_t \ \exists f_1 .. f_t$$

$$
\begin{aligned}
B &= (\overline{d}_1, \overline{e}_1, \overline{f}_1, .., \overline{f}_t) \\
D_i &= (d_i, x_i, \overline{d}_{i+1}, \overline{e}_{i+1}, \overline{f}_1, .., \overline{f}_t) & E_i &= (e_i, \overline{x}_i, \overline{d}_{i+1}, \overline{e}_{i+1}, \overline{f}_1, .., \overline{f}_t) & \text{for } i \in [1..t-1] \\
D_t &= (d_t, x_t, \overline{f}_1, .., \overline{f}_t) & E_t &= (e_t, \overline{x}_t, \overline{f}_1, .., \overline{f}_t) \\
F_i &= (x_i, f_i, \overline{f}_{i+1}, .., \overline{f}_t) & F_i' &= (\overline{x}_i, f_i, \overline{f}_{i+1}, .., \overline{f}_t) & \text{for } i \in [1..t-1] \\
F_t &= (x_t, f_t) & F_t' &= (\overline{x}_t, f_t)
\end{aligned}
$$

Observation 1. For $t > 1$ any member KBKF-lq[t] of the KBKF-lq family is a quantified extended Horn (QE-Horn) formula [11] and QE-Horn formulas are closed under LQ-resolution.

The closure of QE-Horn formulas under LQ-resolution directly follows from their closure under Q-resolution (observe that the resolve$_{\exists L}$ rule does not influence existential literals in the clauses). On the other hand, note that QE-Horn formulas are not closed under QU-resolution. Further, the following three invariants hold for any member of KBKF-lq family.

Lemma 1 (Invariant 1). *Given any LQ-resolution proof Π of a formula KBKF-lq[t], the following holds for any clause $C \in \Pi$: For all $i \in [1..t]$, if $f_i \in C$ then $\text{lit}(x_i) \in C$, and if $\overline{f}_i \in C$ then for any $j \in [i..t]$ either $\overline{f}_j \in C$ or $\text{lit}(x_j) \in C$.*

Proof. First, observe that the invariant holds for any clause in the original clause set of KBKF-lq[t]. Let C be a clause derived from C' by exactly one derivation step, such that $f_i \in C$ and $f_i \in C'$. If $\text{lit}(x_i) \in C'$ then it must hold that $\text{lit}(x_i) \in C$, because resolution on universal variables is forbidden and the presence of f_i disallows the universal reduction of $\text{lit}(x_i)$ in both C' and C. Thus by induction it holds for any clause C that if $f_i \in C$ then $\text{lit}(x_i) \in C$.

Now let C be a clause derived from C' by exactly one derivation step, such that $\overline{f}_i \in C$ and $\overline{f}_i \in C'$. If $\text{lit}(x_j) \in C'$ for some $j \in [i..t]$, then $\text{lit}(x_j) \in C$ for the same reasons as above. If $\overline{f}_j \in C'$ for some $j \in [i..t]$, then either $\text{lit}(x_j) \in C$ (in the case where f_j is the pivot variable, i.e., $C = \text{resolve}(C', f_j, C'')$ with $f_j, \text{lit}(x_j) \in C''$ by the above discussion), or $\overline{f}_j \in C$ (in any other case). Thus by induction it holds for any clause C that if $\overline{f}_i \in C$ then for any $j \in [i..t]$ either $\overline{f}_j \in C$ or $\text{lit}(x_j) \in C$.   □

Lemma 2 (Invariant 2). *Given any LQ-resolution proof Π of a formula KBKF-lq[t] the following holds for any clause $C \in \Pi$: For all $i \in [1..t]$, if $\text{lit}(d_i) \in C$ or $\text{lit}(e_i) \in C$ then $f_j \notin C$ for any $j \in [1..t]$.*

Proof. First, the invariant holds for any clause in the original clause set of KBKF-lq[t]. Now let $C = \text{resolve}(C_1, p, C_2)$, where $\text{lit}(e_i) \in C$ or $\text{lit}(d_i) \in C$, and $\text{lit}(e_i) \in C_1$ or $\text{lit}(d_i) \in C_1$ for some $i \in [1..t]$.

If $\text{lit}(e_k) \in C_2$ or $\text{lit}(d_k) \in C_2$ for some $k \in [1..t]$, then by inductive hypothesis it holds that $f_j \notin C_1$ and $f_j \notin C_2$ for all $j \in [1..t]$. Therefore, by the definition of resolve, it holds that $f_j \notin C$ for all $j \in [1..t]$.

Else, $\text{lit}(e_i) \notin C_2$ and $\text{lit}(d_i) \notin C_2$, thus we are left with $p = f_k$ for some $k \in [1..t]$. By inductive hypothesis, $f_j \notin C_1$ for all $j \in [1..t]$, therefore $\overline{f}_k \in C_1$ and $f_k \in C_2$. By Observation 1 it holds that $f_j \notin C_2$ for all $j \in [1..t]$ with $j \neq k$. Thus for all $j \in [1..t]$ it holds that $f_j \notin C$.

Therefore, by induction it holds for any clause C and for all $i \in [1..t]$ that if $\text{lit}(d_i) \in C$ or $\text{lit}(e_i) \in C$ then $f_j \notin C$ for any $j \in [1..t]$.   □

Lemma 3 (Invariant 3). *Given any LQ-resolution proof Π of a formula KBKF-lq[t] the following holds for any clause $C \in \Pi$: For all $i \in [1..t]$ it holds that if $\text{lit}(d_i) \in C$ or $\text{lit}(e_i) \in C$ then for any $j \in [1..i-1]$ either $\overline{f}_j \in C$ or $\text{lit}(x_j) \in C$.*

Proof. First, note that the invariant holds for any clause of the original clause set of KBKF-lq[t]. Now, let C be a clause retrieved from C' by one derivation step, such that $\text{lit}(e_i) \in C'$ or $\text{lit}(d_i) \in C'$, and $\text{lit}(e_i) \in C$ or $\text{lit}(d_i) \in C$. If for some $j \in [1..i-1]$ it holds that $\text{lit}(x_j) \in C'$, then $\text{lit}(x_j) \in C$ for the same reasons as in the proof of Invariant 1 (recall that $\text{lev}(e_i) = \text{lev}(d_i) > \text{lev}(x_j)$ for $j \in [1..i-1]$, therefore disallowing universal reduction of $\text{lit}(x_j)$ in the presence of either $\text{lit}(e_i)$ or $\text{lit}(d_i)$). If $\overline{f}_j \in C'$ for some $j \in [1..i-1]$, then either $\text{lit}(x_j) \in C$ (in the case where f_j is the pivot variable, i.e., $C = \text{resolve}(C', f_j, C'')$ with $\{f_j, \text{lit}(x_j)\} \in C''$ by Invariant 1), or $\overline{f}_j \in C$ (in any other case).

Therefore by induction it holds for any clause C and for all $i \in [1..t]$ that if $\text{lit}(d_i) \in C$ or $\text{lit}(e_i) \in C$ then for any $j \in [1..i-1]$ either $\overline{f}_j \in C$ or $\text{lit}(x_j) \in C$.   □

Theorem 2. *For $t > 1$ there exists a QU-resolution proof of polynomial size for KBKF-lq[t], but any LQ-resolution proof for KBKF-lq[t] is of exponential size in t (based on Theorem 3.2 in [11]).*

Proof. For $t > 1$, a QU-refutation of polynomial size in t for KBKF-lq[t] can be constructed as follows: The unit clause (f_t) is obtained by the resolution step $\text{resolve}_\forall(F_t, x_t, F_t')$. Then, for each $i \in [1..t-1]$, the unit clause (f_i) is obtained by recursively resolving all previous units $(f_{i+1})..(f_t)$ with the resolvent $\text{resolve}_\forall(F_i, x_i, F_i')$. For $i \in [1..t]$ these unit clauses are used to remove all $\overline{f}_i$ from the clauses D_i, E_i, and B, and the existential literals e_i and d_i are removed one after another by $\text{resolve}_\exists$ over the remaining clauses.

For the remainder of this proof, let Π be an LQ-resolution proof for KBKF-lq[t]. Let the three clauses $C_1 = (A_1, p, X, R_1)$, $C_2 = (A_2, \overline{p}, \overline{X}, R_2)$, and $C =$

(A, X^*, R) be parts of a resolve$_{\exists L}$ step in Π, where X is a set of universal literals, $\overline{X} = \{\overline{x} \mid x \in X\}$, $X^* = \{x^* \mid x \in X\}$, $C = \text{resolve}_{\exists L}(C_1, p, C_2)$ is the resolvent of C_1 and C_2, $A = A_1 \cup A_2$, and $R = R_1 \cup R_2$. Let x_m and x_n be the variables with the lowest, respectively the highest, level among the variables in X. By definition of resolve$_{\exists L}$ it holds that $\text{lev}(p) < \text{lev}(x_m)$. Without loss of generality, for $i \in \{1, 2\}$ let $R_i = \{v \in C_i \mid v \notin X \wedge \text{lev}(v) > \text{lev}(x_m)\}$ and $A_i = \{v \in C_i \mid v \notin (X \cup R_i \cup \{p\})\}$. Therefore, $R = \{v \in C \mid v \notin X^* \wedge \text{lev}(v) > \text{lev}(x_m)\}$ and $A = \{v \in C \mid v \notin (X^* \cup R)\}$. It is important to notice that the existential literals in R have to be removed from successors of C before X^* can be reduced. Further, $f_i \notin R$ for all $i \in [1..t]$ by Invariant 2, and $R_1, R_2 \neq \emptyset$ because otherwise x_m would be reduced before deriving C. Hence $R \subset \{\text{lit}(e_i), \text{lit}(d_i), \text{lit}(x_i) \mid m < i \leq t\} \cup \{\overline{f}_i \mid 1 \leq i \leq t\}$.

We now show by case distinction on the existential variables in R that the clause C can either not contribute to the derivation of $\square$ in Π because at least one of the merged variables can never be reduced, or that the subclause A can be retrieved from C_1, C_2, and the input clauses in a polynomial number of derivation steps in the Q-resolution calculus. Under the assumption that Π is of polynomial size, its polynomial transformation into a Q-resolution proof contradicts with Proposition 1 and Theorem 3.2 in [11], stating that any Q-resolution for any member of KBKF-lq is exponential. Therefore, Π must be exponential.

Case 1. $\overline{f}_n \in R$. To remove $\overline{f}_n$, C has to be resolved with a clause C' containing f_n. By Invariant 1, C' contains $\text{lit}(x_n)$. Thus $\overline{f}_n$ cannot be removed from R due to the level restriction on resolve$_{\exists L}$ steps. Therefore, $C \notin \Pi$.

Case 2. $\text{lit}(d_i) \in R$ or $\text{lit}(e_i) \in R$. Recall that $i > m$, and without loss of generality, let $d_i \in R$. To remove d_i, C has to be resolved with a clause C' containing $\overline{d}_i$. By Invariant 3, C' either contains $\text{lit}(x_m)$ or $\overline{f}_m$. In the first case, the level restriction on resolve$_{\exists L}$ steps forbids the resolution, and in the latter case, Invariant 1 applies to the resolvent similarly as in Case 1. Therefore, $C \notin \Pi$.

Case 3. $\overline{f}_i \in R$ and $i < n$. Similarly to Case 2, to remove $\overline{f}_i$, C has to be resolved with a clause C' containing f_i. By Invariant 1, C' either contains $\text{lit}(x_n)$, which blocks the resolution as in Case 1 and Case 2, or it contains $\overline{f}_n$ and therefore to its resolvent, Case 1 applies. Therefore, $C \notin \Pi$.

Case 4. $\overline{f}_i \in R$ and $i > n$. Assume without loss of generality that $\overline{f}_i \in R_1$. For $j \in [i..t]$ let the set X' contain x_j if $x_j \in R_1$ and contain $\overline{x}_j$ if $x_j \notin R_1$. By applying resolve$_\exists$ over an adequate subset of the clauses $\{F_j, F'_j \mid i \leq j \leq t\}$, the clause (f_i, X') can be obtained in a polynomial number of steps and be resolved with C to eliminate $\overline{f}_i$. This procedure can be applied to eliminate all $\overline{f}_i$ literals from R_1 and thus enable reduce on the variables in X. By applying the same rewriting to C_2 eventually resolve$_{\exists L}(C_1, p, C_2)$ transforms into resolve$_\exists(C_1 \setminus \{X, F_1\}, p, C_2 \setminus \{\overline{X}, F_2\})$, where $F_1 = \{\overline{f}_i \mid i > n \wedge \overline{f}_i \in C_1\}$ and $F_2 = \{\overline{f}_i \mid i > n \wedge \overline{f}_i \in C_2\}$. $\square$

For $\{\textsf{Q}, \textsf{QU}, \textsf{LQ}\}$-resolution, we follow the assumption that universal reduction is performed whenever possible. If one allows postponing the reduction arbitrarily (as in the definition of QU-resolution in [17]), it will generalize the aforementioned proof systems and allow a larger number of sound refutations. In the sequel we call a refutation where the reduction of at least one universal variable has been postponed a *postponed refutation* and a clause that contains a universal variable which could be universally reduced, but is still present in at least one of its child clauses, a *postponed clause*. The following corollary from Proposition 1 shows that postponing cannot lead to shorter refutations in terms of the number of resolutions for any of the {Q, QU, LQ}-resolution proof systems.

Corollary 1. *Given a false QBF Φ, let Π be its shortest {Q, QU, LQ}-refutation, and let Π^* be its shortest postponed {Q, QU, LQ}-refutation. Then $|\Pi^*| \geq |\Pi|$.*

Proposition 1 can be applied to all topologically first postponed clauses in a postponed refutation and therefore, the corollary follows. By Corollary 1, Theorems 1 and 2 hold for postponed QU-refutations as well.

3.2   New Resolution Proof Systems

We propose two additional resolution systems for QBF. The first, LQU-resolution, is defined as an extension of Q-resolution by adding both the resolve$_\forall$ and the resolve$_{\exists L}$ derivation rules. The second, LQU+-resolution, extends LQU-resolution by the new derivation rule resolve$_{\forall L}$ that allows proper long-distance resolutions under universally quantified pivots. The proof for soundness of resolve$_{\forall L}$ is similar to that of resolve$_{\exists L}$ rule in [1]. Note that the index restriction imposed on resolve$_{\forall L}$ cannot be simplified to level restriction as for resolve$_{\exists L}$, since the universal pivot may have the same level as a merged literal in the same proof step. The following example shows that relaxing the index restriction to the level restriction is unsound for resolve$_{\forall L}$.

Ex. 2. Consider the true QBF $\Phi = \forall x \, \forall y \, \exists a. \ (x, y, a)_1 \ (\overline{x}, \overline{y}, a)_2 \ (x, \overline{y}, \overline{a})_3 \ (\overline{x}, y, \overline{a})_4$. The Skolem function $s_a = (x \leftrightarrow y)$ shows that Φ is true. Note that $\text{lev}(x) = \text{lev}(y)$, but $\text{idx}(x) < \text{idx}(y)$. If the index restriction is neglected, then the following unsound proof Π can be built.

$$\Pi = \begin{cases} 1. \ clause_5 = \text{resolve}_{\forall L}(clause_1, x, clause_2) = (y^*, a) \\ 2. \ clause_6 = \text{resolve}_{\forall L}(clause_3, y, clause_4) = (x^*, \overline{a}) \\ 3. \ clause_7 = \text{resolve}_\exists(clause_5, a, clause_6) = (x^*, y^*) \\ 4. \ clause_\emptyset = \text{reduce}(clause_7) = \square \end{cases}$$

Note that the index restriction on x and y would disallow resolve$_{\forall L}$ step 2.

Table 1 compares the five proof systems discussed in this section by listing their derivation rules. In each line, the derivation rules for each proof system are marked by "x". All proof systems are sound and refutationally complete for QBF. The completeness of LQU-resolution and LQU+-resolution follows from the completeness of Q-resolution. The soundness of LQU-resolution and LQU+-resolution is an extension of Theorem 4 in [1] and can be proved similarly. We extend the definition of a {Q,QU,LQ}-resolution proof Π to {LQU,LQU+}-resolution by adding the corresponding derivation rules.

Table 1. Summary of Proof System Rules

	reduce	resolve∃	resolve∀	resolve∃L	resolve∀L
Q-resolution [11]	x	x			
QU-resolution[17]	x	x	x		
LQ-resolution[1]	x	x		x	
LQU-resolution	x	x	x	x	
LQU+-resolution	x	x	x	x	x

3.3   Superiority and Limitation of **LQU-** and **LQU+-resolutions**

KBKF-qu and KBKF-lq families can be combined into a family KBKF-lqu that has exponential smallest proofs in both QU-resolution and LQ-resolution, but polynomial proofs in LQU-resolution. Proposition 2 and Theorem 3 below demonstrate bounds on the shortest proofs for combinations of QBF formulas.

Proposition 2 (cf. [6], Proposition 5). *Given a false QBF $\Phi = \mathcal{P}.\phi$, a literal $e \in V_\exists$, an LQU-resolution proof Π for Φ, both $\Phi_{\lceil\{e\}}$ and $\Phi_{\lceil\{\bar{e}\}}$ are false QBFs and Π can be modified in polynomial time with respect to its size to obtain a new proof $\Pi_{\lceil e}$ (respectively $\Pi_{\lceil\bar{e}}$) deriving $\square$ from $\Phi_{\lceil\{e\}}$ (respectively $\Phi_{\lceil\{\bar{e}\}}$).*

This proposition extends Proposition 5 in [6] by allowing e to have an arbitrary quantifier level and allowing the proof to contain resolve∀ steps. The extension is sound, since the proof in [6] is independent of the quantifier levels of existential variables and can also be incorporated with resolve∀ and resolve∃L rules.[1] The same result for Q-resolution has been proposed in [8]. By Proposition 2 also $\Phi_{\lceil\sigma}$ and $\Pi_{\lceil\sigma}$ for any assignment σ to existential variables of Φ can be constructed.

Theorem 3. *Given two disjoint sets V_1 and V_2 of variables, let $\Phi_1 = \mathcal{P}_1.\phi_1$ and $\Phi_2 = \mathcal{P}_2.\phi_2$ be two false QBFs over V_1 and V_2, respectively. Let Π_1 and Π_2 be their respective shortest LQU-resolution proofs. Let $\Phi = \exists a \mathcal{P}_1 \mathcal{P}_2.(\phi_1 \vee a) \wedge (\phi_2 \vee \bar{a})$, where $a \notin V_1 \cup V_2$, and for $i \in \{1,2\}$, $(\phi_i \vee a)$ stands for $\{C \cup \{a\} \mid C \in \phi_i\}$. Then Φ is false and the size of its shortest LQU-refutation is $|\Pi_1| + |\Pi_2| + 1$.*

Proof. By following the resolution steps of Π_1 on the clauses of $(\phi_1 \vee a)$ we retrieve the clause (a), by following the resolution steps of Π_2, on $(\phi_2 \vee \bar{a})$ we retrieve $(\bar{a})$, and resolving the two unit clauses results in $\square$. Thus an LQU-resolution proof Π with $|\Pi| = |\Pi| = |\Pi_1| + |\Pi_2| + 1$ is constructed. Let Π be any LQU-resolution proof for Φ and let $\Pi_{\lceil\{a\}}$ be the proof generated for $\Phi_{\lceil\{a\}}$ as by Proposition 2 and let n_1 (resp. n_2) be the number of resolution steps in Π under any pivot $p \in V_1$ (resp. any pivot $p \in V_2$). By construction, $\Phi_{\lceil\{a\}} = \Phi_2$ and therefore $n_2 \geq |\Pi_{\lceil\{a\}}| \geq |\Pi_2|$. The dual case holds for $\Pi_{\lceil\{\bar{a}\}}$, resulting in $n_1 \geq |\Pi_{\lceil\{\bar{a}\}}| \geq |\Pi_1|$. Finally, in a derivation of $\square$ from Φ, there must be at least one resolve∃ with pivot variable a. As $V_1 \cap V_2 = \emptyset$ and $a \notin V_1 \cup V_2$ we conclude $|\Pi| \geq |\Pi_1| + |\Pi_2| + 1$. Note that if V_1 and V_2 are not disjoint, then in similar way we can only prove a weaker bound $|\Pi| \geq max(|\Pi_1|, |\Pi_2|) + 1$. $\square$

---

[1] The proof is found in the Appendix of [6], available in the online version of the paper at http://www.kr.tuwien.ac.at/staff/widl/publications/2013/lpar13.pdf

Definition 4 (KBKF-lqu family). *For $t > 1$, let $\mathcal{P}^q.\phi^q$ be the tth member of the KBKF-qu family over variable set V^q, and let $\mathcal{P}^l.\phi^l$ be the tth member of the KBKF-lq family over variable set V^l, where $V^q \cap V^l = \emptyset$. Let a be a fresh variable with $a \notin V^q \cup V^l$. The tth member KBKF-lqu[t] in the KBKF-lqu family is defined as $\exists a \mathcal{P}^q \mathcal{P}^l.(\phi^q \vee a) \wedge (\phi^l \vee \overline{a})$.*

Corollary 2. *For $t > 1$, the smallest proofs for KBKF-lqu[t] are polynomial for LQU-resolution, but are exponential for LQ-resolution and exponential for QU-resolution (based on Theorem 3.2 in [11]).*

Whether the LQU+-resolution calculus has an exponential separation with respect to LQU remains an open problem. The following example, however, shows how LQU+-resolution can be more beneficial than LQU-resolution in some cases.

Ex. 3. Consider the false QBF $\Phi = \exists a \forall x \forall y \exists b.(a, x, b)_1 (\overline{a}, \overline{x}, b)_2 (x, y, \overline{b})_3 (\overline{x}, \overline{y}, \overline{b})_4$.

Notice that Φ is similar as in Ex. 1, that it has Herbrand functions $h_y = h_x = a$, and that an LQU-resolution proof of the falsity of Φ cannot contain any of the steps resolve$_\forall$ and resolve$_{\exists L}$, relevant to the derivation of an empty clause. There exists, however, an LQU+-resolution proof Π, which contains a resolve$_{\forall L}$ step.

$$\Pi = \begin{cases} 1. \ clause_5 = \mathsf{resolve}_{\exists L}(clause_1, a, clause_2) = (x^*, b) \\ 2. \ clause_6 = \mathsf{resolve}_{\forall L}(clause_3, x, clause_4) = (y^*, \overline{b}) \\ 3. \ clause_7 = \mathsf{resolve}_\exists(clause_5, b, clause_6) = (x^*, y^*) \\ 4. \ clause_\emptyset = \mathsf{reduce}(clause_7) = \square \end{cases}$$

4   Certificate Extraction

In this section we examine existing methods for countermodel construction from {Q,LQ}-resolution proofs and extend them for {QU,LQU}-resolution proofs. All the discussions can be dually extended to cube resolution proofs for true QBFs as proposed in [1]. The Algorithm `Countermodel_construct` [1] was proposed to extract Herbrand functions from Q-resolution proofs. We show in the following proposition that this algorithm is also sound for QU-resolution proofs.

Proposition 3. *For a false QBF Φ and a corresponding QU-resolution proof Π, algorithm `Countermodel_construct` of [1] returns a correct countermodel for Φ.*

Proof. Theorem 3 of [1] shows the correctness of `Countermodel_construct` for Q-resolution proofs. Since the way it is proved is not affected by the presence of resolve$_\forall$ steps, it is also sound for QU-resolution proofs.   □

Note that the algorithm `Countermodel_construct` applied to QU-refutations of KBKF[t] proposed in [18] returns countermodel $\mathcal{H} = \bigcup_{i \in [1..t]} h_{xi}$ with $h_{xi} = d_i \wedge \overline{e_i}$, since for each $i \in [1..t]$ the literal $\mathrm{lit}(x_i)$ is universally reduced only twice in the whole proof, namely in clauses (d_i, x_i) and $(e_i, \overline{x}_i)$. It is also worth noticing that KBKF[t] has even simpler Herbrand functions than those constructed by

---

LQU_countermodel_construct
> **input**: a false QBF Φ and its LQ-resolution proof Π
> **output**: Herbrand model $\mathcal{H}$ for Φ
> **begin**
> 00   **let** Σ the set of all assignments to variables V_{P_Π}
> 01   **foreach** assignment $\sigma \in \Sigma$
> 02 $(\Phi^\sigma, \Pi^\sigma) := \mathtt{unmerge}(\Phi, \Pi, \sigma)$;
> 03 $\mathcal{H}^\sigma := \mathtt{Countermodel\_construct}(\Phi^\sigma, \Pi^\sigma)$;
> 04 $\mathcal{H} := \{ \mathsf{h}_x \mid \mathsf{h}_x = \left(\bigvee_{\sigma \in \Sigma} (\mathsf{h}_x^\sigma \wedge \mathtt{cond}(\sigma)) \right)$ for $\mathsf{h}_x^\sigma \in \mathcal{H}^\sigma \}$;
> 05   **return** $\mathcal{H}$;
> **end**

Fig. 1. Algorithm: LQU Countermodel Construction

Countermodel_construct, namely $\mathsf{h}_{xi} = d_i$ for all $i \in [1..t]$. The existence of these simple functions motivates to further investigate the (counter)model extraction from proofs of different resolution systems.

In contrast to QU-resolution proofs, the algorithm Countermodel_construct is unsound for LQ-resolution proofs due to the possible presence of resolve$_{\exists L}$ steps. A conversion of an LQ-resolution proof into a Q-resolution proof in order to apply Countermodel_construct has been proposed [1], but it can result in an exponential blow-up. We propose an algorithm to extract Herbrand functions for a false QBF directly from its LQ-resolution proofs. The algorithm is outlined in Fig. 1. By Proposition 3 it applies LQU-resolution proofs as well. The procedure unmerge(Φ, Π, σ) is central to the algorithm. It transforms an LQU-resolution proof Π into a QU-refutation as follows. Let $V_{P_\Pi} \subseteq V_\exists$ be the exact set of the pivot variables in the resolve$_{\exists L}$ steps of Π. Given an LQU-resolution proof Π for a false QBF $\Phi = \mathcal{P}.\phi$ and an assignment σ to a set V_σ of variables of Φ with $V_{P_\Pi} \subseteq V_\sigma \subseteq V_\exists$, unmerge$(\Phi, \Pi, \sigma)$ traverses Π in a topological order. Whenever it encounters two clauses $C_a = C_1 \cup \{l, p\}$ and $C_b = C_2 \cup \{\bar{l}, \bar{p}\}$ resolving into $C = \mathtt{resolve}_{\exists L}(C_a, p, C_b) = C_1 \cup C_2 \cup \{l^*\}$, it applies the following rewriting rule. Two cases are distinguished by the polarity of the pivot's literal in σ.

$$\frac{(C_1 \cup \{l,p\})\ (C_2 \cup \{\bar{l}, \bar{p}\})}{(C_1 \cup C_2 \cup \{l^*\})} \quad \xrightarrow{p \in \sigma} \quad \frac{\frac{(C_1 \cup \{l,p\})\ (C_1 \cup \{\bar{l}, p\})}{(C_1 \cup \{p\})}\ (C_2 \cup \{\bar{l}, \bar{p}\})}{(C_1 \cup C_2 \cup \{\bar{l}\})}$$

$$\frac{(C_1 \cup \{l,p\})\ (C_2 \cup \{\bar{l}, \bar{p}\})}{(C_1 \cup C_2 \cup \{l^*\})} \quad \xrightarrow{\bar{p} \in \sigma} \quad \frac{(C_1 \cup \{l,p\})\ \frac{(C_2 \cup \{\bar{l}, \bar{p}\})\ (C_2 \cup \{l, \bar{p}\})}{(C_2 \cup \{\bar{p}\})}}{(C_1 \cup C_2 \cup \{l\})}$$

If there are more than one merged literals in C, unmerge is applied several times to eliminate all of them. Intuitively, this procedure adds clauses to ϕ in order to substitute all resolve$_{\exists L}$ steps. It preserves the order of reduce and does not create any new resolve$_{\exists L}$ steps. It never encounters resolve$_{\exists L}$ on two clauses containing merged literals because these literals are removed by the rewriting rule in an earlier iteration. We denote the QBF resulting from unmerge(Φ, Π, σ) by Φ^σ, and the resulting (QU-resolution) proof by Π^σ.

Given a Herbrand model $\mathcal{H}$ and an assignment σ, the Herbrand model $\mathcal{H}_{\lceil\sigma}$ results from replacing each variable v in $\mathcal{H}$ by $\top$ if $v \in \sigma$ and by $\bot$ if $\bar{v} \in \sigma$. The following two observations establish the connection between Φ^σ and $\Phi_{\lceil\sigma}$.

Observation 2. Let $\mathcal{H}$ be a set of Herbrand functions for a false QBF Φ, and σ be an assignment to some existential variables of Φ. Then $\mathcal{H}_{\lceil\sigma}$ is a set of Herbrand functions for the false QBF $\Phi_{\lceil\sigma}$.

Observation 3. For any assignment σ to variables in $V_{P_{\Pi}}$, it holds that $\Pi_{\lceil\sigma} = (\Pi^\sigma)_{\lceil\sigma}$. By Observation 2, if $\mathcal{H}$ is a set of Herbrand functions for Φ^σ, then $\mathcal{H}_{\lceil\sigma}$ is a set of Herbrand functions for $\Phi_{\lceil\sigma}$.

The algorithm LQU_countermodel_construct takes a false QBF Φ and an LQU-resolution proof Π of Φ as input. It then collects the pivots of all resolve$_{\exists L}$ steps in Π into the set $V_{P_{\Pi}}$ and iteratively picks an assignment σ to the variables in $V_{P_{\Pi}}$. For each assignment, a QU-resolution proof is constructed by unmerge in Line 02. Note that unmerge was defined for any set of existential variables containing $V_{P_{\Pi}}$. It however suffices to consider the assignments to $V_{P_{\Pi}}$ only. In Line 03, Countermodel_construct is applied to extract parts of the countermodel for Φ, which are then put together in Line 04. Note that the Herbrand function F_x returned by the algorithm LQU_countermodel_construct for a universal variable x permits its dependency on the universal variables x' with $lvl(x') < lvl(x)$. All occurrences of such x' in F_x should be substituted by the corresponding Herbrand functions $F_{x'}$, resulting into the function that depends only on existential variables.

Theorem 4 below states the soundness of LQU_countermodel_construct. Note that from this theorem also follows the soundness of LQU-resolution.

Theorem 4. *Given a false QBF $\Phi = \mathcal{P}.\phi$ and an LQU-resolution proof Π for Φ, LQU_countermodel_construct returns correct Herbrand functions for Φ.*

Proof. Consider any assignment σ to $V_{P_{\Pi}}$ variables. By construction, $\mathcal{H}_{\lceil\sigma} = \mathcal{H}^\sigma$. Taking in account Observation 3, the Herbrand functions $\mathcal{H}_{\lceil\sigma}$ falsify the formula $\Phi_{\lceil\sigma}$. Thus $\mathcal{H}$ falsifies ϕ under any assignment to existential variables.

It remains to show that for each $x \in V_\forall$, its Herbrand function $h_x \in \mathcal{H}$ respects the variable ordering of $\mathcal{P}$. (As constructed, h_x includes all variables in σ due to the assignment condition cond(σ).) Notice that under a given σ, the constructed h_x^σ is uniquely defined by the ordered set of clauses in Π^σ from which x is removed by universal reduction. By construction, Π^σ has exactly the same universal reduction steps as Π, with the only difference that every literal l^* is replaced by l or $\bar{l}$, depending on σ. For two assignments σ_1 and σ_2, compare clauses $C^{\sigma_1} \in \Pi^{\sigma_1}$ and $C^{\sigma_2} \in \Pi^{\sigma_2}$ that correspond to clause $C \in \Pi$ and result from universal reduction on x. If $(l \in \sigma_1) \wedge (\text{lev}(l) < \text{lev}(x))$ implies $l \in \sigma_2$ for any literal l, then by the definition of unmerge we conclude that $C^{\sigma_1} = C^{\sigma_2}$ and that x was universally reduced as the same literal to get both C^{σ_1} and C^{σ_2}. Thus h_x is independent of any variable in $\{v \in V_{P_{\Pi}} \mid \text{lev}(v) > \text{lev}(x)\}$. □

Table 2. Time and Memory Statistics for KBKF Family of QBF Instances

t	DEPQBF		RESQU			DEPQBF-LQ		RESQU-LQU		
	time	$|\Pi|$	time	memory	verify	time	$|\Pi|$	time	memory	verify
2	0	24	0	1	0	0	27	0	1	0
3	0	50	0	1	0	0	43	0	1	0
4	0	106	0	1	0.1	0	59	0	1	0.1
5	0	230	0	1	0.1	0	75	0	1	0.1
6	0	506	0	1	0.1	0	91	0	1	0.1
7	0	1.1k	0	1	0.1	0	107	0	1	0.1
8	0	2.5k	0	2	0.1	0	123	0	2	0.1
9	0	5.4k	0	3	0.1	0	139	0	2	0.1
10	0.1	11.8k	0.1	7	0.1	0	155	0	4	0.1
11	0.2	25.6k	0.1	14	0.3	0	171	0.1	8	0.1
12	0.5	55.3k	0.3	58	0.7	0	187	0.1	18	0.1
13	1.2	118.8k	0.6	123	2.3	0	203	0.3	37	0.1
14	2.8	254.0k	1.4	261	7.6	0	219	0.7	79	0.1
15	6.8	540.7k	3.0	550	30.5	0	235	1.8	169	0.1
16	16.6	1.15M	6.7	1.2G	-1	0	251	3.9	360	0.8
17	41.0	2.42M	15.1	2.4G	-1	0	267	9.4	767	5.4
18	102.8	5.11M	33.6	5.1G	-1	0	283	20.5	1.6G	40.4
19	261.5	10.75M	74.1	10.7G	-1	0	299	48.8	3.4G	-1
20	674.2	22.54M	175.7	22.5G	-1	0	315	95.1	7.2G	-1

The time complexity of LQU_countermodel_construct is in the worst case exponential in the proof size. In practice, however, it can be more efficient than converting LQ-resolution proofs into Q-resolution proofs [1], as will be evident in Section 5. Note that the algorithm LQU_countermodel_construct is unsound for LQU+-resolution proofs due to the presence of universal variables in V_{P_Π}.

5   Experiments

In this section we evaluate the proposed algorithm LQU_countermodel_construct on members of the KBKF family. To the best of our knowledge, there is currently no tool available to construct QU-resolution proofs (and consequently LQU-resolution proofs). Hence we test LQU_countermodel_construct on LQ-resolution proofs, and compare the results to those obtained by Countermodel_construct [1] from the corresponding Q-resolution proofs. The experiments were conducted on a Linux machine with a Xeon 2.3 GHz CPU and 32 GB RAM.

Table 2 summarizes time and memory statistics for solving, extracting, and verifying Herbrand functions for members of the KBKF family up to $t = 20$. RESQU implements the algorithm Countermodel_construct, RESQU-LQU implements LQU_countermodel_construct, DEPQBF stands for the solver proposed in [13], and DEPQBF-LQ for its extension by LQ-resolution [6]. The column "time" refers to the runtime in seconds, "$|\Pi|$" to the size of the resulting proof, "memory" to the maximal memory consumption (in MB for unit unspecified entries), and "verify" to the time needed by the SAT-solver MINISAT [5] embedded in ABC [3] to verify the certificate where "-1" stands for a timeout with a limit of 1,000s.

The superiority of LQ-resolution compared to Q-resolution is evident in all aspects. Since Q-resolution proofs produced by DEPQBF are exponential in t. RESQU also requires resources exponential in t. On the other hand, LQ-resolution proofs produced by DEPQBF-LQ are linear in t. Despite its exponential worst-case behavior, RESQU-LQU considerably outperforms RESQU in both time and memory consumption, although it still requires exponential resources due to the exponential size of the constructed Herbrand functions.

6   Conclusions and Future Work

We have presented results related to both theoretical and practical aspects of QBF evaluation. On the theoretical side, we have shown the incomparability between two proof systems, QU-resolution and LQ-resolution, from literature. Additionally, we have proposed two new extended proof systems, LQU-resolution and LQU+-resolution, and have shown the two new systems to be exponentially stronger than both of the above. It remains open whether an expo-

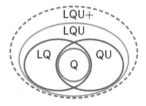

Fig. 2. Relations among the Proof Systems

nential gap exists between the proof complexities of LQU-resolution and LQU+-resolution. Fig. 2 summarizes our results on the relations between the discussed proof systems. Since modern QBF solvers heavily rely on resolution techniques, we expect our theoretical results to inspire future work in the area of QBF solving.

On the practical side, we have designed a new algorithm to extract Herbrand certificates from LQU-resolution proofs. An implementation and experimental evaluation underline its practical applicability and advantage over the certificates from Q-resolution. For future work, a polynomial time algorithm for certificate extraction from LQU-resolution proofs would be very desirable.

References

1. Balabanov, V., Jiang, J.-H.R.: Unified QBF Certification and Its Applications. Formal Methods in System Design 41, 45–65 (2012)
2. Benedetti, M.: sKizzo: A suite to evaluate and certify QBFs. In: Nieuwenhuis, R. (ed.) CADE 2005. LNCS (LNAI), vol. 3632, pp. 369–376. Springer, Heidelberg (2005)
3. Berkeley Logic Synthesis and Verification Group. ABC: A System for Sequential Synthesis and Verification, http://www.eecs.berkeley.edu/$\sim$alanmi/abc/
4. Dershowitz, N., Hanna, Z., Katz, J.: Bounded Model Checking with QBF. In: Bacchus, F., Walsh, T. (eds.) SAT 2005. LNCS, vol. 3569, pp. 408–414. Springer, Heidelberg (2005)
5. Eén, N., Sörensson, N.: An Extensible SAT-Solver. In: Giunchiglia, E., Tacchella, A. (eds.) SAT 2003. LNCS, vol. 2919, pp. 502–518. Springer, Heidelberg (2004)
6. Egly, U., Lonsing, F., Widl, M.: Long-distance resolution: Proof generation and strategy extraction in search-based QBF solving. In: McMillan, K., Middeldorp, A., Voronkov, A. (eds.) LPAR-19 2013. LNCS, vol. 8312, pp. 291–308. Springer, Heidelberg (2013)

7. Giunchiglia, E., Narizzano, M., Tacchella, A.: QuBE++: An Efficient QBF Solver. In: Hu, A.J., Martin, A.K. (eds.) FMCAD 2004. LNCS, vol. 3312, pp. 201–213. Springer, Heidelberg (2004)

8. Goultiaeva, A., Van Gelder, A., Bacchus, F.: A Uniform Approach for Generating Proofs and Strategies for Both True and False QBF Formulas. In: International Joint Conference on Artificial Intelligence (IJCAI), pp. 546–553. AAAI Press (2011)

9. Janota, M., Klieber, W., Marques-Silva, J., Clarke, E.: Solving QBF with Counterexample Guided Refinement. In: Cimatti, A., Sebastiani, R. (eds.) SAT 2012. LNCS, vol. 7317, pp. 114–128. Springer, Heidelberg (2012)

10. Jiang, J.-H.R., Lin, H.-P., Hung, W.-L.: Interpolating Functions from Large Boolean Relations. In: Proc. International Conference on Computer-Aided Design (ICCAD), pp. 779–784. IEEE/ACM (2009)

11. Kleine Büning, H., Karpinski, M., Flögel, A.: Resolution for Quantified Boolean Formulas. Information and Computation 117(1), 12–18 (1995)

12. Lai, C.-F., Jiang, J.-H.R., Wang, K.-H.: BooM: A Decision Procedure for Boolean Matching with Abstraction and Dynamic Learning. In: Design Automation Conference (DAC), pp. 499–504. ACM/IEEE (2010)

13. Lonsing, F., Biere, A.: DepQBF: A Dependency-Aware QBF Solver (System Description). Journal on Satisfiability, Boolean Modeling and Computation 7, 71–76 (2010)

14. Marques Silva, J.P., Lynce, I., Malik, S.: Conflict-Driven Clause Learning SAT Solvers. In: Handbook of Satisfiability, pp. 131–153. IOS Press (2009)

15. Rintanen, J.: Asymptotically Optimal Encodings of Conformant Planning in QBF. In: National Conference on Artificial Intelligence (AAAI), pp. 1045–1050. AAAI Press (2007)

16. Staber, S., Bloem, R.: Fault localization and correction with QBF. In: Marques-Silva, J., Sakallah, K.A. (eds.) SAT 2007. LNCS, vol. 4501, pp. 355–368. Springer, Heidelberg (2007)

17. Van Gelder, A.: Input Distance and Lower Bounds for Propositional Resolution Proof Length. In: Bacchus, F., Walsh, T. (eds.) SAT 2005. LNCS, vol. 3569, pp. 282–293. Springer, Heidelberg (2005)

18. Van Gelder, A.: Contributions to the Theory of Practical Quantified Boolean Formula Solving. In: Milano, M. (ed.) CP 2012. LNCS, vol. 7514, pp. 647–663. Springer, Heidelberg (2012)

19. Zhang, L., Malik, S.: Conflict Driven Learning in a Quantified Boolean Satisfiability Solver. In: Proc. International Conference on Computer-Aided Design (ICCAD), pp. 442–449. ACM (2002)

20. Zhang, L., Malik, S.: The Quest for Efficient Boolean Satisfiability Solvers. In: Brinksma, E., Larsen, K.G. (eds.) CAV 2002. LNCS, vol. 2404, pp. 17–36. Springer, Heidelberg (2002)

Unified Characterisations
of Resolution Hardness Measures[*]

Olaf Beyersdorff[1] and Oliver Kullmann[2]

[1] School of Computing, University of Leeds, UK
[2] Computer Science Department, Swansea University, UK

Abstract. Various "hardness" measures have been studied for resolution, providing theoretical insight into the proof complexity of resolution and its fragments, as well as explanations for the hardness of instances in SAT solving. In this paper we aim at a unified view of a number of hardness measures, including different measures of width, space and size of resolution proofs. Our main contribution is a unified game-theoretic characterisation of these measures. As consequences we obtain new relations between the different hardness measures. In particular, we prove a generalised version of Atserias and Dalmau's result on the relation between resolution width and space from [5].

1 Introduction

Arguably, resolution is the best understood among all propositional proof system, and at the same time it is the most important one in terms of applications. To understand the complexity of resolution proofs, various *hardness measures* have been defined and investigated. Historically the first and most studied measure is the *size of resolution proofs*, with the first lower bounds dating back to Tseitin [56] and Haken [34]. A number of ingenious techniques have been developed to show lower bounds for the size of resolution proofs, among them feasible interpolation [42], which applies to many further systems. In their seminal paper [13], Ben-Sasson and Wigderson showed that resolution size lower bounds can be elegantly obtained by showing lower bounds to the *width* of resolution proofs. Indeed, the discovery of this relation between width and size of resolution proofs was a milestone in our understanding of resolution. Around the same time *(tree) resolution space* was investigated, and first lower bounds were obtained [54,24,25,26,55]. The primary method to obtain lower bounds on resolution space is based on width, and the general bound was shown in the fundamental paper by Atserias and Dalmau [5]. Since then the relations between size, width and space have been intensely investigated, resulting in particular in sharp trade-off results [11,8,12,49,50,9]. Independently, in [43,46,47] the concept of "hardness" has been introduced, with an algorithmic focus (as shown in [43], equivalent to tree resolution space; one can also say "tree-hardness"), together with a generalised form of width, which we call "asymmetric width" in this paper.

[*] Research supported by a grant from the John Templeton Foundation.

C. Sinz and U. Egly (Eds.): SAT 2014, LNCS 8561, pp. 170–187, 2014.

One of the prime motivations to understand these measures is their close correspondence to SAT solving. In particular, resolution size and space relate to the running time and memory consumption, respectively, of executions of SAT solvers on unsatisfiable instances. However, size and space are not the only measures which are interesting with respect to SAT solving, and the question what constitutes a good hardness measure for practical SAT solving is a very important one (cf. [4,37] for discussions).

The aim of this paper is to review different hardness measures defined in the literature, and to provide *unified characterisations* for these measures in terms of Prover-Delayer games and sets of partial assignments satisfying some consistency conditions. These unified characterisations allow elegant proofs of basic relations between the different hardness measures. Unlike in the works [8,12,49], our emphasis here is not on trade-off results, but on *exact relations* between the different measures. For a clause-set F we will consider the following measures: *(i) size measures:* the depth $\mathrm{dep}(F)$ and the hardness $\mathrm{hd}(F)$ (of best resolution refutations of F); *(ii) width measures:* the symmetric and asymmetric width $\mathrm{wid}(F)$ and $\mathrm{awid}(F)$; *(iii) clause-space measures:* semantic space $\mathrm{css}(F)$, resolution space $\mathrm{crs}(F)$ and tree-resolution space $\mathrm{cts}(F)$.

Game-theoretic methods have a long tradition in proof complexity, as they provide intuitive and simplified methods for lower bounds in resolution, e.g. for Haken's exponential bound for the pigeonhole principle in dag-like resolution [51], or the optimal bound in tree resolution [14], and even work for very strong systems [10]. Inspired by the Prover-Delayer game of Pudlák and Impagliazzo [52], we devise a game that characterises the hardness measure $\mathrm{hd}(F)$. In contrast to [52] it also works for satisfiable formulas (Theorem 12), due to elimination of the communication between Prover and Delayer. We then explain a more general game, allowing the Prover to also *forget* some information. This game tightly characterises the asymmetric width hardness $\mathrm{awid}(F)$ (Theorem 23); and restricting this game by disallowing forgetting yields the hd-game (Lemma 24).

Characterisations by partial assignments provide an alternative combinatorial description of the hardness measures. In [5] such a characterisation is obtained for $\mathrm{wid}(F)$. Taking this as a starting point, we devise a hierarchy of consistency conditions for sets of partial assignments which serve to characterise asymmetric width $\mathrm{awid}(F)$ (k-consistency, Theorem 22), hardness $\mathrm{hd}(F)$ (weak k-consistency), and depth $\mathrm{dep}(F)$ (bare k-consistency).

Relations between these measures can be easily obtained by exploiting the above characterisations. We obtain a generalised version of Atserias and Dalmau's connection between width and resolution space from [5], replacing symmetric width by the stronger notion of asymmetric width (handling long clauses now), and resolution space by the possibly tighter semantic space (Theorem 27). The full picture is presented in the following diagram, where $F \in \mathcal{CLS}$ has n variables, minimal clause length p, and maximal length q of necessary clauses:

$$p \longrightarrow \mathrm{awid}(F) \longrightarrow \mathrm{css}(F) \xrightarrow{\sim *3} \mathrm{crs}(F) \longrightarrow \mathrm{cts}(F) \xrightarrow{=-1} \mathrm{hd}(F) \Longrightarrow \mathrm{dep}(F) \longrightarrow n$$

$$q \longrightarrow \mathrm{wid}(F)$$

An arrow "$h(F) \to h'(F)$" means $h(F) \le h'(F)$, and there is a sequence (F_n) of clause-sets with bounded $h(F_n)$ but unbounded $h'(F_n)$; in case of an undirected edge no such separation is possible (crs differs from css at most by a factor of 3, while cts -1 = hd). The separation awid $\to$ css is shown in [50], crs $\to$ cts in [37], hd $\to$ dep and wid $\to$ dep use unsatisfiable Horn 3-clause-sets, and dep $\to n$ uses unsatisfiable clause-sets which are not minimally unsatisfiable.

These measures do not just apply to unsatisfiable clause-sets, but are extended to **satisfiable clause-sets**, taking a *worst-case approach* over all unsatisfiable sub-instances obtained by applying partial assignments (instantiations). For a fixed bound these measures allow for polynomial-time SAT solving via "oblivious" SAT algorithms — certain basic steps, applied in an arbitrary manner, are guaranteed to succeed. The sets $\mathcal{UC}_k$ of all clause-sets F with $\mathrm{hd}(F) \le k$ yield the basic hierarchy, and we have SAT decision in time $O(n(F)^{2\,\mathrm{hd}(F)-2} \cdot \ell(F))$. The special case $\mathcal{UC}_1 = \mathcal{UC}$ was introduced in [58] for the purpose of Knowledge Compilation (KC), and in [29,32] it is shown that $\mathcal{UC} = \mathcal{SLUR}$ holds, where $\mathcal{SLUR}$ is the class introduced in [53] as an umbrella class for polynomial-time SAT solving. By [6,32] we get that membership decision for $\mathcal{UC}_k$ with $k \ge 1$ is coNP-complete.

Perhaps the main aim of measuring the complexity of satisfiable clause-sets is to obtain *SAT representations of boolean functions* of various quality ("hardness") and sizes; see [30,33] for investigations into XOR-constraints. The goal is to obtain "good" representations F of boolean functions (like cardinality or XOR-constraints) in the context of a larger SAT problem representations. "Good" means not "too big" and of "good" inference power. The latter means (at least), that all unsatisfiable instantiations of F should be easy for SAT solvers, motivating the worst-case approach (over all unsatisfiable sub-instances). In the diagram above, having low $\mathrm{dep}(F)$ is the strongest condition, having low $\mathrm{awid}(F)$ the weakest. The KC aspects, concerning size-hardness trade-offs, are further investigated in [31]; see Corollary 29 for an application. This study of the "best" choice of a representation, considering size (number of clauses) and hardness (like hd, awid or css) among *all (logically) equivalent* clause-sets, likely could not be carried out using (symmetric) width, but requires *asymmetric width*, so that unbounded clause length can be handled. The traditional method to bound the clause-length, by breaking up clauses via auxiliary variables, introduces unnecessary complexity, and can hardly be applied if we only want to consider (logically) equivalent clause-sets (without auxiliary variables).

This paper is organised as follows. After fixing notation in Sect. 2, we define all hardness measures in Sect. 3 and prove some first results. Our main results then follow in Sect. 4, where we prove the combinatorial characterisations of the measures and infer basic connections. We conclude in Sect. 5 with a discussion and some open questions.

2   Preliminaries

We use the general notions as in [39], but also define all notations.

Clause-Sets. $\mathcal{VA}$ is the (infinite) set of variables, while $\mathcal{LIT}$ is the set of literals, where every literal is either a variable v or a complemented (negated) variable $\overline{v}$. For a set $L \subseteq \mathcal{LIT}$ of literals we use $\overline{L} := \{\overline{x} : x \in L\}$. A clause is a finite $C \subset \mathcal{LIT}$ with $C \cap \overline{C} = \emptyset$ (i.e., without conflicting literals), the set of all clauses is $\mathcal{CL}$. A clause-set is a finite set of clauses, the set of all clause-sets is $\mathcal{CLS}$. For $k \in \mathbb{N}_0$ we define k–$\mathcal{CLS}$ as the set of all $F \in \mathcal{CLS}$ where every clause $C \in F$ has length (width) at most k, i.e., $|C| \leq k$. We use var : $\mathcal{LIT} \to \mathcal{VA}$ for the underlying variable of a literal, while $\mathrm{var}(C) := \{\mathrm{var}(x) : x \in C\}$ for a clause C, and $\mathrm{var}(F) := \bigcup_{C \in F} \mathrm{var}(C)$ for a clause-set F. Measures for $F \in \mathcal{CLS}$ are $n(F) := |\mathrm{var}(F)| \in \mathbb{N}_0$ (number of variables) and $c(F) := |F| \in \mathbb{N}_0$ (number of clauses). A special clause is the empty clause $\perp := \emptyset \in \mathcal{CL}$, a special clause-set is the empty clause-set $\top := \emptyset \in \mathcal{CLS}$.

A partial assignment is a map $\varphi : V \to \{0, 1\}$ for some finite $V \subset \mathcal{VA}$, the set of all partial assignments is $\mathcal{PASS}$; we use $\mathrm{var}(\varphi) := V$, and the number of variables in a partial assignment is denoted by $n(\varphi) := |\mathrm{var}(\varphi)|$. For a clause C we denote by $\varphi_C \in \mathcal{PASS}$ the partial assignment which sets precisely the literals in C to 0; furthermore we use $\langle x \to \varepsilon \rangle \in \mathcal{PASS}$ for a literal x and $\varepsilon \in \{0, 1\}$, while $\langle\rangle \in \mathcal{PASS}$ denotes the empty partial assignment. The natural partial order on $\mathcal{PASS}$ is given by inclusion $\varphi \subseteq \psi$, that is, $\mathrm{var}(\varphi) \subseteq \mathrm{var}(\psi)$ and φ, ψ are compatible (do not assign different values to the same variable). The application (instantiation) of φ to $F \in \mathcal{CLS}$ is denoted by $\varphi * F \in \mathcal{CLS}$, obtained by first removing satisfied clauses $C \in F$ (i.e., containing a literal $x \in C$ with $\varphi(x) = 1$), and then removing all falsified literals from the remaining clauses.

The set of satisfiable clause-sets is $\mathcal{SAT} := \{F \in \mathcal{CLS} \mid \exists \varphi \in \mathcal{PASS} : \varphi * F = \top\}$, while $\mathcal{USAT} := \mathcal{CLS} \setminus \mathcal{SAT}$ is the set of unsatisfiable clause-sets. For $F, F' \in \mathcal{CLS}$ the implication-relation is defined as usual: $F \models F' :\Leftrightarrow \forall \varphi \in \mathcal{PASS} : \varphi * F = \top \Rightarrow \varphi * F' = \top$. We write $F \models C$ for $F \models \{C\}$. A clause C with $F \models C$ is an *implicate* of F, while a *prime implicate* is an implicate C such that no $C' \subset C$ is also an implicate; $\mathrm{prc}_0(F)$ is the set of prime implicates of F. Finally, by $\mathrm{r}_1 : \mathcal{CLS} \to \mathcal{CLS}$ we denote unit-clause propagation, which is defined recursively by $\mathrm{r}_1(F) := \{\perp\}$ if $\perp \in F$, $\mathrm{r}_1(F) := F$ if F does not contain a unit-clause, while otherwise choose $\{x\} \in F$ and set $\mathrm{r}_1(F) := \mathrm{r}_1(\langle x \to 1\rangle * F)$.

Resolution. Two clauses C, D are resolvable if $|C \cap \overline{D}| = 1$, i.e., they clash in exactly one variable. For two resolvable clauses C and D the resolvent $C \diamond D := (C \cup D) \setminus \{x, \overline{x}\}$ for $C \cap \overline{D} = \{x\}$ is the union of the two clauses minus the resolution literals. $\mathrm{var}(x)$ is called the resolution variable. The closure of $F \in \mathcal{CLS}$ under resolution has $\mathrm{prc}_0(F)$ as its subsumption-minimal elements.

The set of nodes of a tree T is denoted by $\mathrm{nds}(T)$, the set of leaves by $\mathrm{lvs}(T) \subseteq \mathrm{nds}(T)$. The height $\mathrm{ht}_T(w) \in \mathbb{N}_0$ of a node $w \in \mathrm{nds}(T)$ is the height of the subtree of T rooted at w (so $\mathrm{lvs}(T) = \{w \in \mathrm{nds}(T) : \mathrm{ht}_T(w) = 0\}$). A resolution tree is a pair $R = (T, C)$ such that T is an ordered rooted tree, where every inner node has exactly two children, and where the set of nodes is denoted by $\mathrm{nds}(T)$ and the root by $\mathrm{rt}(T) \in \mathrm{nds}(T)$, while $C : \mathrm{nds}(T) \to \mathcal{CL}$ labels every node with a clause such that the label of an inner node is the resolvent of the labels of its two parents. We use $\mathrm{ax}(R) := \{C(w) : w \in \mathrm{lvs}(T)\} \in \mathcal{CLS}$ for the "axioms" (or "premisses") of R, $C(R) := C(\mathrm{rt}(T)) \in \mathcal{CL}$ for the "conclusion", and $\mathrm{cl}(R) := \{C(w) : w \in \mathrm{nds}(T)\} \in \mathcal{CLS}$ for the set of all clauses in R.

A resolution proof R of a clause C from a clause-set F, denoted by $R : F \vdash C$, is a resolution tree $R = (T, C)$ such that $\mathrm{ax}(R) \subseteq F$ and $C(R) = C$. We use $F \vdash C$ if there exists a resolution proof R of some $C' \subseteq C$ from F (i.e., $R : F \vdash C'$). A resolution refutation of a clause-set F is a resolution proof deriving $\bot$ from F. The tree-resolution complexity $\mathrm{Comp}_R^*(R) \in \mathbb{N}$ is the number of leaves in R, that is, $\mathrm{Comp}_R^*(R) := |\mathrm{lvs}(T)|$. The resolution complexity $\mathrm{Comp}_R(R) \in \mathbb{N}$ is the number of *distinct* clauses in R, that is $\mathrm{Comp}_R(R) := c(\mathrm{cl}(R))$. Finally, for $F \in \mathcal{USAT}$ we set $\mathrm{Comp}_R^*(F) := \min\{\mathrm{Comp}_R^*(R) \mid R : F \vdash \bot\} \in \mathbb{N}$ and $\mathrm{Comp}_R(F) := \min\{\mathrm{Comp}_R(R) \mid R : F \vdash \bot\} \in \mathbb{N}$.

3　Hardness Measures

In this section we define the hardness measures $\mathrm{hd}, \mathrm{dep}, \mathrm{wid}, \mathrm{awid}, \mathrm{css}, \mathrm{crs}, \mathrm{cts}$ ("hardness, depth, width, asymmetric width, semantic/resolution/tree space") that we investigate in this article, and observe some first connections.

First we discuss a *general method for extending measures* h_0 for unsatisfiable clause-sets to measures h for arbitrary clause-sets. The basic idea is to consider the hardness of unsatisfiable sub-instances, obtained by partial instantiations. In a probabilistic setting this has been considered e.g. in [1,3]. We however consider the worst-case, which yields precise measurements. The special case of extension of "hardness" was first mentioned (as one of two possibilities) by [4]. Our motivation was that the extension of clause-sets falsifiable by unit-clause propagation yields precisely the class $\mathcal{SLUR}$ ([29,32]).

A measure $h_0 : \mathcal{USAT} \to \mathbb{N}_0$, which is not increased by applying partial assignments, is extended to $h : \mathcal{CLS} \to \mathbb{N}_0$ by $h(\top) := \min_{F \in \mathcal{USAT}} h_0(F)$, while for $F \in \mathcal{CLS} \setminus \{\top\}$ we define $h(F)$ as the maximum of $h_0(\varphi * F)$ for $\varphi \in \mathcal{PASS}$ with unsatisfiable $\varphi * F$. So also h is not increased by applying partial assignments, and $h(F) = h_0(F)$ for $F \in \mathcal{USAT}$, while for $h_0 \le h_0'$ we get $h \le h'$. Note that for the computation of $h(F)$, as the maximum of $h_0(\varphi * F)$ for unsatisfiable $\varphi * F$, one only needs to consider minimal φ (since application of partial assignments can not increase the measure), that is, φ_C for $C \in \mathrm{prc}_0(F)$; so for $F \in \mathcal{CLS} \setminus \{\top\}$ we have $h(F) = \max_{C \in \mathrm{prc}_0(F)} h_0(\varphi * F)$. In the following we will define the hardness measure only for unsatisfiable clauses and then extend them via the above method.

3.1　Tree-Hardness

We start with what in our opinion is one of the central hardness measures for resolution, which is why we simply call it *hardness* (but for differentiation it might be called *tree-hardness*, then written "thd"). This concept was reinvented in the literature several times. Intuitively, hardness measures the height of the biggest full binary tree which can be embedded into each tree-like resolution refutation of the formula. This is also known as the Horton-Strahler number of a tree (see

[59,23]). In the context of resolution this measure was first introduced in [43,47]. In a more loose sense, based on reduction rules, "hardness classes" are mentioned in [27,28], based on an unpublished manuscript of Stålmarck from 1994.[1] The equivalent approach via tree-resolution space was introduced in [54,25,26,55]. These approaches concern only unsatisfiable clause-sets; the extension to satisfiable clause-sets considered in [43,47] generalises the reduction-rules-based approach, and is essentially different from the general extension process as discussed above; the extension as in this paper was first considered in [4].

Definition 1. *For* $F \in \mathcal{USAT}$ *let* $\mathbf{hd}(F) \in \mathbb{N}_0$ *be the minimal* $k \in \mathbb{N}_0$ *such that a resolution tree* $T : F \vdash \bot$ *exists, where the Horton-Strahler number of* T *is at most* k, *that is, for every node in* T *there exists a path to some leaf of length at most* k. *For* $k \in \mathbb{N}_0$ *let* $\mathcal{UC}_k := \{F \in \mathcal{CLS} : \text{hd}(F) \leq k\}$.

See [43,47,29,32] for equivalent descriptions in this setting, where especially the algorithmic approach, via generalised unit-clause propagation r_k, is notable: hardness is the minimal level k of generalised unit-clause propagation needed to derive a contradiction under any instantiation. As shown in [43, Corollary 7.9], and more generally in [47, Theorem 5.14], we have

$$2^{\text{hd}(F)} \leq \text{Comp}^*_{\text{R}}(F) \leq (n(F)+1)^{\text{hd}(F)}$$

for $F \in \mathcal{USAT}$.[2] A simpler measure is the minimum depth ([57,20,21]):

Definition 2. *For* $F \in \mathcal{USAT}$ *let* $\mathbf{dep}(F) \in \mathbb{N}_0$ *be the minimal height of a resolution tree* $T : F \vdash \bot$.

Obviously hd$(F) \leq$ dep(F) for all $F \in \mathcal{CLS}$. For $k \in \mathbb{N}_0$ the class of $F \in \mathcal{CLS}$ with dep$(F) \leq k$ is called CANON(k) in [19,7]; by definition CANON$(0) = \mathcal{UC}_0$. See Subsection 7.2 in [32] and Subsection 9.2 in [31] for further results.

3.2 Asymmetric Width

The standard resolution-width of an unsatisfiable clause-set is the minimal k such that a resolution refutation using only clauses of length at most k exists:

Definition 3. *For* $F \in \mathcal{USAT}$ *the* **symmetric width** $\mathbf{wid}(F) \in \mathbb{N}_0$ *is the smallest* $k \in \mathbb{N}_0$ *such that there is* $T : F \vdash \bot$ *with* cl$(T) \in k\text{-}\mathcal{CLS}$.

Based on the notion of "k-resolution" introduced in [38], the "asymmetric width" was introduced in [43,46,47] (and further studied in [32,31,33]).[3] Different from the symmetric width, only *one parent clause* needs to have size at most k (while there is no restriction on the other parent clause nor on the resolvent):

---

[1] We have never seen these fragments called "A proof theoretic concept of tautological hardness", but the ideas circulated amongst some researchers from the formal methods community.

[2] Our motivation for the lower bound came from [22]. A similar lower bound is mentioned in [27,28], based on the manuscript of Stålmarck. An equivalent bound is shown in [52] (see Subsection 4.1). In [24,25] the lower bound $2^{\text{crs}(F)-1}$ is shown.

[3] In [32] the notation "whd" was used, to emphasise that we have an extension of "hardness"; but now we consider the relation to "width" as more important.

Definition 4. *For a resolution tree T its **asymmetric width** $\text{awid}(T) \in \mathbb{N}_0$ is defined as 0 if T is trivial (i.e., $|\text{nds}(T)| = 1$), while otherwise for left and right children w_1, w_2 with subtrees T_1, T_2 we define $\text{awid}(T)$ as the maximum of $\min(|C(w_1)|, |C(w_2)|)$ and $\max(\text{awid}(T_1), \text{awid}(T_2))$.*

We write $\boldsymbol{R : F \vdash^k C}$ if $R : F \vdash C$ and $\text{awid}(R) \leq k$. Now for $F \in \mathcal{USAT}$ we define $\boldsymbol{\text{awid}(F)} := \min\{k \in \mathbb{N}_0 \mid F \vdash^k \perp\}$. Finally for $k \in \mathbb{N}_0$ let $\boldsymbol{\mathcal{WC}_k} := \{F \in \mathcal{CLS} : \text{awid}(F) \leq k\}$.

The asymmetric width is a natural, but less known generalisation of symmetric width, and these measures can be very different. Namely, for an unsatisfiable Horn clause-set F holds $\text{awid}(F) \leq 1$, since unit-clause resolution (i.e., asymmetric width at most 1) is sufficient to derive unsatisfiability. But $\text{wid}(F)$ is unbounded: if F is minimally unsatisfiable, then $\text{wid}(F)$ equals the maximal clause-length of F. For general minimally unsatisfiable F, the maximal clause-length is a lower bound for $\text{wid}(F)$, but is unrelated to $\text{awid}(F)$. For bounded clause-length of F however, wid and awid can be considered asymptotically equivalent by Lemma 5 below.

In a seminal paper, Ben-Sasson and Wigderson [13] observe a fundamental relation between symmetric width and proof size for resolution refutations, thereby establishing one of the main methods to prove resolution lower bounds. We recall that in [43, Theorem 8.11] and [47, Theorem 6.12, Lemma 6.15] this size-width relation is indeed strengthened to asymmetric width:

$$e^{\frac{1}{8} \frac{\text{awid}(F)^2}{n(F)}} < \text{Comp}_{\text{R}}(F) < 6 \cdot n(F)^{\text{awid}(F)+2}$$

for $F \in \mathcal{USAT} \setminus \{\{\perp\}\}$, where $e^{\frac{1}{8}} = 1.1331484\ldots$ Note that compared to [13] the numerator of the exponent does not depend on the maximal clause-length of F. In [43, Lemma 8.13] it is shown that the partial ordering principle has asymmetric width the square-root of the number of variables, while having a polysize resolution refutation. Comparing asymmetric width to (tree-)hardness, we have $\mathcal{WC}_0 = \mathcal{UC}_0$ and $\mathcal{WC}_1 = \mathcal{UC}_1$, while for all $F \in \mathcal{CLS}$ holds $\text{awid}(F) \leq \text{hd}(F)$. The latter is shown in [47, Lemma 6.8] (for unsatisfiable F), and in Corollary 25 below we provide an alternative proof.

It is an open problem whether for (fixed) $k \geq 3$ we can decide "$F \vdash^k \perp$" in polynomial time. For $k = 1$ there is a linear-time algorithm (since $F \vdash^1 \perp$ iff $r_1(F) = \{\perp\}$), and for $k = 2$ there is a quartic-time algorithm by [18]. See the underlying report [17] for some partial results. In [43,47] a stronger system was considered (which allows polynomial-time decision). It uses the closure under input resolution, where only the conclusion is restricted to length $\leq k$. Using this system, [43, Lemma 8.5] obtains the connections $\text{wid}(F) - \max(p, \text{awid}(F)) \leq \text{awid}(F)$ for $F \in \mathcal{USAT} \cap p\text{-}\mathcal{CLS}$ (see [47, Lemma 6.22] for a generalisation). We give a freestanding proof in the underlying report [17]:

Lemma 5. *For $F \in p\text{-}\mathcal{CLS}, p \in \mathbb{N}_0$, holds $\text{wid}(F) \leq \text{awid}(F) + \max(p, \text{awid}(F))$.*

3.3 Space Complexity

The last measures that we discuss in this paper relate to space complexity. We consider three measures: *semantic space*, *resolution space* and *tree space* (all counting clauses to be stored, under different rules). Semantic space was introduced in [2]; a slightly modified definition follows.

Definition 6. *Consider $F \in \mathcal{CLS}$ and $k \in \mathbb{N}$. A* **semantic k-sequence for** F *is a sequence $F_1, \ldots, F_p \in \mathcal{CLS}$, $p \in \mathbb{N}$, fulfilling the following conditions:*

1. *For all $i \in \{1, \ldots, p\}$ holds $c(F_i) \leq k$.*
2. *$F_1 = \top$, and for $i \in \{2, \ldots, p\}$ either holds $F_{i-1} \models F_i$ (inference), or there is $C \in F$ with $F_i = F_{i-1} \cup \{C\}$ (axiom download).*

A semantic sequence is called **complete** *if $F_p \in \mathcal{USAT}$. For $F \in \mathcal{USAT}$ the* **semantic-space complexity** *of F, denoted by $\mathbf{css}(F) \in \mathbb{N}$ ("c" for "clause"), is the minimal $k \in \mathbb{N}$ such there is a complete semantic k-sequence for F.*

Different from [2], the elimination of clauses ("memory erasure") is integrated into the inference step, since we want our bound awid $\leq$ css to be as tight as possible, and the tree-space, a special case of semantic space, shall fulfil cts = hd +1. By definition we have $\mathrm{css}(\varphi * F) \leq \mathrm{css}(F)$ for $F \in \mathcal{USAT}$ and $\varphi \in \mathcal{PASS}$, and thus $\mathrm{css}(F)$ is naturally defined for all $F \in \mathcal{CLS}$.

We come to the notion of resolution space originating in [40,41] and [54,25]. This measure was intensively studied during the last decade (cf. e.g. [12,49]).

Definition 7. *Consider $F \in \mathcal{CLS}$ and $k \in \mathbb{N}$. A* **resolution k-sequence for** F *is a sequence $F_1, \ldots, F_p \in \mathcal{CLS}$, $p \in \mathbb{N}$, fulfilling the following conditions:*

1. *For all $i \in \{1, \ldots, p\}$ holds $c(F_i) \leq k$.*
2. *$F_1 = \top$, and for $i \in \{2, \ldots, p\}$ either holds $F_i \setminus F_{i-1} = \{C\}$, where C is a resolvent of two clauses in F_i (removal of clauses and/or addition of one resolvent), or there is $C \in F$ with $F_i = F_{i-1} \cup \{C\}$ (axiom download).*

A resolution k-sequence is **complete** *if $\bot \in F_p$. For $F \in \mathcal{USAT}$ the* **resolution-space complexity** *of F, denoted by $\mathbf{crs}(F) \in \mathbb{N}$, is the minimal $k \in \mathbb{N}$ such there is a complete resolution k-sequence for F.*

We can also define a variant of space for tree-like resolution refutations.

Definition 8. *A* **tree k-sequence for** F *is a resolution k-sequence for F, such that in case of adding an inferred clause via $F_i \setminus F_{i-1} = \{R\}$, for $R = C \diamond D$ with $C, D \in F_{i-1}$, we always have $C, D \notin F_i$. For $F \in \mathcal{USAT}$ the* **tree-resolution space complexity** *of F, denoted by $\mathbf{cts}(F) \in \mathbb{N}$, is the minimal $k \in \mathbb{N}$ such there is a complete tree k-sequence for F.*

Both measures crs, cts are again not increased by applying partial assignments. By definition we have $\mathrm{css}(F) \leq \mathrm{crs}(F) \leq \mathrm{cts}(F)$ for $F \in \mathcal{CLS}$. We recall a basic connection between tree space and hardness ([43, Subsection 7.2.1]):

Lemma 9 ([43]). *For $F \in \mathcal{CLS}$ holds $\mathrm{cts}(F) = \mathrm{hd}(F) + 1$.*

We remarked earlier that by definition we have $\operatorname{css}(F) \leq \operatorname{crs}(F)$. In fact, the two measures are the same up to a linear factor, as shown by [2]. Our factor is different from [2]; see the underlying report [17] for the proof:

Proposition 10. *For $F \in \mathcal{CLS}$ we have $\operatorname{crs}(F) \leq 3\operatorname{css}(F) - 2$.*

See the conclusions (Section 5) for further discussions.

4 Combinatorial Characterisations

In this section we come to the main topic of this article: the characterisations of the hardness measures introduced in the previous section by Prover-Delayer games and sets of partial assignments.

4.1 Game Characterisations for Hardness

The game of Pudlák and Impagliazzo [52] is a well-known and classic Prover-Delayer game, which serves as one of the main and conceptually very simple methods to obtain resolution lower bounds for unsatisfiable formulas in CNF. The game proceeds between a Prover and a Delayer. The Delayer claims to know a satisfying assignment for an unsatisfiable clause-set, while the Prover wants to expose his lie and in each round asks for a variable value. The Delayer can either choose to answer this question by setting the variable to 0/1, or can defer the choice to the Prover. In the latter case, Delayer scores one point. This game provides a method for showing lower bounds for tree resolution. Namely, Pudlák and Impagliazzo [52] show that exhibiting a Delayer strategy for a CNF F that scores at least p points against every Prover implies a lower bound of 2^p for the proof size of F in tree resolution. This can now be understood through hardness; by Lemma 9 we know that for unsatisfiable clause-set F holds $\operatorname{cts}(F) = \operatorname{hd}(F)+1$, while in [26] it was shown that the optimal value of the above game plus one equals $\operatorname{cts}(F)$, and thus $\operatorname{hd}(F)$ is the optimal value of that game for F. We remark that thus the game of Pudlák and Impagliazzo does not characterise tree resolution size *precisely*; in [16,14] a modified (asymmetric) version of the game is introduced, which precisely characterises tree resolution size ([15]). We present now the generalised hardness game, also handling satisfiable clause-sets. First we need to determine how hardness is affected when assigning one variable:

Lemma 11. *For clause-sets $F \in \mathcal{CLS}$ and $v \in \operatorname{var}(F)$ either there is $\varepsilon \in \{0,1\}$ with $\operatorname{hd}(\langle v \to \varepsilon \rangle * F) = \operatorname{hd}(F)$ and $\operatorname{hd}(\langle v \to \bar{\varepsilon} \rangle * F) \leq \operatorname{hd}(F)$, or we have $\operatorname{hd}(\langle v \to 0 \rangle * F) = \operatorname{hd}(\langle v \to 1 \rangle * F) = \operatorname{hd}(F) - 1$. If F is unsatisfiable and $\operatorname{hd}(F) > 0$, then there is a variable $v \in \operatorname{var}(F)$ and $\varepsilon \in \{0,1\}$ with $\operatorname{hd}(\langle v \to \varepsilon \rangle * F) < \operatorname{hd}(F)$.*

Proof. The assertion on the existence of v and ε follows by definition. Assume now that neither of the two cases holds, i.e., that there is some $\varepsilon \in \{0,1\}$ such that $\operatorname{hd}(\langle v \to \varepsilon \rangle * F) \leq \operatorname{hd}(F) - 1$ and $\operatorname{hd}(\langle v \to \bar{\varepsilon} \rangle * F) \leq \operatorname{hd}(F) - 2$. Consider a partial assignment φ such that $\varphi * F \in \mathcal{USAT}$ and $\operatorname{hd}(\varphi * F) = \operatorname{hd}(F)$ (recall Definition 1). Then $v \notin \operatorname{var}(\varphi)$ holds. Now $\operatorname{hd}(\langle v \to \varepsilon \rangle * (\varphi * F)) \leq \operatorname{hd}(F) - 1$ and $\operatorname{hd}(\langle v \to \bar{\varepsilon} \rangle * (\varphi * F)) \leq \operatorname{hd}(F) - 2$, so by definition of hardness for unsatisfiable clause-sets we have $\operatorname{hd}(\varphi * F) \leq \operatorname{hd}(F) - 1$, a contradiction. □

We are ready to present the new game, which characterises $\mathrm{hd}(F)$ for arbitrary F. A feature of this game, not shared by the original game, is that there is just one "atomic action" for both players, the choice of a variable *and* a value, and the rules are just about how this choice can be employed.

Theorem 12. *Consider $F \in \mathcal{CLS}$. The following game is played between Prover and Delayer, where the partial assignments θ all fulfil $\mathrm{var}(\theta) \subseteq \mathrm{var}(F)$:*

1. *The two players play in turns, and Delayer starts. Initially $\theta := \langle \rangle$.*
2. *A move of Delayer extends θ to $\theta' \supseteq \theta$.*
3. *A move of Prover extends θ to $\theta' \supseteq \theta$ with $\theta' * F = \top$ or $n(\theta') = n(\theta) + 1$.*
4. *The game ends as soon $\bot \in \theta * F$ or $\theta * F = \top$. In the first case Delayer gets as many points as variables have been assigned by Prover. In the second case Delayer gets zero points.*

Now there is a strategy of Delayer which can always achieve $\mathrm{hd}(F)$ many points, while Prover can always avoid that Delayer gets $\mathrm{hd}(F) + 1$ or more points.

Proof. The strategy of Prover is: If $\theta * F$ is satisfiable, then extend θ to a satisfying assignment. Otherwise choose $v \in \mathrm{var}(F)$ and $\varepsilon \in \{0,1\}$ s.t. $\mathrm{hd}(\langle v \to \varepsilon \rangle * F)$ is minimal. The strategy of Delayer is: Initially extend $\langle \rangle$ to some θ such that $\theta * F \in \mathcal{USAT}$ and $\mathrm{hd}(\theta * F)$ is maximal. For all other moves, and also for the first move as an additional extension, as long as there are variables $v \in \mathrm{var}(\theta * F)$ and $\varepsilon \in \{0,1\}$ with $\mathrm{hd}(\langle v \to \varepsilon \rangle * (\theta * F)) \le \mathrm{hd}(\theta * F) - 2$, choose such a pair (v, ε) and extend θ to $\theta \cup \langle v \to \overline{\varepsilon} \rangle$. The assertion follows by Lemma 11. $\square$

The game of Theorem 12 can be extended to handle asymmetric width (Theorem 23): Delayer in both cases just extends the current partial assignment, while Prover for awid can additionally "forget" assignments.

4.2   Characterising Hardness and Depth by Partial Assignments

We now provide an alternative characterisation of hardness of clause-sets F by sets $\mathbb{P}$ of partial assignments, complementing the games. The "harder" F is, the better $\mathbb{P}$ "approximates" satisfying F. The minimum condition is:

Definition 13. *A set $\mathbb{P} \subseteq \mathcal{PASS}$ is **minimal consistent for** $F \in \mathcal{CLS}$ if $\mathrm{var}(\mathbb{P}) = \bigcup_{\varphi \in \mathbb{P}} \mathrm{var}(\varphi) \subseteq \mathrm{var}(F)$, for all $\varphi \in \mathbb{P}$ holds $\bot \notin \varphi * F$, and $\mathbb{P} \ne \emptyset$.*

$\mathbb{P}$ is a partially ordered set (by inclusion). Recall that a *chain* K is a subset constituting a linear order, while the *length* of K is $|K| - 1 \in \mathbb{Z}_{\ge -1}$, and a **maximal chain** is a chain which can not be extended without breaking linearity.

Definition 14. *For $k \in \mathbb{N}_0$ and $F \in \mathcal{USAT}$ let a **weakly k-consistent set of partial assignments for** F be a minimally consistent set $\mathbb{P}$ for F, such that the minimum length of a maximal chain in $\mathbb{P}$ is at least k, and for every non-maximal $\varphi \in \mathbb{P}$, every $v \in \mathrm{var}(F) \setminus \mathrm{var}(\varphi)$ and every $\varepsilon \in \{0,1\}$ there is $\varphi' \in \mathbb{P}$ with $\varphi \cup \langle v \to \varepsilon \rangle \subseteq \varphi'$.*

There can be gaps between $\varphi \subset \varphi'$ for $\varphi, \varphi' \in \mathbb{P}$, corresponding to the moves of Delayer in Theorem 12, who needs to prevent all "bad" assignments at once.

Proposition 15. *For all $F \in \mathcal{USAT}$ we have $\mathrm{hd}(F) > k$ if and only if there is a weakly k-consistent set for partial assignments for F.*

Proof. If there is a weakly k-consistent $\mathbb{P}$, then Delayer from Theorem 12 has a strategy achieving at last $k+1$ points by choosing a minimal $\theta' \in \mathbb{P}$ extending θ, and maintaining in this way $\theta \in \mathbb{P}$ as long as possible. And a weakly $(\mathrm{hd}(F)-1)$-consistent $\mathbb{P}$ for $\mathrm{hd}(F) > 0$ is given by the partial assignments obtained from those $\varphi \in \mathcal{PASS}$ with $\bot \notin \varphi * F$ by extending φ to $\varphi' := \varphi \cup \langle v \to \varepsilon \rangle$ for such $v \in \mathrm{var}(F) \setminus \mathrm{var}(\varphi)$ and $\varepsilon \in \{0,1\}$ with $\mathrm{hd}(\varphi' * F) = \mathrm{hd}(\varphi * F)$, and repeating this extension as long as possible. $\square$

A similar characterisation can also be given for the depth-measure $\mathrm{dep}(F)$ (cf. Definition 2). For this we relax the concept of weak consistency:

Definition 16. *For $k \in \mathbb{N}_0$ and $F \in \mathcal{USAT}$ let a **barely k-consistent set of partial assignments for** F be a minimally consistent $\mathbb{P}$ for F such that $\langle \rangle \in \mathbb{P}$, and for every $\varphi \in \mathbb{P}$ with $n(\varphi) < k$ and all $v \in \mathrm{var}(F) \setminus \mathrm{var}(\varphi)$ there is $\varepsilon \in \{0,1\}$ with $\varphi \cup \langle v \to \varepsilon \rangle \in \mathbb{P}$.*

By [57, Theorem 2.4] we get the following characterisation (we provide a proof due to technical differences):

Proposition 17. *For all $F \in \mathcal{USAT}$ we have $\mathrm{dep}(F) > k$ if and only if there is a barely k-consistent set for partial assignments for F.*

Proof. If F has a resolution proof T of height k, then for a barely k'-consistent $\mathbb{P}$ for F we have $k' < k$, since otherwise starting at the root of T we follow a path given by extending $\langle \rangle$ according to the extension-condition of $\mathbb{P}$, and we arrive at a $\varphi \in \mathbb{P}$ falsifying an axiom of T, contradicting the definition of $\mathbb{P}$. On the other hand, if $\mathrm{dep}(F) > k$, then there is a barely k-consistent $\mathbb{P}$ for F as follows: for $j \in \{0, \ldots, k\}$ put those partial assignments $\varphi \in \mathcal{PASS}$ with $\mathrm{var}(\varphi) \subseteq \mathrm{var}(F)$ and $n(\varphi) = j$ into $\mathbb{P}$ which do not falsify any clause derivable by a resolution tree of depth at most $k - j$ from F. Now consider $\varphi \in \mathbb{P}$ with $j := n(\varphi) < k$, together with $v \in \mathrm{var}(F) \setminus \mathrm{var}(\varphi)$. Assume that for both $\varepsilon \in \{0,1\}$ we have $\varphi \cup \langle v \to \varepsilon \rangle \notin \mathbb{P}$. So there are clauses C, D derivable from T by a resolution tree of depth at most $k - j - 1$, with $v \in C, \overline{v} \in D$, and $\varphi * \{C, D\} = \{\bot\}$. But then $\varphi * \{C \diamond D\} = \{\bot\}$, contradicting the defining condition for φ. $\square$

4.3   Characterising Symmetric Width by Partial Assignments

We now turn to characterisations of the width-hardness measures, starting with the symmetric width measure wid. It is instructive to review the characterisation for $\mathrm{wid}(F)$ for $F \in \mathcal{USAT}$ from [5], using a different formulation.

Definition 18. *Consider $F \in \mathcal{CLS}$ and $k \in \mathbb{N}_0$. A **symmetrically k-consistent set of partial assignments for** F is a minimally consistent $\mathbb{P}$ for F, such that for all $\varphi \in \mathbb{P}$, all $v \in \mathrm{var}(F) \setminus \mathrm{var}(\varphi)$, and all $\psi \subseteq \varphi$ with $n(\psi) < k$ there exists $\varepsilon \in \{0,1\}$ and $\varphi' \in \mathbb{P}$ with $\psi \cup \langle v \to \varepsilon \rangle \subseteq \varphi'$.*

Note that a symmetrically k-consistent set is also barely k-consistent. For the (simple) proof of the following lemma see the underlying report [17].

Lemma 19. *Consider $F \in \mathcal{USAT}$ and $k \in \mathbb{N}_0$. Then Duplicator wins the Boolean existential k-pebble game on F in the sense of [5] if and only if there exists a symmetrically k-consistent set of partial assignment for F.*

By [5, Theorem 2] (there "$F \in r$-$\mathcal{CLS}$" is superfluous):

Corollary 20. *For $F \in \mathcal{USAT}$ and $k \in \mathbb{N}_0$ holds $\text{wid}(F) > k$ if and only if there exists a symmetrically k-consistent set of partial assignments for F.*

4.4 Characterising Asymmetric Width by Partial Assignments

Similar to Definition 18, we characterise asymmetric width — the only difference is, that the extensions must work for both truth values.

Definition 21. *Consider $F \in \mathcal{CLS}$ and $k \in \mathbb{N}_0$. A k-consistent set of partial assignments for F is a minimally consistent $\mathbb{P}$ for F, such that for all $\varphi \in \mathbb{P}$, all $v \in \text{var}(F) \setminus \text{var}(\varphi)$, all $\psi \subseteq \varphi$ with $n(\psi) < k$ and for both $\varepsilon \in \{0,1\}$ there is $\varphi' \in \mathbb{P}$ with $\psi \cup \langle v \to \varepsilon \rangle \subseteq \varphi'$.*

Similarly to [5, Theorem 2], where the authors provide a characterisation of symmetric width, we obtain a characterisation of asymmetric width:

Theorem 22. *For $F \in \mathcal{USAT}$ and $k \in \mathbb{N}_0$ holds $\text{awid}(F) > k$ if and only if there exists a k-consistent set of partial assignments for F.*

Proof. First assume $\text{awid}(F) > k$. Let $F' := \{C \in \mathcal{CL} \mid \exists R : F \vdash^k C\}$. Note that by definition $F \subseteq F'$, while by assumption we have $\bot \notin F'$. Let $\mathbb{P}$ be the set of maximal $\varphi \in \mathcal{PASS}$ with $\text{var}(\varphi) \subseteq \text{var}(F)$ and $\bot \notin \varphi * F'$. We show that $\mathbb{P}$ is a k-consistent set of partial assignments for F. Consider $\varphi \in \mathbb{P}$, $v \in \text{var}(F) \setminus \text{var}(\varphi)$ and $\psi \subseteq \varphi$ with $n(\psi) < k$. Due to maximality of φ there are $C, D \in F'$ with $v \in C$, $\overline{v} \in D$ and $(\varphi \cup \langle v \to 0 \rangle) * \{C\} = (\varphi \cup \langle v \to 1 \rangle) * \{D\} = \{\bot\}$. Assume that there is $\varepsilon \in \{0,1\}$, such that for $\psi' := \psi \cup \langle v \to \varepsilon \rangle$ there is no $\varphi' \in \mathbb{P}$ with $\psi' \subseteq \varphi'$. Thus there is $E \in F'$ with $\psi' * \{E\} = \{\bot\}$; so we have $v \in \text{var}(E)$ and $|E| \leq k$. Now E is resolvable with either C or D via k-resolution, and for the resolvent $R \in F'$ we have $\varphi * \{R\} = \{\bot\}$, contradicting the definition of $\mathbb{P}$.

For the other direction, assume that $\mathbb{P}$ is a k-consistent set of partial assignments for F. For the sake of contradiction assume there is $T : F \vdash^k \bot$. We show by induction on $\text{ht}_T(w)$ that for all $w \in \text{nds}(T)$ and all $\varphi \in \mathbb{P}$ holds $\varphi * \{C(w)\} \neq \{\bot\}$, which at the root of T (where the clause-label is $\bot$) yields a contradiction. If $\text{ht}_T(w) = 0$ (i.e., w is a leaf), then the assertion follows by definition; so assume $\text{ht}_T(w) > 0$. Let w_1, w_2 be the two children of w, and let $C := C(w)$ and $C_i := C(w_i)$ for $i \in \{1,2\}$. W.l.o.g. $|C_1| \leq k$. Note $C = C_1 \diamond C_2$; let v be the resolution variable, where w.l.o.g. $v \in C_1$. Consider $\varphi \in \mathbb{P}$; we have to show $\varphi * \{C\} \neq \{\bot\}$, and so assume $\varphi * \{C\} = \{\bot\}$. By induction hypothesis we know $\bot \notin \varphi * \{C_1, C_2\}$, and thus $v \notin \text{var}(\varphi)$. Let $\psi := \varphi | (\text{var}(C_1) \setminus \{v\})$, and $\psi' := \psi \cup \langle v \to 0 \rangle$. There is $\varphi' \in \mathbb{P}$ with $\psi' \subseteq \varphi'$, thus $\psi' * \{C_1\} = \{\bot\}$, contradicting the induction hypothesis. □

4.5   Game Characterisation for Asymmetric Width

The characterisation of asymmetric width by partial assignments from the previous subsection will now be employed for a game-theoretic characterisation; in fact, the k-consistent sets of partial assignments will directly translate into strategies for Delayer (while a strategy of Prover is given by a resolution refutation). We only handle the unsatisfiable case here — the general case can be handled as in Theorem 12. This game extends (in a sense) the Prover-Delayer game from [55] for symmetric width (but again without communication).

Theorem 23. *Consider* $F \in \mathcal{USAT}$. *The following game is played between Prover and Delayer (as in Theorem 12, always* $\text{var}(\theta) \subseteq \text{var}(F)$ *holds):*

1. *The two players play in turns, and Delayer starts. Initially* $\theta := \langle \rangle$.
2. *Delayer extends* θ *to* $\theta' \supseteq \theta$.
3. *Prover chooses some* θ' *compatible with* θ *such that* $|\text{var}(\theta') \setminus \text{var}(\theta)| = 1$.
4. *If* $\perp \in \theta * F$, *then the game ends, and Delayer gets the maximum of* $n(\theta')$ *chosen by Prover as points (0 if Prover didn't make a choice).*
5. *Prover must play in such a way that the game is finite.*

We have the following:

1. *For a strategy of Delayer, which achieves* $k \in \mathbb{N}$ *points whatever Prover does, we have* $\text{awid}(F) \geq k$.
2. *For a strategy of Prover, which guarantees that Delayer gets at most* $k \in \mathbb{N}_0$ *points in any case, we have* $\text{awid}(F) \leq k$.
3. *There is a strategy of Delayer which guarantees at least* $\text{awid}(F)$ *many points (whatever Prover does).*
4. *There is a strategy of Prover which guarantees at most* $\text{awid}(F)$ *many points for Delayer (whatever Delayer does).*

Proof. W.l.o.g. $\perp \notin F$. Part 1 follows by Part 4 (if $\text{awid}(F) < k$, then Prover could guarantee at most $k-1$ points), and Part 2 follows by Part 3 (if $\text{awid}(F) > k$, then Delayer could guarantee at least $k+1$ points).

Let now $k := \text{awid}(F)$. For Part 3, a strategy of Delayer guaranteeing k many points (at least) is as follows: Delayer chooses a $(k-1)$-consistent set $\mathbb{P}$ of partial assignment (by Theorem 22). The move of Delayer is to choose some $\theta' \in \mathbb{P}$. If Prover then chooses some θ' with $n(\theta') \leq k-1$, then the possibility of extension is maintained for Delayer. In this way the empty clause is never created. Otherwise the Delayer has reached his goal, and might choose anything.

It remains to show that Prover can force the creation of the empty clause such that Delayer obtains at most k many points. For that consider a resolution refutation $R : F \vdash \perp$ which is a k-resolution tree. The strategy of Prover is to construct partial assignments θ' (from θ as given by Delayer) which falsify some clause C of length at most k in R, where the height of the node is decreasing — this will falsify finally some clause in F, finishing the game. The Prover considers initially (before the first move of Delayer) just the root. When Prover is to move, he considers a path from the current clause to some leaf, such that only clauses of length at most k are on that path. There must be a first clause C (starting from the falsified clause, towards the leaves) on that path not falsified by θ (since θ does not falsify any axiom). It must be the case that θ falsifies all literals in C besides one literal $x \in C$, where $\text{var}(x) \notin \text{var}(\theta)$. Now Prover chooses θ' as the restriction of θ to $\text{var}(C) \setminus \{\text{var}(x)\}$ and extended by $x \to 0$. $\qquad \square$

We already remarked in Section 3 that always $\text{awid}(F) \leq \text{hd}(F)$. Based on the game characterisations shown here, we provide an easy alternative proof for this fundamental fact for $F \in \mathcal{USAT}$:

Lemma 24. *Consider the game of Theorem 23, when restricted in such a way that Prover must always choose some θ' with $n(\theta') > n(\theta)$. This game is precisely the game of Theorem 12.*

Corollary 25. *For all $F \in \mathcal{CLS}$ we have $\text{awid}(F) \leq \text{hd}(F)$.*

4.6   Width Hardness versus Semantic Space

We have already seen in Corollary 25, that our game-theoretic characterisations allow quite easy and elegant proofs on tight relations between different hardness measures. Our next result also follows this paradigm. It provides a striking relation between asymmetric width and semantic space. We recall that Atserias and Dalmau [5, Theorem 3] have shown $\text{wid}(F) \leq \text{crs}(F) + r - 1$, where $F \in \mathcal{USAT} \cap r\text{-}\mathcal{CLS}$ (all $r \geq 0$ are allowed; note that now we can drop the unsatisfiability condition). We generalise this result in Theorem 27 below, replacing resolution space $\text{crs}(F)$ by the tighter notion of semantic space $\text{css}(F)$. More important, we eliminate the additional $r - 1$ in the inequality, by changing symmetric width $\text{wid}(F)$ into asymmetric width $\text{awid}(F)$ (cf. Lemma 5 for the relation between these two measures). First a lemma similar to [5, Lemma 5]:

Lemma 26. *Consider $F \in \mathcal{CLS}$, a k-consistent set $\mathbb{P}$ of partial assignments for F ($k \in \mathbb{N}_0$), and a semantic k-sequence $(F_1, \ldots, F_p)$ for F (recall Definition 6). Then there exist $\varphi_i \in \mathbb{P}$ with $\varphi_i * F_i = \top$ for each $i \in \{1, \ldots, p\}$.*

Proof. Set $\varphi_1 := \langle \rangle \in \mathbb{P}$. For $i \in \{2, \ldots, p\}$ the partial assignment φ_i is defined inductively. If $\varphi_{i-1} * F_i = \top$, then $\varphi_i := \varphi_{i-1}$; this covers the case where F_i is obtained from F_{i-1} by addition of inferred clauses and/or removal of clauses. So consider $F_i = F_{i-1} \cup \{C\}$ for $C \in F \setminus F_{i-1}$ (thus $c(F_i) < k$), and we assume $\varphi_{i-1} * F_i \neq \top$. So there is a literal $x \in C$ with $\text{var}(x) \notin \varphi_{i-1}$, since φ_{i-1} does not falsify clauses from F. Choose some $\psi \subseteq \varphi_{i-1}$ with $n(\psi) \leq c(F_{i-1})$ such that $\psi * F_{i-1} = \top$.[4] There is $\varphi_i \in \mathbb{P}$ with $\psi \cup \langle x \to 1 \rangle \subseteq \varphi_i$, whence $\varphi_i * F_i = \top$. $\qquad \square$

We can now show the promised generalisation of [5, Theorem 3]:

Theorem 27. *For $F \in \mathcal{CLS}$ holds $\text{awid}(F) \leq \text{css}(F)$.*

---

[4] For every partial assignment φ and every clause-set F with $\varphi * F = \top$ there exists $\psi \subseteq \varphi$ with $n(\psi) \subseteq c(F)$ and $\psi * F = \top$; see for example Lemma 4 in [5], and see Corollary 8.6 in [48] for a generalisation.

Proof. Assume $F \in \mathcal{USAT}$ and $\mathrm{awid}(F) > \mathrm{css}(F)$; let $k := \mathrm{css}(F)$. By Theorem 22 there is a k-consistent set $\mathbb{P}$ for F. Let $(F_1, \ldots, F_p)$ be a complete semantic k-sequence for F according to Definition 6. Now for the sequence $(\varphi_1, \ldots, \varphi_p)$ according to Lemma 26 we have $\varphi_p * F_p = \top$, contradicting $F_p \in \mathcal{USAT}$.   □

We are now in a position to order most of the hardness measures that we investigated here (cf. also the diagram in the introduction):

Corollary 28. $\mathrm{awid}(F) \leq \mathrm{css}(F) \leq \mathrm{crs}(F) \leq \mathrm{cts}(F) = \mathrm{hd}(F) + 1$ *for* $F \in \mathcal{CLS}$.

We conclude by an application of the extended measures $\mathrm{cts}, \mathrm{css} : \mathcal{CLS} \to \mathbb{N}_0$. In [31] it is shown that for every k there are clause-sets in $\mathcal{UC}_{k+1}$ where every (logically) equivalent clause-set in $\mathcal{WC}_k$ is exponentially bigger. This implies, in the language of representing boolean functions via CNFs, that allowing the tree-space to increase by 2 over semantic space allows for an exponential saving in size (regarding logical equivalence):

Corollary 29. *For all constant* $k \in \mathbb{N}$ *there are sequences* (F_n) *of clause-sets with* $\mathrm{cts}(F_n) \leq k + 2$ *for all* n, *where all equivalent sequences* (F'_n) *with* $\mathrm{css}(F'_n) \leq k$ *(for all* n*) are exponentially bigger.*

5   Conclusion and Open Problems

In this paper we aimed at *unified characterisations* for the main hardness measures for resolution, thereby obtaining precise relations between these measures. Continuing this programme, a deeper understanding of the three space measures is required. In terms of the game-theoretic characterisations, the main question left open is whether $\mathrm{crs}, \mathrm{css} : \mathcal{CLS} \to \mathbb{N}$ can be characterised in a similar spirit by games and/or partial assignments (for cts we provided such characterisations).

A further block of questions concerns the *exact relationship between the measures*. We believe that Theorem 27 can be improved:

Conjecture 30. $\mathrm{awid}(F) + 1 \leq \mathrm{css}(F)$ for $F \in \mathcal{CLS}$.

Then in Corollary 29 the "+2" could be replaced by "+1". Note that Corollary 29 shows that such small measurement differences actually matter! Concerning the space measures, it is conceivable that $\mathrm{crs} = \mathrm{css}$ could hold; if not then there could be substantial differences between crs and css regarding expressive power, that is, regarding the power to represent boolean functions. Concerning the precise relation between symmetric and asymmetric width, it appears that Lemma 5 could be improved to $\mathrm{wid}(F) \leq \mathrm{awid}(F) + p - 1$ for $F \in p\text{-}\mathcal{CLS}, p \geq 1$ (then Theorem 27 would precisely imply [5, Theorem 3]).

The question on the *expressive power* of the various classes (measures) seems very relevant, and can also be raised when allowing new variables for the representation of boolean functions; see [31] for a thorough discussion of these issues.

Also for practical SAT solving the influence of *blocked clauses elimination/addition* (introduced in [44,45], a generalisation of Extended Resolution) on hardness measures needs to be studied (see [35,36] for recent developments).

Finally, a general theory of *hardness measures* might be possible (which might also be applicable to other proof systems than resolution).

References

1. Achlioptas, D., Beame, P., Molloy, M.: A sharp threshold in proof complexity yields lower bounds for satisfiability search. Journal of Computer and System Sciences 68, 238–268 (2004)
2. Alekhnovich, M., Ben-Sasson, E., Razborov, A.A., Wigderson, A.: Space complexity in propositional calculus. SIAM Journal on Computing 31(4), 1184–1211 (2002)
3. Alekhnovich, M., Hirsch, E.A., Itsykson, D.: Exponential lower bounds for the running time of DPLL algorithms on satisfiable formulas. Journal of Automated Reasoning 35(1-3), 51–72 (2005)
4. Ansótegui, C., Bonet, M.L., Levy, J., Manyà, F.: Measuring the hardness of SAT instances. In: Fox, D., Gomes, C. (eds.) Proceedings of the 23th AAAI Conference on Artificial Intelligence (AAAI 2008), pp. 222–228 (2008)
5. Atserias, A., Dalmau, V.: A combinatorial characterization of resolution width. Journal of Computer and System Sciences 74, 323–334 (2008)
6. Babka, M., Balyo, T., Čepek, O., Gurský, Š., Kučera, P., Vlček, V.: Complexity issues related to propagation completeness. Artificial Intelligence 203, 19–34 (2013)
7. Balyo, T., Gurský, Š., Kučera, P., Vlček, V.: On hierarchies over the SLUR class. In: Twelfth International Symposium on Artificial Intelligence and Mathematics, ISAIM 2012 (January 2012),
   http://www.cs.uic.edu/bin/view/Isaim2012/AcceptedPapers
8. Beame, P., Beck, C., Impagliazzo, R.: Time-space tradeoffs in resolution: superpolynomial lower bounds for superlinear space. In: STOC 2012 Proceedings of the 44th Symposium on Theory of Computing, pp. 213–232 (2012)
9. Beck, C., Nordström, J., Tang, B.: Some trade-off results for polynomial calculus: extended abstract. In: Proceedings of the Forty-fifth Annual ACM Symposium on Theory of Computing (STOC 2013), pp. 813–822 (2013)
10. Ben-Sasson, E., Harsha, P.: Lower bounds for bounded depth Frege proofs via Buss-Pudlák games. ACM Transactions on Computational Logic 11(3) (2010)
11. Ben-Sasson, E., Nordström, J.: Short proofs may be spacious: An optimal separation of space and length in resolution. In: Proc. 49th IEEE Symposium on the Foundations of Computer Science, pp. 709–718 (2008)
12. Ben-Sasson, E., Nordström, J.: Understanding space in proof complexity: Separations and trade-offs via substitutions. In: Innovations in Computer Science, ICS 2010, January 7-9, pp. 401–416. Tsinghua University, Beijing (2011)
13. Ben-Sasson, E., Wigderson, A.: Short proofs are narrow - resolution made simple. Journal of the ACM 48(2), 149–169 (2001)
14. Beyersdorff, O., Galesi, N., Lauria, M.: A lower bound for the pigeonhole principle in tree-like resolution by asymmetric prover-delayer games. Information Processing Letters 110(23), 1074–1077 (2010)
15. Beyersdorff, O., Galesi, N., Lauria, M.: A characterization of tree-like resolution size. Information Processing Letters 113(18), 666–671 (2013)
16. Beyersdorff, O., Galesi, N., Lauria, M.: Parameterized complexity of DPLL search procedures. ACM Transactions on Computational Logic 14(3) (2013)
17. Beyersdorff, O., Kullmann, O.: Hardness measures and resolution lower bounds. Tech. Rep. arXiv:1310.7627v2 [cs.CC], arXiv (February 2014)
18. Buro, M., Kleine Büning, H.: On resolution with short clauses. Annals of Mathematics and Artificial Intelligence 18(2-4), 243–260 (1996)
19. Čepek, O., Kučera, P., Vlček, V.: Properties of SLUR formulae. In: Bieliková, M., Friedrich, G., Gottlob, G., Katzenbeisser, S., Turán, G. (eds.) SOFSEM 2012. LNCS, vol. 7147, pp. 177–189. Springer, Heidelberg (2012)

20. Chan, S.M.: Just a pebble game. In: 2013 IEEE Conference on Computational Complexity (CCC), pp. 133–143 (2013)
21. Chan, S.M.: Just a pebble game. Tech. Rep. TR13-042, Electronic Colloquium on Computational Complexity (ECCC), 41 pages (March 2013), http://eccc.hpi-web.de/report/2013/042/
22. Clegg, M., Edmonds, J., Impagliazzo, R.: Using the Groebner basis algorithm to find proofs of unsatisfiability. In: Proceedings of the 28th ACM Symposium on Theory of Computation, pp. 174–183 (1996)
23. Esparza, J., Luttenberger, M., Schlund, M.: A brief history of Strahler numbers. In: Dediu, A.-H., Martín-Vide, C., Sierra-Rodríguez, J.-L., Truthe, B. (eds.) LATA 2014. LNCS, vol. 8370, pp. 1–13. Springer, Heidelberg (2014)
24. Esteban, J.L., Torán, J.: Space bounds for resolution. In: Meinel, C., Tison, S. (eds.) STACS 1999. LNCS, vol. 1563, pp. 551–560. Springer, Heidelberg (1999)
25. Esteban, J.L., Torán, J.: Space bounds for resolution. Information and Computation 171(1), 84–97 (2001)
26. Esteban, J.L., Torán, J.: A combinatorial characterization of treelike resolution space. Information Processing Letters 87(6), 295–300 (2003)
27. Groote, J.F., Warners, J.P.: The popositional formula checker HeerHugo. Tech. Rep. SEN-R9905, Centre for Mathematics and Computer Science (CWI), Amsterdam, The Netherlands (January 1999)
28. Groote, J.F., Warners, J.P.: The propositional formula checker HeerHugo. Journal of Automated Reasoning 24(1-2), 101–125 (2000)
29. Gwynne, M., Kullmann, O.: Generalising and unifying SLUR and unit-refutation completeness. In: van Emde Boas, P., Groen, F.C.A., Italiano, G.F., Nawrocki, J., Sack, H. (eds.) SOFSEM 2013. LNCS, vol. 7741, pp. 220–232. Springer, Heidelberg (2013)
30. Gwynne, M., Kullmann, O.: On SAT representations of XOR constraints. Tech. Rep. arXiv:1309.3060v4 [cs.CC], arXiv (December 2013)
31. Gwynne, M., Kullmann, O.: Trading inference effort versus size in CNF knowledge compilation. Tech. Rep. arXiv:1310.5746v2 [cs.CC], arXiv (November 2013)
32. Gwynne, M., Kullmann, O.: Generalising unit-refutation completeness and SLUR via nested input resolution. Journal of Automated Reasoning 52(1), 31–65 (2014)
33. Gwynne, M., Kullmann, O.: On SAT representations of XOR constraints. In: Dediu, A.-H., Martín-Vide, C., Sierra-Rodríguez, J.-L., Truthe, B. (eds.) LATA 2014. LNCS, vol. 8370, pp. 409–420. Springer, Heidelberg (2014)
34. Haken, A.: The intractability of resolution. Theoretical Computer Science 39, 297–308 (1985)
35. Järvisalo, M., Biere, A., Heule, M.: Simulating circuit-level simplifications on CNF. Journal of Automated Reasoning 49(4), 583–619 (2012)
36. Järvisalo, M., Heule, M.J.H., Biere, A.: Inprocessing rules. In: Gramlich, B., Miller, D., Sattler, U. (eds.) IJCAR 2012. LNCS, vol. 7364, pp. 355–370. Springer, Heidelberg (2012)
37. Järvisalo, M., Matsliah, A., Nordström, J., Živný, S.: Relating proof complexity measures and practical hardness of SAT. In: Milano, M. (ed.) CP 2012. LNCS, vol. 7514, pp. 316–331. Springer, Heidelberg (2012)
38. Kleine Büning, H.: On generalized Horn formulas and k-resolution. Theoretical Computer Science 116, 405–413 (1993)
39. Kleine Büning, H., Kullmann, O.: Minimal unsatisfiability and autarkies. In: Biere, A., Heule, M.J., van Maaren, H., Walsh, T. (eds.) Handbook of Satisfiability. Frontiers in Artificial Intelligence and Applications, vol. 185, ch. 11, pp. 339–401. IOS Press (February 2009)

40. Kleine Büning, H., Lettmann, T.: Aussagenlogik: Deduktion und Algorithmen. Leitfäden und Monographen der Informatik, B.G. Teubner Stuttgart (1994)
41. Kleine Büning, H., Lettmann, T.: Propositional Logic: Deduction and Algorithms. Cambridge University Press (1999)
42. Krajíček, J.: Interpolation theorems, lower bounds for proof systems and independence results for bounded arithmetic. The Journal of Symbolic Logic 62(2), 457–486 (1997)
43. Kullmann, O.: Investigating a general hierarchy of polynomially decidable classes of CNF's based on short tree-like resolution proofs. Tech. Rep. TR99-041, Electronic Colloquium on Computational Complexity (ECCC) (October 1999), http://eccc.hpi-web.de/report/1999/041/
44. Kullmann, O.: New methods for 3-SAT decision and worst-case analysis. Theoretical Computer Science 223(1-2), 1–72 (1999)
45. Kullmann, O.: On a generalization of extended resolution. Discrete Applied Mathematics 96-97, 149–176 (October 1999)
46. Kullmann, O.: An improved version of width restricted resolution. In: Electronical Proceedings of Sixth International Symposium on Artificial Intelligence and Mathematics, 11 pages (January 2000), http://rutcor.rutgers.edu/~amai/aimath00/AcceptedCont.htm
47. Kullmann, O.: Upper and lower bounds on the complexity of generalised resolution and generalised constraint satisfaction problems. Annals of Mathematics and Artificial Intelligence 40(3-4), 303–352 (2004)
48. Kullmann, O.: Constraint satisfaction problems in clausal form I: Autarkies and deficiency. Fundamenta Informaticae 109(1), 27–81 (2011)
49. Nordström, J.: Pebble games, proof complexity, and time-space trade-offs. Logical Methods in Computer Science 9(3), 1–63 (2013)
50. Nordström, J., Håstad, J.: Towards an optimal separation of space and length in resolution. Theory of Computing 9(14), 471–557 (2013)
51. Pudlák, P.: Proofs as games. American Math. Monthly, 541–550 (2000)
52. Pudlák, P., Impagliazzo, R.: A lower bound for DLL algorithms for k-SAT (preliminary version). In: SODA, pp. 128–136. ACM/SIAM (2000)
53. Schlipf, J.S., Annexstein, F.S., Franco, J.V., Swaminathan, R.: On finding solutions for extended Horn formulas. Information Processing Letters 54, 133–137 (1995)
54. Torán, J.: Lower bounds for space in resolution. In: Flum, J., Rodríguez-Artalejo, M. (eds.) CSL 1999. LNCS, vol. 1683, pp. 362–373. Springer, Heidelberg (1999)
55. Torán, J.: Space and width in propositional resolution (Column: Computational Complexity). Bulletin of the European Association of Theoretical Computer Science (EATCS) 83, 86–104 (2004)
56. Tseitin, G.: On the complexity of derivation in propositional calculus. In: Seminars in Mathematics, vol. 8. V.A. Steklov Mathematical Institute, Leningrad (1968), English translation: Slisenko, A.O. (ed.) Studies in mathematics and mathematical logic, Part II, pp. 115–125 (1970)
57. Urquhart, A.: The depth of resolution proofs. Studia Logica 99, 349–364 (2011)
58. del Val, A.: Tractable databases: How to make propositional unit resolution complete through compilation. In: Proceedings of the 4th International Conference on Principles of Knowledge Representation and Reasoning (KR 1994), pp. 551–561 (1994)
59. Viennot, X.G.: Trees everywhere. In: Arnold, A. (ed.) CAAP 1990. LNCS, vol. 431, pp. 18–41. Springer, Heidelberg (1990)

Community Branching
for Parallel Portfolio SAT Solvers

Tomohiro Sonobe[1,2], Shuya Kondoh[3], and Mary Inaba[3]

[1] National Institute of Informatics, Japan
[2] JST, ERATO, Kawarabayashi Large Graph Project, Japan
[3] Graduate School of Information Science and Technology,
University of Tokyo, Japan
tominlab@gmail.com

Abstract. Portfolio approach for parallel SAT solvers is known as the standard parallelisation technique. In portfolio, diversification is one of the important factors in order to enable workers (solvers) to conduct a vast search. The diversification is implemented by setting different parameters for each worker in the state-of-the-art parallel portfolio SAT solvers. However, it is difficult to combine the search parameters properly in order to avoid overlaps of search spaces between the workers For this issue, we propose a novel diversification technique, called community branching. In this method, we assign a different set (or sets) of variables (called a community) to each worker and force them to select these variables as decision variables in early decision levels. In this manner, we can avoid the overlaps of the search spaces between the workers more vigorously than the existing method. We create a graph, where a vertex corresponds to a variable and an edge stands for a relation between two variables in a same clause, and we apply a modularity-based community detection algorithm to it. The variables in a community have strong relationships, and a distributed search for different communities can benefit the whole search. Experimental results show that we could speedup an existing parallel SAT solver with our proposal.

Keywords: Parallel SAT solver, Portfolio, Diversification, Decision, Community detection.

1   Introduction

The state-of-the-art SAT solvers are based on the Davis-Putnam-Logemann-Loveland (DPLL) algorithm [5]. The so-called Conflict-Driven Clause Learning (CDCL) SAT solvers consist of various techniques such as clause learning, non-chronological backtracking, Variable State Independent Decaying Sum (VSIDS) [14] decision heuristic, and restart upon DPLL algorithm.

On the other hand, state-of-the-art parallel SAT solvers are also based on CDCL solvers. Solvers based on the divide-and-conquer approach prevailed first, such as PSATO [17], MiraXT [12], and PMSat [7]. The divide-and-conquer solvers often have a difficulty for selecting splitting variables of the search space,

C. Sinz and U. Egly (Eds.): SAT 2014, LNCS 8561, pp. 188–196, 2014.

and that causes a difficult load-balancing. In recent years, the portfolio approach [9] [11] has become the mainstream of parallel SAT solvers. In the portfolio, workers (solvers) conduct a competitive and cooperative search without the splitting of the search space between other workers, and the answer (SAT or UNSAT) of the fastest solver is adopted. The advantages of the portfolio are that no load-balancing is needed and no synchronisation of the end of the search for unsatisfiable instances (because all the workers conduct the search for the whole search space). In order to fully utilise the advantages of the portfolio, diversification [10] is needed. The diversification is a differentiation of workers' search to avoid overlaps between the workers, and is achieved by differentiating search parameters of the workers, such as decision heuristics, learning schemes, and frequency of restart.

However, it is difficult to combine the search parameters properly in order to avoid overlaps of search spaces between the workers. In addition, there is no guaranty that different parameter sets can prevent two or more solvers from searching for the same spaces. This is because the existing diversification only focuses on differentiating "search activity" of the SAT solver. For this issue, we propose a new diversification technique, *community branching*. Community branching forces each worker in the portfolio to branch on a specific set (or sets) of the variables in early decision levels, and prompt the workers to conduct searching in the different search spaces. This method focuses on differentiating "search space" of the SAT solver as compared to the existing diversification. In order to achieve effective branchings, we used a community detection algorithm to make the sets of variables.

Community detection algorithms [4,3,16] gathered attention in analysing real-world networks, such as Internet [6] and social networks [8]. These algorithms are used in uncovering structures of graphs by gathering vertices into a community, where a high density of edges within the community and low density of the edges between the communities. The recent algorithms are designed to handle over one million vertices within a reasonable computational time, and they are based on modularity [4] optimisation. The modularity value stands for a relative value compared with random partition of the vertices. The higher the modularity value is, the more structured the graph is. As for SAT problem, [1] points out that Variable Incidence Graph (VIG) made from a real-world SAT instance can have a high modularity value. In VIG, the vertices correspond to the Boolean variables and the edges correspond to relations between two variables in a same clause. The variables in the community have close relationships (strongly constrained) to each other. Hence an intensive search for them in parallel can benefit the whole search. A similar attempt was evaluated for MAXSAT problems in [13].

In community branching, a graph is made from the given CNF and learnt clauses, and a community detection algorithm is performed to the graph. Then the detected communities are assigned to each worker. Each worker conducts intensive searches for the given communities, which means that each worker selects the variables in the communities as decision variables in the early phase of the search. We implemented community branching to PeneLoPe [2] with Graph Folding Algorithm [3] as community detection algorithm and conducted experiments.

In Section 2, we detail community branching. We show the experimental results in Section 3 and conclude the paper in Section 4.

2   Community Branching

In this section, we detail the procedure of community branching.

First, we create a graph in the VIG format from the given CNF. In the VIG, the vertices correspond to the Boolean variables and the edges correspond to the relations between the variables in the same clause. A clause "c" generates $(|c| \times |c-1|/2)$ edges between every pair of the variables. The weight of each edge is $(1/(|c| \times |c-1|/2))$, therefore, the sum of the weights of the edges added by each clause is always 1. There are various modularity-based community detection algorithms such as fast greedy algorithm [4], label propagation algorithm [15], and Graph Folding Algorithm (GFA) [3]. Note that a vertex is not included in two or more communities in these algorithms. As stated in [1], the SAT instances from the real-world applications can show high modularity values. In this paper, we used the GFA because it exhibited a good performance for execution time in our preliminary experiments. After the community detection, the communities are assigned to the workers. Each worker conducts intensive searches for the variables in the given community. In particular, the VSIDS scores of the variables are increased periodically. In this manner, each worker searches in the different search spaces and the diversification of search space can be achieved.

Figure 1 shows the pseudo code of community branching. The function "assign_communities" conducts the community detection for the VIG made from the given CNF and the learnt clauses. The reason why the learnt clauses are included is that this function can be called multiple times during the search. We should reconstruct the communities along with the transformation of the graph caused by the learnt clauses, which is also mentioned in [1]. The function "convert_CNF_to_VIG" converts the CNF including the learnt clauses into the VIG and "community_detection" returns detected communities. The variable "coms" is a set of the communities (a two-dimensional array). Then the communities in "coms" are sorted by each size in descending order. We should conduct the search for larger communities preferentially because they can be core parts of the given instance, and these cores should be distributed to each worker. Finally, each community is assigned to each worker (the variable "worker_num" stands for the number of the workers). This procedure is conducted only by a master thread (thread ID 0), and the learnt clauses only in the master thread are used. Note that the number of the communities depends on the community detection algorithm. If the number of the communities is greater than the number of the workers, two or more communities are assigned to one worker. In the reverse case, some workers have no communities (for example in the case of 8 workers and 6 detected communities, 2 workers have no communities and they conduct the search without community branching).

```
assign_communities() {
  coms = community_detection(convert_CNF_to_VIG(given_CNF + learnt_clauses));
  sort_by_size_in_descending_order(coms);
  for(i = 0; i < coms.size(); i++) {
    worker_id = i % worker_num;
    assign(coms[i], worker_id);
  }
}

run_count = 0;
community_branching() {
  if (run_count++ % INTERVAL > 0)
    return;
  [choose one of the assigned communities in rotation]
  for each var in the chosen community
    bump_VSIDS_score(var, BUMP_RATIO);
}
```

Fig. 1. Pseudo code of community branching

After the assignment, each worker calls the function "community_branching" in Figure 1 for every restart. This function chooses a community from the assigned communities and increases the VSIDS scores of the variables in the chosen community. The variable "run_count" counts the number of executions of this function, and the main part of this function is executed for every "INTERVAL" restarts. We need this constant number because we should change the speed of the switching of the branching communities appropriately. In the main part of this function, the community is chosen from the assigned communities in rotation and the VSIDS scores of the variables in the chosen community are increased by proportional to "BUMP_RATIO". In general, "BUMP_RATIO" is set to 1 at clause learning (the VSIDS scores are bumped at clause learning). In order to force the variables in the community to be selected as decision variables right after the restart, "BUMP_RATIO" should be a large value, such as 100. In the following section, we set "INTERVAL" and "BUMP_RATIO" as certain values by considering preliminary experiments. In addition, we set an interval for reconstruction of the communities, an interval for the function "assign_communities". We call it "Community Reconstruction Interval (CRI)" in this paper. In particular, the communities are reconstructed for every "CRI" restarts.

3   Experimental Results

In order to confirm the effectiveness of our proposal, we conducted experiments by using a state-of-the-art parallel portfolio SAT solver PeneLoPe [2]. We utilised the default settings of PeneLoPe, but we turned on the deterministic mode for reproducibility. We used 300 instances from the application category of the SAT Competition 2011. The experiments were conducted on a Linux machine with two Intel Xeon six-core CPUs, running at 3.33 GHz and 144 GB of RAM. Timeout was set to 5000 seconds (wall-clock time) for each instance. The number of running threads was set to eight.

From our preliminary experiments, we found that good performances could be achieved when "INTERVAL" = 1 and "BUMP_RATIO" = 100 for Pene-LoPe. On the basis of this result, we compared three types of parameter settings: no community branching (no_cb), community branching with CRI = 500, INTERVAL = 1, and BUMP_RATIO = 100 (cri500_int1_bump100), community branching with CRI = 3000, INTERVAL = 1, and BUMP_RATIO = 100 (cri3000_int1_bump100).

Figure 2 and Figure 3 show the cactus plots of the results. Figure 2 shows 105 satisfiable instances and Figure 3 shows 126 unsatisfiable instances. Instances that were not solved by the three solvers were not included. Table 1 details the results. From these results, we can see that community branching improved the performance of PeneLoPe, especially for the unsatisfiable instances. Eventually, community branching with CRI = 3000, INTERVAL = 1, and BUMP_RATIO = 100 could solve eleven more instances than the base solver within a shorter runtime. However, we could not find suitable CRIs in this experiment. Although longer CRI (cri3000) exhibited a better performance, we should reconstruct the communities appropriately, as stated in [1].

Figure 4 and Figure 5 show the scatter plots of the relation between the number of the detected communities (made only from the given CNF) and speedup (seconds) achieved by community branching (the runtime difference no_cb - cri3000_int1_bump1, instances that were not solved within the time limit are calculated as 5000) for each instance. Figure 4 shows 105 satisfiable instances and Figure 4 shows 126 unsatisfiable instances. There were only three satisfiable instances and four unsatisfiable instances where the number of the detected communities was less than the number of the workers (eight). From these results, we can find that instances in which the number of the detected communities was less than around 100 could make impacts on the performances, even though a few instances were exceptional in the satisfiable instances. For instances where the number of the detected communities is large compared to the number of the workers (e.g. 100 communities and 8 workers), each worker is forced to conduct the search for various communities. It may cause an excessively diversified search, and the performance can be deteriorated. As future work, we are going to merge the detected communities, for example adjacent communities, into one community, with consideration for the number of the workers. We are also planning to change the parameters of community branching dynamically according to the number of the detected communities and their size.

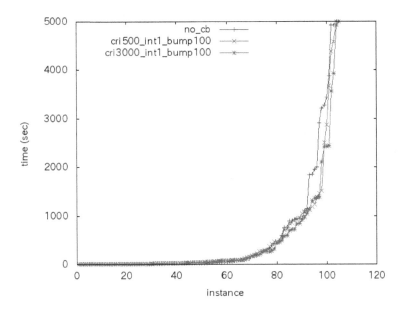

Fig. 2. The experimental results of 105 satisfiable instances

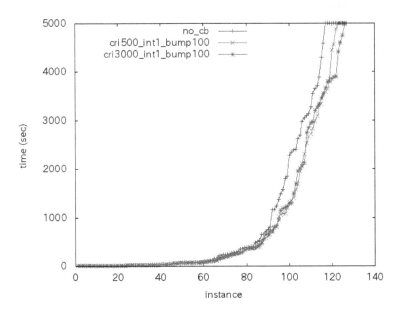

Fig. 3. The experimental results of 126 unsatisfiable instances

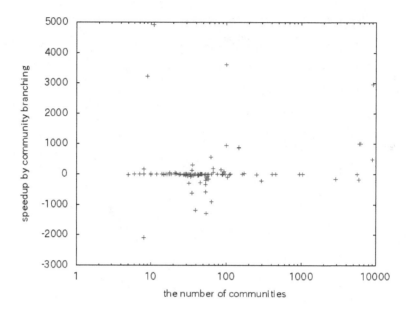

Fig. 4. The relation between the number of communities and speedup (seconds) by cri3000_int1_bump1 for 105 satisfiable instances

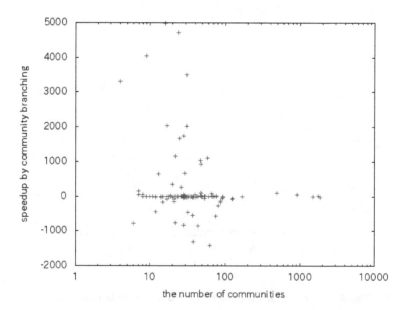

Fig. 5. The relation between the number of communities and speedup (seconds) by cri3000_int1_bump1 for 126 unsatisfiable instances

Table 1. The details of the results: 231 (SAT: 105, UNSAT: 126) instances could be solved at least one solver. The instances that could not be solved within 5000 seconds are calculated as 5000.

	SAT (105)	UNSAT (126)	total	total time for 231 instances
no_cb	103	116	219	385220
cri500_int1_bump1	103	122	225	313450
cri3000_int1_bump1	104	126	230	310090

4   Conclusion

In this paper, we proposed community branching for making more diversification in the parallel portfolio SAT solvers. Compared with the existing diversification technique that focuses on differentiating the search activity, community branching can achieve diversification of the search spaces. Experimental results indicate that out proposal can improve the performance of the base solver. In addition, we expect that community branching is effective for higher degree of parallelisation, more than eight workers. As future work, we have to tune the parameters of community branching. We are also planning to change them dynamically according to the given instance.

Acknowledgment. We appreciate the insightful comments from the reviewers.

References

1. Ansótegui, C., Giráldez-Cru, J., Levy, J.: The Community Structure of SAT Formulas. In: Cimatti, A., Sebastiani, R. (eds.) SAT 2012. LNCS, vol. 7317, pp. 410–423. Springer, Heidelberg (2012)
2. Balint, A., Belov, A., Heule, M.J.H., Jrvisalo, M.: Proceedings of SAT Competition 2013; Solver and Benchmark Descriptions (2013)
3. Blondel, V.D., Guillaume, J., Lambiotte, R., Lefebvre, E.: Fast unfolding of communities in large networks. Journal of Statistical Mechanics: Theory and Experiment 2008(10), 10008 (2008)
4. Clauset, A., Newman, M.E.J., Moore, C.: Finding community structure in very large networks. Physical Review E 70(6), 066111 (2004)
5. Davis, M., Logemann, G., Loveland, D.: A machine program for theorem-proving. Communications of the ACM 5(7), 394–397 (1962)
6. Eriksen, K.A., Simonsen, I., Maslov, S., Sneppen, K.: Modularity and Extreme Edges of the Internet. Physical Review 90, 148701 (2003)
7. Gil, L., Flores, P.F., Silveira, L.M.: PMSat: A parallel version of MiniSAT. JSAT 6(1-3), 71–98 (2009)
8. Girvan, M., Newman, M.E.J.: Community Structure in Social and Biological Networks. Proceedings of the National Academy of Sciences 99(12), 7821–7826 (2002)
9. Gomes, C.P., Selman, B.: Algorithm Portfolio Design: Theory vs. Practice. In: Proceedings of the 13th Conference on Uncertainty in Artificial Intelligence, UAI 1997, pp. 190–197 (1997)

10. Guo, L., Hamadi, Y., Jabbour, S., Sais, L.: Diversification and Intensification in Parallel SAT Solving. In: Proceedings of the 16th International Conference on Principles and Practice of Constraint Programming, CP 2010, pp. 252–265 (2010)
11. Hamadi, Y., Jabbour, S., Sais, L.: ManySAT: A Parallel SAT Solver. JSAT 6(4), 245–262 (2009)
12. Lewis, M., Schubert, T., Becker, B.: Multithreaded SAT Solving. In: Proceedings of the 2007 Asia and South Pacific Design Automation Conference, ASP-DAC 2007, pp. 926–931 (2007)
13. Martins, R., Manquinho, V., Lynce, I.: Community-Based Partitioning for MaxSAT Solving. In: Järvisalo, M., Van Gelder, A. (eds.) SAT 2013. LNCS, vol. 7962, pp. 182–191. Springer, Heidelberg (2013)
14. Moskewicz, M.W., Madigan, C.F., Zhao, Y., Zhang, L., Malik, S.: Chaff: Engineering an Efficient SAT Solver. In: Proceedings of the 38th Annual Design Automation Conference, DAC 2001, pp. 530–535 (2001)
15. Raghavan, U.N., Albert, R., Kumara, S.: Near linear time algorithm to detect community structures in large-scale networks. Physical Review E 76(3), 036106 (2007)
16. Rosvall, M., Bergstrom, C.T.: Maps of random walks on complex networks reveal community structure. Proceedings of the National Academy of Sciences 105(4), 1118–1123 (2008)
17. Zhang, H., Bonacina, M.P., Hsiang, J.: PSATO: A Distributed Propositional Prover and its Application to Quasigroup Problems. J. Symb. Comput. 21, 543–560 (1996)

Lazy Clause Exchange Policy for Parallel SAT Solvers

Gilles Audemard[1,*] and Laurent Simon[2,**]

[1] Univ. Lille-Nord de France, CRIL/CNRS UMR 8188, Lens.
[2] Univ. Bordeaux, LABRI, Bordeaux

Abstract. Managing learnt clauses among a parallel, memory shared, SAT solver is a crucial but difficult task. Based on some statistical experiments made on learnt clauses, we propose a simple parallel version of Glucose that uses a lazy policy to exchange clauses between cores. This policy does not send a clause when it is learnt, but later, when it has a chance to be useful locally. We also propose a strategy for clauses importation that put them in "probation" before a potential entry in the search, thus limiting the negative impact of high importation rates, both in terms of noise and decreasing propagation speed.

1 Introduction

The success story of SAT solving is one of the most impressive in recent computer science history. The theoretical and practical progresses observed in the area had a direct impact in a number of connected areas. SAT solvers are nowadays used in many critical applications (BMC [5], Bio-informatics [17] ...) by direct encodings of problems to propositional logic (often leading to huge formulas), or by using SAT solvers on an abstraction level only.

However, if until now measured progresses are quite impressive, the recent trends in computer architecture are forcing the community to study new efficient frameworks, by considering the native parallel (and sometimes massively parallel) architecture of current and upcoming computers. CPU speed is stalling, but the number of cores is increasing. Computers with one shared memory and a large number of cores are the norm today. A few specialized cards even allow more than two hundreds threads on the same board. Thus, designing efficient and scalable parallel SAT solvers is now a crucial challenge for the community [11]. Existing approaches can be roughly partitioned in two. Firstly, the "portfolio" approach tries to launch in parallel a set of solvers on the same formula. This can be trivially done by running the best known solvers without any communications [19] or, more interestingly, with communication between threads. This communication is generally limited to learnt clauses sharing [10,1,6]. Secondly, the divide and conquer approach tries first to reduce the whole formula in smaller ones and then solve them [2,13,12] with or without any communications. Note that some attempts have been made on combination of portfolio and divide and conquer approaches [7].

In our approach, we would like to consider CDCL solvers as clauses producer engines and thus, in this case, the current divide and conquer paradigm may not be ideal

---

* This work has been supported by CNRS and OSEO, under the ISI project "Pajero".
** This work was possible thanks to the BIRS meeting 14w5101.

C. Sinz and U. Egly (Eds.): SAT 2014, LNCS 8561, pp. 197–205, 2014.

because the division is made on the variable search space, not on the proof space. When designing a clause sharing parallel CDCL solver, an important question arises: which clauses to export and import? The more is not the best. Importing too many clauses will completely paralyze the considered thread. If we have N threads sending its clause with probability p then, after C conflicts, each thread will have on average C learnt clauses and $p \times (N - 1) \times C$ imported clauses. Thus, keeping p as low as possible is critical. Too many imported clauses by too many distinct threads will destroy the effort of the current thread to focus on a subproblem. This problem is not new and was already mentioned and partially answered by one of the first parallel SAT solver: MANYSAT [10,8].

Let us summarize the original contributions of our approach. First, we try to carefully identify clauses that have a chance to be useful locally (even locally, just a part of produced clauses are in fact really useful). These clauses will be detected "lazily" before exporting them. Secondly, we don't directly import clauses. We put them in "probation" before adding them to the clauses database, thus limiting their impact on the current search. In the following section, we review the different approaches proposed for parallelizing Modern SAT solvers. Then, in section 3, we detail the principal "Lazy" clause exchange policy proposed in GLUCOSE-SYRUP, our parallel version of GLUCOSE. Then, we propose some experiments (section 5) and conclude.

2   Preliminaries and Previous Works

We assume the reader familiar with the essentials of propositional logic, SAT solving and CDCL solvers. These solvers are branching on literals and, at any step of the search, ensure that all the unit clauses w.r.t the current partial assignment are correctly "propagated" until an empty clause is found (a backtrack is then fired, or the unsatisfiability is proven) or a total assignment is reached (the formula is SAT). Each time a conflict occurs, a clause (called *asserting clause*) is learnt and is used to force backtracking, leading to new propagations. Of course, many additional ingredients are essential but reviewing all of them will clearly be beyond the scope of this paper.

The important point to be emphasized here, even for the non specialist, is that solvers are learning a lot of clauses (more than 5000 per second), partially guided in its search by previous learnt clauses forcing new unit propagations. The management of the clauses database was firstly pointed out as an essential ingredient with the design of GLUCOSE [4]. Indeed, keeping too many learnt clauses will slow down the unit propagation process, while deleting too many of them will break the overall learning benefit. Consequently, identifying good learnt clauses – relevant to the (future) proof derivation – is clearly an important challenge. The first proposed quality measure followed the success of the activity-based "VSIDS" heuristic [18]. More precisely, a learnt clause is considered relevant (in the future) to the proof, if it is involved more often in recent conflicts, *i.e.* used to derive asserting clauses by resolution. This deletion strategy supposes that a useful clause in the past would be useful in the future. In [4], the authors proposed a more accurate measure called LBD (*Literal Block Distance*) to estimate the quality of a learnt clause. This measure is based on the number of distinct decision levels occurring in a learnt clause and computed when the clause is learnt. Intensive experiments demonstrated that clauses with small LBD values are used more often than those of higher LBD ones. This measure is important here because it seems to offer a good prediction for the quality of a clause, and it seems to be a good starting point if we may want to build a good clause exchange policy between threads. However, as we will see, the LBD measure seems essentially relevant to the current search only. Even if it has been proven to be more relevant than size (see below the PLINGELING strategy), a small LBD clause may not be of great interest for another thread.

2.1 About Clause-Sharing Parallel Approaches

A number of previous works have been proposed around Portfolio approaches, with more or less cooperation/diversification between threads. The main idea is to exploit the complementarity between different sequential CDCL strategies to let them compete on the same formula with more or less cooperation between them [10,6,16,1]. Each thread deals with the whole formula and cooperation is achieved through the exchange of learnt clauses. Non-Portfolio approaches are mostly based on the divide-and-conquer paradigm [2,13]. We here focus on Portfolio parallel approaches in this section.

As mentioned in the introduction, the size of the learnt clause database is crucial for sequential solvers. This is not only true for maintaining a good unit propagation speed, but this seems also essential to guide the solver to the best possible proof it can build. These observations are even more crucial when many threads are cooperating. For a parallel portfolio SAT solver, it is not desirable to share as many clauses as possible and, obviously, each of the clause-sharing portfolio approaches mentioned above had to develop its own strategy to carefully select the clauses to share. A first and quite natural solution to limit the number of exported clauses is to simply share the smallest clauses according to their size. This was the strategy adopted in [10]. Based on the observation that small clauses appear less and less during the search, authors of MANYSAT proposed a very nice dynamic clause sharing policy using pairwise size limits to control the exchange between threads [9]. In the same paper, they also anecdotally proposed another dynamic policy based on the activity of variables according to the VSIDS heuristic. However, such weighting function is highly fickle and, unfortunately, the top-ranked variable (according to VSIDS) may not be of any interest only 0.1s later (according to the same VSIDS). It is thus hopeless to try to directly rely on this highly dynamic strategy to measure the quality of imported clauses. In Penelope [1], authors use the freezing strategy to manage learnt clauses [3] allowing to share much more clauses without a high overhead. Finally, PLINGELING shares all clauses with a size less than 40 and LBD less than 8 [6].

2.2 Portfolio or Not Portfolio

We chose to focus, in this paper, on a Clause-Sharing Parallel Approach of GLUCOSE engines. This approach is often misleadingly called "Portfolio" because, in general, each engine must have its own configuration to ensure orthogonal searches between threads. However, considering that GLUCOSE is seen as an efficient proof producer, this idea of "orthogonal" search on variables assignments is not clear. We would like to keep the word "portfolio" only to approaches that tries to take advantage of running many distinct solvers, each of them specialized on a subset of problems, and thus focusing on the competition between each configuration strengths. In our approach, we will restrict the "orthogonal" search to its minimum. Thus, we rather see our approach as a simple parallelization effort of the same engine rather than a "portfolio" one. Our final goal is to see all solvers working together to produce a single proof, as short as possible.

3 Lazy Clause Exchange

This section describes the strategy we propose for parallelizing GLUCOSE. This strategy is only based on how to identify "good" clauses to export and to import. The parallel solver is called "GLUCOSE-SYRUP" (i.e. *a lot of glucose*).

3.1 Identifying Useless Clauses?

The identification of useless clauses is still an open question for sequential solvers. We thus do not pretend to fully answer it in this section. However, let us take some time in this section to report a set of experiments on sequential solvers that motivated the strategies proposed in GLUCOSE-SYRUP.

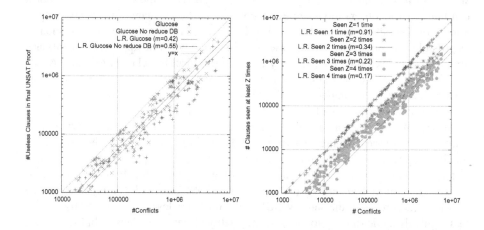

Fig. 1. (Left) Useless clauses in the final UNSAT proof w.r.t the total number of generated clauses. Glucose no reduce DB is a hacked version of GLUCOSE that do not perform any learnt clause removal. Experiments are done on a set of 250 UNSAT problems from competitions 2011 and 2013. Only successful run are collected. "L.R." stands for "Linear Regression" with $y = m \times x + n$. **(Right)** Scatter plot of the number of conflicts (X axis) against the number of clauses seen at least Z={1, 2, 3, 4} times for all the successful launches (on SAT 2011, satelited, problems).

Viewing CDCL solvers as clauses producers is not the mainstream approach. However, as it was reported in [15], this may be one of the key points for an efficient parallelization. This work suggested the following very simple experiment. When an instance is UNSAT (proven by GLUCOSE), we identify clauses that occur in the proof, and measure how many of them are *useless* for the proof. The term "useless" is now clearly defined: it does not occur in the final proof for UNSAT. However, it should be noticed here that this definition can be misleading. A "useless" clause may be crucial at some stage of the search to update the heuristic or to propagate a literal earlier in the search tree. Thus, this notion should be used with caution.

This being said, Figure 1-Left shows a surprisingly large number of useless clauses. On the original GLUCOSE solver, 45% of the learnt clauses are, on average, not useful. This result is not the only surprise here. When GLUCOSE keeps all its clauses (called "Glucose No Reduce DB" on the figure), this number is even more important (55%!). This mean that keeping all the clauses, even if the final number of conflicts is smaller than the original GLUCOSE, leads to a larger proportion of useless clauses. We could have expected the opposite to happen: the aggressive clause database reduction in GLUCOSE throws away many clauses. Those clauses will not have a chance to occur in the proof afterwards. As a very short conclusion on this figure, we see that even for a single engine, considering the usefulness of a clause is not an easy task. Moreover, around half of the generated clauses are not useful. Sending them to other threads may not be the right move to do.

3.2   Lazy Exportation of Interesting Clauses

Before presenting the export strategy in GLUCOSE-SYRUP, let us focus now on Figure 1-Right. This figure shows how many clauses are seen at least $Z=\{1,2,3,4\}$ times during all conflicts analysis of GLUCOSE (not propagations). The figure already shows that 91% of the clauses, on all the problems, are seen at least once. However, when $Z=2$, this ratio drops to 34%, then 22% and 17% for $Z=4$. This suggests to export only clauses seen at least Z times, and to fix $Z>1$. We propose to simply fix $Z=2$ to already get rid of 66% (100-34%) of the locally generated clauses, and to send only clauses seen two times. The strategy is called "*lazy*" because we do not try to guess in advance the usefulness of a clause. Instead, we simply (and somehow "lazily") wait for the clause to be seen twice in the conflict analysis.

To refine the above observations, we conducted two more experiments. Firstly, we supposed that, in many cases, a learnt clause had a high probability to be seen in the very next conflict analysis, because the clause is immediately propagated and is clearly at one of the deepest levels of the search tree. However, as clearly shown Figure 2-Left, only 41% of the learnt clauses are immediately used for conflict analysis (remember that 91% of the clauses are seen at least once). This means that we really need a lazy strategy to identify interesting clauses: we cannot suppose any locality (in the number of conflicts) for identifying when a clause will be used for the second time.

As we already pointed out in the second section, PLINGELING is using a fixed strategy to filter the exported clauses. It restricts the exportations to clauses of LBD smaller than 8 and size smaller than 40. We use a more flexible limit in our approach. We automatically adapt the LBD and size thresholds according to the characteristics of the current clauses in the learnt clause database. Each time a clause database reduction is fired, the current median LBD value and the average size of learnt clauses are updated.

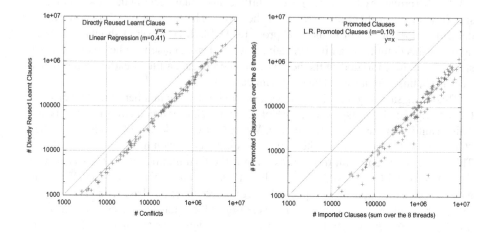

Fig. 2. Left Scatter plot of the number of conflicts (X axis) against the number of clauses directly reused in the next conflict analysis (on a single engine GLUCOSE). **Right** Scatter plot of the number of imported clauses (1-Watched clauses) against promoted clauses (clauses found empty, and pushed to the 2-Watched literal scheme) on successful run of GLUCOSE-SYRUP with 8 threads on the SAT'11 competition benchmarks (after SatElite).

In most of the cases, we observed that the dynamic values are more relaxed than the fixed values of PLINGELING.

Of course, unit clauses and binary clauses are exported without any restriction, as soon as they are learnt. We extended this strategy to all glue clauses (clauses of LBD=2). Thus the above strategy is for clauses of at least size 3 that are not glues.

3.3 Lazy Importation of Clauses

We have seen how carefully choosing which clauses to send can be crucial for limiting the communication overhead. The extra work for importing clauses can also be limited if we can consider the following four points. (1) Importing clauses after each conflict may have an important impact (and a negative one) on all solvers strategies (need to check the trail, and to backjump when necessary, breaking the solver locality, ...) ; (2) Importing clauses may have an important impact on the cleaning strategies of GLUCOSE (the learnt clause database will increase much more faster) ; (3) The more is not the best. A *"bad"* clause may force the solver to propagate a literal in the wrong direction, i.e. it will not be able to properly explore the current subproblem ; (4) There is no simple way of computing the LBD of an incoming clause. This measure cannot be used because LBD is relative to the current search of each solver, and a good clause for one thread may not be good for another one. So, before adding a clause to the solver database, one has to carefully ensure that the clause is useful for the current search.

For the first point (1), we decided to import clauses (unary, binary and others) only when the solver is at decision level 0, *i.e.* right after a restart or right after it learnt a unit clause. This event occurs a lot during the search (many times per second) and this

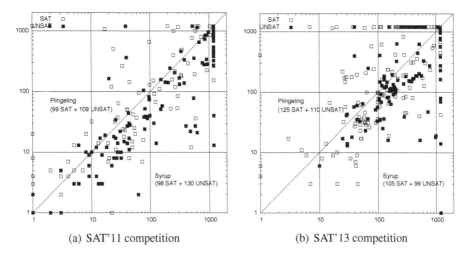

(a) SAT'11 competition   (b) SAT'13 competition

Fig. 3. GLUCOSE-SYRUP VS PLINGELING on SAT competitions benchmarks, application track. Each dot represents an instance. A dot below the diagonal indicates an instance solved faster with GLUCOSE-SYRUP. On SAT'11 Problems (a): PLINGELING: 208 / GLUCOSE-SYRUP: 228. On SAT'13 Problems(b): PLINGELING: 235 / GLUCOSE-SYRUP: 204.

rate is clearly sufficient for a good collaboration. For the second point, (2) we import clauses in another set of clauses, that can have its own cleaning rules and will not pollute GLUCOSE cleaning strategies on its learnt clauses.

The main "Lazy" solution we propose is a solution for the two last items, (3) and (4). It is somehow related to the PSM strategy [3]: when a clause is imported, we watch it only by one literal. This watching scheme does not guarantee anymore that all unit propagations are performed after each decision. However, this is sufficient to ensure that any conflicting clause will be detected during unit propagation. The interest of this technique is twofold. Firstly, the clause will not pollute the current search of the solver, except when it is falsified. Secondly, the cost for handling the set of imported clause is heavily reduced. This technique can be viewed as a more reactive PSM strategy.

Figure 2-Right shows that only 10% of the imported clauses are falsified at some point. It demonstrates how well founded is our strategy for the importation: 90% of the clauses are never falsified by the solver strategy. Technically speaking, as soon as an imported clause is falsified, it is "promoted", i.e. we watch it with 2 literals like an internal learnt clause. The clause is then part of the solver search strategy, and can be used for propagation.

4 Experiments

In this section, we compare PLINGELING [6], the winner of the SAT'13 competition with our first version of GLUCOSE-SYRUP. Let us notice here that this version is still preliminary in the sense that absolutely no tuning has been conducted on the set of parameters we chose for the 8 threads. All threads are identical except their parameter playing on the VSIDS scoring scheme. One thread is however configured like GLU-COSE 2.0 version.

Figure 3 compares PLINGELING [6], the winner of the SAT'13 competition against GLUCOSE-SYRUP using the classical scatter plots. We use two test sets of problems, because each solver has its own strengths and weaknesses. Let us start with the SAT'11 competition, application track, problems. For this test set, GLUCOSE-SYRUP is able to solve 228 instances (98 SAT and 130 UNSAT) whereas PLINGELING is able to solve 208 instances (99 SAT and 109 UNSAT). Note that the sequential version of GLUCOSE only solves 187 instances (87 SAT and 100 UNSAT). GLUCOSE-SYRUP clearly extends the efficiency of GLUCOSE on UNSAT problems to the parallel case. Moreover, many points are below the diagonal indicating that, in many cases, our solver is faster (even for SAT instances) than PLINGELING. This result may be partially explained by all the inprocessing [14] techniques embedded in PLINGELING, that may not be efficient enough with the time we fixed. More importantly, we think the big gap observed for UNSAT instances can come from the ability of GLUCOSE-SYRUP to share and exploit promising clauses. It is also fair to notice that GLUCOSE, the underlying sequential engine is quite good in solving those UNSAT instances. Let us continue with instances coming from SAT'13 competition. Here, the picture is totally inverted. Results are clearly in favor of PLINGELING. It solves 235 instances (125 SAT and 110 UNSAT) whereas GLUCOSE-SYRUP *is only able* (GLUCOSE the underlying solver can only solve 173 instances (93 SAT and 80 UNSAT)) to solve 204 (105 SAT and 99 UN-SAT). This is a big difference. However, if we study this result more deeply, we can observe that, here again, in many cases GLUCOSE-SYRUP is faster than PLINGELING. The differences between the two approaches arise after 500 seconds: PLINGELING is able to solve difficult and/or particular problems, by exploiting *inprocessing* techniques (xor-reasoning, equivalence checking, ...). The SAT'13 competition is indeed the first competition to contain so many problems with xor chains and counters. On this set of problem, *inprocessings* techniques are mandatory. However, for genuine parallel CDCL solvers, we showed that GLUCOSE-SYRUP is clearly a new parallel approach offering very good performances.

5   Conclusion

Clauses sharing among threads in a parallel SAT solver remains a difficult task. By experimentally studying learnt clauses usefulness, we propose a lazy policy for clauses exports. Furthermore, we propose a new scheme for clauses importations by putting them in probation before adding them to the clauses database. Experiments based on our solver GLUCOSE shows that this method for parallelization allows a very good scaling up of the underlying sequential engines.

References

1. Audemard, G., Hoessen, B., Jabbour, S., Lagniez, J.-M., Piette, C.: Revisiting clause exchange in parallel SAT solving. In: Cimatti, A., Sebastiani, R. (eds.) SAT 2012. LNCS, vol. 7317, pp. 200–213. Springer, Heidelberg (2012)

2. Audemard, G., Hoessen, B., Jabbour, S., Piette, C.: An effective distributed D&C approach for the satisfiability problem. In: 22nd Euromicro International Conference on Parallel, Distributed and Network-Based Processing (PDP 2014) (February 2014)
3. Audemard, G., Lagniez, J.-M., Mazure, B., Saïs, L.: On freezing and reactivating learnt clauses. In: Sakallah, K.A., Simon, L. (eds.) SAT 2011. LNCS, vol. 6695, pp. 188–200. Springer, Heidelberg (2011)
4. Audemard, G., Simon, L.: Predicting learnt clauses quality in modern SAT solvers. In: Proceedings of IJCAI, pp. 399–404 (2009)
5. Biere, A., Cimatti, A., Clarke, E., Zhu, Y.: Symbolic Model Checking without BDDs. In: Cleaveland, W.R. (ed.) TACAS 1999. LNCS, vol. 1579, pp. 193–207. Springer, Heidelberg (1999)
6. Biere, A.: Lingeling, plingeling and treengeling entering the sat competition 2013. In: Proceedings of SAT Competition 2013; Solver and Benchmark Descriptions, p. 51 (2013), http://fmv.jku.at/lingeling
7. Gebser, M., Kaufmann, B., Schaub, T.: Multi-threaded asp solving with clasp. TPLP 12(4-5), 525–545 (2012)
8. Guo, L., Hamadi, Y., Jabbour, S., Sais, L.: Diversification and intensification in parallel SAT solving. In: Cohen, D. (ed.) CP 2010. LNCS, vol. 6308, pp. 252–265. Springer, Heidelberg (2010)
9. Hamadi, Y., Jabbour, S., Sais, L.: Control-based clause sharing in parallel SAT solving. In: Proceedings of IJCAI, pp. 499–504 (2009)
10. Hamadi, Y., Jabbour, S., Saïs, L.: Manysat: A parallel SAT solver. Journal on Satisfiability, Boolean Modeling and Computation 6, 245–262 (2009)
11. Hamadi, Y., Wintersteiger, C.M.: Seven challenges in parallel SAT solving. AI Magazine 34(2), 99–106 (2013)
12. Heule, M.J.H., Kullmann, O., Wieringa, S., Biere, A.: Cube and conquer: Guiding CDCL SAT solvers by lookaheads. In: Eder, K., Lourenço, J., Shehory, O. (eds.) HVC 2011. LNCS, vol. 7261, pp. 50–65. Springer, Heidelberg (2012)
13. Hyvärinen, A.E.J., Junttila, T., Niemelä, I.: Partitioning SAT instances for distributed solving. In: Fermüller, C.G., Voronkov, A. (eds.) LPAR-17. LNCS, vol. 6397, pp. 372–386. Springer, Heidelberg (2010)
14. Järvisalo, M., Heule, M.J.H., Biere, A.: Inprocessing rules. In: Gramlich, B., Miller, D., Sattler, U. (eds.) IJCAR 2012. LNCS (LNAI), vol. 7364, pp. 355–370. Springer, Heidelberg (2012)
15. Katsirelos, G., Sabharwal, A., Samulowitz, H., Simon, L.: Resolution and parallelizability: Barriers to the efficient parallelization of SAT solvers. In: AAAI 2013 (2013)
16. Kottler, S., Kaufmann, M.: SArTagnan - A parallel portfolio SAT solver with lockless physical clause sharing. In: Pragmatics of SAT (2011)
17. Lynce, I., Marques-Silva, J.: SAT in bioinformatics: Making the case with haplotype inference. In: Biere, A., Gomes, C.P. (eds.) SAT 2006. LNCS, vol. 4121, pp. 136–141. Springer, Heidelberg (2006)
18. Moskewicz, M., Madigan, C., Zhao, Y., Zhang, L., Malik, S.: Chaff: Engineering an efficient SAT solver. In: Proceedings of DAC, pp. 530–535 (2001)
19. Roussel, O.: ppfolio, http://www.cril.univ-artois.fr/~roussel/ppfolio

Ultimately Incremental SAT

Alexander Nadel[1], Vadim Ryvchin[1,2], and Ofer Strichman[2]

[1] Design Technology Solutions Group, Intel Corporation, Haifa, Israel
[2] Information Systems Engineering, IE, Technion, Haifa, Israel
alexander.nadel@intel.com, rvadim@tx.technion.ac.il,
ofers@ie.technion.ac.il

Abstract. Incremental SAT solving under assumptions, introduced in Minisat, is in wide use. However, Minisat's algorithm for incremental SAT solving under assumptions has two main drawbacks which hinder performance considerably. First, it is not compliant with the highly effective and commonly used preprocessor SatELite. Second, all the assumptions are left in the formula, rather than being represented as unit clauses, propagated, and eliminated. Two previous attempts to overcome these problems solve either the first or the second of them, but not both. This paper remedies this situation by proposing a comprehensive solution for incremental SAT solving under assumptions, where SatELite is applied and all the assumptions are propagated. Our algorithm outperforms existing approaches over publicly available instances generated by a prominent industrial application in hardware validation.

1   Introduction

Modern backtrack search-based SAT solvers are indispensable in a broad variety of applications [2]. In the classical SAT interface, the solver receives one formula in Conjunctive Normal Form (CNF) and is required to decide whether it is satisfiable or unsatisfiable. However, many practical applications [18–20, 10, 4, 11, 8] require solving a sequence of related SAT formulas.

Rather than solving related formulas separately, modern solvers attempt to solve them *incrementally*, that is, to propagate information gathered during the solving process to future instances in the sequence. Initially there was *Clause Sharing (CS)* [19, 20], in which relevant learned clauses are propagated between instances. Then came Minisat [9], whose widely-used incremental interface induces the problem of *incremental SAT solving under assumptions*. In incremental SAT solving under assumptions the user may invoke the solving function multiple times, each time with a different set of *assumption literals* and, possibly, additional clauses. The solver then checks the satisfiability of all the clauses provided so far, while enforcing the values of the current assumptions only. This paper aims to improve the algorithm to solve the problem of incremental SAT solving under assumptions.

In Minisat's approach to solving the problem, the same SAT solver instance solves the entire sequence internally, and hence the state of the solver is

C. Sinz and U. Egly (Eds.): SAT 2014, LNCS 8561, pp. 206–218, 2014.
© Springer International Publishing Switzerland 2014

preserved between incremental steps, which means that it shares not only the relevant learned clauses, but also the scores that guide decision, restart, and clause-deletion heuristics. The assumptions are modeled as first decision variables; all inferred clauses that depend on some of the assumptions include their negation, which means that they are *pervasive*, i.e., they are implied by the formula regardless of the assumptions and can therefore be reused in subsequent steps. We refer to this approach as *Minisat-Alg* .

Independently of advances in incremental SAT solving, a breakthrough in *non-incremental* SAT solving was achieved with the SatELite [7] preprocessor. SatELite applies non-increasing *variable elimination* (existentially quantifying variables by resolution as long as this does not increase the size of the formula), subsumption and self-subsumption until fixed-point, before the SAT solver begins the search. Currently, SatELite is the most dominant and useful preprocessing algorithm [1].

Combining Minisat-Alg with SatELite could clearly make incremental SAT solving faster. Unfortunately, however, these two methods are incompatible. The problem is that one cannot eliminate a variable that may reappear in future steps. In [15] we suggested a solution to this problem. We will briefly explain this solution, which we call *incremental SatELite* , in Section 2.2.

Another problem with combining Minisat-Alg and preprocessing relates to the assumptions. Minisat-Alg uses assumptions as first decision variables and hence cannot enjoy the benefits of modeling them as unit clauses. This is necessary for forcing the inferred clauses to be pervasive. However, modeling them instead as unit clauses typically leads to major simplifications of the formula with BCP and SatELite, and this can have a crucial impact on the size of the formula and the performance of the solver.

In [13] the first two authors suggested a method that we call here *assumption propagation*, which enjoys both worlds, at the price of using multiple SAT instances. It models assumptions as units, but then it transforms *temporary* (i.e., assumption-dependent) clauses to pervasive clauses after each invocation using the so-called *Temporary2Pervasive (T2P)* transformation, i.e., the assumptions upon which each inferred clause depends are added to it, negated. Assumption propagation can be combined with SatELite [13]. Assumption propagation is inspired by CS, which would use multiple SAT instances, encode assumptions as units, and reuse the set of pervasive conflict clauses to solve our problem. CS, however, is neither compliant with SatELite nor does it transform temporary clauses to pervasive.

Our new approach, which we call *Ultimately Incremental SAT (UI-SAT)*, combines the advantages of the approaches we have presented. We summarize the differences between UI-SAT and the other algorithms in Table 1. The columns represent the following:

1. *Algorithm:* the algorithm name and origin.
2. *Instances:* the number of SAT instances used. Recall that a single instance is preferable because of score sharing.

3. *Assumptions as units:* indicates whether assumptions are modeled as unit clauses, which, via SatELite and BCP, simplifies the formula.
4. *SatELite:* indicates the way SatELite is applied. CS and Minisat-Alg cannot apply SatELite at all, while assumption propagation applies full SatELite for each incremental step. Incremental SatELite and UI-SAT carry out a more efficient procedure while enjoying the same effect: they apply SatELite incrementally, which means that they handle only the newly-provided clauses and assumptions for each step.
5. *Assumption-dependent clauses:* the way in which assumption-dependent conflict clauses are treated. CS marks all the conflict clauses that depend on assumptions and discards them before the next incremental step. Potentially useful information is lost this way. Minisat-Alg and incremental SatELite reuse all such clauses across all the incremental steps, where the negation of the assumption literals are part of the clauses. Assumption propagation operates similarly to CS, except that it applies *T2P*. Such an approach allows one to reuse the conflict clauses for the next step instead of discarding them, while still enjoying the benefits of modeling assumptions as unit clauses. Finally, UI-SAT introduces an incremental version of *T2P*: before step i it only adds a literal $\neg l$ if l was an assumption literal in step $i-1$ and is not an assumption at step i.

Table 1. A comparison of different approaches to incremental SAT solving under assumptions

Algorithm	Instances	Assumptions as units	SatELite	Assumption-dep. Clauses
Clause Sharing (CS) [19, 20]	Multiple	Yes	No	Discard
Minisat-Alg [9]	One	No	No	Keep all
Assumption prop.[13]	Multiple	Yes	Full	*T2P*
Incremental SatELite [15]	One	No	Incremental	Keep all
UI-SAT (this paper)	One	Yes	Incremental	Incremental *T2P*

As is clearly evident from the table, our new approach, UI-SAT, is the most comprehensive solution to the problem of incremental SAT solving under assumptions to date. Note that we solve the most general problem. First, we do not make any a priori suppositions about the use of assumptions. The assumption literals can be part of the original problem and/or be selector variables; they can be flipped between instances or discarded, and new assumption literals can be introduced. Second, we do not assume that look-ahead information is available. Indeed, in various applications information about future instances is not available:

- Some applications require interactive communication with the user for determining the next portion of the problem:

- One example is an article from IBM [5] that shows that using an *incremental* SAT-based model checker, based on an incremental SAT solver, is critical for speeding-up regression verication, where a new version of a hardware design is re-verified with respect to the same (or very similar) specication. The changes in the design and the specification are not known a-priory.
- The benchmarks used in this paper spring from another such application in formal verification. Assume that a verification engineer needs to formally verify a set of properties in some circuit up to a certain bound. Formal verification techniques cannot scale to large modern circuits, hence the engineer needs to select a sub-circuit and mimic the environment of the larger circuit by imposing assumptions [12]. The engineer then invokes an *incremental* bounded model checker to verify a property under the assumptions. If the result is satisfiable, then either the environment is not set correctly, that is, assumptions are incorrect or missing, or there is a real bug. In practice the first reason is much more common than the second. To discover which of the possibilities is the correct one, the engineer needs to analyze the counterexample. If the reason for satisfiability lies in incorrect modeling of the environment, the assumptions must be modified and BMC invoked again. When one property has been verified, the engineer can move on to another. Practice shows that most of the validation time is spent in this process, which is known as the *debug loop*.

– In other applications the calculation of the next portion of the problem depends on the results of the previous invocation of the SAT solver. For example, various tasks in MicroCode validation [11] are solved by using a symbolic execution engine to explore the paths of the program. The generated proof obligations are solved by an incremental SAT-based SMT solver. In this application, the next explored path of the program is determined based on the result of the previous computation.

In the experimental results section, we show the superiority of UI-SAT to the other algorithms over the same instances used in [13]; these were generated by Intel's incremental bounded model checker and are publicly available.

We continue in the following section with formalization and various preliminaries, which are mostly similar to those in [15]. In Section 3 we present our algorithm, and in Section 4 we summarize our empirical evaluation. We conclude in Section 5.

2   Preliminaries

Let φ be a CNF formula. We denote by $vars(\varphi)$ the variables used in φ. For a clause c we write $c \in \varphi$ to denote that c is a clause in φ. For $v \in vars(\varphi)$ we define $\varphi_v = \{c \mid c \in \varphi \wedge v \in c\}$ and $\varphi_{\bar{v}} = \{c \mid c \in \varphi \wedge \bar{v} \in c\}$ (a somewhat abusing notation, as we refer here to v as both a variable and a literal). When a literal is given as an assumption and in a future step it stops being an assumption, we say that it is *invalidated*.

At step 0 of incremental SAT solving under assumptions the SAT solver is given a CNF formula φ^0 and a set of assumption literals A_0. It has to solve φ^0 under A_0, i.e., to determine whether $\varphi^0 \wedge A_0$ is satisfiable. At each step i for $i > 0$, additional clauses Δ^i and a different set of assumption literals A_i are provided to the solver. We denote by φ^i all the input clauses available at step i: $\varphi^i \equiv \varphi^0 \wedge \bigwedge_{j=1..i} \Delta^j$. At step i, the solver should solve $\varphi^i \wedge A_i$.

2.1 Preprocessing

The three preprocessing techniques that are implemented in SatELite are:

Variable Elimination
Input: formula φ and a variable $v \in vars(\varphi)$.
Output: formula φ' such that $v \notin vars(\varphi')$ and φ' and φ are equisatisfiable. Typically this technique is applied for a variable v only if the number of clauses in φ' is not larger than in φ.

Subsumption
Input: $\varphi \wedge (l_1 \vee \cdots \vee l_i) \wedge (l_1 \vee \cdots \vee l_i \vee l_{i+1} \vee \cdots \vee l_j)$.
Output: $\varphi \wedge (l_1 \vee \cdots \vee l_i)$.

Self-Subsumption
Input: $\varphi \wedge (l_1 \vee \cdots \vee l_i \vee l) \wedge (l_1 \vee \cdots \vee l_i \vee l_{i+1} \vee \cdots \vee l_j \vee \bar{l})$.
Output: $\varphi \wedge (l_1 \vee \cdots \vee l_i \vee l) \wedge (l_1 \vee \cdots \vee l_i \vee l_{i+1} \vee \cdots \vee l_j)$.

2.2 Preprocessing in an Incremental Setting

The problem in combining variable elimination and incremental SAT is that a variable that is eliminated at step i can later reappear in Δ^j for some $j > i$. Since the elimination at step i may remove *original* clauses that contain that variable, soundness is possibly lost. For example, suppose that a formula contains the two clauses $(v_1 \vee v), (v_2 \vee \bar{v})$. Eliminating v results in removing these two clauses and adding the resolvent $(v_1 \vee v_2)$. Suppose, now, that in the next iteration the clauses $(\bar{v}), (\bar{v_1})$ are added, which clearly contradict $(v_1 \vee v)$. Yet since we erased that clause and since there is no contradiction between the resolvent and the new clauses, the new formula is satisfiable.

The solution we proposed in [15], which is also the basis for the generalization we propose in the next section, appears in pseudo-code in Alg. 1. The algorithm is applied at the beginning of each incremental step. Consider the pseudo-code. *SubsumptionQ* is a global queue of clauses.

For each $c \in SubsumptionQ$, and each $c' \in \varphi^i$, RemoveSubsumptions executes these two conditional steps:

Algorithm 1. Preprocessing in an incremental SAT setting (without assumptions propagation), as it appeared in [15].

```
 1: function PREPROCESS-INC(int i)                        ▷ preprocessing of φⁱ
 2:     SubsumptionQ = {c | ∃v. v ∈ c ∧ v ∈ vars(Δⁱ)};
 3:     REMOVESUBSUMPTIONS ();
 4:     for (j = 0 … |ElimVarQ| − 1) do                 ▷ scanning eliminated vars in order
 5:         v = ElimVarQ[j].v;
 6:         if |φ_v^i| = |φ_v̄^i| = 0 then continue;
 7:         REELIMINATE-OR-REINTRODUCE(j, i);
 8:     while SubsumptionQ ≠ ∅ do
 9:         for each variable v ∉ ElimVarQ do            ▷ scanning the rest
10:             SubsumptionQ = ELIMINATEVAR-INC(v, i);
11:             REMOVESUBSUMPTIONS ();
12:         SubsumptionQ = {c | vars(c) ∩ TouchedVars ≠ ∅};
13:         TouchedVars = ∅;
```

1. if $c \subset c'$ it performs subsumption;
2. else if c self-subsumes c' then it performs self-subsumption.

The rest of the algorithm relies on information that we maintain as part of the process of variable elimination. $ElimVarQ$ is a queue of pairs $\langle v, num \rangle$ where v is a variable that was eliminated (or reeliminated as we explain below) in step $i - 1$, and num is the number of resolvent clauses resulting from eliminating v (this number is later used for deciding whether to reintroduce v). In addition, we save the original clauses that are removed when eliminating v in sets S_v and $S_{\bar{v}}$ according to $v's$ phase in those clauses. If v later reappears in some Δ^i, we either *reeliminate* or *reintroduce* v based on these sets. These sets do not contain *all* the original clauses that contain v, because some of them might have been removed from the formula earlier, when another variable, which shares a clause with v was eliminated. We showed in [15] that as long as the order in which the variables are reeliminated or reintroduced remains fixed, soundness is preserved. This order is indeed enforced in lines 4–7.

We now explain how REELIMINATE-OR-REINTRODUCE works by returning to the example above. Suppose that v was eliminated in the base iteration and v_1 was not. We have $S_v = (v_1 \vee v)$, $S_{\bar{v}} = (v_2 \vee \bar{v})$, and $\Delta^1 = \{(\bar{v}), (\bar{v_1})\}$. We can now decide to do one of the following for each variable that is currently eliminated (only v in our case):

- Reeliminate v. For this we need to compute the resolvents that would have been generated if v was not already eliminated. Let $\varphi_v^1 = \{\}$ and $\varphi_{\bar{v}}^1 = \{(\bar{v})\}$ denote the subset of clauses in φ^1 that contain v and $\bar{v}$, respectively.[1] We compute

$$\text{RESOLVE}(\varphi_v^1, \varphi_{\bar{v}}^1) \cup \text{RESOLVE}(\varphi_v^1, S_{\bar{v}}) \cup \text{RESOLVE}(S_v, \varphi_{\bar{v}}^1) =$$
$$\{\} \cup \{\} \cup \text{RESOLVE}((v_1 \vee v), (\bar{v})) = (v_1) \,,$$

---

[1] These sets may contain not only new clauses from Δ^i as in this example, but also clauses that were retrieved when variables earlier-in-the-order were reintroduced.

add it to the formula and remove all clauses containing v or $\bar{v}$. This leaves us with $(v_1 \vee v_2) \wedge (v_1) = (v_1)$. Now adding $(\bar{v}_1)$ leads to the desired contradiction.
- Reintroduce v. For this we only need to add $S_v, S_{\bar{v}}$ back to the formula, which leaves us with $(v_1 \vee v_2), (v_1 \vee v), (v_2 \vee \bar{v}), (\bar{v}), (\bar{v}_1)$, which is again contradictory, as it should be.

A reasonable criterion for choosing between these two options is the expected size of the resulting formula after this step.

Variables not in $ElimVarQ$ are preprocessed in the loop starting in line 8. It is basically the same procedure that appears in SatELite, iterating between variable elimination and subsumption, with a small difference in the implementation of the variable elimination, as can be seen in ELIMINATEVAR-INC of Alg. 2: in addition to elimination, it also updates $ElimVarQ$ and the S sets. We have ignored in this short description various optimizations and subtleties related to difference between original and conflict clauses. A correctness proof of this algorithm can be found in a technical report [16].

Algorithm 2. Variable elimination for φ^i, where the eliminated variable v was *not* eliminated in φ^{i-1}.

1: **function** ELIMINATEVAR-INC(var v, int i)
2:     clauseset $Res = $ RESOLVE $(\varphi_v^i, \varphi_{\bar{v}}^i)$;   ▷ Resolve all. Remove tautologies.
3:     **if** $|Res| > |\varphi_v^i| + |\varphi_{\bar{v}}^i|$ **then return** $\emptyset$;   ▷ no variable elimination
4:     $S_v = \varphi_v^i; S_{\bar{v}} = \varphi_{\bar{v}}^i$;   ▷ Save for possible reintroduction
5:     $ElimVarQ$.push($\langle v, |Res| \rangle$);   ▷ Save #resolvents in queue
6:     $\varphi^i = (\varphi^i \cup Res) \setminus (\varphi_v^i \cup \varphi_{\bar{v}}^i)$;
7:     CLEARDATASTRUCTURES (v);
8:     $TouchedVars = TouchedVars \cup vars(Res)$;   ▷ used in Alg. 1
9:     **return** Res;

---

3   Adding Assumption Propagation

We now describe an incremental version of the preprocessing algorithm that supports assumption propagation. As mentioned in Section 1, we are targeting the general case, where assumptions may appear in both polarities in the formula, and where in each increment the set of assumption variables that is used is arbitrary, e.g., Δ^i may include assumption variables that have been used before. Assuming that the algorithm described in Sec. 2.2 is correct (as was proven in [16]), to maintain correctness while adding assumption propagation we maintain three invariants about the formula ψ being solved at step i:

1. $\varphi^i \implies \psi$. Specifically, this implies that every conflict clause that was learned in previous instances and appears in ψ is implied by φ^i.

2. All the original clauses of φ^i that were subsumed in previous iterations owing to assumptions that are now invalidated, appear in ψ.

3. All the literals in the original clauses of φ^i, which were self-subsumed in previous iterations owing to assumptions that are now invalidated, are restored to their original clauses in ψ.

We say that a clause is *temporary* if it is either an assumption or was derived from one or more assumptions, and *pervasive* otherwise. We mark a conflict clause as temporary during conflict analysis if one of its antecedents is marked that way. For each temporary clause c we maintain a list $SubsumedClauses[c]$ of clauses that were subsumed by c. As can be seen in Alg. 3, our subsumption function is similar to that of SatELite [7], with the difference that we update $SubsumedClauses[c]$ with the subsumed clause in line 6. This allows us to retrieve these clauses in future instances, in which the root assumptions of c are no longer valid (i.e., the assumptions upon which c depends).

Algorithm 3. Eliminating the subsumed clauses, and saving on the side those that depend on assumptions (line 6). Without this line this algorithm is equivalent to that used in SatELite. In future instances where some of the root assumptions of c are not valid, we retrieve the clauses from $SubsumedClauses[c]$.

1: **function** SUBSUME(Clause c)
2:     Pick the literal p in c that has the shortest occur list;
3:     **for** each $c' \in occur(p)$ **do** $\triangleright occur(p) = \{c \mid p \in c, c \in \varphi\}$
4:         **if** c is a subset of c' **then**
5:             Remove c' from the clauses database;
6:                 **if** c is temporary **then** $SubsumedClauses[c].Add(c')$;

---

self-subsumption is presented in Alg. 4. Whereas in SatELite self-subsumption is done simply by erasing a literal, here we distinguish between temporary and pervasive clauses. For pervasive clauses we indeed erase the literal (line 9), but for temporary clauses we perform resolution. This results in adding the same clause as in the case of pervasive clauses (c'', which is added in line 8, is equivalent to the subsumed clause c' minus the literal p), but this resolution is recorded explicitly in the resolution DAG, and consequently we are able to update $SubsumedClauses[c'']$ with the subsumed clause c'.

The net result of the modifications to the SUBSUME and SELFSUBSUME functions is that we have a resolution DAG with all the information needed to undo the effect of an assumption literal, hence maintaining the three invariants. Indeed, let us now shift our focus to UNDOASSUMPTIONS in Alg. 5. This function computes, for each clause c, the set $A(c)$ of invalidated assumption literals upon which it depends. Our implementation does so via BFS traversal of the resolution graph starting from each assumption literal in $A_{i-1} \setminus A_i$. It then performs *T2P* in line 7, namely it adds to c the negation of the literals in $A(c)$, so it can be reused regardless of the values of those assumption literals, and hence invariant #1 is maintained. Note that this is an improvement over [13] because

Algorithm 4. Self-subsumption. We perform resolution (rather than just eliminate a literal as in SatELite) and save the eliminated clause, because we will need to retrieve it in future instances in which some of the root assumptions of c are invalidated.

```
1: function SELFSUBSUME(Clause c)
2:     for each p ∈ c do
3:         for each c′ subsumed by c[p := p̄] do
4:             if c is temporary then
5:                 c″ = res(c, c′);
6:                 SubsumedClauses[c].Add(c′);
7:                 Remove c′ from clause database;
8:                 AddClause(c″);
9:             else remove p̄ from c′;
```

it is incremental: it only adds those assumption literals that were invalidated at the current step, whereas in [13] the entire set of assumptions upon which the clause depends is recomputed each time.

Algorithm 5. Undoing the effect of invalidated assumptions. Line 5 refers also to deleted clauses. In line 7 we do not consider a clause c if it has been deleted (this is a heuristic decision).

```
1: function UNDOASSUMPTIONS
2:     For each clause c, A(c) = ∅;
3:     RootAssumptions = assumptions in A_{i−1} \ A_i;
4:     for each a ∈ RootAssumptions do
5:         for each clause c descendant of a on the resolution graph do
6:             A(c) = A(c) ∪ {a};
7:     for each clause c for which A(c) ≠ ∅ do
8:         Replace c with ((⋁_{a∈A(c)} ā) ∨ c);                    ▷ T2P
9:         for each clause c′ ∈ SubsumedClauses[c] do ADDCLAUSE(c′);
```

In addition, in line 9 the function restores the clauses that were subsumed by c, and hence maintains invariant #2 and, indirectly, also invariant #3 because of the way self-subsumption is done, as was shown in Alg. 4. There are several subtleties in implementation of this algorithm that are not mentioned explicitly in the pseudo-code. First, the clauses visited in line 5 also include learned clauses that have been deleted, because such clauses may have subsumed other clauses, and may have participated in inferring other (temporary) clauses. Hence, when the deletion strategy of the solver deletes a learned clause, we do not remove it from the resolution DAG. Whether or not to turn such a deleted clause into a pervasive one in line 7 is a matter of heuristics. In our implementation we do not. Finally, observe that line 8 replaces each unit clause c corresponding to an assumption literal $a \in RootAssumptions$ with a tautology $(\bar{a} \vee a)$. Our implementation therefore simply erases such clauses.

Given this ability to undo the effect of assumptions, we can now integrate this function with the PREPROCESS-INC function of Alg. 1. Consider PREPROCESS-INC-WITH-ASSUMPTIONS in Alg. 6. It undoes the effect of the invalidated assumptions by calling UNDOASSUMPTIONS, adds as unit clauses the new assumptions in A_i, and finally calls PREPROCESS-INC. This is the algorithm that we call at the beginning of each new instance.

Note that, unlike in Minisat-Alg, CS, and incremental SatELite, one needs to store and maintain the resolution derivation in order to implement our algorithm. This may have a negative impact on performance. To mitigate this problem, we store only a partial resolution derivation that is sufficient to implement our approach: only clauses that are backward reachable from the assumptions are stored. Using partial resolution was initially proposed in [17] in the context of unsatisfiable core extraction. It was first used in the context of incremental SAT solving under assumptions in the assumption propagation algorithm [13].

Algorithm 6. Preprocessing in an incremental SAT setting with assumptions propagation.

1: **function** PREPROCESS-INC-WITH-ASSUMPTIONS(int i)   ▷ preprocessing of φ^i
2:    UNDOASSUMPTIONS();
3:    $\forall a \in A_i \setminus A_{i-1}$ ADDCLAUSE$((a))$;   ▷ Assumptions as units;
4:    PREPROCESS-INC(i);   ▷ See Alg. 1

---

4   Experimental Results

This section analyzes the performance of our new algorithm UI-SAT against the existing algorithms for incremental SAT solving under assumptions listed in Table 1 in Section 1.

We used instances generated by Intel's formal verification flow applied to the *incremental* formal verification of hardware, where the model checker is allowed to be re-used incrementally each time with a different set of assumptions. Such a mode is essential for ensuring fast and effective application of formal verification in industrial settings. Further details about the application are provided in Section 2 of [13]. We used the same instance set as in [13]. Overall, 210 instances were used, but the results below refer only to the 186 of them that were fully solved by at least one solver within the time limit of 3600 seconds. The instances are characterized by a large number of assumptions, all of which being part of the original formula (and not selector variables). The assumptions mimic the environment of a sub-circuit that is verified with incremental BMC in Intel, as described in Section 1. All of the instances are publicly available at [14].

We implemented all the algorithms in Intel's new SAT solver Fiver. Note that the implementation in [13] was made in another SAT solver (Eureka), hence the results presented in this paper are not directly comparable to the results presented in [13].

Table 2. Performance results (the run-time is in seconds)

	CS	Minisat-Alg	Assump. Prop.	Incr. SatELite	UI-SAT
Run-time	223,424	159,423	182,530	209,781	**64,176**
Unsolved	28	14	24	16	**1**

For the experiments we used machines running Intel® Xeon® processors with 3Ghz CPU frequency and 32Gb of memory.

Table 2 and Figure 1 present the results of all the algorithms. Note that Minisat-Alg refers to Minisat's algorithm as implemented in Fiver, rather than to the Minisat SAT solver.

UI-SAT is clearly the fastest approach. UI-SAT outperforms the next best algorithm Minisat-Alg by 2.5x and solves 13 more instances. CS is the slowest algorithm, as expected. Interestingly, Minisat-Alg outperforms both assumption propagation and incremental SatELite. In [13] assumption propagation is faster than Minisat-Alg in the presence of step look-ahead (that is, a limited form of look-ahead information), while incremental SatELite is faster than Minisat-Alg in [15] on a different instance set having considerably fewer assumptions. One can see that the integration of incremental SatELite and assumption propagation is critical for performance.

Figure 2 provides a direct comparison between UI-SAT and Minisat-Alg. One can see that UI-SAT outperforms Minisat-Alg on the majority of difficult instances. Overall, UI-SAT is faster than Minisat-Alg by at least 2 sec. on 110 instances, while Minisat-Alg is faster than UI-SAT by at least 2 sec. on 58 instances. However, when the run-time of both solvers is at least 100 sec., the dif-

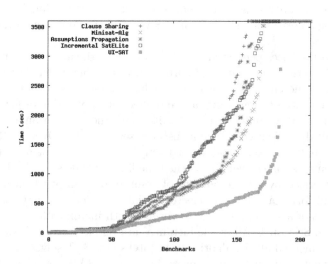

Fig. 1. A comparison of the different techniques over 186 industrial instances generated by a bounded model checker of hardware

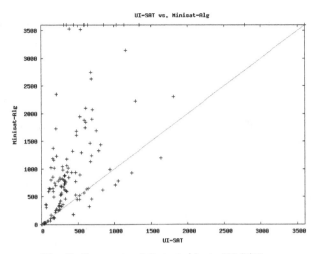

Fig. 2. Comparing Minisat-Alg to UI-SAT

ference in performance is more substantial, as UI-SAT is faster than Minisat-Alg on 107 instances, while Minisat-Alg is faster than UI-SAT on only 17 instances.

5 Conclusion

We have introduced a new algorithm for incremental SAT solving under assumptions, called Ultimately Incremental SAT (UI-SAT). UI-SAT ensures that SatELite is applied and all the assumptions propagated, that is, represented as unit clauses. We have shown that UI-SAT outperforms existing approaches, including clause sharing, Minisat's algorithm, incremental SatELite (without assumption propagation), and assumption propagation (without incremental SatELite) on a publicly available set of 186 instances generated by an incremental bounded model checker. UI-SAT is 2.5x faster than the second best approach, and solves 13 more instances.

Acknowledgments. The authors would like to thank Daher Kaiss for supporting this work and Paul Inbar for editing the paper.

References

1. Balint, A., Manthey, N.: Boosting the performance of SLS and CDCL solvers by preprocessor tuning. In: Pragmatics of SAT (2013)
2. Biere, A., Heule, M.J.H., van Maaren, H., Walsh, T. (eds.): Handbook of Satisfiability. Frontiers in Artificial Intelligence and Applications, vol. 185. IOS Press (February 2009)
3. Bloem, R., Sharygina, N. (eds.): Proceedings of 10th International Conference on Formal Methods in Computer-Aided Design, FMCAD 2010, Lugano, Switzerland, October 20-23. IEEE (2010)

4. Cabodi, G., Lavagno, L., Murciano, M., Kondratyev, A., Watanabe, Y.: Speeding-up heuristic allocation, scheduling and binding with SAT-based abstraction/refinement techniques. ACM Trans. Design Autom. Electr. Syst. 15(2) (2010)
5. Chockler, H., Ivrii, A., Matsliah, A., Moran, S., Nevo, Z.: Incremental formal verification of hardware. In: Bjesse, P., Slobodová, A. (eds.) FMCAD, pp. 135–143. FMCAD Inc. (2011)
6. Cimatti, A., Sebastiani, R. (eds.): SAT 2012. LNCS, vol. 7317. Springer, Heidelberg (2012)
7. Eén, N., Biere, A.: Effective preprocessing in SAT through variable and clause elimination. In: Bacchus, F., Walsh, T. (eds.) SAT 2005. LNCS, vol. 3569, pp. 61–75. Springer, Heidelberg (2005)
8. Eén, N., Mishchenko, A., Amla, N.: A single-instance incremental SAT formulation of proof- and counterexample-based abstraction. In: Bloem, Sharygina (eds.) [3], pp. 181–188
9. Eén, N., Sörensson, N.: An extensible SAT-solver. In: Giunchiglia, E., Tacchella, A. (eds.) SAT 2003. LNCS, vol. 2919, pp. 502–518. Springer, Heidelberg (2004)
10. Eén, N., Sörensson, N.: Temporal induction by incremental SAT solving. Electr. Notes Theor. Comput. Sci. 89(4) (2003)
11. Franzén, A., Cimatti, A., Nadel, A., Sebastiani, R., Shalev, J.: Applying SMT in symbolic execution of microcode. In: Bloem, Sharygina (eds.) [3], pp. 121–128
12. Khasidashvili, Z., Kaiss, D., Bustan, D.: A compositional theory for post-reboot observational equivalence checking of hardware. In: FMCAD, pp. 136–143. IEEE (2009)
13. Nadel, A., Ryvchin, V.: Efficient SAT solving under assumptions. In: Cimatti, Sebastiani (eds.) [6], pp. 242–255
14. Nadel, A., Ryvchin, V., Strichman, O.: UI-SAT benchmark set: https://copy.com/osV4myggyNRa
15. Nadel, A., Ryvchin, V., Strichman, O.: Preprocessing in incremental SAT. In: Cimatti, Sebastiani (eds.) [6], pp. 256–269
16. Nadel, A., Ryvchin, V., Strichman, O.: Preprocessing in incremental SAT. Technical Report IE/IS-2012-02, Technion (2012), http://ie.technion.ac.il/~ofers/publications/sat12t.pdf
17. Ryvchin, V., Strichman, O.: Faster extraction of high-level minimal unsatisfiable cores. In: Sakallah, K.A., Simon, L. (eds.) SAT 2011. LNCS, vol. 6695, pp. 174–187. Springer, Heidelberg (2011)
18. Silva, J.P.M., Sakallah, K.A.: Robust search algorithms for test pattern generation. In: FTCS, pp. 152–161 (1997)
19. Shtrichman, O.: Pruning techniques for the SAT-based bounded model checking problem. In: Margaria, T., Melham, T.F. (eds.) CHARME 2001. LNCS, vol. 2144, pp. 58–70. Springer, Heidelberg (2001)
20. Whittemore, J., Kim, J., Sakallah, K.A.: SATIRE: A new incremental satisfiability engine. In: DAC, pp. 542–545. ACM (2001)

A SAT Attack on the Erdős Discrepancy Conjecture

Boris Konev and Alexei Lisitsa

Department of Computer Science
University of Liverpool, United Kingdom

Abstract In 1930s Paul Erdős conjectured that for any positive integer C in any infinite ± 1 sequence (x_n) there exists a subsequence $x_d, x_{2d}, x_{3d}, \ldots, x_{kd}$, for some positive integers k and d, such that $| \sum_{i=1}^{k} x_{id} | > C$. The conjecture has been referred to as one of the major open problems in combinatorial number theory and discrepancy theory. For the particular case of $C = 1$ a human proof of the conjecture exists; for $C = 2$ a bespoke computer program had generated sequences of length 1124 of discrepancy 2, but the status of the conjecture remained open even for such a small bound. We show that by encoding the problem into Boolean satisfiability and applying the state of the art SAT solver, one can obtain a discrepancy 2 sequence of length 1160 and a *proof* of the Erdős discrepancy conjecture for $C = 2$, claiming that no discrepancy 2 sequence of length 1161, or more, exists. We also present our partial results for the case of $C = 3$.

1 Introduction

Discrepancy theory is a branch of mathematics dealing with irregularities of distributions of points in some space in combinatorial, measure-theoretic and geometric settings [1,2,3,4]. The paradigmatic combinatorial discrepancy theory setting can be described in terms of a hypergraph $\mathcal{H} = (U, S)$, that is, a set U and a family of its subsets $S \subseteq 2^U$. Consider a colouring $c : U \to \{+1, -1\}$ of the elements of U in *blue* ($+1$) and *red* (-1) colours. Then one may ask whether there exists a colouring of the elements of U such that in every element of S colours are distributed uniformly or a discrepancy of colours is always inevitable. Formally, the discrepancy (deviation from a uniform distribution) of a hypergraph $\mathcal{H}$ is defined as $\min_c(\max_{s \in S} | \sum_{e \in s} c(e)|)$. Discrepancy theory also has practical applications in computational complexity [2], complexity of communication [5] and differential privacy [6].

One of the oldest problems of discrepancy theory is the discrepancy of hypergraphs over the set of natural numbers with the subsets (hyperedges) forming arithmetical progressions over this set [7]. Roth's theorem [8], one of the main results in the area, states that for the hypergraph formed by the arithmetical progressions in $\{1, \ldots, l\}$, that is $\mathcal{H}_l = (U_l, S_l)$, where $U_l = \{1, 2, \ldots, l\}$ and elements of S_l being of the form $(ai + b)$ for arbitrary a, b, the discrepancy grows at least as $\frac{1}{20}l^{1/4}$.

Surprisingly, for the more restricted case of *homogeneous* arithmetic progressions of the form (ai), the question of the discrepancy bounds is open for more than eighty years. In 1930s Paul Erdős conjectured [9] that the discrepancy is unbounded. This conjecture became known as the Erdős discrepancy problem (EDP) and its proving or disproving has been referred to as one of the major open problems in combinatorial number theory and discrepancy theory [1,4,10].

C. Sinz and U. Egly (Eds.): SAT 2014, LNCS 8561, pp. 219–226, 2014.
© Springer International Publishing Switzerland 2014

The problem can be naturally described in terms of sequences of $+1$ and -1 (and this is how Erdős himself introduced it). Then Erdős's conjecture states that for any $C > 0$ in any infinite ± 1 sequence (x_n) there exists a subsequence $x_d, x_{2d}, x_{3d}, \ldots, x_{kd}$, for some positive integers k and d, such that $| \sum_{i=1}^{k} x_{id} | > C$. The general definition of discrepancy given above can be specialised as follows. The discrepancy of a finite ± 1 sequence $\bar{x} = x_1, \ldots, x_l$ of length l can be defined as $\max_{d=1,\ldots,l}(| \sum_{i=1}^{\lfloor \frac{l}{d} \rfloor} x_{id} |)$. For an infinite sequence (x_n) its discrepancy is the supremum of discrepancies of all its initial finite fragments.

For random ± 1 sequences of length l the discrepancy grows as $l^{1/2+o(1)}$ and the explicit constructions of a sequence with slowly growing discrepancy at the rate of $\log_3 l$ have been demonstrated [11,12]. By considering cases, one can see that any ± 1 sequence of length 12 or more has discrepancy at least 2; that is, Erdős's conjecture holds for the particular case $C = 1$ (also implied by the stronger result in [13]). For all other values of C the status of the conjecture remained unknown. Although widely believed not to be the case, there was still a possibility that an infinite sequence of discrepancy 2 existed.

The EDP has attracted renewed interest in 2009-2010 as it became a topic of the Polymath project [14], a widely publicised endeavour in collective math initiated by T. Gowers [15]. As part of this activity (see discussion in [14]) an attempt has been made to attack the problem using computers. A purposely written computer program had successfully found ± 1 sequences of length 1124 having discrepancy 2; however, it failed to produce a discrepancy 2 sequence of a larger length and it has been claimed that "given how long a finite sequence can be, it seems unlikely that we could answer this question just by a clever search of all possibilities on a computer" [14].

In this paper we settle the status of the EDP for $C = 2$. We show that by encoding the problem into Boolean satisfiability and applying the state of the art SAT solvers, one can obtain a sequence of length 1160 of discrepancy 2 and a proof of the Erdős discrepancy conjecture for $C = 2$, claiming that no sequence of length 1161 and discrepancy 2 exists. We also present our partial results for the case of $C = 3$ and demonstrate the existence of a sequence of length 13 900 of discrepancy 3.

2   SAT Encoding

Checking that a ± 1 sequence of length l has discrepancy C is quite straightforward and so for the existence claims the specific encoding details are of limited interest and could be left as an exercise to the reader. The negative results (that is, our claim that no infinite discrepancy 2 sequence exists), however, require us to give a short description of our SAT encoding of the EDP. The encoding in full for all cases discussed in this paper and the program generating the encoding of the EDP for arbitrary given values of C and l can be found in [16].

We employ the automata based approach similar to the encoding of temporal formulae for bounded model checking [17]. In Figure 1 we give an automaton that accepts a ± 1 word of length m if, and only if, the word represents a ± 1 sequence $y_1, \ldots, y_m$ such that $\sum_{i=1}^{m} | y_i | > C$ (and for all $m' < m$ it holds $\sum_{i=1}^{m} | y_i | \leq C$). Notice that if a subsequence $x_d, x_{2d}, \ldots, x_{kd}$ of $\bar{x} = x_1, \ldots, x_l$ contains less than C

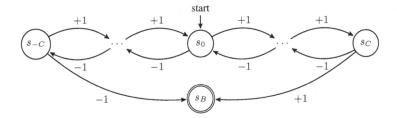

Figure 1. Automaton $\mathcal{A}_C$

elements, then the discrepancy of $\bar{x}$ cannot exceed C. It should be clear then that if for every $d : 1 \leq d \leq \lfloor \frac{l}{C+1} \rfloor$ the automaton $\mathcal{A}_C$ *does not* accept the subsequence $x_d, x_{2d}, \ldots, x_{kd}$, where $k = \lfloor \frac{l}{d} \rfloor$, then the discrepancy of the sequence $\bar{x}$ does not exceed C.

The trace of the automaton $\mathcal{A}_C$ on the subsequence $x_d, x_{2d}, \ldots, x_{kd}$ can be encoded by a Boolean formula in the obvious way. To explain representation details, first consider

$$\phi_{(l,C,d)} = s_0^{(1,d)} \bigwedge_{i=1}^{\lfloor \frac{l}{d} \rfloor} \left[\bigwedge_{-C \leq j < C} \left(s_j^{(i,d)} \wedge p_{id} \to s_{j+1}^{(i+1,d)} \right) \wedge \right.$$
$$\bigwedge_{-C < j \leq C} \left(s_j^{(i,d)} \wedge \neg p_{id} \to s_{j-1}^{(i+1,d)} \right) \wedge \qquad (1)$$
$$\left(s_C^{(i,d)} \wedge p_{id} \to B \right) \wedge$$
$$\left. \left(s_{-C}^{(i,d)} \wedge \neg p_{id} \to B \right) \right],$$

where the intended meaning is that proposition $s_j^{(i,d)}$ is true if, and only if, the automaton $\mathcal{A}_C$ is in the state s_j having read first $(i - 1)$ symbols of the input word, and proposition p_i is true if, and only if, the i-th symbol of the input word is $+1$.

Let

$$\phi_{(l,C)} = \neg B \wedge \bigwedge_{d=1}^{\lfloor \frac{l}{C+1} \rfloor} \phi_{(l,C,d)} \wedge \mathsf{frame}_{(l,C)},$$

where $\mathsf{frame}_{(l,C)}$ is a Boolean formula encoding that the automaton state is correctly defined, that is, exactly one proposition from each of the sets $\{ s_j^{(i,d)} \mid -C \leq j \leq C \}$, for $d = 1, \ldots, \lfloor \frac{l}{C+1} \rfloor$ and $1 \leq i \leq \lfloor \frac{l}{d} \rfloor$, is true in every model of $\phi_{(l,C)}$.

The following statement can be easily proved by an investigation of models of $\phi_{(l,C)}$ and the traces of $\mathcal{A}_C$. Notice that although $\phi_{(l,C)}$ encodes the traces of $\mathcal{A}_C$ on all subsequences of $\bar{x}$ they all share the same proposition B—as soon as the automaton accepts any of these subsequences, the entire sequence should be rejected.

Proposition 1. *The formula $\phi_{(l,C)}$ is satisfiable if, and only if, there exists a ±1 sequence $\bar{x} = x_1, \ldots, x_l$ of length l of discrepancy at most C. Moreover, if $\phi_{(l,C)}$ is satisfiable, the sequence $\bar{x} = x_1, \ldots, x_l$ of discrepancy C is uniquely identified by the assignment of truth values to propositions $p_1, \ldots p_l$.*

The encoding described above, albeit very natural, is quite wasteful: the size of formula $\mathrm{frame}_{(l,C)}$ is quadratic in the number of states. To reduce the size, in our implementation we use a slightly different encoding of the traces of $\mathcal{A}_C$. Namely, we replace in (1) every occurrence of $s_j^{(i,d)}$ with a conjunction of propositions representing the numerical value of j in binary, where the most significant bit encodes the sign of j and the other bits encode an unsigned number $0 \ldots C$ in the usual way. We denote the resulting formula $\phi_{(l,C,d)}^b$.

For example, for $C = 2$ the values $-C \ldots C$ can be represented in binary by 3 bits. Then $\phi_{(l,C,d)}^b$ contains, for example,

$$\left(\neg b_2^{(i,d)} \wedge \neg b_1^{(i,d)} \wedge \neg b_0^{(i,d)}\right) \wedge p_{id} \rightarrow \neg b_2^{(i+1,d)} \wedge \neg b_1^{(i+1,d)} \wedge b_0^{(i+1,d)}$$

encoding the transition from s_0 to s_1 having read $+1$.

We also exclude by a formula $\mathrm{frame}_{(l,C)}^b$ all combinations of bits that do not correspond to any states of $\mathcal{A}_C$. For example, for $C = 2$ we have

$$\mathrm{frame}_{(l,C)}^b = \bigwedge_{d=1}^{\lfloor\frac{l}{C+1}\rfloor} \bigwedge_{i=1}^{\lfloor\frac{l}{d}\rfloor+1} \left[\neg(b_2^{(i,d)} \wedge \neg b_1^{(i,d)} \wedge \neg b_0^{(i,d)}) \wedge \right.$$
$$\neg(\neg b_2^{(i,d)} \wedge b_1^{(i,d)} \wedge b_0^{(i,d)}) \wedge$$
$$\left. \neg(b_2^{(i,d)} \wedge b_1^{(i,d)} \wedge b_0^{(i,d)}) \right].$$

The first conjunct disallows the binary value 100, a 'negated zero', the other two encode that $\mathcal{A}_C$, for $C = 2$, does not have neither s_3 nor s_{-3}. The following statement is a direct consequence of Proposition 1.

Proposition 2. *The formula* $\phi_{(l,C)}^b = \neg B \wedge \bigwedge_{d=1}^{\lfloor\frac{l}{C+1}\rfloor} \phi_{(l,C,d)}^b \wedge \mathrm{frame}_{(l,C)}^b$ *is satisfiable if, and only if, there exists a* ± 1 *sequence* $\bar{x} = x_1, \ldots, x_l$ *of length* l *of discrepancy at most* C. *Moreover, if* $\phi_{(l,C)}$ *is satisfiable, the sequence* $\bar{x} = x_1, \ldots, x_l$ *of discrepancy* C *is uniquely identified by the assignment of truth values to propositions* $p_1, \ldots p_l$.

3   Results

In our experiments we used the Lingeling SAT solver [18], the winner of the *SAT-UNSAT* category of the SAT'13 competition [19] and the Glucose solver [20] version, the winner of the *certified UNSAT* category of the SAT'13 competition [19]. All experiments were conducted on PCs equipped with an Intel Core i5-2500K CPU running at 3.30GHz and 16GB of RAM.

By iteratively increasing the length of the sequence, we establish precisely that the maximal length of a ± 1 sequence of discrepancy 2 is 1160. On our system it took Plingeling, the parallel version of the Lingeling solver, about 800 seconds[1] to find a satisfying assignment. One of the sequences of length 1160 of discrepancy 2 can be found in Appendix A for reader's amusement.

---

[1] The time taken by the solver varies significantly from experiment to experiment; in one run it took the solver just 166.8 seconds to find a satisfying assignment.

Proposition 3. *There exists a sequence of length* 1160 *of discrepancy* 2.

When we increased the length of the sequence to 1161, Plingeling reported unsatisfiability. In order to corroborate this statement, we also used Glucose. It took the solver about 21 500 seconds to compute a Delete Reverse Unit Propagation (DRUP) certificate of unsatisfiability, which is a compact representation of the resolution refutation of the given formula [21]. The correctness of the unsatisfiability certificate has been independently verified with the drup-trim tool [22]. The size of the certificate is about 13 GB[2], and the time needed to verify the certificate was comparable with the time needed to generate it. Combined with Proposition 2, we obtain a computer proof of the following statement.

Theorem 1. *Every sequence of length* 1161 *has discrepancy at least* 3.

So we conclude.

Corollary 1. *The Erdős discrepancy conjecture holds true for* $C = 2$.

In an attempt to better understand this result, we looked at the smaller unsatisfiable subset of $\phi^b_{(l,C)}$ identified by the drup-trim tool. It turned out that the encoding of some automata traces is not present in the subset. A further manual minimisation showed that, although $\lfloor \frac{1161}{3} \rfloor$ is 387, to show unsatisfiability it suffices to consider subsequences of $x_1, \ldots, x_{1161}$ of the form $x_d, \ldots, x_{kd}$ for the values of d ranging from 1 to 358. It remains to be seen whether or not this observation can be helpful for a human proof of the conjecture

We also applied our methodology to identify sequences of discrepancy 3, however, we did not manage to prove the conjecture. Having spent 8 days, 21 hours and 55 minutes (or 770122.2 seconds total), on the encoding of the problem using 380 404 variables and 4 641 640 clauses Plingeling has successfully identified a sequence of length 13 900 with discrepancy 3. The encoding and the generated sequence can be found in [16].

Proposition 4. *There exists a sequence of length* 13 900 *of discrepancy* 3.

However, the general question of existence of a finite bound on the length of ± 1 sequences of discrepancy 3 (that is, the Erdős conjecture for $C = 3$) remains open.

4   Discussion

We have demonstrated that SAT-based methods can be used to tackle the longstanding mathematical question on the discrepancy of ± 1 sequences. For EDP with $C = 2$ we have identified the exact boundary between satisfiability and unsatisfiability, that is, we found the longest discrepancy 2 sequence and proved that no larger sequence of

---

[2] It is possible to reduce the size of the certificate by exploiting symmetry. Notice that if a sequence $x_1, \ldots, x_n$ has discrepancy C so does the sequence $-x_1, \ldots, -x_n$. Fixing the value of one of x_i to 1 reduces the search space and hence the size of the certificate. Moreover, one should fix the value of x_i occurring in many subsequences of $x_1, \ldots, x_n$. E.g. with x_{60} set to 1, the size of the DRUP certificate becomes 3.6 GB.

discrepancy 2 exists. There is, however, a noticeable asymmetry between these findings. The fact that a sequence of length 1160 has discrepancy 2 can be easily checked either by a straightforward computer program or even manually. The negative witness, that is, the DRUP unsatisfiability certificate, is probably one of longest proofs of a non-trivial mathematical result ever produced[3], so one may have doubts about to which degree this can be accepted as a proof of a mathematical statement.

But this is the best we can get for the moment. Essentially, the unsatisfiability proof corresponds to the verification that the search in a huge search space has been done correctly and completed without finding a satisfying assignment. It is a challenging problem to produce a compact proof more amenable for human comprehension.

Finally notice that apart from the obtained results the proposed methodology can be used to further experimentally explore variants of the Erdős problem as well as more general discrepancy theory problems.

Acknowledgement. We thank Armin Biere, Marijn Huele, Pascal Fontaine, Donald Knuth, Laurent Simon and Laurent Théry for helpful comments and discussions on the preliminary version of this paper.

References

1. Beck, J., Chen, W.W.L.: Irregularities of Distribution. Cambridge University Press (1987)
2. Chazelle, B.: The Discrepancy Method: Randomness and Complexity. Cambridge University Press, New York (2000)
3. Matousek, J.: Geometric Discrepancy: An Illustrated Guide. Algorithms and combinatorics, vol. 18. Springer (1999)
4. Beck, J., Sós, V.T.: Discrepancy theory. In: Graham, R.L., Grötschel, M., Lovász, L. (eds.) Handbook of Combinatorics, vol. 2, pp. 1405–1446. Elsivier (1995)
5. Alon, N.: Transmitting in the n-dimensional cube. Discrete Applied Mathematics 37/38, 9–11 (1992)
6. Muthukrishnan, S., Nikolov, A.: Optimal private halfspace counting via discrepancy. In: Proceedings of the 44th Symposium on Theory of Computing, STOC 2012, pp. 1285–1292. ACM, New York (2012)
7. Matousek, J., Spencer, J.: Discrepancy in arithmetic progressions. Journal of the American Mathematical Society 9, 195–204 (1996)
8. Roth, K.F.: Remark concerning integer sequence. Acta Arithmetica 9, 257–260 (1964)
9. Erdős, P.: Some unsolved problems. The Michigan Mathematical Journal 4(3), 291–300 (1957)
10. Nikolov, A., Talwar, K.: On the hereditary discrepancy of homogeneous arithmetic progressions. CoRR abs/1309.6034v1 (2013)
11. Gowers, T.: Erdős and arithmetic progressoins. In: Erdős Centennial Conference (2013), http://www.renyi.hu/conferences/erdos100/program.html, (accessed January 29, 2014)
12. Borwein, P., Choi, S.K.K., Coons, M.: Completely multiplicative functions taking values in $\{1, -1\}$. Transactions of the American Mathematical Society 362(12), 6279–6291 (2010)

---

[3] In the preliminary version of this paper [23] we even compared the original unoptimised 13 GB proof with the size of the whole Wikipedia.

13. Mathias, A.R.D.: On a conjecture of Erdős and Čudakov. In: Combinatorics, Geometry and Probability (1993)
14. Erdős discrepancy problem: Polymath wiki, `http://michaelnielsen.org/polymath1/index.php?title=The_Erd` (accessed January 29, 2014)
15. Gowers, T.: Is massively collaborative mathematics possible, `http://gowers.wordpress.com/2009/01/27/is-massively-collaborative-mathematics-possible/` (accessed January 29 , 2014)
16. Konev, B., Lisitsa, A.: Addendum to: A SAT attack on the Erdős discrepancy conjecture, `http://www.csc.liv.ac.uk/~konev/SAT14`
17. Biere, A.: Bounded model checking. In: Handbook of Satisfiability. Frontiers in Artificial Intelligence and Applications, vol. 185, pp. 457–481. IOS Press (2009)
18. Biere, A.: Lingeling, Plingeling and Treengeling entering the SAT Competition 2013. In: Proceedings of SAT Competition 2013, pp. 51–52. University of Helsinki (2013)
19. Balint, A., Belov, A., Heule, M.J.H., Järvisalo, M. (eds.): Proceedings of SAT competition 2013. University of Helsinki (2013)
20. Audemard, G., Simon, L.: Glucose 2.3 in the SAT 2013 Competition. In: Proceedings of SAT Competition 2013, pp. 42–43. University of Helsinki (2013)
21. Goldberg, E.I., Novikov, Y.: Verification of proofs of unsatisfiability for CNF formulas. In: Proceedings of Design, Automation and Test in Europe Conference and Exposition (DATE 2003), Munich, Germany, March 3-7, pp. 10886–10891 (2003)
22. Heule, M.J.H.: DRUP checker, `http://www.cs.utexas.edu/~marijn/drup/` (accessed January 29 , 2014)
23. Konev, B., Lisitsa, A.: A sat attack on the erdos discrepancy conjecture. CoRR abs/1402.2184 (2014)

A   One of the Sequences of Length 1160 Having Discrepancy 2

We give a graphical representation of one of the sequences of length 1160 obtained from the satisfying assignment computed with the Plingeling solver. Here + stands for +1 and − for −1, respectively.

```
- + + - + - - + + - + + - + - - + - - + + - + - - + - - +
+ - + - - + + - + + - + - + + - - + + - + - - - + - + + -
+ - - + - - + + + + - - - + - - + + - + - - + + - + + - - -
- + + - + + - + - + + - - + + - + - + - - - - + + - + - - +
+ - + + - + - - + + - + - - + - - - + - + + - + - - + + + -
+ + - + - - + - - + + - + + - + - - + + - + - - + + + - +
- + - - - - + + + - + - - + - - + + + - - - + + - + + - +
- - + - - + + + - - + - + - + - - + - + + + - + + - + - -
+ - - + + - + - - + + - + + - + - - + - - + + - - + + + -
- - + + + - + - - - + + - + - - + + - - + - + - - + - + +
+ - + - - + + - + + - + - - + + - + - - + - - + + - + - -
+ + - - + - + + - + - + - - + - + - + + - + - - + + - + -
- + - - + + - + - + - + + - + - + - + + - - - + - + - - +
+ + + - - + - - - + + - + - + + - + - - + + - + - - + - -
+ + - + - - + + + + - - + - - - + - + + + + - - + - - + +
- + + - + - - + + - + - - + - - + + - + - - + + - + + - +
- - + + - + - - + - - + + - + + - + - - - - + + + - + - -
+ + - - + + + - - - + - + + - + - - + - + + - - - + - + +
- + + - + - - + - - + + - - + + + + - + - - + - - + - - +
+ + + - - + - - + + + - - - + + - + + - + - - + + - - + -
+ - - + - - + + - + + - + - - + - - + - + + + - + + - + -
- + - - + + - - + - + + - + + - + - - + - - + - - + + - +
+ - + - - + + + - - - + + - + - - + + - + + - - - + + + -
- - + + - + + - - - - + + + - - + - + + - + - - + - - + +
- + - - + + - + + + - + - + + - - + + - - + - - - + + +
- + + - - + + - - - - + + - + + + - - + + - - - + + + - -
- - + - + - + + - + + - + + - + - + - - - - + + + - - + +
- + - - + + - + + - + - - + - - + - - + + - + - - + + - +
+ - + - - + + - - + - + - - + - + - + - + + + + - - - + -
+ - + + - - + - - + - + - + - + + - + - + + + - - + - + -
- + - - + - + + + - - + - + + + - - - + + - + - - + - - +
+ - + + - - + + - - - + + - + - + + - - + + - + - - - + -
+ + - + - - + - + + - - + + - + - - + + - + - - + - + + +
- + - - + + - - + - + - + + + - - + - + - - + + - + + - +
- - + - - + - + + - - - + - + + - + - + + - - + + - + - -
+ + + - + - - - - + + - - + - + + - + - + + - - + + - + -
- + + - + - + + - - + + - + - - - + - + + - + - - + + + -
- - - + - + - + + - - + + - + - - + + - + + - + + - + - -
+ - - + - - + + + + - - - + + - - - + - + - + + - + - + +
+ - - + - + + - - + - + - - + - + - + + - - - + + + - + +
```

Dominant Controllability Check
Using QBF-Solver and Netlist Optimizer

Tamir Heyman[1], Dan Smith[2], Yogesh Mahajan[2], Lance Leong[2],
and Husam Abu-Haimed[3]

[1] NVIDIA Inc Beaverton, OR, 97006
[2] NVIDIA Inc Santa Clara, CA, 95050
[3] Atrenta Inc Santa Clara, CA, 95050

Abstract. This paper presents an application of formal methods to the verification of hardware power management modules. The property being verified is called Dominant Controllability and is a property of a netlist node and a subset of the inputs. The property holds if there exists an assignment to the subset of the inputs such that it sets the node to 0/1 regardless of the values at the rest of the inputs. Verification of power management modules in recent CPU and GPU designs includes hundreds of such properties. Two approaches are described for verifying such properties: netlist optmization and QBF solving. In the latter case, a QBF preprocessor is used, requiring partial model reconstruction. Each method can be used independently or combined into a third algorithm that heuristically selects a method based on its performance on a design. Experimental results for these methods are presented and discussed.

Keywords: Controllability, QBF, QBF Preprocessor, Netlist Optimizer.

1 Introduction

Power management in hardware is becoming a major challenge in recent designs. Power management is significant because the ratio between performance and power consumption is very important for many applications like mobile devices, servers and gaming consoles. The main power consumers in these applications are the CPU and GPU. In order to save power in a very large hardware design, a number of components can be turned on and off by a power management module. This module has a set of inputs that controls a power switch (clamp signal, sleep signal) to each component individually. The design specifications define the set of power switches and a set of inputs that controls them. The challenge is that these switches are controlled by other modules as well. In order for the power module to function correctly, the power module should have dominant control (i.e., have priority) over other modules connected to the switch. Each switch should be turned on and off by an assignment to the power module inputs regardless of any other controlling modules. Existence of such assignments means that the power module has dominant control over the switch.

This paper presents the use of formal methods for checking dominant controllability. This check calls for formal methods due to the large search space. First,

C. Sinz and U. Egly (Eds.): SAT 2014, LNCS 8561, pp. 227–242, 2014.

the power management module has a large number of inputs that control multiple power switches. Second, many switches are controlled by a large number of control modules other than the power management module. In such cases, random search may fail to find a controlling assignment even when it exists. Neither is enumerating all possible controlling assignments feasible. Formal methods are needed to efficiently verify whether the power module has priority over all the other modules in controlling the power switch.

This paper describes two dominant controllability check methods. The first is called Const-controllability and uses netlist optimization. For each node representing a power switch, this method tries different assignments to the set of specified inputs. For each assignment tried, the check applies fast netlist optimization to identify assignments which force the node to a constant regardless of the other inputs. If such an assignment is found, it is a controlling assignment.

The second method is called QBF-controllability and uses a QBF solver. This method creates a QBF formula which is true if and only if the node is controllable. This check uses a QBF solver that not only returns SAT/UNSAT/ABORT but also returns a controlling assignment for SAT cases. This assignment is used by the netlist optimizer to verify the correctness of the QBF results. This way QBF-controllability utilizes the technology of the QBF solver and the reliability of the netlist optimization. To the best of our knowledge this is the first time QBF is used for this application.

This paper also describes the use of a QBF preprocessor. During our work, we found that using a QBF preprocessor reduces runtimes for QBF solving and reduces the number of ABORT cases. However, additional work was needed to reconstruct the assignment after preprocessing the formula.

Finally, this paper describes Alternating-controllability, an algorithm which uses both QBF-controllability and Const-controllability. In four out of six designs its runtime is close to the best runtime of either method.

2   Background

The hardware design is given as a Boolean *netlist* that is a directed acyclic graph (DAG) with nodes corresponding to logic gates and directed edges corresponding to wires connecting the gates. The present work is concerned with combinational netlists only. *Fanin* edges drive a node. A *fanin cone* of a node n is the subset of all nodes of the netlist reachable transitively from n through fanin edges. A *primary input* (PI) is a node without *fanin* edges. $PIs(n)$ is the set of all the PI nodes in the fanin cone of n. PIs is the union of all $PIs(n)$. An *assignment* $A(v_i, p_i)$ forces a value $v_i \in \{false, true\}$ on a primary input p_i. An *assignment* $A(\boldsymbol{v}, \boldsymbol{p})$ forces a vector $\boldsymbol{v}$ of Boolean values on the same size vector $\boldsymbol{p}$ of primary inputs. Note: there are $2^{|\boldsymbol{p}|}$ different assignments to $\boldsymbol{p}$. Given an assignment $A(\boldsymbol{v}, \boldsymbol{p})$ and a node n, we denote by $val(n, A(\boldsymbol{v}, \boldsymbol{p}))$ the value $true$ or $false$ of n under the assignment $A(\boldsymbol{v}, \boldsymbol{p})$. Note: it is enough to assign a value only to the primary inputs in $PIs(n)$ for the evaluation of n to be $false$ or $true$.

A *partial* assignment forces values on a subset $P_1 \subset PIs$ of the primary inputs and leaves the reset of the inputs unassigned. Under such partial assignment

$A(\boldsymbol{v_1}, \boldsymbol{p_1})$, the value of $val(n, A(\boldsymbol{v_1}, \boldsymbol{p_1}))$ is not always determined. In such cases we say that $val(n, A(\boldsymbol{v_1}, \boldsymbol{p_1})) = unknown$.

3   Dominant Controllability

Large hardware designs often include a power management module. This module includes a set C of distinguished primary inputs that can be set externally by the software which uses the design. C controls a set PS of power switches in the design. A power switch may also be controlled by other modules. In order for the power module to function correctly, the power module should have dominant control of the switch over the other modules. This allows the software to manage the power consumption depending on the actual usage.

Part of the hardware specification is the set C and the set of power switches PS. A power switch in the design is a node n in the *netlist* that can be *on* or *off* depending on the node's evaluation. The evaluation of a node n is only influenced by $PIs(n)$. Therefore, for each node $n \in PS$, only the subset $C_n = C \cap PIs(n)$ needs to be analyzed.

Let $\boldsymbol{C_n}$ denote a vector of all the elements of C_n and $\boldsymbol{O_n}$ denote a vector of all elements of $PIs(n) \setminus C_n$. Recall that $\oplus$ denotes vector concatenation.

Definition 1. Dominant Controllability *A node n is dominant controllable by $C_n \subseteq C$ if and only if $\forall v \exists v \forall u.(val(n, A(\boldsymbol{v}, \boldsymbol{C_n}) \oplus A(\boldsymbol{u}, \boldsymbol{O_n})) == v)$, where $\boldsymbol{v}$ and $\boldsymbol{u}$ are vectors of Boolean variables.*

In other words, the set of inputs C_n has dominant controllability over n if there exist two assignments $A_0 = A(\boldsymbol{v_0}, \boldsymbol{C_n})$ and $A_1 = A(\boldsymbol{v_1}, \boldsymbol{C_n})$ that set n to *false* and *true* respectively, regardless of the other inputs' values. In the rest of this paper, we use controllable in the sense of Dominant Controllability.

4   Controllability Check Using Netlist Optimization

This approach uses an optimizer which implements fast (linear runtime) netlist optimization algorithms that were introduced for SAT sweeping [5,10,13]. Given a node n and an assignment $A_1 = A(\boldsymbol{v}, \boldsymbol{C_n})$, the optimizer returns a reduced cone of n denoted as $opt(A_1, n)$. The reduced netlist $opt(A_1, n)$ is computed recursively. First, the inputs which appear in A_1 are assigned to corresponding constant values and other inputs are left as free variables. Then every node n with a fanin $\langle n_1, n_2 \ldots \rangle$ is computed as $optimizer(n, \langle opt(A, n_1), opt(A, n_2) \ldots \rangle)$. For example if n is an output of an AND gate with fanins $\langle n_1, n_2 \rangle$ and $opt(A_1, n_1) = false$, then $optimizer$ returns the node $false$.

The optimizer has the following property: $val(opt(A, n), A) \iff val(n, A)$. Informally, the optimizer does not change the value of the node n. Let $\boldsymbol{C_n}$ be a vector of all the elements of C_n and $\boldsymbol{O_n}$ be a vector of all elements of $PIs \setminus C_n$. In addition let $\boldsymbol{F} = A(\boldsymbol{v}, \boldsymbol{C_n}) \oplus A(\boldsymbol{u}, \boldsymbol{O_n})$ be a full assignment to PIs. Then, $val(opt(A(\boldsymbol{v}, \boldsymbol{C_n}), n), A(\boldsymbol{v}, \boldsymbol{C_n})) \iff val(opt(\boldsymbol{F}, n), \boldsymbol{F})$, for any set of values $\boldsymbol{u}$

forced on O_n. Consequently, if a partial assignment forces a node to a constant value, any partial assignment to the rest of the inputs will not change the node's value. This observation is what enables the use of the optimizer in the Const-controllability check. The algorithm searches for two assignments $A_0=A(v_0, C_n)$ and $A_1=A(v_1, C_n)$ which force n to 0 and 1 respectively.

Const-Controllability checks if a node n is controllable by the subset C_n. Its runtime is controlled by a parameter $MaxTries$. The check uses the netlist optimizer to determine if an assignment $A(v, C_n)$ demonstrates controllability. The complexity of the check is linear in the size of n's cone, times $MaxTries$. The optimizer does not always return the smallest most optimized cone of n, since it uses fast linear algorithms and is not NP-complete like SAT. This means that while Const-Controllability is fast, it is incomplete.

Const-controllability search adapts to the size of C_n. If $|C_n|$ is very small (≤ 5), the algorithm enumerates all possible assignments to C_n. For larger $|C_n|$ it first tries two frequently used assignments, all zeros ($Zeros$) and all ones ($Ones$). Our experimental results show that when these assignments demonstrate controllability, Const-controllability runtime is only a few seconds. If A_0, A_1 are still not found after these two assignments, the check continues with randomly picked assignments from the range $0 \ldots 2^{|C_n|} - 1$.

If Const-Controllability finds both A_0 and A_1, it terminates, the controllability check passes and returns SAT. If the number of tries reaches $MaxTries$ and $MaxTries < 2^{|C_n|} - 1$ the check terminates and returns $ABORT_t$. If all possible assignments $0 \ldots 2^{|C_n|} - 1$ have been checked but A_0 or A_1 have not been found then the check fails and returns $ABORT_a$. The difference between $ABORT_t$ and $ABORT_a$ is that $ABORT_t$ results can change with larger $MaxTries$. In contrast, $ABORT_a$ remains the same for larger $MaxTries$. Note that since the optimizer is incomplete, Const-Controllability returns $ABORT_a$ and not UNSAT.

Const-Controllability check (presented in Fig. 1) uses two Boolean flags $found_0$ and $found_1$ to indicate whether A_0 and A_1 have been found. The check iterates until it reaches the smaller of $MaxTries$, $2^{|C_n|} - 1$ or until both $found_0$ and $found_1$ are set. In each iteration, the check sets some values (as discussed previously) on the inputs in C_n and calls the optimizer. The optimizer analyzes the cone of n and returns an optimized cone. If the optimized cone of n is a constant ($false, true$), the check sets the flags $found_0, found_1$ accordingly.

5   Controllability Check Using QBF-Solver

QBF-Controllability($C_n, n, MaxTries$) is presented in Fig. 2. It finds each of the assignments A_1 and A_0 in two separate steps. The parameter $MaxTries$ is used to control the resources used by the procedure and limits the maximum number of decisions made by the QBF solver. The QBF solver is modified to return the controlling assignment and will be presented in more detail in Section 6. When both A_1 and A_0 are found, the QBF-Controllability check passes and returns SAT. If a proof is found that either assignment does not exist, the check fails and returns UNSAT. In other cases the search terminates inconclusively and

function **Const-Controllability** $(C, n, MaxTries)$
1   $i = 0$, $found_0 = False$, $found_1 = False$, $A_0 = \emptyset$, $A_1 = \emptyset$
2   While $(i < 2^{|C|} \wedge i < MaxTries \wedge (\neg found_0 \vee \neg found_1))$
3   If $(2^{|C|} < MaxTries)$ Set $A(\boldsymbol{i}, \boldsymbol{C})$ // $\boldsymbol{i}$ is bitvector representation of i
4   Else
5   If $(i = 0)$ Set $A(\langle 0, 0, \dots, 0 \rangle, \boldsymbol{C})$ /* Zeros */
6   If $(i = 1)$ Set $A(\langle 1, 1, \dots, 1 \rangle, \boldsymbol{C})$ /* Ones */
7   If $(i > 1)$ Set $A(Random(2^{|C|}, \boldsymbol{C}))$
8   If $Eval(\emptyset, opt(A, n)) = False$
9       $found_0 = True$, $A_0 = A$
10   If $Eval(\emptyset, opt(A, n)) = True$
11      $found_1 = True$, $A_1 = A$
12    $i = i + 1$
13 If $found_0 \wedge found_1$ Return (SAT, A_0, A_1)
14 If $2^{|C|} < MaxTries$ Return $(ABORT_a, A_0, A_1)$
15 Return $(ABORT_t, A_0, A_1)$

Fig. 1. Pseudo–code Const-Controllability

returns ABORT. The check is complete (given enough time it returns SAT if and only if the node n is controllable) assuming the QBF solver itself is complete.

In order to test dominant controllability of node n to value $true$, the check uses a QBF solver to test if $\exists \boldsymbol{v} \forall \boldsymbol{u}.(val(n, A(\boldsymbol{v}, \boldsymbol{C_n}) \oplus A(\boldsymbol{u}, \boldsymbol{O_n})) == true)$ is satisfiable. If the QBF solver returns UNSAT, then we can conclude that there is no controlling assignment. If the QBF solver returns ABORT, we learn nothing.

function **QBF-Controllability**$(C_n, n, MaxTries)$
1   $(result, A_1) = $ QBF-Controllability-Value$(C_n, n, MaxTries, True)$
2   If $result == SAT$
3   $(result, A_0) = $ QBF-Controllability-Value$(C_n, n, MaxTries, False)$
4   Return $(result, A_1, A_0)$

Fig. 2. Pseudo–code QBF-Controllability

The function QBF-Controllability-Value$(C_n, n, MaxTries, val)$ is described in Fig. 3. It searches for a controlling assignment to the inputs in C_n that sets the node n to the value val. First, it calls GenQDIMACS(F) to translate the formula F which is a quantified netlist expression to a QDIMACS [1] formatted formula F_{qbf}. The generated QDIMACS instance is of the form $F_{qbf} = \exists V_0 \forall V_1 \exists V_2.CNF(n)$ where V_0 maps to C_n, V_1 maps to $PIs(n) \setminus C_n$ and V_2 has variables that do not correspond to primary inputs. The GenQDIMACS$(\dots)$ procedure has some implementation details which are outside the scope of this paper: it bitblasts the netlist, runs constant propagation and does some simple bitlevel optimizations during the translation.

Since we want to call the QBF solver on F_{qbf}, and then use the partial model returned by the QBF solver to infer an assignment to primary inputs in C_n, Gen-QDIMACS($\dots$) also returns a one-one mapping $M_{qbf} : C_n \to V_0$. If a variable $v \in V_0$ is assigned to *true* by the QBF solver, then the corresponding primary input $p = M_{qbf}^{-1}(v)$ is assigned to *true*, and so on.[1]

QBF-Solver($\dots$) returns SAT, UNSAT or ABORT. If F_{qbf} is SAT, the QBF solver also returns an existential prefix $A_{outer\_\exists}$ of the satisfying model. In this case, the optimizer evaluates $opt(A, n)$ to verify the returned controlling assignment. This verification step guards against bugs in the translation process from quantified netlist expression to QDIMACS and bugs in the QBF solver itself.

function **QBF-Controllability-Value** $(C_n, n, MaxTries, value)$
1    $F = \exists\boldsymbol{v}\forall\boldsymbol{u}.(val(n, A(\boldsymbol{v}, \boldsymbol{C_n}) \oplus A(\boldsymbol{u}, \boldsymbol{O_n})) == value)$
2    $(F_{qbf}, C_{qbf}) = \text{GenQDIMACS}(F)$
3    $(result, A_{outer\_\exists}) = \text{QBF-Solver}(F_{qbf}, MaxTries)$
4    $A = \emptyset$
5   If $(result == \text{SAT})$
6   foreach $(val, var\_id) \in A_{outer\_\exists}$
7          $A = A \cup (val, M_{qbf}^{-1}(var\_id))$
8   If $(opt(A, n) \neq val)$
9          $result = \text{ABORT}$
10   Return $(result, A)$

Fig. 3. Pseudo–code QBF-Controllability-Value

6   QBF-Solver Providing Existential Prefix Models

For our application, we need to check the satisfiablity of the formula $F = \exists S_0 \forall S_1 \exists S_2.\phi_{cnf}$ and if a satisfying model exists, we must return a Boolean assignment to all the variables in S_0 which can be extended to a model of F. (It's possible that a subset of S_0 is adequate for purposes of controllability, but we ignore that for sake of simplicity.)

The formula $\exists S_0 \forall S_1 \exists S_2.\phi_{cnf}$ is a quantified Boolean formula (QBF) in *prenex* form, with $\exists S_0 \forall S_1 \exists S_2$ being the quantifier prefix of F while propositional formula ϕ_{cnf} forms the *matrix* of F. This formula is *closed*, with S_0, S_1, S_2 being a partition of the set of all the Boolean variables that occur in ϕ_{cnf}. The variables in S_i happen to be at scope i. Here, scope 0 is the *outermost* scope, while scope 2 is the *innermost* scope. (While scanning the quantifier prefix from left to right, the scope increases whenever the quantifier type changes from *existential* $\exists$ to *universal* $\forall$ or vice versa.) The concepts of scope and *nesting* of scopes are useful for interpreting QBF formulas, and this can be done recursively. We illustrate

---

[1] The mapping M_{qbf} is used by GenQDIMACS($\dots$) during construction of F_{qbf} to add the quantifier prefix portion of the QDIMACS formula.

this informally here for prenex formulas. If $Q = \exists v.Q'$ ($Q = \forall v.Q'$) is a closed QBF formula and v does not occur in Q', then Q is satisfiable iff Q' is satisfiable. If v does occur in Q', then let $Q'|_{v=0}$ ($Q'|_{v=1}$) denote the result of *substituting* 0 (1) for the *free* occurrence of v in Q'. Then, if $Q = \exists v.Q'$ ($Q = \forall v.Q'$) is a closed QBF formula and v occurs in Q', we say that Q is satisfiable iff either (both) of $Q'|_{v=0}$ and $Q'|_{v=1}$ are satisfiable. Lastly, note that both $Q'|_{v=0}$ and $Q'|_{v=1}$ are closed prenex QBF formulas and the procedure can be repeated.

The above informal description of the semantics suggests the existence of *tree-like models* which witness the satisfiability of a closed prenex QBF formula. A tree-like model is a tree where the child nodes are variable assignments, and its structure is such that it demonstrates why the formula is satisfiable. In order to do so, a tree-like model satisfies the following constraints. First, the variable assignments along any path from the root to any leaf must form a satisfying assignment for the matrix ϕ_{cnf}. Second, any node corresponding to an assignment to a universal variable must have a sibling with the same variable assigned to the opposite value. Third, the scope of the variable associated with a node must be $\leq$ the scope of variables associated with all its descendants.

A more detailed presentation of preliminaries of QBF formulas, including scopes, may be found in [11]. Tree-like models are introduced and discussed in more detail in [15].

In this paper, when we use the word *model*, we have tree-like models in mind. We now introduce the concept of existential prefix model, as motivated by our application.

Definition 2. Existential Prefix Model *If* $F = \exists S_0 \forall S_1 \exists S_2.\phi_{cnf}$, *then we define an* existential prefix model *for* F *to be any assignment* A *to all variables in* S_0 *which can be extended to a model for* F. *We use the notation* $EPM(A, F)$ *to denote that* A *is an existential prefix model for* F.

Our QBF solver provides existential prefix models for QDIMACS instances. It combines a search/learning based QBF solver (depqbf-1.0 [11]) with a modified QBF preprocessor (bloqqer [3]) while maintaining the necessary bookkeeping information to calculate existential prefix models in case of a SAT result.

bloqqer is a QDIMACS preprocessor that transforms formula F to F' while preserving satisfiability. It does not guarantee that a model for F' is a model for F. In case F' is UNSAT, there is no existential prefix model for F because bloqqer preserves satisfiability. In case F' is SAT, then we can use an existential prefix model for F' together with knowledge of the transformations applied by the preprocessor to reconstruct an existential prefix model for F. We view the execution of bloqqer as a sequence of transformations of formulas $F = Q_0 \to Q_1 \to \ldots \to Q_{n-1} \to Q_n = F'$. As the transformations are applied, information needed to reconstruct the existential prefix model is pushed onto a stack. depqbf is then called on F'. If depqbf declares SAT, the reconstruction procedure queries the solver state in depqbf (collects the most recent assignments to variables in S_0) to determine an existential prefix model for $F' = Q_n$. With that initial seed, the reconstruction procedure then pops the reconstruction information from the stack to reconstruct an existential prefix model for

Q_{n-1} and continues in this fashion until the stack is empty and an existential prefix model for F is determined.

The key observation about depqbf is that when it declares a SAT result for $F = \exists S_0 \forall S_1 \exists S_2 . \phi_{cnf}$, it retains the most recent assignments to S_0, and these assignments can be extended to a model for F. Put differently, the solver terminates with SAT result when it derives a satisfied cube under the most recent assignment to S_0. Note that the solver maintains the most recent assignments to variables that have been eliminated from the matrix, for example by applications of pure literal rule or unit clause rule. As a result, no modifications were needed to depqbf itself since its API provides the necessary functions to inspect solver state.

The modifications needed to bloqqer were more extensive, and are related to partial model reconstruction. In the following, we review individual transformations $Q_{pre} \rightarrow Q_{post}$ applied by bloqqer and summarize the information that needs to be saved in order to reconstruct the existential prefix model. Note that the initial QBF formula is in prenex form with the matrix in CNF form, and each transformation applied by bloqqer maintains this form. To make the presentation easier, we introduce the following notation.

Definition 3. Outer(C,l) & Inner(C,l): *Given a clause C and a literal $l \in C$, we use the notation $Outer(C, l)$ for the set of all literals in $(C \setminus l)$ which have a scope outside $(\leq)$ the scope of l. The notation $Inner(C, l)$ denotes $C \setminus (Outer(C, l) \cup l)$.*

For example, in $\exists x \forall y \exists z . C_1 \wedge C_2$, where $C_1 = \{x, y, z\}$ and $C_2 = \{y, z\}$, we have $Outer(C_1, z) = \{x, y\}$ and $Inner(C_2, z) = \{\}$. Note that we treat the clause C as either a disjunction of literals or set of literals, based on the context.

1. *Unused Variable Removal:* A variable which is never used in ϕ_{cnf} can be dropped from the quantifier list. No information needs to be saved for reconstruction. If an unassigned variable is encountered during any reconstruction steps, we assign it to a concrete value.[2]
2. *Pure Literal Rule:* If an existential (universal) variable v occurs in ϕ_{cnf} in only one phase as literal l, then all clauses containing l (occurrences of l) can be removed from ϕ_{cnf}. If $v \in S_0$, we push the constraint l on the stack.
3. *Forall Reduction:* If the literal l with the greatest scope in a clause C is universal, then C can be replaced by $(C \setminus l)$. This is model preserving and hence no reconstruction is needed.
4. *Unit Clause Rule:* If a forall reduced clause C has only one (existential) literal l, then all occurrences of $\neg l$ are removed from ϕ_{cnf}. This is model preserving. (It is followed by pure literal rule application.)
5. *Subsumption Based Clause Deletion:* If clauses $C_1, C_2 \in \phi_{cnf}$ are such that $C_2 \subset C_1$, then clause C_1 can be deleted. This is model preserving.
6. *Hidden Literal Addition:* For clauses $C_1, C_2 \in \phi_{cnf}$, if there is a literal $l \in C_2$ such that $(C_2 \setminus l) \subset C_1$, then C_1 can be replaced with $(C_1 \cup \neg l)$. Since $C_1 \wedge C_2 = (C_1 \cup \neg l) \wedge C_2$ under these conditions, this is model preserving.

---

[2] If we dont assign a concrete value, the next time we see the variable during reconstruction, we may interpret the variable inconsistently.

7. *Forward/Backward Clause Strengthening:* This is the reverse direction of hidden literal addition and is model preserving.[3]

8. *Tautology Removal:* If a clause $C \in \phi_{cnf}$ contains both a literal l and its negation $\neg l$, then C can be deleted. This is model preserving.

9. *Covered Literal Addition:* With l being an existential literal, define N_l as being the set $\{C' \in \phi_{cnf} \mid \neg l \in C'\}$. Any clause $C \in \phi_{cnf}$ containing the existential literal l can then be replaced with $C \cup N_l^{outer}$, where $N_l^{outer} = \{\bigcap_{C' \in N_l} Outer(C', \neg l)\}$.

   Given a model[4] for Q_{post}, consider the subtrees of the model rooted at occurrences of $var(l)$. For each such subtree, the path from the root of the model to the root of the subtree contains assignments to all the variables in $Outer(C, l) \cup N_l^{outer}$. If N_l^{outer} is true under these assignments, then set $var(l)$ at the root of this subtree such that l is *true* and make C satisfied along all paths through the subtree. If N_l^{outer} is *false* under these assignments, then $Outer(C, l)$ and hence C must already be *true* along all paths of the model through this subtree. This process gives a model for Q_{pre}. For reconstruction of $EPM(Q_{pre})$, we push the constraint $N_l^{outer} \Rightarrow l$ on the stack when $var(l) \in S_0$.

10. *Blocked Clause Elimination:* With $l \in C$ being an existential literal, define N_l as being the set $\{C' \in \phi_{cnf} \mid \neg l \in C'\}$. If $Outer(C, l) \cup Outer(C', \neg l)$ is a tautology for each $C' \in N$, then the clause C can be deleted.

    Given a model for Q_{post}, consider the subtrees of the model rooted at occurrences of $var(l)$. For each such subtree, the path from the root of the model to the root of the subtree contains assignments to all the variables in $Outer(C, l)$. If $Outer(C, l)$ evaluates to *false* under this assignment, then set $var(l)$ such that l is *true*. Note that all paths through the subtree also satisfy all clauses in N_l. This process gives a model for Q_{pre}. Since we are interested in reconstructing $EPM(Q_{pre})$, we push the constraint $\neg Outer(C, l) \Rightarrow l$ on the stack when $var(l) \in S_0$.

11. *Hidden Tautology Removal:* This is a sequential application of one or more steps of hidden literal addition and covered literal addition, followed by tautology removal. Reconstruction information is pushed on the stack for each (speculative) covered literal addition during the hidden tautology check. If the clause is not found to be redundant, the speculative additions to the reconstruction stack are popped, though this is not needed for correctness.

12. *Hidden Blocked Clause Elimination:* This is a sequential application of one or more steps of hidden literal addition and covered literal addition, followed by blocked clause elimination. Similar to the case of hidden tautology removal, reconstruction information for covered literal additions is pushed speculatively on to the stack. If blocked clause elimination succeeds, then the reconstruction information for the blocked clause elimination is also pushed

---

[3] "forward" and "backward" refer to the relative chronological ages of the clauses involved. There are separate procedures in **bloqqer** for each direction.

[4] We really want a model such that all children of $var(l)$ have a strictly deeper scope. We get this by reordering variables in the same scope as $var(l)$ in a model for Q_{post}.

on the stack. Otherwise, the speculative additions to the stack are popped, though this is not needed for correctness.

13. *Equivalent literal removal:* If $\phi_{cnf} \Rightarrow (x == y)$, the scope of $var(x)$ is outside ($\leq$) the scope of $var(y)$, and $var(y)$ is an existential variable, then the literal substitutions $[y \mapsto x]$ and $[\neg y \mapsto \neg x]$ are applied to ϕ_{cnf}. If $var(y) \in S_0$, then the constraints $x \Rightarrow y$ and $\neg x \Rightarrow \neg y$ are pushed on the stack.

14. *Existential Variable Elimination:* With l being an existential literal, define P_l as being the set $\{C \in \phi_{cnf} \mid l \in C\}$ and N_l as being the set $\{C \in \phi_{cnf} \mid \neg l \in C\}$. Suppose that for each $(C_p, C_n) \in P_l \times N_l$, either $(Outer(C_p, l) \cup Outer(C_n, \neg l))$ is a tautology or $(Inner(C_p, l) \cup Inner(C_n, \neg l))$ is the empty set. Under these conditions, $P_l \cup N_l$ can be deleted from ϕ_{cnf} and replaced with $R = \{(C_p \setminus l \cup (C_n \setminus \neg l) \mid (C_p, C_n) \in P_l \times N_l\}$.

    Given a model for Q_{post}, consider the subtrees rooted at occurrences of $var(l)$. For each such subtree, the assignments along the paths through the subtree satisfy all clauses in R. Let A be the partial assignment implied by the path from the root of the model to the root of the subtree. If $Outer(C_p, l)$ is false under A for any of the $C_p \in P_l$, then modify the assignment to $var(l)$ at the root of this subtree such that l is *true*. Otherwise, modify the assignment to $var(l)$ at the root of this subtree such that l is *false*. This makes all paths through the subtree satisfy all clauses in $P_l \cup N_l$. This process gives a model for Q_{pre}. If $var(l) \in S_0$, push the constraint $(\bigwedge_{C_p \in P_l} Outer(C_p, l)) \Leftrightarrow \neg l$ on the stack.

15. *Forall Variable Expansion:* This transforms formula $Q_{pre} = \exists S_0 \forall S_1 \exists S_2.\phi_{cnf}$ to $Q_{post} = \exists S_0 \forall (S_1 \setminus v) \exists (S_2 \cup S_2^{copy}).\phi_{cnf} [v \mapsto 1] \wedge \phi_{cnf} [v \mapsto 0, S_2 \mapsto S_2^{copy}]$ where $v \in S_1$ and S_2^{copy} is a fresh set of newly introduced existential variables. An existential prefix model for Q_{post} restricted to S_0 is an existential prefix model for Q_{pre} and no reconstruction information needs to be saved.

16. *Clause Splitting:* Clause splitting is used by the preprocessor to split long clauses into shorter clauses by introducing new existential variables at the deepest scope of the formula. Given a model for Q_{post}, deleting the part corresponding to the newly added variables gives a model for Q_{pre}.

To summarize, we need to push reconstruction information on the stack for the following transformations - *Pure Literal Rule, Covered Literal Addition, Blocked Clause Elimination, Equivalent Literal Removal* and *Existential Variable Elimination*.

Finally, we conclude the section with a note on how we tested the QBF solver implementation. We used a fuzzer (Qbfuzz) [4] to generate a small random QDI-MACS instance F with 30–100 variables. In case the QBF solver returns UNSAT on F, the test harness confirms that depqbf does not return SAT on F. In case the QBF solver returns SAT and an existential prefix model A, the test harness confirms that depqbf returns SAT on F'' where F'' is F with unit clauses added to force the assignment A.

7   Controllability Check Using an Alternating Approach

Alternating-Controllability(C, PS) depicted below in Figure 4 checks that each power switch node in the set PS is controllable by a set of primary inputs C. This algorithm harnesses the benefits of the two methods: QBF-Controllability and Const-Controllability. It runs two rounds: a fast round with a time budget $TimeLimit$ for each method. Based on the winner of the first round, a second round is executed without a time limit ($None$).

In the first round, using Const-Controllability first is better because QBF-Controllability has a higher overhead. The first round limits the number of tries per node to $FastTries$ to avoid spending all the method's time budget on a single node. In the second round the order in which the methods are run is determined by the results of the first round. In this round there is no time limit. In addition a larger number of tries $MaxTries$ is used per node by each method.

The algorithm maintains the set PS_l of nodes left to verify. A node is removed from the set PS_l when one of the methods returns SAT or UNSAT for it. The algorithm reports ABORT for the remaining nodes in this set after the two rounds end.

function **Alternating-Controllability**(C, PS)
1   $Order = Const\_Controllability, QBF\_Controllability$
2   $PS_l = PS$
3   Foreach method in the sequence $Order$
4     $D[method] = $ Timed-Controllability$(method, C, PS_l, FastTries, TimeLimit)$
5     $PS_l = PS \setminus D[method]$
6   If $|D[Const\_Controllability]| < |D[QBF\_Controllability]|$
7     $Order = QBF\_Controllability, Const\_Controllability$
8   Foreach method in the sequence $Order$
9     $D[method] = $ Timed-Controllability$(method, C, PS_l, MaxTries, None)$
10    $PS_l = PS \setminus D[method]$
11 Report ABORT for all nodes in PS_l

Fig. 4. Pseudo–code Alternating-Controllability

Timed-Controllability$(method, C, PS, MaxTries, TimeLimit)$ depicted   in Figure 5 runs Const-Controllability or QBF-Controllability with a time budget. It returns all the nodes that are found to be SAT or UNSAT.

8   Related Work

Controllability is a property of a node in the netlist [2]. In combinational logic, controllability is the ability to set a node to a specified value by an assignment to the primary inputs. In Automatic Test Generation (ATG), controllability is a measure of the difficulty to set each signal to *true* and *false*. A precise measure of such controllability is hard to define/compute and random controllability is

function **Timed-Controllability**$(method, C, PS, MaxTries, TimeLimit)$
1   Iterate n over all PS or until $runtime > TimeLimit$
2   If $method = Const\_Controllability$
3   Const-Controllability$(C, n, MaxTries)$
4   Else
5   QBF-Controllability$(C, n, MaxTries)$
6   Return all nodes $n \in PS$ that resulted in SAT or $UNSAT$

Fig. 5. Pseudo–code Timed-Controllability

used for some applications. Random controllability looks at the probability of setting a signal to $true$ or $false$ using random assignments to the primary inputs. Our work focuses on a specific concept : Dominant-controllability. This notion of controllability emphasizes the element of dominance of a subset of inputs over other inputs to the controllability concept. In recent complicated designs such as GPU and CPU the specification of the power control module includes the dominance element in order to allow the software running the design to control the power consumption. Our work applies formal methods to verify Dominant-controllability.

Three valued simulation can also be used for Const-Controllability instead of the optimizer. If the value of n is a constant, then the requested assignment to C_n is found. However, three valued simulation may be too pessimistic (the value of n is X, even though it should be a constant) and may result in missing a valid assignment. For example, an AND gate with two inputs In and $\neg In$ may evaluate to X when it is actually $false$. In addition, when the number of inputs in C_n is very large it is not practical to enumerate all possibilities.

In our work, we use the QBF solver `depqbf` and the QBF preprocessor `bloqqer`. For search/learning based QBF solvers such as `depqbf`, there is work related to certificate generation for QBF results [14,16]. There is also related work that involves model/proof reconstruction after steps including preprocessing, such as [7,8,12]. These efforts are aimed at a more general setting than our application, which is limited to the significantly simpler problem of reconstructing existential prefix models for satisfiable QBF formulas. Specifically, we don't need to generate a full certificate since we are only interested in an existential prefix model. Our verification of SAT results is done using a constant propagation and optimization process which is very fast and efficient. Since we accept ABORT results for our application, we did not feel the cost of certificate generation and verification was justified for the case of UNSAT results. Since our work is more focused on existential prefix model reconstruction after preprocessing by `bloqqer`, we would also like to point to what we feel is the closest related work which is actually for solution reconstruction after preprocessing of CNF formulas in the domain of SAT solvers [9]. We also note that some of the specific techniques used within `bloqqer` have been introduced in other preprocessors, for example, existential variable elimination in [6], and analogous procedures for existential prefix model reconstruction should apply.

9   Experimental Results

Table 1 presents the experimental results for Const-Controllability on six industrial designs. $D1$, $D2$, $D3$ and $D4$ are previously verified designs with no known bugs. $D5$ and $D6$ are newer designs with potential bugs. Each design specifies a control set C. Const-Controllability checks Dominant Controllability of each node (power switch) in the power management module. We controlled the runtime of Const-Controllability by limiting $MaxTries$ to 100 for each node.

The SAT column shows the number of controllable nodes. $ABORT_a$ shows the number of nodes that are not controllable even after the algorithm tests all $2^{|C|}$ possible assignments. The $ABORT_t$ column shows the number of nodes for which the algorithm found only one or none of the necessary assignments (A_0 or A_1) before hitting the threshold of 100 random assignments. The $Zeros$ ($Ones$) assignment sets all inputs to 0 (1). $1stRand$ is the first random assignment tried. The columns $Zeros$, $Ones$ and $1stRand$ show how many times these assignments were the controlling assignments A_0 or A_1. The column $\widetilde{PIs}(n)$ is the median size of $PIs(n)$, with the minimum and maximum values in parentheses. The column $\widetilde{C_n}$ is the median size of C_n, with the minimum and maximum values in parentheses. The averages are slightly higher, but the median represents the strength of the algorithm on the smaller examples. The last column is the total runtime for checking all nodes in seconds. Note: Since each node requires two assignments, the sum of $Zeros$, $Ones$ and $1stRand$ may be larger than the total number of nodes.

Our first observation is that the performance of Const-Controllability improves as the percentage of uncontrollable nodes decreases. The reason is that Const-Controllability is unable to conclude that a node is uncontrollable. Therefore, each node that is uncontrollable consumes runtime without contributing to the number of nodes resolved.

The second observation is that $Ones$ assignment is very useful in some designs like $D3$ where almost all of the power switches tend to use this assignment! (The $Ones$ assignment forces some nodes to 1 and others to 0).

The third observation is that many of the assignments are easy to guess. In D3, the first random assignment forces the node to 1 or 0 in 93% of the cases.

Last observation is that in newer designs the size of the controlling set is very small. For example in $D5$ the median size of C_n is only seven – for nodes with six control inputs, the algorithm terminates after enumerating all 64 assignments.

Table 2 presents the experimental results of QBF-Controllability. The table shows the results for the same six designs as Table 1. In the modified QBF solver, we run **bloqqer** preprocessor with no time limit and limit the number of decisions made by **depqbf** to 3,000. We tested all the SAT results using our optimizer and all of them were verified. We choose to limit the runtime using number of decisions to get deterministic results.

In all six designs, using the preprocessor reduced the total runtime and the number of timeouts. For example, without the preprocessor $D1$ has 30 timeouts and $D2$ has 27 timeouts.

Table 1. Const-Controllability

	SAT	$ABORT_a$	$ABORT_t$	Zeros	Ones	1stRand	$\widetilde{PIs}(n)$	$\widetilde{C_n}$	Runtime
D1	538	0	9	0	538	93	53(53–108)	33(33–74)	178
D2	195	0	0	0	195	76	30(30–280)	21(21–154)	126
D3	4,518	0	0	0	4,517	4,203	94(58–305)	58(37–98)	12,044
D4	508	0	26	0	488	272	88(34–449)	59(25–236)	13,226
D5	0	0	805	1	0	0	36(7–74)	7(2–46)	147,506
D6	0	114	293	1	0	0	73(15–98)	34(2–61)	64,549

Table 2. QBF-Controllability

	SAT	UNSAT	ABORT	#clauses	$|\exists|$	$|\forall|$	Runtime
D1	547	0	0	592(489–2,435)	39(33–74)	23(19–31)	755
D2	194	0	1	667(174–4,005)	36(21–154)	23(9–124)	356
D3	3,810	0	708	1k(0.9k–4k)	64(37–98)	60(32–203)	24,215
D4	352	0	182	1.5k(0.2k–7.5k)	46(25–236)	79(9–402)	3,281
D5	0	789	16	7k(0.4k–244k)	13(2–46)	281(16–9,645)	27,347
D6	0	399	8	4k(0.4k–111k)	14(2–61)	217(16–4,854)	8,818

The columns *SAT*, *UNSAT* and *ABORT* show the distribution of QBF-solver results. *#clauses* is the average number of clauses for each example. The column $|\exists|$ has the average number of variables in the outermost existential quantifier, with the min and max values in parentheses. The column $|\forall|$ has the average number of variables in the outermost universal quantifier, with the min and max values in parentheses. *Runtime* is the total runtime of checking all the nodes in seconds.

Our first observation is that the performance of QBF-Controllability improves as the percentage of bugs increases. In these cases, the solver quickly realizes that there is no way to satisfy the formula and returns UNSAT.

Our second observation is that on simple cases, QBF-Controllability is slower than Const-Controllability. There are a few possibilities which explain this. First is that Const-controllability is using the *Ones* assignment which has a high rate of success on these designs. Second, the QBF-Controllability makes two calls to the modified solver per node, while Const-Controllability runs a single loop to look for both assignments. Lastly, QBF-Controllability incurs an overhead for generating the QDIMACS instance while Const-Controllability operates natively on the netlist.

Table 3 presents the experimental results of Alternating-Controllability. The table shows the results for the same six designs as Table 1. The columns *SAT*, *UNSAT* and *ABORT* show the distribution of the final results at the end of both rounds. The *FirstRound* and *SecondRound* columns show the number of SAT and UNSAT results found by QBF-Controllability and Const-controllability during each round. *Runtime* is the total runtime in seconds for checking all the nodes.

Our first observation is that the winner in the first round is indeed the faster method for all designs. For example, Table 1 and Table 2 show that $D1$ is faster when using Const-controllability compared to QBF-controllability, and Const-controllability is indeed the winner of the first round in this case.

Our second observation is that the total runtime of Alternating-Controllability is closer to the best of Const-Controllability and QBF-Controllability than to the sum of them. For example the runtime for $D5$ is 67k seconds. It is closer to the best runtime 27k seconds (QBF-Controllability) than to the sum (174k).

Table 3. Alternating-Controllability

	SAT	*UNSAT*	*ABORT*	First Round		Second Round		*Runtime*
				QBF	Const	QBF	Const	
D1	547	0	0	9	538	0	0	262
D2	195	0	0	1	194	0	0	157
D3	4,518	0	0	68	92	0	4,358	13,221
D4	511	0	26	0	14	0	497	15,115
D5	0	789	16	10	0	779	0	67,660
D6	0	399	8	19	0	280	0	15,292

10 Future Work

A promising future direction is to use a SAT solver instead of the netlist optimizer in Const-Controllability. When testing if a node n is forced to *true* (*false*) by a partial assignment to the set C_n we can query a SAT solver to check if there is any assignment to the rest of the free inputs in $PIs(n)$ which sets n to *false* (*true*). If the SAT solver returns UNSAT then we will conclude that n is forced by this assignment to *true* (*false*). Since the SAT instances for each node will be closely related we will look into using an incremental SAT solver.

Another future direction is to improve the resource control and reduce the overhead of the translation from netlist to QDIMACS. A circuit based QBF solver which is capable of providing existential prefix models can avoid the translation overhead. Another possibility is to evaluate the performance of QBF solvers which can utilize non-prenex and non-CNF formula representations.

References

1. QDIMACS Standard Version 1.1 (released on December 21, 2005)
2. Abramovici, M., Breuer, M.A., Friedman, A.D.: Digital Systems Testing and Testable Design. Computer Science Press (1990)
3. Biere, A., Lonsing, F., Seidl, M.: Blocked Clause Elimination for QBF. In: Bjørner, N., Sofronie-Stokkermans, V. (eds.) CADE 2011. LNCS (LNAI), vol. 6803, pp. 101–115. Springer, Heidelberg (2011)

4. Brummayer, R., Lonsing, F., Biere, A.: Automated Testing and Debugging of SAT and QBF Solvers. In: Strichman, O., Szeider, S. (eds.) SAT 2010. LNCS, vol. 6175, pp. 44–57. Springer, Heidelberg (2010)
5. Darringer, J.A., Joyner Jr., W.H., Berman, C.L., Trevillyan, L.: Logic Synthesis Through Local Transformations. IBM Journal of Research and Development 25(4), 272–280 (1981)
6. Giunchiglia, E., Marin, P., Narizzano, M.: sQueezeBF: An Effective Preprocessor for QBFs based on Equivalence Reasoning. In: Strichman, O., Szeider, S. (eds.) SAT 2010. LNCS, vol. 6175, pp. 85–98. Springer, Heidelberg (2010)
7. Goultiaeva, A., Van Gelder, A., Bacchus, F.: A Uniform Approach for Generating Proofs and Strategies for Both True and False QBF Formulas. In: Proceedings of the Twenty-Second International Joint Conference on Artificial Intelligence, IJCAI 2011, vol. 1, pp. 546–553. AAAI Press (2011)
8. Janota, M., Grigore, R., Marques-Silva, J.: On QBF Proofs and Preprocessing. CoRR abs/1310.2491 (2013)
9. Järvisalo, M., Biere, A.: Reconstructing Solutions After Blocked Clause Elimination. In: Strichman, O., Szeider, S. (eds.) SAT 2010. LNCS, vol. 6175, pp. 340–345. Springer, Heidelberg (2010)
10. Kuehlmann, A.: Dynamic Transition Relation Simplification for Bounded Property Checking. In: ICCAD, pp. 50–57. IEEE Computer Society / ACM (2004)
11. Lonsing, F., Biere, A.: Integrating Dependency Schemes in Search-based QBF Solvers. In: Strichman, O., Szeider, S. (eds.) SAT 2010. LNCS, vol. 6175, pp. 158–171. Springer, Heidelberg (2010)
12. Marin, P., Miller, C., Lewis, M., Becker, B.: Verification of Partial Designs Using Incremental QBF Solving. In: Design, Automation Test in Europe Conference Exhibition DATE 2012, pp. 623–628 (March 2012)
13. Mishchenko, A., Chatterjee, S., Brayton, R.K., Eén, N.: Improvements to Combinational Equivalence Checking. In: ICCAD, pp. 836–843 (2006)
14. Niemetz, A., Preiner, M., Lonsing, F., Seidl, M., Biere, A.: Resolution-based Certificate Extraction for QBF. In: Cimatti, A., Sebastiani, R. (eds.) SAT 2012. LNCS, vol. 7317, pp. 430–435. Springer, Heidelberg (2012)
15. Samulowitz, H., Davies, J., Bacchus, F.: Preprocessing QBF. In: Benhamou, F. (ed.) CP 2006. LNCS, vol. 4204, pp. 514–529. Springer, Heidelberg (2006)
16. Yu, Y., Malik, S.: Validating the Result of a Quantified Boolean Formula (QBF) Solver: Theory and Practice. In: Tang, T. (ed.) ASP-DAC, pp. 1047–1051. ACM Press (2005)

Fast DQBF Refutation

Bernd Finkbeiner and Leander Tentrup

Saarland University

Abstract. Dependency Quantified Boolean Formulas (DQBF) extend QBF with Henkin quantifiers, which allow for non-linear dependencies between the quantified variables. This extension is useful in verification problems for incomplete designs, such as the partial equivalence checking (PEC) problem, where a partial circuit, with some parts left open as "black boxes", is compared against a full circuit. The PEC problem is to decide whether the black boxes in the partial circuit can be filled in such a way that the two circuits become equivalent, while respecting that each black box only observes the subset of the signals that are designated as its input. We present a new algorithm that efficiently refutes unsatisfiable DQBF formulas. The algorithm detects situations in which already a subset of the possible assignments of the universally quantified variables suffices to rule out a satisfying assignment of the existentially quantified variables. Our experimental evaluation on PEC benchmarks shows that the new algorithm is a significant improvement both over approximative QBF-based methods, where our results are much more accurate, and over precise methods based on variable elimination, where the new algorithm scales better in the number of Henkin quantifiers.

1 Introduction

Dependency Quantified Boolean Formulas (DQBF) are an extension of QBF which allows for non-linear dependencies between quantified variables. Non-linear dependencies occur naturally in verification problems for incomplete designs, such as the *partial equivalence checking* (PEC) problem [10], where a partial circuit, with some parts left open as "black boxes", is compared against a full circuit. The inputs to the circuit are modeled as universally quantified variables and the outputs of the black boxes as existentially quantified variables. Since the output of a black box should only depend on the inputs that are actually visible to the black box, we need to restrict the dependencies of the existentially quantified variables to subsets of the universally quantified variables.

There has been some success in extending standard techniques of QBF solving to DQBF [3,9,10], but, generally, it has proven very difficult to scale the classic algorithms to larger DQBF problems. Fröhlich et al. conclude, based on experiments with various techniques from DPLL-based SAT/QBF-solving, including unit propagation, pure literal reduction, clause learning, selection heuristics and watched literal schemes, that "it does not perform very well" [9].

A much faster alternative to such precise methods is to *approximate* the dependencies, such that all dependencies become linear and DQBF thus simplifies

C. Sinz and U. Egly (Eds.): SAT 2014, LNCS 8561, pp. 243–251, 2014.

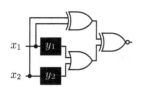

Fig. 1. Example of a partial equivalence checking (PEC) problem. A partial design, consisting here of the two black boxes and the OR gate, is compared to the reference circuit above, here consisting of a single XOR gate. The output of the complete circuit is 1 iff the completion of the partial design and the reference circuit compute the same result.

to QBF. For the PEC problem, an overapproximation of the dependencies is still useful to find errors (if the black box cannot be implemented with additional inputs, then it can, for sure, not be implemented according to the original design), but it significantly decreases the accuracy, because errors that result precisely from the incomparable dependencies of the black boxes are no longer detected. Consider, for example, the toy PEC problem shown in Fig. 1, where we ask whether it is possible to implement the XOR gate at the top as an OR of the two black boxes below, which *each* only see *one* of the two inputs x_1 and x_2. This is obviously not possible; however, the three overapproximating linearizations $\forall x_1 \forall x_2 \exists y_1 \exists y_2$, $\forall x_1 \exists y_1 \forall x_2 \exists y_2$ and $\forall x_1 \exists y_2 \forall x_2 \exists y_1$ all result in a positive answer, because an output that depends on both x_1 and x_2 can compute $x_1 \oplus x_2$, which gives the correct result, assuming that the other black box simply outputs constant 0.

In this paper, we present a new algorithm for DQBF that combines the efficiency of the QBF abstraction with the accuracy of the classic methods [3,9,10]. We focus on the *refutation* of DQBF, because this corresponds to the identification of errors in the PEC problem. Our algorithm identifies situations in which already a subset of the possible assignments of the universally quantified variables suffices to rule out a satisfying assignment of the existentially quantified variables. We call assignments to the universal variables *paths*. In the PEC example from Fig. 1, there are 4 possible paths $(x_1 x_2, x_1 \overline{x_2}, \overline{x_1} x_2, \overline{x_1} \overline{x_2})$. However, already 3 paths[1], $x_1 x_2, x_1 \overline{x_2}$, and $\overline{x_1} \overline{x_2}$, suffice to rule out a satisfying assignment for the existential variables: Since y_1 does not depend on x_2, its value must be the same for $x_1 x_2$ and for $x_1 \overline{x_2}$; likewise, the value of y_2 must be the same for $x_1 \overline{x_2}$ and $\overline{x_1} \overline{x_2}$. For $x_1 x_2$, both y_1 and y_2 must be 0, because $1 \oplus 1 = 0$. However, if $y_1 = 0$ for $x_1 x_2$, then $y_1 = 0$ also for $x_1 \overline{x_2}$, which leads to a contradiction, because y_2 must be equal to 1 for $x_1 \overline{x_2}$ because $1 \oplus 0 = 1$ and, at the same time, equal to 0, because $0 \oplus 0 = 0$. In our algorithm, we specify the existence of such a set of paths as a QBF formula. We iteratively increase the number of paths to be considered and terminate as soon as a satisfying assignment is ruled out.

The proofs can be found in the full version of this paper [8].

Related Work. DQBF was first defined by Peterson and Reif [15] and gained more attention recently [1,10]. The first investigation of practical methods for DQBF solving is a DPLL-based approach due to Fröhlich et al. [9]. In addition, there is an expansion-based solver [10]. It is also possible to reduce DQBF

---

[1] Later in the paper, we present an optimization that further reduces the number of paths required in this example to just two.

to SAT, using skolemization methods similar to those originally developed for
QBF [3]. To the best of our knowledge, there is, however, so far no publicly available DQBF solver. The idea of partial expansions was used in previous work,
e.g., there is a two-phase proof system for QBF that expands certain paths and
than refutes the formula by propositional resolution [11]. The verification of in-
completely specified circuits has received significant attention (cf. [14, 16]); the
connection between DQBF and the PEC problem was first pointed out by Gitina
et al. [10]. On a more general level, the verification of partial designs is related to
the synthesis problems for reactive systems with incomplete information and for
distributed systems (cf. [6, 12]). In previous work, we have proposed an efficient
method for disproving the existence of distributed realizations of specifications
given in linear-time temporal logic (LTL) [7] that bounds, similar to the approach
of this paper, the number of paths under consideration.

2   DQBF

Let $\mathcal{V}$ be a finite set of propositional variables. We use the convention to denote
the set of all universal variables $\mathcal{X}$, an element $x \in \mathcal{X}$, and a subset $X \subseteq \mathcal{X}$ ($y \in$
$Y \subseteq \mathcal{Y}$ for existential variables, respectively). The standard form of DQBF [9] is

$$\forall x_1. \forall x_2. \forall x_3 \ldots \exists_{H_1} y_1. \exists_{H_2} y_2. \exists_{H_3} y_3 \ldots \varphi \ , \tag{1}$$

that is, formulas beginning with universal quantified variables followed by the
existentially quantified *Henkin quantifier* and the quantifier-free *matrix* φ. A
Henkin quantifier $\exists_H y$ explicitly states the dependency for variable y by its
support set $H \subseteq \mathcal{X}$, which is the difference to QBF, where the preceding universal quantification determines the dependency of an existential variable. For
the matrix φ we allow negation $\neg$, disjunction $\vee$, conjunction $\wedge$, implication $\rightarrow$,
equivalence $\leftrightarrow$, exclusive or $\oplus$, and the abbreviations *true* $\top$ and *false* $\bot$.

   A DQBF formula Φ is satisfiable, if there exists a *Skolem function* f_y for each
existential variable $y \in \mathcal{Y}$, such that for all possible assignments of the universal
variables $\mathcal{X}$, the Skolem functions evaluated on these assignments satisfy the
matrix. An assignment is a function $\alpha : \mathcal{V} \rightarrow \{0, 1\}$ and a Skolem function
$f_y : (H_y \rightarrow \{0, 1\}) \rightarrow (\{y\} \rightarrow \{0, 1\})$ maps assignments of the dependencies to
an assignment of y. We identify an assignment α by a set $\{v \in \mathcal{V} \mid \alpha(v) = 1\} \subseteq$
$2^{\mathcal{V}}$ and f_y by a function $2^{H_y} \rightarrow 2^{\{y\}}$. We represent a function f_y as a binary
decision tree (BDT), where the branching of the tree represents the assignment
of the dependencies and f_y serves as the labeling function for the leaves, see
Fig. 2(a)–(c) for examples of BDTs. As a notation for paths, we use sequences
of (possibly negated) variables $x \in \mathcal{X}$, e.g., the path $x_1 \overline{x_2}$ is a shorthand for
the assignment $\{x_1\}$. A path $X \subseteq \mathcal{X}$ of a BDT satisfies a propositional formula
φ, if the assignment $X \cup f_y(X)$, i.e., the joint assignment of this path and the
respective labeling, satisfies φ. A *model* $\mathcal{M}$ of a satisfiable formula Φ is a binary
decision tree over $\mathcal{X}$ such that (1) every path in the tree satisfies the matrix
and (2) the labels of the leaves are *consistent* according to the dependencies of
the existential variables, i.e., there exists a decomposition of the decision tree

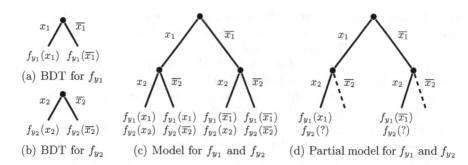

(a) BDT for f_{y_1}

(b) BDT for f_{y_2} (c) Model for f_{y_1} and f_{y_2} (d) Partial model for f_{y_1} and f_{y_2}

Fig. 2. The figure shows the binary decision trees for f_{y_1} (a) and f_{y_2} (b), their composition (c), and a partial model (d) for the PEC problem in Fig. 1

into individual Skolem functions f_y for each $y \in \mathcal{Y}$. For example, the Skolem functions of a satisfiable DQBF formula

$$\forall x_1, x_2. \exists_{\{x_1\}} y_1. \exists_{\{x_2\}} y_2. \varphi \tag{2}$$

are the unary functions f_{y_1} and f_{y_2}, depicted in Fig. 2(a) and (b), respectively. Figure 2(c) shows the corresponding model, that is the composition of f_{y_1} and f_{y_2}. In this representation, the incomparable dependencies become visible: Despite the branching of the tree by both variables x_1 and x_2, the results of the Skolem functions f_{y_1} and f_{y_2} must be equal on paths that cannot be distinguished according to the dependencies, e.g., as y_2 does not depend on x_1, the paths $x_1 x_2$ and $\overline{x_1} x_2$ are indistinguishable for f_{y_2} and the result on both paths is $f_{y_2}(x_2)$. A DQBF formula is *unsatisfiable* if there does not exist a model, i.e., for all *candidate models* always at least one path violates the matrix φ.

QBF Approximation. Given a DQBF formula Φ, a QBF formula Ψ with the same matrix is an approximation of Φ, written $\Phi \preceq \Psi$ if for all existential variables $y \in \mathcal{Y}$ it holds that $H_y \subseteq X_y$, where H_y is the support set of y and $X_y \subseteq \mathcal{X}$ is the dependency set of y in the QBF formula. Given two QBF approximations Ψ and Ψ', we call Ψ *stronger* than Ψ', written $\Psi \preceq \Psi'$, if for all $y \in \mathcal{Y}$ it holds that $X_y \subseteq X'_y$ [10]. In (2), y_1 and y_2 have *incomparable* dependencies as neither $\{x_1\} \subseteq \{x_2\}$ nor $\{x_2\} \subseteq \{x_1\}$. Hence, in all strongest QBF approximations, that is $\forall x_1 \exists y_1 \forall x_2 \exists y_2$ and $\forall x_2 \exists y_2 \forall x_1 \exists y_1$, at least one existential variable has more dependencies than before. The resulting inaccuracy was already highlighted in the introduction on the PEC problem from Fig. 1, which corresponds to formula (2) with matrix $\varphi = (y_1 \vee y_2) \leftrightarrow (x_1 \oplus x_2)$. All QBF abstractions of (2) are satisfiable despite the DQBF formula being unsatisfiable.

Variable Elimination. The expansion based method for converting a DQBF formula Φ into a logically equivalent QBF formula Ψ [1, 5] uses the idea of unrolling the binary decision tree, e.g., expanding (2) by x_1 gives us:

$$\forall x_2. \exists_{\{x_2\}} y_2. \exists_\emptyset y_1, y'_1. \varphi|_{x_1=0} \wedge \varphi'|_{x_1=1} , \tag{3}$$

where $\varphi|_{x=b}$ denotes the formula φ where all occurrences of x are substituted by b and φ' is the formula obtained from φ by replacing all occurrences of y_1 by y_1'. In the expansion of x_1, only variable y_1 is duplicated to represent the different choices of the Skolem function f_{y_1} on the paths that differ in the assignment of x_1. Likewise, variable y_2 is not duplicated. After the expansion of all universal variables, the resulting existential QBF formula can be solved by a SAT solver.

3   Bounded Unsatisfiability

Instead of expanding the whole binary decision tree it is often possible to determine unsatisfiability with only a subset of the assignments to the universal variables. We introduce the notion of partial models that are decision trees which contain only a subset of the original paths. Formally, a *partial model* $\mathcal{P}$ of a DQBF formula Φ is a decision tree over $\mathcal{X}$ consisting of paths $P \subseteq 2^{\mathcal{X}}$ such that (1) every path in the tree satisfies the matrix and (2) the labels of the leaves are *consistent* according to the dependencies of the existential variables and the selected paths P. As partial models are weaker than models, the existence of a partial model does not imply the existence of a model, but from the non-existence of a partial model follows the non-existence of a model.

Lemma 1. *Given a DQBF formula Φ and a set of paths $P \subseteq 2^{\mathcal{X}}$. Φ is unsatisfiable if there does not exist a partial model over P.*

We turn the idea of non-existing partial models into the bounded unsatisfiability problem that limits the number of paths under consideration in order to show that no partial model exists. For a $k \geq 1$, a DQBF formula Φ is *k-bounded unsatisfiable* if there exists a set of paths $P \subseteq 2^{\mathcal{X}}$ with $|P| \leq k$ such that there does not exist a partial model over P.

Theorem 2. *A DQBF formula Φ is unsatisfiable iff it is k-bounded unsatisfiable for some $k \geq 1$.*

4   From DQBF to QBF

We give an encoding of the k-bounded unsatisfiability problem to QBF for a fixed bound k. Before presenting the general encoding, we show the basic steps on formula (2) $\forall x_1, x_2. \exists_{\{x_1\}} y_1. \exists_{\{x_2\}} y_2. \varphi$. The formula is unsatisfiable iff for all candidate models, there exists a path that violates the matrix φ. Instead of expanding all four paths, we restrict the binary decision tree on two paths (but do not choose which one) and encode the search for the paths as QBF formula

$$\exists x_1^1, x_1^2, x_2^1, x_2^2. \forall y_1^1, y_1^2. \forall y_2^1, y_2^2. \neg\varphi^1 \vee \neg\varphi^2 \ , \tag{4}$$

that asserts that either path violates φ. This, however, does not accurately represent the incomparable dependencies of y_1 and y_2. For the assignment depicted in Fig. 2(d), where only x_1 has a different assignment on the two paths, y_2^1 and

y_2^2 can have different assignment as well, despite the fact that y_2 does not depend on x_1. To fix this inaccuracy, we introduce a *consistency condition* that ensures the restricted choices across multiple paths. For example, the consistency condition for y_2 in (4) would be $(y_2^1 \leftrightarrow y_2^2) \vee (x_2^1 \leftrightarrow x_2^2)$, i.e., either the assignment of y_2 is equal on both paths, or the assignment of the dependency x_2 is different on both paths. In the following, we describe the general encoding.

We build a QBF formula Ψ that encodes the k-bounded unsatisfiability problem, i.e., for a given bound k, the satisfaction of Ψ implies that Φ is unsatisfiable. In the encoding, we introduce k copies of the existential and universal variables in the DQBF formula Φ. Moreover, we specify a consistency condition that enforces that the universal variables can only act according to the assignment of the dependencies given by the support sets.

$$\text{bunsat}(\Phi, k) := \exists x_1^1, \ldots, x_m^1, x_1^2, \ldots, x_m^2, \ldots, x_m^k. \, \forall y_1^1, \ldots, y_n^1, y_1^2, \ldots, y_n^2, \ldots, y_n^k.$$
$$\text{consistent}(\{y_1, \ldots, y_n\}, k) \rightarrow \bigvee_{1 \leq i \leq k} \neg \varphi^k \, , \tag{5}$$

where φ^k denotes the formula φ where every variable v is replaced by v^k. The consistency condition is given by the formula

$$\text{consistent}(Y, k) := \bigwedge_{y \in Y} \bigwedge_{(i,j) \in \{1, \ldots, k\}^2} \left((y^i \leftrightarrow y^j) \vee \left(\bigvee_{x \in H_y} x^i \leftrightarrow x^j \right) \right) . \tag{6}$$

Theorem 3. *A DQBF formula Φ is unsatisfiable iff there exists a bound $k \geq 1$ such that the QBF formula bunsat(Φ, k) is satisfiable.*

Proposition 4. *Let Φ be a DQBF formula. For $k \geq 1$, the QBF formula bunsat(Φ, k) has $k \cdot |\mathcal{X}|$ existential and $k \cdot |\mathcal{Y}|$ universal variables, respectively, and the matrix is of size $\mathcal{O}(|\mathcal{Y}| \cdot k^2 \cdot \max_{y \in \mathcal{Y}} |H_y| + k \cdot |\varphi|)$.*

Reducing the Bound. One critical observation in the QBF encoding in (5) is that many paths are not needed for proving the unsatisfiability, but for enforcing consistency across the labels in the partial model. The reason for this is that in (5) we used the weakest QBF approximation of Φ. By using a stronger QBF abstraction, the QBF formula itself takes care for a part of the consistency condition. The stronger the QBF abstraction, the better is the dependency modeling and the fewer paths must be chosen.

Example. Consider again the PEC example from Fig. 1. As we have seen, there exist two strongest QBF approximations, but both are satisfiable due to overapproximation. However, we prove unsatisfiability by using a strongest QBF abstraction together with a bound of two: The formula

$$\exists x_1^1, x_1^2. \, \forall y_1^1, y_1^2. \, \exists x_2^1, x_2^2. \, \forall y_2^1, y_2^2. \, ((y_2^1 \leftrightarrow y_2^2) \vee (x_1^1 \leftrightarrow x_1^2)) \rightarrow (\neg \varphi^1 \vee \neg \varphi^2) \tag{7}$$

is satisfiable (choose arbitrary assignment α with $\alpha(x_1^1) \neq \alpha(x_1^2)$ and $\alpha(x_2^1) = \alpha(x_2^2)$). As the assignment of y_2 must be the same on both paths (otherwise it violates the consistency condition), it holds that either $\alpha(y_2) \neq \alpha(x_1^1)$ or $\alpha(y_2) \neq \alpha(x_1^2)$, hence the matrix is violated on either path.

Table 1. Results of the bounded unsatisfiability method on PEC examples

circuit	BBs	unsat.	bound 1 / 2		time	circuit	BBs	unsat.	bound 1 / 2		time
multi-plier	1	950	100%	100%	27.5%	multi-plexer	1	931	100%	100%	57.1%
	3	927	97.6%	100%	22.2%		3	908	97.9%	99.8%	48.0%
	5	924	87.6%	99.6%	17.9%		5	906	95.7%	98.5%	41.9%
	7	912	67.9%	95.9%	13.8%		7	896	92.5%	97.0%	35.5%
	9	870	30.9%	76.2%	16.2%		9	889	88.9%	94.7%	28.9%
adder	1	962	100%	100%	10.8%	look-ahead	1	999	100%	100%	4.5%
	3	959	100%	100%	9.0%		3	997	98.2%	100%	3.4%
	5	959	99.9%	100%	8.9%		5	996	97.1%	100%	3.3%
	7	951	99.5%	100%	7.0%		7	996	94.2%	99.9%	0.7%
	9	957	98.5%	99.9%	6.8%		9	986	84.4%	99.1%	0.8%

The table shows the approx. ratio of the bounded-path prototype using a bound ≤ 2 and the median running time relative to the expansion solver (timeout after 5min per instance). For every number of black boxes, we generated 1000 random instances.

5   Experimental Results

We have implemented our new method as a prototype that uses the bounded un-satisfiability method together with a strongest QBF abstraction. In this section, we report on experiments carried out on a 2.6 GHz Opteron system. The QBF instances generated by our reduction are solved using a combination of the QBF preprocessor Bloqqer [4] in version 031 and the QBF solver DepQBF [13] in version 2.0. As a base of comparison, we have also implemented an expansion-based DQBF solver using the BDD library CUDD in version 2.4.2[2].

Table 1 shows the performance of our solver on a number of PEC benchmarks, including the arithmetic circuits *multiplier* (4-bit) and *adder* (32-bit), a 32-bit *lookahead* arbiter implementation, and a 32-bit *multiplexer* [17]. The PEC instances are created as follows: Starting with a circuit, we exchange a variable number of gates by black boxes and use one copy of the original circuit as the specification. Random faults are inserted by replacing exactly one gate with a different gate. With only one exception (instances with more than 7 black boxes of the *multiplier* instances), more than 94% of the instances were solved correctly with bound two, while the number of correctly solved instances by the QBF abstraction drops as low as 84.4%. The running times in Table 1 are given relative to the running time of the expansion-based solver. Our solver outperforms the expansion-based solver significantly, especially on benchmarks with a large number of black boxes. For example, with 9 black boxes, the difference ranges from 37% faster (*adder*) to more than 5 times faster (*lookahead*).

Table 1 indicates that an increase of the bound from 1, which corresponds to the plain QBF approximation, to 2 already results in a significant improvement of accuracy. Table 2 analyzes the impact of the bound on the approximation quality in more detail. Here, the benchmark is a circuit family from [10], depicted for two

---

[2] The source code of the benchmarks and tools are available at
http://react.uni-saarland.de/tools/sat14/

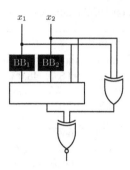

Fig. 3. XOR template

Table 2. Result of the XOR random function example

BBs	total	sat.	unsat.	bounded unsatisfiable			
				1	2	3	> 3
2	65536	32377	33159	22687	10472	0	0
			100%	68.4%	31.6%	0%	0%
3	50000	9273	40727	11257	26169	3115	186
			100%	27.6%	64.3%	7.6%	0.5%
4	50000	190	49810	5002	43781	1015	12
			100%	10.0%	87.9%	2.0%	< 0.1%

The table shows the approximation quality of the bounded-path prototype with respect to the number of black boxes.

black boxes in Fig. 3. The circuit uses the XOR of the inputs as specification and a random Boolean function as implementation, where black boxes with pairwise different dependencies serve as inputs to this function. For two black boxes, we created all $65536 = 2^{16}$ instances of Boolean functions with four inputs. For more than two, we selected a random subset of 50000 instances.

Table 2 shows an interesting correlation between the plain QBF approximation and the bounded-path method: The less effective the plain approximation, the more effective the bounded-path method. With an increasing number of black boxes, the number of solved instances by the plain QBF approximation (bound 1) decreases by more than a half with every black box. At the same time, the relative number of instances that are solved with a bound of at most two is always larger than 90%. With a bound of at most three, nearly all unsatisfiable instances are detected (> 99%). While the QBF approximation alone thus does not lead to satisfactory results, a comparatively small bound suffices to solve almost all instances.

6 Conclusion

We have presented a method for DQBF refutation that significantly outperforms the previous expansion-based approach on PEC benchmarks. The new method is based on an improved approximation of the DQBF formula within QBF, based on evaluating the formula on multiple paths and expressing dependency constraints as consistency conditions. Our experiments show that considering multiple paths significantly improves the accuracy, especially with an increasing number of black boxes. Compared to an expansion-based solver, the running time of our prototype implementation scales better with the number of black boxes. In the future, we plan to extend the method to sequential circuits and to integrate QBF certification [2] in order to identify faulty components.

Acknowledgements. This work was partially supported by the German Research Foundation (DFG) as part of SFB/TR 14 AVACS. We thank Christoph Scholl for comments on an earlier version of this paper.

References

1. Balabanov, V., Chiang, H.-J.K., Jiang, J.-H.R.: Henkin quantifiers and Boolean formulae. In: Cimatti, A., Sebastiani, R. (eds.) SAT 2012. LNCS, vol. 7317, pp. 129–142. Springer, Heidelberg (2012)
2. Balabanov, V., Jiang, J.H.R.: Unified QBF certification and its applications. Formal Methods in System Design 41(1), 45–65 (2012)
3. Benedetti, M.: Evaluating QBFs via symbolic skolemization. In: Baader, F., Voronkov, A. (eds.) LPAR 2004. LNCS (LNAI), vol. 3452, pp. 285–300. Springer, Heidelberg (2005)
4. Biere, A., Lonsing, F., Seidl, M.: Blocked clause elimination for QBF. In: Bjørner, N., Sofronie-Stokkermans, V. (eds.) CADE 2011. LNCS (LNAI), vol. 6803, pp. 101–115. Springer, Heidelberg (2011)
5. Bubeck, U., Kleine Büning, H.: Dependency quantified horn formulas: Models and complexity. In: Biere, A., Gomes, C.P. (eds.) SAT 2006. LNCS, vol. 4121, pp. 198–211. Springer, Heidelberg (2006)
6. Finkbeiner, B., Schewe, S.: Uniform distributed synthesis. In: LICS, pp. 321–330. IEEE Computer Society (2005)
7. Finkbeiner, B., Tentrup, L.: Detecting unrealizable specifications of distributed systems. In: Ábrahám, E., Havelund, K. (eds.) TACAS 2014. LNCS. vol. 8413, pp. 78–92. Springer, Heidelberg (2014)
8. Finkbeiner, B., Tentrup, L.: Fast DQBF refutation. Reports of SFB/TR 14 AVACS 97, SFB/TR 14 AVACS (2014), http://www.avacs.org, ISSN: 1860-9821
9. Fröhlich, A., Kovásznai, G., Biere, A.: A DPLL algorithm for solving DQBF. In: Proc. POS 2012 (2012)
10. Gitina, K., Reimer, S., Sauer, M., Wimmer, R., Scholl, C., Becker, B.: Equivalence checking of partial designs using dependency quantified Boolean formulae. In: ICCD, pp. 396–403. IEEE (2013)
11. Janota, M., Marques-Silva, J.: On propositional QBF expansions and Q-resolution. In: Järvisalo, M., Van Gelder, A. (eds.) SAT 2013. LNCS, vol. 7962, pp. 67–82. Springer, Heidelberg (2013)
12. Kupferman, O., Vardi, M.Y.: Synthesis with incomplete information. In: ICTL (1997)
13. Lonsing, F., Biere, A.: DepQBF: A dependency-aware QBF solver. JSAT 7(2-3), 71–76 (2010)
14. Nopper, T., Scholl, C., Becker, B.: Computation of minimal counterexamples by using black box techniques and symbolic methods. In: IEEE/ACM International Conference on Computer-Aided Design, ICCAD 2007, pp. 273–280 (November 2007)
15. Peterson, G.L., Reif, J.: Multiple-person alternation. In: 20th Annual Symposium on Foundations of Computer Science 1979, pp. 348–363 (October 1979)
16. Scholl, C., Becker, B.: Checking equivalence for partial implementations. In: Proceedings of the 38th Annual Design Automation Conference, DAC 2011, pp. 238–243. ACM, New York (2001)
17. William, J., Dally, R.C.H.: Digital Design, A Systems Approach. Cambridge University Press (2012)

Impact of Community Structure
on SAT Solver Performance

Zack Newsham[1], Vijay Ganesh[1],
Sebastian Fischmeister[1], Gilles Audemard[2], and Laurent Simon[3]

[1] University of Waterloo, Waterloo, Ontario, Canada
[2] Laboratoire Bordelais de Recherche en Informatique, Bordeaux Cedex, France
[3] Université Lille-Nord de France, CRIL - CNRS UMR 8188, Artois, F-62307 Lens

Abstract. Modern CDCL SAT solvers routinely solve very large industrial SAT instances in relatively short periods of time. It is clear that these solvers somehow exploit the structure of real-world instances. However, to-date there have been few results that precisely characterise this structure. In this paper, we provide evidence that the community structure of real-world SAT instances is correlated with the running time of CDCL SAT solvers. It has been known for some time that real-world SAT instances, viewed as graphs, have natural *communities* in them. A community is a sub-graph of the graph of a SAT instance, such that this sub-graph has more internal edges than outgoing to the rest of the graph. The community structure of a graph is often characterised by a quality metric called Q. Intuitively, a graph with high-quality community structure (high Q) is *easily separable* into smaller communities, while the one with low Q is not. We provide three results based on empirical data which show that community structure of real-world industrial instances is a better predictor of the running time of CDCL solvers than other commonly considered factors such as variables and clauses. First, we show that there is a strong correlation between the Q value and Literal Block Distance metric of quality of conflict clauses used in clause-deletion policies in Glucose-like solvers. Second, using regression analysis, we show that the the number of communities and the Q value of the graph of real-world SAT instances is more predictive of the running time of CDCL solvers than traditional metrics like number of variables or clauses. Finally, we show that randomly-generated SAT instances with $0.05 \leq Q \leq 0.13$ are dramatically harder to solve for CDCL solvers than otherwise.

1 Introduction

In the last few years, we have witnessed impressive improvements in the performance of conflict-driven clause-learning (CDCL) Boolean SAT solvers over real-world industrial SAT instances, despite the fact that the Boolean satisfiability problem is known to be NP-complete and the worst-case time complexity of our best solvers is exponential in the size of the formula. What is even more impressive is that these solvers perform extremely well even for never-before-seen classes of large industrial instances, where the biggest instances may have

C. Sinz and U. Egly (Eds.): SAT 2014, LNCS 8561, pp. 252–268, 2014.

upwards of 10 million clauses and millions of variables in them. In other words, one cannot reasonably argue that these solvers are being hand-tuned for every class of real-world instances. It is all but clear that CDCL solvers employ a very general class of techniques, that have been robustly implemented and continuously tested for many applications ranging from software engineering to AI. All of this begs the question why CDCL solvers are so efficient, and whether they are exploiting some structural features of real-world instances. It is this question that we address in this paper.

In this paper, we present three results that show that there is correlation between the presence of natural communities [6] in real-world SAT instances [3,4] and the running time of MiniSAT CDCL solver [7] (by extension many other CDCL SAT solvers that are either built using MiniSAT code or use the most important techniques employed by CDCL SAT solvers). Informally, a community [6] in a SAT formula, when viewed as the variable-incidence graph [1], is a sub-graph that has more edges internal to itself than going out to the remainder of the graph. There is previous work pointing to some correlation between community structure in SAT instances and performance of CDCL solvers [2]. However, we provide much stronger evidence as discussed in the Contributions sub-section below.

We characterise the structure of SAT instances through a well-known metric called the *Q value* [6] and the number of communities present in its graph. The Q value is a widely-accepted quality metric that measures whether the communities in a graph are *easily separable*. In particular, formulas [2] with high Q tend to have few inter-community edges relative to the number of communities, while those with low Q have lots of inter-community edges.

Contributions:

1. We show that there is a strong correlation between Literal Block Distance (LBD), introduced in a paper [5] by some of the authors on learnt clause quality, and number of communities. This correlation fits better and better as the search progresses. In their original paper [5], the authors suggested that the quality of a learnt clause can be measured using the LBD metric. I.e. the lower the LBD the better the learnt clause. They also suggested a learnt clause deletion policy, wherein clauses with high LBD were marked for deletion. The result we found in this paper suggests that low LBD clauses also are shared by very few communities.

2. We performed a regression analysis of the performance of the MiniSAT [7] solver over SAT 2013 competition instances [1], using a variety of factors that characterise Boolean formulas including number of variables, number of clauses, number of communities, Q and even ratios between some of these

---

[1] A variable-incidence graph of a Boolean SAT formula is one where the variables of the formula are nodes and there is an edge between two nodes if the corresponding variables occur in the same clause.

[2] In the rest of the paper, we do not distinguish a formula from its variable-incidence graph.

factors. We found that the number of communities and Q were more correlated with the running time of MiniSAT over these instances (real-world, hard combinatorial, and random) than the traditional factors like number of variables, clauses or the clause-variable ratio.

3. Additionally, we generated approximately 500,000 random Boolean formulas and made the surprising finding that MiniSAT finds it hard to solve instances with Q value lying in the range from 0.05 to 0.13, whereas it was able to easily solve the ones outside this range. While previous work [11] has shown that the clause-to-variable ratio is predictive of solver run time on randomly-generated instances (phase transition at clause-to-variable ratio of 4.2 [11]), this metric is not predictive at all of solver efficiency on real-world instances [15]. By contrast, according to our experiments, Q and number of communities measure for both real-world and random instances are correlated with the running time of MiniSAT (and by extension all solvers that are significantly similar to it algorithmically) on these instances.

2   Background

In this Section, we provide some background on regression analysis, the concept of the community structure of graphs and how it relates to SAT formulas.

2.1   Community Structure of SAT Formulas

The idea of decomposing graphs into *natural communities* [6,17] arose in the study of complex networks such as the *graph of biological systems* or the Internet. Informally, a network or graph is said to have community structure, if the graph can be decomposed into sub-graphs where the sub-graphs have more internal edges than outgoing edges. Each such sub-graph (aka module) is called a community. Modularity is a measure of the quality of the community structure of a graph. The idea behind this measure is that graphs with high modularity have dense connections between nodes within sub-graph but have few inter-module connections. It is easy to see informally that maximising modularity is one way to detect the optimal community structure inherent in a graph. Many algorithms [6,17] have been proposed to solve the problem of finding an optimal community structure of a graph, the most well-known among them being the one from Girvan and Newman [6]. The quality measure for optimal community structure is often referred to as the Q value, and we will continue to call it similarly. There are many different ways of computing the Q value and we refer the reader to these paper [6,17,12].

We experimented with two different algorithms the Clauset-Neuman-Moore (CNM) algorithm [6] and the online community detection algorithm (OL) [17]. While we did find that the CNM algorithm resulted in a better community structure — evidenced by fewer communities with few links between them — we chose the OL algorithm because of its vastly superior run time. This was of particular importance due to the sheer size and number of the SAT instances we processed. Our initial experiments were conducted with an implementation of the CNM algorithm, then repeated with the OL algorithm. The results we present in Section 3.4 were observed, regardless of the choice of algorithm.

2.2   Linear Regression

In this paper we make use of linear regression techniques for the result that correlates the Q value and number of communities with the running time of the MiniSAT CDCL SAT solver. In case the reader is not familiar with this topic, we provide a very brief description of the basic ideas involved.

Given multiple independent factors and a single dependent variable, linear regression can be used to determined the relationship between the factors and the variables based on a provided model. For the scope of this paper the dependent variable will always be $log(time)$, while the independent factors (such as Q value, number of communities, variables, and clauses) will be appropriately specified for each experiment in Section 3.4. This model can either look only at the main effects of the factors specified, or at both the effects of factors and the interactions between them.

We provide a few important definitions below:

ANOVA stands for analyses of variance. In the scope of this paper, it is generated by the linear regression, and used to understand the influence that specific factors and interactions between factors have on the dependent variable.

R^2 represents the amount of variability in the data that has been accounted for by the model and is used to measure the *goodness of fit* of the model. It ranges from zero to one with one representing a perfect model. Due to the nature of the calculation, the R^2 value will increase when additional factors are added to the model. In this paper, we refer to the *Adjusted R^2* which is modified to only increase if an added factor contributes positively to the model.

Confidence Levels are used to specify a certain level of confidence that a given statement is true. They can be used to calculate the likelihood of a given set of input values resulting in a given output (for example time), or they can be used to estimate the likelihood that a factor in a model is significant. They are measured in percent, typical values are 99.9, 99 and 95. Any result with a confidence level below 95% is considered unimportant in the context of this paper.

Confidence Intervals are used to provide a range for a value at a given confidence level, which is usually set to 95% or 99%. They show that with a given percentage probability, an estimated value will lie between a certain range.

Kolmogorov–Smirnov test is used to provide quantitative assurances that a provided sample belongs to a specified distribution, it results in a value between zero and one, with values approaching zero indicating that the provided sample does belong to the specified distribution.

Residuals is the difference between a fitted dependent variable and the corresponding provided dependent variable. It represents the amount of error for a given set of input factors when calculating the output.

3   Experimental Results

In this Section we describe our experimental results that correlate Q value/the number of communities with the running time of two CDCL solvers we considered in our experiments, namely, MiniSAT [8] and Glucose [5].

3.1   Correlation between LBD and Community Structure

In this section, we propose to link the number of communities contained in a clause with an efficient measure used in recent CDCL (Conflict-Driven Clause-Learning) solvers to score learnt clause usefulness. Interestingly, this measure is used in most of today's best SAT solvers.

We assume that the reader is familiar with the basic concepts and techniques relevant to CDCL SAT solvers. Briefly, these solvers branch by making decisions on the value of literals, and at any step of the search, ensure that all the unit clauses w.r.t the current partial assignment are correctly *propagated* until either an empty clause is found (the input formula is UNSAT) or a complete assignment is found (the input formula is SAT). The important point to be emphasised is that solvers learn clauses at a fast rate (generally around 5000 per seconds), which can overwhelm the memory of the system unless some steps are taken to routinely get rid of some of them. Such learnt clause deletion policies have come to be recognised as crucial to the efficiency of solvers. The trick to the success of such deletion policies is that somehow the solver has to differentiate *good* clauses from the ones that are not so good.

Prior to the 2009 version of the Glucose solver [5], deletion policies were primarily based on the past VSIDS activity of the clauses, i.e., learnt clauses with low VSIDS scores were deleted. However in their paper [5], the authors proposed that a better scoring mechanism for learnt clauses is to rank them by the number of distinct decision levels the variables in these clauses belonged to. This measure is called the Literal Block Distance (LBD) of a clause. The smaller the LBD score of a clause, the higher its rank. The intuition behind this scoring mechanism is the following: The lower the LBD score of a clause, the fewer the number of decision levels needed for this clause to be unit propagated or falsified again during the search. Clauses with LBD score of 2 are called glue clauses, because it allows for merging (glue) of two *blocks of propagations* with the aim of removing one decision variable (this notion in turn inspired the name of the Glucose solver). It is important to note that this behaviour is more likely to happen when using the phase saving heuristic for branching [14], because all variables are set to their last propagated value when possible.

The hypothesis we test in this sub-section is that the notion of *blocks* of propagations (i.e. a decision variable and all the propagated variables by unit propagation at the same decision level) is highly correlated to the idea of communities. To be more precise, if an input instance has high-quality community structure (communities with few inter-community edges) then we hypothesise that the conflict clauses that are shared between fewer communities are likely to cause more propagation per decision and hence are likely to have a lower LBD score. We verify our hypothesis, namely, that there is indeed a strong relationship between the number of communities of a clause (initial or learnt) and its LBD score computed by Glucose.

Intuitively, we consider clauses that are shared between very few communities as higher quality than the ones shared between many communities, because such clauses are localised to a small set of communities possibly enabling the solver to in-effect partition the problem into many "small set of communities" and solve them one at a time.

Experiment Set Up: We limit our study to the set of industrial instances of the SAT 2013 competition (Applications category). This is in line with our observation that SAT instances obtained from real-world applications have good community structure, and consequently the notion of LBD scoring will likely have the biggest impact on performance for such instances than otherwise. Put it differently, if their is indeed a relationship between LBD and community structure of SAT instances, we hope to characterise it in this set of problems first. For our experiment, we store, for each formula of the 189 instances (SAT'13 Competition, Application Track), the value of the LBD and the number of different communities of each learnt clauses. (There are 300 application instances in the SAT'13, however we were able to compute the communities of only 189 of them due to resource constraints.)

We would like to emphasise few points here: (1) First, all instances were pre-processed using the SatELite simplifier before any experiments were conducted for this study. All CDCL solvers have pre-processing simplifying routines in them. It is very likely that these kind of simplified formulas are representative of the inherent structure of formulas CDCL solvers are efficient on. (2) Second, we computed the community structure using the Newman algorithm [12] (aka CNM algorithm) on the variable-incidence graph representation of the formula. Results were stored in a separate file once for all the experiments. For each SAT instance, the corresponding communities were first labeled. Then, for every instance we maintained a map from variables occurring in that instance to the label of the corresponding community the variable belonged to. (3) Third, Glucose was launched on these instances without any information regarding their community structure computed in step (2). Glucose computed the LBD values for the input instances, and stored them in a large trace file that was analysed later: for each instance, for each learnt clause, we compared the LBD value of that learnt clause and computed, thanks to (2), the number of distinct communities the learnt clause contained. Finally, note that we used a maximum number of conflicts for our study, not a time out. In Section 3.2, the maximum conflicts studied is set to the first 20,000 conflicts. In the Section 3.3, it is set to 100,000. Those values were chosen w.r.t. the statistical tools we used.

3.2 Observing the Clear Relationship between Communities and LBD by Heatmaps

It is not trivial to express a relationship between thousands of values, following unknown distribution functions, on hundreds of problems. Hence, to show the

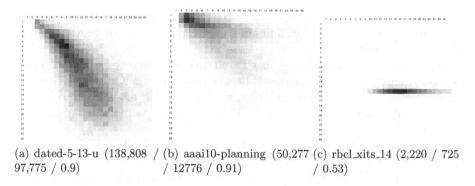

(a) dated-5-13-u (138,808 / 97,775 / 0.9)   (b) aaai10-planning (50,277 / 12776 / 0.91)   (c) rbcl_xits_14 (2,220 / 725 / 0.53)

Fig. 1. Relationship between Literal Block Distance (LBD) and communities for a selection of instances. The x-axis corresponds to LBD, the y-axis to communities. Blue intensity represents the frequency of learnt clauses with the considered LBD and community value. For each instance, we provide in parenthesis, the number of variables, the number of communities, and the quality value (Q). The figure is analysed in Section 3.2.

general trend, we chose to build one heatmap per instance: we compute the number of occurrences of each couple $(LBD, Community)$ on the considered problem and assign the intensity of a colour with respect to this number of occurrences. The result is shown for some characteristic instances Figure 1. As we can see, there is an obvious and clear relationship on the first two instances (a, b) which are the most frequent cases. Intensive colours follow the diagonal: many clauses have approximatively the same LBD and the same number of communities. This behaviour appears in most cases. All heatmaps are available on the web (a temporary url for the reviewing process available on request), with more or less a strong diagonal shape.

From these figures we can conclude that there has to be a strong relationship between LBD and number of communities. Small LBD learnt clauses add stronger constraints between fewer communities. This may allow the solver to focus its search on a smaller part of the search tree, avoiding *scattering*, the phenomena where the solver jumps between lots of communities creating learnt clauses that link these communities together thus making the structure of the SAT instance worse and consequently harder to solve. We think this study gives a new point of view of LBD and provides a new explanation of its efficiency. Of course, there exists a few cases that do not exhibit such a relationship. This is the case, for example, for the last example provided in the Figure 1.c. In this instance, it seems that in many cases all learnt clauses involve 15 communities and more than ten decision levels.

We do not yet have strong evidence that correlates classes of instance, and the average number of communities that a learnt clause belongs to. Our suspicion is that all the original problems for which CDCL solvers were designed, the BMC problems, may exhibit a particularly good relationship.

3.3   How Close are Communities and LBD?

If the LBD seems heavily related to communities, the question is now how close to the LBD is it? In particular, on some extreme cases, the simple size of a clause could be a good predictor for its number of communities (clearly, the larger the clause is, the bigger the number of communities can be). Thus, we also have to ensure that (1) the LBD score is more accurate than the size of the clause for predicting the number of communities and (2) the more we update the LBD, the closest we are to the number of communities (the solver is consistent along its search).

In answer to the above question we present two figures: 2.a and 2.b. For both figures we computed each problem instance X, for each learnt clause c during the first 100,000 conflicts of Glucose solving X, the standard deviation δ_X of the values:

1. $|LBD(c) - \#Com(c)|$, representing the dispersion of the differences between the LBD of a clause c and its number of communities, shown as "LBD and communities" in the figure;
2. $|size(c) - \#Com(c)|$, representing the dispersion of the differences between the size of a clause c and its number of communities, shown as "Size and communities";
3. $|LBD(c) - size(c)|$, representing the dispersion of the differences between the LBD of a clause c and its size, shown as "Size and LBD";

In the first experiment, we try to see how close LBD and size are to the number of communities. We represent in Figure 2 the cumulative distribution function of all the δ_X for each of the three cases. This figure clearly highlights that the relationship between LBD and number of communities is much more accurate (it

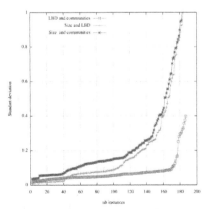

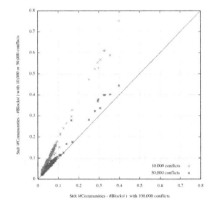

(a) Cumulative distribution function of standard deviation between different measures

(b) Evolution of the standard deviation during the search

Fig. 2. Some standard deviations. This figure is analysed Section 3.3.

is less than 0.1 for a large majority of instances) than the relationship between size and LBD or between size and communities. Thus, as a first conclusion, we see that, in the large majority of the cases, the LBD is really close to the number of communities. The only hypothesis here is that the standard deviation has some meaning over the analysed data, which seems to be a plausible hypothesis.

The last experiment we conducted studies the evolution of LBD scores during the execution of Glucose. We focus now on the values of δ_X accounting for the dispersion of values $|LBD(c) - \#Com(c)|$ for each instance X. We compare the values obtained after 10,000 conflicts with the values obtained after 100,000 (shown as "10,000 conflicts" on the figure) and the values obtained after 50,000 conflicts with, again, the values obtained after 100,000 (shown as "50,000 conflicts"). The comparison is done by the two scatter plots in Figure 2.b. Two conclusions can be drawn from this. It seems clear that the longer the solver is running, the more accurate it is at estimating the number of communities of clauses by the LBD. This may also be explained by the fact that, the longer the solver is running, the longer it is working on fewer communities/LBD, thus focusing on small subparts of the problem. However, which one of these two hypothesis is more accurate is still an ongoing work.

3.4 Experimental Setup: Correlation between Solve Time and Community Structure

In this section, we present the hypothesis that it is possible to correlate the characteristics of SAT instance C and the running time of CDCL SAT solvers in solving C. Previous attempts in this direction have largely focused on characterising the hardness of solving SAT instances in terms of number of variables, clauses or the clause-variable ratio [9]. In our experiments, we go beyond variables, clauses, and their ratio to also considering number of communities and the Q value (modularity). To test this hypothesis we performed two experiments. The first experiment we did was to correlate the above-mentioned characteristics and the running time of the MiniSAT solver over all instances in the SAT 2013 competition [1]. The second experiment we performed was a controlled one, wherein, we randomly-generated instances varying a subset of their characteristics such as number of variables, clauses, Q value, and number of communities and keeping the rest constant. We then ran MiniSAT on these randomly generated instances and recorded its running time. We then plotted the running time against changing Q to see how the two are correlated. All the data for these experiments is available at [13].

3.5 Community Structure and SAT 2013 Competition Instances

We performed the following steps in this experiment. First, we attempted to calculate the community structure of every SAT instance from all categories (hard combinatorial, random, and application) the SAT 2013 competition [1]. For this we used the OL algorithm [17]. Due to the size of some of the formula it was not possible to get this information for every instance. As such we were only able to run the community structure analysis on approximately 800 instances. The generated results were then aggregated with the solve time of the MiniSAT solver (from the SAT 2013 competition website [1]) for each instance.

The analysis was performed on the $log(time)$ rather than raw recorded time due to the presence of a large number of timeouts, which would have a skewed distribution. Having said that, our results are similarly strong without this constraint. In addition to this, we standardised our data to have a mean of zero and a standard deviation of one. This is standard practice when performing regression on factors that have large differences in scale, and ensures that importance is not falsely reported based only on scale.

After formatting the data as described, we fitted a linear regression model to it using a stepwise regression technique to choose the best model, this was identified as:

$$log(time) = |V| \oplus |CL| \oplus Q \oplus |CO| \oplus \text{QCOR} \oplus \text{VCLR}$$

Where V is the set of all variables in a formula, CL is the set of all clauses, CO is the set of all communities, QCOR is the ratio of $\frac{Q}{|CO|}$, and VCLR is the ratio of $\frac{|V|}{|CL|}$. In this model the $\oplus$ operator denotes that the factor and all interactions with all other factors were to be considered. Performing the regression resulted in a residual vs fitted plot, shown in Figure 3(b) where the x-axis shows the fitted values and the y-axis shows the residuals. As well as a normal quantile plot shown in Figure 3(a) where the x-axis shows the standardised residuals plotted against a randomly generated normally distributed sample with the same mean and standard deviation on the y-axis. In the normal quantile plot, the presence of a slight curve in the line is indicative that the distribution of the data may be bimodal, as such we are unable to measure confidence intervals for the accuracy of the model. In addition to this, the residual plot shows that the data may be biased, but at least has relatively even variance. Unfortunately, the presence of the timeout results has played a role in the biased nature of the experiment, however dropping them from the results entirely leads to a bias in the opposite direction.

The adjusted R^2 of our model is 0.5159. While this relatively low R^2 value indicates that there is some factor we have not considered, our model is far better than any previous model, which relied only on number of variables and clauses. This model, which takes the form of

$$log(time) = |V| \oplus |CL| \oplus \text{VCLR}$$

is also given for comparison and results in an adjusted R^2 of 0.3148 — making it significantly less predictive than our model. In addition to this, the more distinct S shape, and presence of sharp curves in Figure 4(a) shows that the distribution of the data is less normal. This result is confirmed when using the Kolomogorov-Sminov (KS) method to test *goodness of fit*. Our model results in a KS value of 0.1283 compared with the previously available model which gave a KS value of 0.3154, this lack of normality makes it impossible to estimate

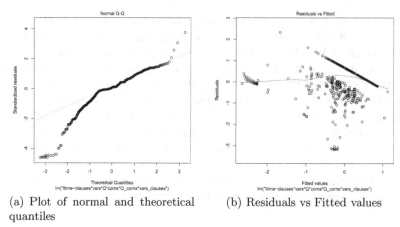

(a) Plot of normal and theoretical
quantiles

(b) Residuals vs Fitted values

Fig. 3. Plots for the model including community metrics $R^2 = 0.5159$

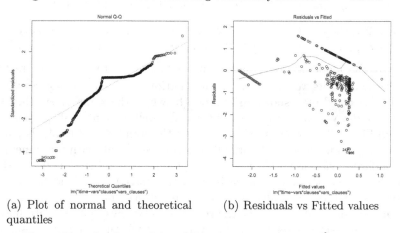

(a) Plot of normal and theoretical
quantiles

(b) Residuals vs Fitted values

Fig. 4. Plots for the model without community metrics $R^2 = 0.3148$

confidence intervals for the results. However, it is possible to rank the factors by importance (because the data was standardised prior to regression). The results in Figures 3(a) and 3(b) show that our model, while not perfect, is a major step towards being a predictor for solve time. This is confirmed when viewing the results of the regression shown in the Table 1 (This table can also be viewed from our website [13]).

Bottomline Result: The main result we found from the regression is that the Q factor is involved in every one of the significant interactions at a 99.9% confidence level. In addition to this we found that $|V|$ (number of variables) alone is not significant, and $|CL|$ (number of clauses) alone is only marginally significant. Furthermore, $|CO|$ (number of communities) proved to be the most predictive effect, as well as being involved with numerous other interactions that are also significant.

3.6   Community Structure and Random Instances

In this Section, we describe the experiments, where we ran MiniSAT on a large set of randomly-generated SAT instances, to better understand the effects of varying the various factors of the input formulas in a controlled fashion. We ran a controlled experiment in which approximately 550,000 formulas were generated and executed. In performing this experiment we discovered that there is a large increase in average solution time when the $0.05 \leq Q \leq 0.13$. This can be seen clearly in Figure 5(a).

The formulas were generated by varying the number of variables from 500 to 2000 in increments of 100, the number of clauses from 2000 to 10,000 in increments of 1000, the desired number of communities from 20 to 400 in increments of 20, and the desired Q value, from zero to one in increments of 0.01. Each individual trial was repeated three times with the same characteristics. This was necessary due to the non-deterministic nature of the generation technique. The resulting experiments were ran in a random order for several hours to generate a large volume of data.

To generate a specific instance we perform the following actions: Let us assume that the set of variables be denoted as $V = \{V_i : 0 \leq i < n_v\}$ where n_v is the desired number of variables. Similarly, let the set of groups be $G = \{G_x : 0 \leq x < n_g\}$ where n_g is the desired number of groups. A group is a rough estimate of a community, and is used only to guide the generator in producing a structured problem.

First, we assign variables to groups such that each group $G_x = \{V_y : y = r_v * |G| + x; 0 \leq r_v < \frac{|V|}{|G|}\}$, where r_v is randomly selected. Next, we generate the set of clauses $C = \{C_z : 0 < z < n_c\}$ as follows, where n_c is the desired number of clauses such that $C_z = \{V_{z1} \vee V_{z2} \vee V_{z3}\}$. Each clause is constructed as follows: First, a group G_x and a variable $V_{z1} \in G_x$ are randomly selected. This is followed by a selection of another variable V_{z2} from either G_x or V with probability of q of being selected from G_x. Finally, a third variable V_{z3} is selected from either G_x or V with probability of q of being chosen from G_x. The value q (lies between 0 and 1) can be used as a rough

Table 1. List of the factors with 99.9% significance level. $\odot$ indicates an interaction between two or more factors and *Sig* stands for Significance. The full table is listed in Appendix Table 2.

Factor	Estimate	Std. Error	t value	$Pr(> \lvert t \rvert)$	Sig
$\lvert CO \rvert$	-1.237e+00	3.202e-01	-3.864	0.000121	***
$\lvert CL \rvert \odot Q \odot QCOR$	-4.226e+02	1.207e+02	-3.500	0.000492	***
$\lvert CL \rvert \odot Q$	-2.137e+02	6.136e+01	-3.483	0.000523	***
$\lvert CL \rvert \odot Q \odot \lvert CO \rvert \odot QCOR \odot VCLR$	-1.177e+03	3.461e+02	-3.402	0.000702	***
$\lvert CL \rvert \odot Q \odot \lvert CO \rvert$	-6.024e+02	1.774e+02	-3.396	0.000719	***
$Q \odot QCOR$	3.415e+02	1.023e+02	3.339	0.000881	***
Q	1.726e+02	5.200e+01	3.318	0.000947	***

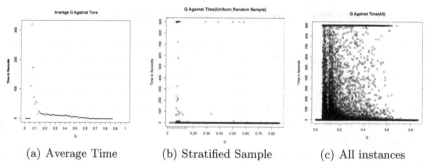

(a) Average Time   (b) Stratified Sample   (c) All instances

Fig. 5. A plot of Q against time, for three different data sets

estimator of Q, the modularity of the formula and is provided as input to the generator. The result is a randomly-generated 3-CNF formula.

During our analyses of the results, we discovered that our random generation technique resulted in far more results in the $0.05 \leq Q \leq 0.12$ range than in any other range. To ensure that the results seen were not because of this discrepancy, we re-ran many of the trials, focusing on data outside of this range, and generated approximately another 307,000 formula.

To ensure an unbiased analyses we also performed basic analyses on a stratified random sample taken uniformly across the range of Q. From the 545,000 results 2250 were randomly sampled, with 250 results taken from each range of 0.1, as there were no results with $Q > 0.9$ this range was not included in the sample. This process ensured their was no bias in the results based purely on frequency. The resulting sample is shown in Figure 5(b) which shows that when $0.05 \leq Q \leq 0.12$, the formula take far longer to solve. while this range is slightly different from the results of the full dataset ($0.05 \leq Q \leq 0.13$) This can be explained by the reduced dataset.

Figure 5(b) appears to shows that for all values of $Q < 0.05$ *or* $Q > 0.12$, almost all the formulas finish in approximately zero seconds, however this is only because of the scale, in reality while a large number of them do complete very quickly, in less than one second, numerous other results take varying amounts of time anywhere between zero and 900 seconds (the timeout).

The data collected from each result is as follows: number of variables, number of clauses, number of communities, Q metric, result and time, Prior to analyses we ensured the quality of the experiments by checking that we had a good distribution of results in both the SAT and UNSAT categories, and that in both there was a reasonable distribution across time. While we did discover a more even distribution of time in the case of the UNSAT vs SAT formula, it is not enough to affect the results. Similarly, while the majority of the results are in the lower end of the scale, this confirms the result that for the majority of the range of Q, most SAT instances are relatively easy to solve. Once the distribution of SAT vs UNSAT was determined, Q was plotted against time for all results Figure 5(c). From this plots we see a very clear trend in the relationship between Q and solution time, namely that when $0.05 \leq Q \leq 0.13$ there is a significant increase in average solution time. A more clear representation of these results is Figure 5(a), which plots Q on the x-axis against the average execution time of the formulas on the y-axis.

In addition to the result showing that when $0.05 \leq Q \leq 0.13$ the formula is hard, we also noticed several interesting features of these graphs. From these graphs we made several observations; Firstly, when looking at the full dataset as shown in Figure 5(c), it can be seen that none of the 2500 formula with a $Q < 0.05$ had a solve time of $> 100ms$. Secondly, while not immediately clear from looking at Figure 5(c) we discovered that none of the SAT formula had a $Q < 0.1096$, while we think it would be possible to generate a satisfiable formula that had a Q value in this range, it does not typically occur in randomly generated instances.

Bottomline Result: The basic takehome message here is that, when we accounted for potential bias in the generation process and eliminated instances that were quick to solve, we got the following result: randomly-generated instances with Q values in the range $0.05 \leq Q \leq 0.12$ (for the reduced set of instances uniformly binned for Q values ranging from 0 to 0.9 in increments of 0.1) were unusually hard for MiniSAT to solve compared to instances outside this range, and this result only depends on the Q value and not on any other factor such as number of variables or clauses.

4   Related Work

In [3] Levy et al introduced the concept of SAT problems having community structure. The paper showed that numerous problems in the SAT 2010 race contained very high modularity compared to graphs of any other nature. It was also suggested in this paper that SAT solvers are able to exploit this *hidden structure* in order to achieve good solve times. However, the paper was unable to explain what characteristics of community structure leads to poor or good solve times. Also, in [4] the authors state that while SAT solvers have shown improvements in solve times for numerous industrial applications, their has been less success in improving the solve time of randomly generated instances. They posit that this is due to the lack of structure present in randomly generated instances.

In [16] Xu et al describe a SAT solver that chooses its algorithms based on 48 features of the input formula. While they did use certain graph theoretic concepts, such as node-degree statistics, they did not consider the concept of communities as a feature of the input. The list of 48 features could be used in a more comprehensive model than the ones used in our regression. In [10] Habet et al present an empirical study of the effect of conflict analyses on solution time in CDCL solvers.

In [2] the authors present a the notion of fractal dimensions in SAT solvers, they have discovered that as the SAT solver progresses the fractal dimension increases when new learnt clauses are added to the formula. They have also discovered that learnt clauses do not connect distant parts of the formula (ones with long shortest paths between nodes), as one would expect. This is interesting when combined with the work we present stating that clauses which are comprised of variables in a small number of communities are more useful to the solver. This means that even when a learnt clause that does connect distant variable in the formula is added, it is not as useful as a clause that connects locally occurring variables.

5 Conclusions

In this paper we presented evidence that the community structure present in real world SAT instances is correlated with solution time of CDCL SAT solvers. First, we highlighted a relationship between Literal Block Distance (LBD), a measure indicating the importance of a learnt clause in CDCL solver, and community structure. In particular, learnt clauses that are shared by few communities are highly correlated with high-quality learnt clauses with low LBD scores. In other words, we have a new measure (number of communities shared by a learnt clause) of quality of learnt clauses that correlates with a very successful existing one (LBD). This result provides new insights into the efficiency of the LBD measure and should be considered to improve solver performance. Second, we introduced a model, that while not perfect, is a first step towards a predictive model for the solution time of SAT instances. Finally, we presented a result showing that randomly generated instances are particularly difficult to solve, regardless of number of clauses or variables, when their modularity is between 0.05 and 0.13.

6 Future Work

As mentioned in Section 3.4, our regression is one of the early predictive model for solve time of a CDCL solver based on community structure. However, there are many other factors not discussed in this paper that may play an important role in determining solve time, factors such as: median/mean size of clauses, the number of clauses that feature a subset of variables that appear together in another clause, the number of unique pairs of variables appearing in a clause, or the size of the largest clique in the variable graph. Any or all of these features may play a role in determining solution time. In the future we intend to explore as many of these, and as many of the 48 features from [16] as necessary to improve our model. Another potential for research is implementing a solver that takes advantage of community structure to improve solve time, this could be implemented in several ways, one of which is to create a clause deletion heuristic based on the community structure (as opposed to LBD deletion policy). Another could be to implement a decision heuristic that chooses variables that appear in learnt clauses with variables from very few number of communities. The idea being the more local a conflict clause is to a community, the higher its quality.

References

1. 2013 sat competition, http://satcompetition.org/2013/ (accessed: January 31, 2014)
2. Ansótegui, C., Bonet, M.L., Giráldez-Cru, J., Levy, J.: The fractal dimension of sat formulas. arXiv preprint arXiv:1308.5046 (2013)

3. Ansótegui, C., Giráldez-Cru, J., Levy, J.: The community structure of sat formulas. In: Cimatti, A., Sebastiani, R. (eds.) SAT 2012. LNCS, vol. 7317, pp. 410–423. Springer, Heidelberg (2012)
4. Ansótegui, C., Levy, J.: On the modularity of industrial sat instances. In: CCIA, pp. 11–20 (2011)
5. Audemard, G., Simon, L.: Predicting learnt clauses quality in modern SAT solvers. Proceedings of IJCAI, 399–404 (2009)
6. Clauset, A., Newman, M.E.J., Moore, C.: Finding community structure in very large networks. Physical review E 70(6), 66111 (2004)
7. Eén, N., Sörensson, N.: An extensible sat-solver. In: Giunchiglia, E., Tacchella, A. (eds.) SAT 2003. LNCS, vol. 2919, pp. 502–518. Springer, Heidelberg (2004)
8. Een, N., Sörensson, N.: Minisat: A sat solver with conflict-clause minimization. In: SAT, vol. 5 (2005)
9. Ian, P.: Gent and Toby Walsh. The sat phase transition. In: ECAI, pp. 105–109. PITMAN (1994)
10. Habet, D., Toumi, D.: Empirical study of the behavior of conflict analysis in cdcl solvers. In: Schulte, C. (ed.) CP 2013. LNCS, vol. 8124, pp. 678–693. Springer, Heidelberg (2013)
11. Mitchell, D., Selman, B., Levesque, H.: Hard and easy distributions of sat problems. In: AAAI, vol. 92, pp. 459–465. Citeseer (1992)
12. Newman, M.E.J.: Fast algorithm for detecting community structure in networks. Physical review E 69(6), 66133 (2004)
13. Newsham, Z., Ganesh, V., Fischmeister, S., Audemard, G., Simon, L.: Community Structure of SAT Instances Webpage with Data and Code, https://ece.uwaterloo.ca/~vganesh/satcommunitystructure.html
14. Pipatsrisawat, K., Darwiche, A.: A lightweight component caching scheme for satisfiability solvers. In: Marques-Silva, J., Sakallah, K.A. (eds.) SAT 2007. LNCS, vol. 4501, pp. 294–299. Springer, Heidelberg (2007)
15. Vardi, M.: Phase transition and computation complexity (2012), http://www.lsv.ens-cachan.fr/Events/fmt2012/SLIDES/moshevardi.pdf
16. Xu, L., Hutter, F., Hoos, H.H., Leyton-Brown, K.: Satzilla: Portfolio-based algorithm selection for sat. J. Artif. Intell. Res. (JAIR) 32, 565–606 (2008)
17. Zhang, W., Pan, G., Wu, Z., Li, S.: Online community detection for large complex networks. In: Proceedings of the Twenty-Third International Joint Conference on Artificial Intelligence, pp. 1903–1909. AAAI Press (2013)

7 Appendix

Table 2. List of all significant effects, three stars indicates the highest level of confidence that the effect is important. $\odot$ indicates an interaction between two or more factors and *Sig* stands for Significance.

Factor	Estimate	Std. Error	t value	$Pr(> |t|)$	Sig
$|CO|$	-1.237e+00	3.202e-01	-3.864	0.000121	***
$|CL| \odot Q \odot QCOR$	-4.226e+02	1.207e+02	-3.500	0.000492	***
$|CL| \odot Q$	-2.137e+02	6.136e+01	-3.483	0.000523	***
$|CL| \odot Q \odot |CO| \odot QCOR \odot VCLR$	-1.177e+03	3.461e+02	-3.402	0.000702	***
$|CL| \odot Q \odot |CO|$	-6.024e+02	1.774e+02	-3.396	0.000719	***
$Q \odot QCOR$	3.415e+02	1.023e+02	3.339	0.000881	***
Q	1.726e+02	5.200e+01	3.318	0.000947	***
$Q \odot |CO| \odot QCOR$	9.451e+02	2.927e+02	3.229	0.001292	**
$Q \odot |CO|$	4.839e+02	1.503e+02	3.220	0.001335	**
$|V| \odot QCOR$	-3.177e+01	1.004e+01	-3.164	0.001617	**
$|CL| \odot |V| \odot VCLR$	-1.263e+01	4.503e+00	-2.805	0.005163	**
$|CL| \odot |V| \odot QCOR \odot VCLR$	-2.521e+01	9.008e+00	-2.798	0.005263	**
$|V|$	-1.376e+01	4.947e+00	-2.782	0.005526	**
$QCOR$	-1.057e+01	3.912e+00	-2.701	0.007065	**
$|CL| \odot |V| \odot QCOR$	2.096e+01	7.894e+00	2.656	0.008073	**
$(Intercept)$	-4.949e+00	1.950e+00	-2.538	0.011327	*
$|CL| \odot QCOR$	9.486e+00	3.792e+00	2.502	0.012556	*
$|CL| \odot |V|$	9.641e+00	3.933e+00	2.451	0.014456	*
$QCOR \odot VCLR$	9.035e+00	3.789e+00	2.385	0.017323	*
$VCLR$	4.452e+00	1.892e+00	2.353	0.018845	*
$|CL|$	4.299e+00	1.894e+00	2.270	0.023507	*
$|V| \odot QCOR \odot VCLR$	1.700e+01	7.556e+00	2.250	0.024755	*
$|V| \odot VCLR$	8.059e+00	3.811e+00	2.115	0.034769	*
$|CL| \odot |V| \odot Q \odot QCOR$	-4.680e+02	2.298e+02	-2.036	0.042060	*
$|CL| \odot |V| \odot Q$	-2.373e+02	1.167e+02	-2.034	0.042268	*
$|CL| \odot |V| \odot Q \odot |CO|$	-6.594e+02	3.315e+02	-1.989	0.047042	*
$|CL| \odot |V| \odot Q \odot |CO| \odot QCOR$	-1.286e+03	6.469e+02	-1.988	0.047160	*

Variable Dependencies and Q-Resolution*

Friedrich Slivovsky and Stefan Szeider

Institute of Information Systems, Vienna University of Technology, Vienna, Austria
fs@kr.tuwien.ac.at, stefan@szeider.net

Abstract. We propose $Q(D)$-resolution, a proof system for Quantified
Boolean Formulas. $Q(D)$-resolution is a generalization of Q-resolution
parameterized by a dependency scheme D. This system is motivated by
the generalization of the QDPLL algorithm using dependency schemes
implemented in the solver DepQBF. We prove soundness of $Q(D)$-res-
olution for a dependency scheme D that is strictly more general than
the standard dependency scheme; the latter is currently used by Dep-
QBF. This result is obtained by proving correctness of an algorithm
that transforms $Q(D)$-resolution refutations into Q-resolution refutations
and could be of independent practical interest. We also give an alternative characterization of resolution-path dependencies in terms of directed
walks in a formula's implication graph which admits an algorithmically
more advantageous treatment.

1 Introduction

The satisfiability problem of *Quantified Boolean Formulas* (QBFs) is a canonical PSPACE-complete decision problem [17]. QBFs offer a convenient language
for representing problems from domains such as model checking, planning, or
knowledge representation and reasoning. In practice, QBF solvers are expected
to generate certificates witnessing the satisfiability or unsatisfiablity of input
formulas. These certificates serve a dual purpose:

1. Certificates encode information that is valuable in applications settings (e.g.,
   a plan, or a counterexample in model checking).
2. Certificates can be used to verify that the answer given by the solver is
   correct.

Search-based QBF solvers implementing the QDPLL algorithm [7] typically use
variants of *Q-resolution* [5] as a certificate language [13]. When assigning variables during the search process, the QDPLL algorithm observes the order induced by the nesting of quantifiers in the input formula. This is often needlessly
restrictive, in particular for formulas in the common QCNF format which places
all quantifiers in a single, linear quantifier prefix. An appealing approach to
dealing with this restriction is the generalization of QDPLL by means of *dependency schemes* implemented in DepQBF [2,12]. A dependency scheme maps

---

* This research was supported by the ERC (COMPLEX REASON, 239962).

each QCNF formula to a binary relation on its variables that represents potential variable dependencies [14,15]. This relation is used by DepQBF to gain additional freedom in decision making and in the definition of more powerful rules for constraint learning and unit propagation. The latter correspond to a generalization of the *forall-reduction* rule of Q-resolution [2]. Since certificates produced by DepQBF may involve applications of this more general rule, they do not correspond to ordinary Q-resolution proofs. With respect to the two purposes of certificates mentioned above, this has the following consequences.

1. The canonical algorithm for extracting Skolem/Herbrand models from Q-resolution refutations [1] does not work for certificates generated by DepQBF.
2. It is unclear whether certificates can serve as *proofs* of truth or falsity, that is, whether the underlying proof system is sound.

In this paper, we introduce $Q(D)$-*resolution* to study the proof system used by DepQBF to generate proofs. We define the *reflexive resolution-path dependency scheme* D^{rrs} and prove correctness of an algorithm that transforms $Q(D^{rrs})$-resolution refutations into Q-resolution refutations, thus establishing soundness of $Q(D^{rrs})$-resolution. Since D^{rrs} is strictly more general than the *standard dependency scheme* D^{std} currently implemented in DepQBF, the soundness result carries over to $Q(D^{std})$-resolution. We also provide an alternative characterization of *resolution-path dependencies* [16,18] in terms of directed walks in a formula's *implication graph* which admits an algorithmically more advantageous treatment in terms of strongly connected components. The dependency scheme D^{rrs} is a variant of the *resolution path dependency scheme* D^{res} [16]. We justify our working with D^{rrs} instead of D^{res} by demonstrating that $Q(D^{res})$-resolution is unsound.

2   Preliminaries

Sequences. We write ε for the empty sequence. If $s = s_1 \ldots s_k$ and $r = r_1 \ldots r_l$ are sequences then by $s * r$ we denote the sequence $s_1 \ldots s_k r_1 \ldots r_l$. Let r be a sequence such that $r = p * s$. Then p is a *prefix* of r and s is a *suffix* of r. If t is a prefix (suffix) of r then t is a *proper* prefix (suffix) of r if $t \neq r$. The *length* of a sequence $s_1 \ldots s_k$ is k. A *shortest* sequence with property P is a sequence of minimum length with property P. If p_i is a sequence for every $i \in \{1, \ldots, k\}$ we use $\langle p_i \rangle_{i=1}^k$ as a shorthand for the sequence $p_1 * p_2 * \cdots * p_k$. At the cost of introducing some ambiguity we write a for both the singleton set $\{a\}$ and the sequence containing only a.

Trees. A *rooted binary tree* is a directed graph G such that (a) there exists a vertex $r \in V(G)$ (called the *root* of G) such that for every $w \in V(G)$ there is a unique walk from r to w in G, and (b) for every $v \in V(G)$ there are at most two distinct vertices $u, w \in V(G)$ such that $(v, u) \in E(G)$ and $(v, w) \in E(G)$. A *labelled rooted binary tree* is a triple $T = (V(T), E(T), \lambda)$ where $(V(T), E(T))$ is a rooted binary tree and λ is a function with domain $V(T) \cup E(T)$. We say $x \in V(T) \cup E(T)$ is *labelled with* $\lambda(x)$.

Formulas. We consider quantified Boolean formulas in *quantified conjunctive normal form* (QCNF). A QCNF formula is a pair QF, where Q is a (quantifier) *prefix* and F is a *CNF formula*, called the *matrix* of $\mathcal{F}$. A CNF formula is a finite set of *clauses*, where each clause is a finite set of *literals*. Literals are negated or unnegated propositional *variables*. If x is a variable, we put $\overline{x} = \neg x$ and $\overline{\neg x} = x$, and let $var(x) = var(\neg x) = x$. If X is a set of literals, we write $\overline{X}$ for the set $\{\,\overline{a} : a \in X\,\}$. For a clause C, we let $var(C)$ be the set of variables occuring (negated or unnegated) in C. For a QCNF formula $\mathcal{F}$ with matrix F, we put $var(\mathcal{F}) = var(F) = \bigcup_{C \in F} var(C)$. We call a clause *tautological* if it contains the same variable negated as well as unnegated. We assume that the matrix of a formula contains only non-tautological clauses. The prefix of a QCNF formula $\mathcal{F}$ is a sequence $Q_1 x_1 \ldots Q_n x_n$, where $x_1 \ldots x_n$ is a permutation of $var(\mathcal{F})$ and $Q_i \in \{\forall, \exists\}$ for $1 \le i \le n$. We define a total order $<_\mathcal{F}$ on $var(\mathcal{F})$ by letting $x_i <_\mathcal{F} x_j$ if and only if $i < j$. The sets of *existential* and *universal* variables occurring in $\mathcal{F}$ are given by $var_\exists(\mathcal{F}) = \{\,x_i : 1 \le i \le n, Q_i = \exists\,\}$ and $var_\forall(\mathcal{F}) = \{\,x_i : 1 \le i \le n, Q_i = \forall\,\}$. Relative to $\mathcal{F}$, a literal a is existential (universal) if $var(a)$ is existential (universal). We define $R_\mathcal{F} = \{\,(x, y) : x <_\mathcal{F} y\,\}$ and let $R_\mathcal{F}(x) = \{\,y \in var(\mathcal{F}) : (x, y) \in R_\mathcal{F}\,\}$ for $x \in var(\mathcal{F})$. Moreover, we let $D_\mathcal{F}^{\mathrm{trv}} = \{\,(x_i, x_j) \in R_\mathcal{F} : Q_i \ne Q_j\,\}$. If F is a CNF formula and a is a literal then $F[a] = \{\,C \setminus \overline{a} : C \in F, a \notin C\,\}$. We extend this to QCNF formulas $\mathcal{F} = QF$ by letting $\mathcal{F}[a] = Q' F[a]$, where Q' is obtained from Q by deleting $var(a)$ and its associated quantifier. Let $\mathcal{F}$ be a QCNF formula. If $var(\mathcal{F}) = \emptyset$ then $\mathcal{F}$ is *true* (or *satisfiable*) if $F = \emptyset$. Otherwise, let $\mathcal{F} = Q_x x\, \mathcal{F}'$. If $Q_x = \forall$ then $\mathcal{F}$ is *true* if both $\mathcal{F}'[x]$ and $\mathcal{F}'[\neg x]$ are true. If $Q_x = \exists$ then $\mathcal{F}$ is true if at least one of $\mathcal{F}'[x]$ and $\mathcal{F}'[\neg x_1]$ is true. If $\mathcal{F}$ is not true then $\mathcal{F}$ is *false* (or *unsatisfiable*).

3   Q(D)-Resolution

A *proto-dependency scheme* is a mapping D that associates each QCNF formula $\mathcal{F}$ with a binary relation $D_\mathcal{F} \subseteq D_\mathcal{F}^{\mathrm{trv}}$. The *trivial dependency scheme* is the mapping $D^{\mathrm{trv}} : \mathcal{F} \mapsto D_\mathcal{F}^{\mathrm{trv}}$. A proto-dependency scheme D is *tractable* if the relation $D(\mathcal{F})$ can be computed in polynomial time for each QCNF formula $\mathcal{F}$.

We will represent $Q(D)$-resolution derivations as labelled rooted binary trees constructed by means of the following operations.[1]

- If C is a clause then $\triangle(C)$ denotes a labelled rooted binary tree consisting of a single (root) vertex labelled with C.
- Let $T_1 = (V_1, E_1, \lambda_1)$ and $T_2 = (V_2, E_2, \lambda_2)$ be labelled rooted binary trees. For every literal a, we define the operation $\odot_a$ as follows. We assume without loss of generality that V_1 and V_2 are disjoint. Let r_1 and r_2 denote the roots of T_1 and T_2, respectively, and let $C_1 = \lambda_1(r_1)$ and $C_2 = \lambda_2(r_2)$. Then $T_1 \odot_a T_2$ denotes the labelled rooted binary tree obtained by taking the union of T_1 and T_2, adding a new vertex r labelled with $C = (C_1 \setminus a) \cup (C_2 \setminus \overline{a})$, and making r the root by adding edges (r, r_1) and (r, r_2) labelled with a and $\overline{a}$, respectively.

---

[1] Our notation is inspired by [3].

– Let $T = (V, E, \lambda)$ be a labelled rooted binary tree with root r and $\lambda(r) = C$. For a literal a we construct the labelled rooted binary tree $T\|_a$ starting from T by adding a fresh vertex r' labelled with $C \setminus a$ and an edge (r', r) labelled with a.

Definition 1 (Tree-like $Q(D)$-resolution derivation). *Let D be a proto-dependency scheme and let $\mathcal{F} = QF$ be a QCNF formula. A tree-like $Q(D)$-resolution derivation from $\mathcal{F}$ is a labelled rooted binary tree that can be constructed using the following rules.*

1. *If $C \in F$ then $\triangle(C)$ is a tree-like $Q(D)$-resolution derivation from $\mathcal{F}$.*
2. *Let T_1, T_2 be $Q(D)$-resolution derivations from $\mathcal{F}$ whose roots are labelled with C_1 and C_2, respectively. If $a \in C_1$ is an existential literal such that $\overline{a} \in C_2$, and $(C_1 \setminus a) \cup (C_2 \setminus \overline{a})$ is non-tautological then $T_1 \odot_a T_2$ is a tree-like $Q(D)$-resolution derivation from $\mathcal{F}$.*
3. *Let T be a $Q(D)$-resolution derivation with root label C. If $a \in C$ is a universal literal and there is no existential literal $b \in var(C)$ such that $(var(a), var(b)) \in D(\mathcal{F})$ then $T\|_a$ is a tree-like $Q(D)$-resolution derivation from $\mathcal{F}$.*

Rules 2 and 3 are known as *resolution* and *forall-reduction*, respectively. We will usually refer to tree-like $Q(D)$-resolution derivations as $Q(D)$-*derivations* (or simply as *derivations* if D is clear from the context). Let $T = (V, E, \lambda)$ be a $Q(D)$-derivation from $\mathcal{F}$ with root r. We say that T is a *derivation of* $\lambda(r)$ or that T *derives* $\lambda(r)$, and call $\lambda(r)$ the *conclusion* of T. We call T a *refutation* of $\mathcal{F}$ if T derives the empty clause $\emptyset$. If T_1 and T_2 are derivations from $\mathcal{F}$ of clauses C_1 and C_2 such that $C_1 \subseteq C_2$ we say T_1 *subsumes* T_2. The *size* of T, denoted $|T|$, is defined to be $|V|$. A *position* of T is either a sequence of edge labels occurring on a walk in T starting from r or the empty sequence ε. Let π be a position of T. We let $T[\pi]$ denote the *subderivation of T at π* defined as

$$T[\pi] = \begin{cases} T & \text{if } \pi = \varepsilon, \\ T_1[\sigma] & \text{if } T = T_1 \odot_\ell T_2 \text{ and } \pi = \ell * \sigma, \\ T'[\sigma] & \text{if } T = T'\|_\ell \text{ and } \pi = \ell * \sigma. \end{cases}$$

If $T[\pi] = T_1 \odot_\ell T_2$ we refer to $T[\pi]$ as a *resolution step* (on $var(\ell)$); if $T[\pi] = S\|_a$ then $T[\pi]$ is a *forall-reduction step* (on ℓ). Let $\mathcal{F}$ be a QCNF formula, let x be a universal variable of $\mathcal{F}$, and let y be an existential variable of $\mathcal{F}$. Then $(x, y) \in D^{\mathrm{trv}}(\mathcal{F})$ if and only if $x <_{\mathcal{F}} y$. It follows that the forall-reduction rule of "ordinary" Q-resolution [5] corresponds to forall-reduction in $Q(D^{\mathrm{trv}})$-resolution. Accordingly, we define Q-resolution as follows.

Definition 2 (Q-resolution). *Let $\mathcal{F}$ be a QCNF formula. A Q-resolution derivation from $\mathcal{F}$ is a $Q(D^{\mathrm{trv}})$-derivation from $\mathcal{F}$, and a Q-resolution refutation of $\mathcal{F}$ is a $Q(D^{\mathrm{trv}})$-refutation of $\mathcal{F}$.*

In spite of its simplicity, Q-resolution is a sound and complete proof system for unsatisfiable QCNF formulas.

Fact 1 ([5]). *Let $\mathcal{F}$ be a QCNF formula. There exists a Q-resolution refutation of $\mathcal{F}$ if and only if $\mathcal{F}$ is unsatisfiable.*

Let $\mathcal{F}$ be a QCNF formula, let a be a universal literal, and let b be an existential literal. We say that b *blocks* a (relative to $\mathcal{F}$) if $var(a) <_{\mathcal{F}} var(b)$. We extend this to clauses C and say that C *blocks* a if there is a literal $b \in C$ such that b blocks a. If $T = S\|_a$ is a $Q(D)$-derivation from $\mathcal{F}$ and C blocks a we say that C *blocks* T. In $Q(D)$-derivations, forall-reduction can be applied even in the presence of blocking literals. We refer to such occurrences of forall-reduction as *D-reductions*.

Definition 3 (D-reduction). *Let D be a proto-dependency scheme, let $\mathcal{F}$ be a QCNF formula, and let T be a $Q(D)$-derivation from $\mathcal{F}$. Let $T[\pi] = S\|_a$ be a derivation of clause C. If C blocks $S\|_a$ then $S\|_a$ is a D-reduction (of T). If, in addition, there is no D-reduction $R\|_b$ of T such that $var(b) <_{\mathcal{F}} var(a)$, then $T[\pi]$ is an* outermost *D-reduction (of T).*

A $Q(D)$-derivation that does not contain D-reductions is already a Q-resolution derivation.

4   Dependency Schemes and $Q(D)$-Resolution

In the literature, there are two definitions of *Dependency Scheme* that refine our abstract notion of a proto-dependency scheme:

1. Dependency schemes and so-called *cumulative dependency schemes* that characterize truth-value preserving permutations of a formula's prefix [15].
2. Dependency schemes based on a semantic notion of *independence* [14].

It turns out these notions are too weak to characterize soundness of $Q(D)$-resolution. For dependency schemes of type 2, this can be shown using the following formula:

$$\mathcal{F} = \forall x \forall y \exists z \; \{\{x, y, \neg z\}, \{\neg x, \neg y, z\}\}$$

The formula $\mathcal{F}$ is true if the value assigned to z matches the value assigned to x or the value assigned to y. Accordingly, both $f_z(x, y) = x$ and $f'_z(x, y) = y$ are *models* [6] of $\mathcal{F}$. This implies that z is both independent of x and independent of y in the sense of [14]. Let D be the proto-dependency scheme such that $D(\mathcal{F}) = \emptyset$ and $D(\mathcal{G}) = D^{\mathrm{trv}}(\mathcal{G})$ for every QCNF formula $\mathcal{G} \neq \mathcal{F}$. Then D is a dependency scheme of type 2, but $Q(D)$-resolution is unsound: since $(x, z) \notin D(\mathcal{F})$ and $(y, z) \notin D(\mathcal{F})$, forall-reduction derives the clauses $\{x\}$ and $\{\neg x\}$ which yield a $Q(D)$-resolution refutation of $\mathcal{F}$.

The dependency scheme D in the above example is constructed with the counterexample in mind. For dependency schemes of type 1, we do not have to come up with artificial proto-dependency schemes. We will now show that $Q(D^{\mathrm{res}})$-resolution is unsound, where D^{res} is the *resolution-path dependency scheme* [16,18] (we give a simplified but equivalent version of the definition in [16]).

Definition 4 (Resolution Path). *Let $\mathcal{F} = QF$ be a QCNF formula and let $X \subseteq var_{\exists}(\mathcal{F})$. A resolution path (from a_1 to a_{2k}) via X (in $\mathcal{F}$) is a sequence $p = a_1 \ldots a_{2k}$ of literals satisfying the following conditions.*

1. *There is a $C_i \in F$ such that $a_{2i-1}, a_{2i} \in C_i$, for all $i \in \{1, \ldots, k\}$.*
2. *$var(a_{2i-1}) \neq var(a_{2i})$ for all $i \in \{1, \ldots, k\}$.*
3. *$a_{2i}, a_{2i+1} \in X \cup \overline{X}$ for all $i \in \{1, \ldots, k-1\}$.*
4. *$\overline{a_{2i}} = a_{2i+1}$ for all $i \in \{1, \ldots, k-1\}$.*

For every $i \in \{1, \ldots, k\}$ we say that p goes through $var(a_{2i})$.

Definition 5 (Resolution-Path Dependency Pair). *Let $\mathcal{F}$ be a QCNF formula and let $X \subseteq var(\mathcal{F})$. We call (x, y) a resolution-path dependency pair of $\mathcal{F}$ with respect to X if there are literals a and b such that $var(a) = x$ and $var(b) = y$ and there exist resolution paths from a to b and from $\overline{a}$ to $\overline{b}$ via X.*

Definition 6 (Resolution-Path Dependency Scheme). *The resolution path dependency scheme D^{res} maps each QCNF formula $\mathcal{F}$ to the relation $D^{res}_{\mathcal{F}} = \{ (x, y) \in R_{\mathcal{F}} : (x, y)$ is a resolution-path dependency pair of $\mathcal{F}$ with respect to $(R_{\mathcal{F}}(x) \cap var_{\exists}(\mathcal{F})) \setminus y \}$.*

The resolution-path dependency scheme is a cumulative dependency scheme [16]. However, the following example (taken from [14]) demonstrates that $Q(D^{res})$-resolution is unsound.

Example 7. Let $\mathcal{F} = \forall x \exists z \forall u \exists y\ F$, where F contains the clauses $\{x, u, \neg y\}$, $\{\neg x, \neg u, \neg y\}$, $\{z, u, y\}$, $\{\neg z, u, \neg y\}$, $\{\neg z, \neg u, y\}$, and $\{z, \neg u, \neg y\}$.

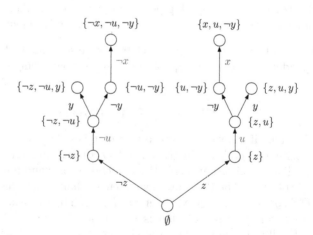

Fig. 1. $Q(D^{res})$-refutation of $\mathcal{F}$ from Example 7

The formula $\mathcal{F}$ is satisfiable, but Figure 4 shows a $Q(D^{res})$-refutation of $\mathcal{F}$. It is straightforward to verify that the pair (x, y) is not in $D^{res}(\mathcal{F})$ since every resolution path from x or $\neg x$ to y goes through y. As a consequence, one can derive

Definition 4 (Resolution Path). *Let $\mathcal{F} = QF$ be a QCNF formula and let $X \subseteq var_\exists(\mathcal{F})$. A resolution path (from a_1 to a_{2k}) via X (in $\mathcal{F}$) is a sequence $p = a_1 \ldots a_{2k}$ of literals satisfying the following conditions.*

1. *There is a $C_i \in F$ such that $a_{2i-1}, a_{2i} \in C_i$, for all $i \in \{1, \ldots, k\}$.*
2. *$var(a_{2i-1}) \neq var(a_{2i})$ for all $i \in \{1, \ldots, k\}$.*
3. *$a_{2i}, a_{2i+1} \in X \cup \overline{X}$ for all $i \in \{1, \ldots, k-1\}$.*
4. *$\overline{a_{2i}} = a_{2i+1}$ for all $i \in \{1, \ldots, k-1\}$.*

For every $i \in \{1, \ldots, k\}$ we say that p goes through $var(a_{2i})$.

Definition 5 (Resolution-Path Dependency Pair). *Let $\mathcal{F}$ be a QCNF formula and let $X \subseteq var(\mathcal{F})$. We call (x, y) a resolution-path dependency pair of $\mathcal{F}$ with respect to X if there are literals a and b such that $var(a) = x$ and $var(b) = y$ and there exist resolution paths from a to b and from $\overline{a}$ to $\overline{b}$ via X.*

Definition 6 (Resolution-Path Dependency Scheme). *The resolution path dependency scheme D^{res} maps each QCNF formula $\mathcal{F}$ to the relation $D^{\mathrm{res}}_{\mathcal{F}} = \{ (x, y) \in R_{\mathcal{F}} : (x, y)$ is a resolution-path dependency pair of $\mathcal{F}$ with respect to $(R_{\mathcal{F}}(x) \cap var_\exists(\mathcal{F})) \setminus y \}$.*

The resolution-path dependency scheme is a cumulative dependency scheme [16]. However, the following example (taken from [14]) demonstrates that $Q(D^{\mathrm{res}})$-resolution is unsound.

Example 7. Let $\mathcal{F} = \forall x \exists z \forall u \exists y \, F$, where F contains the clauses $\{x, u, \neg y\}$, $\{\neg x, \neg u, \neg y\}$, $\{z, u, y\}$, $\{\neg z, u, \neg y\}$, $\{\neg z, \neg u, y\}$, and $\{z, \neg u, \neg y\}$.

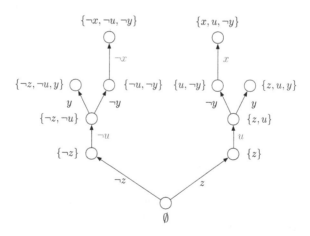

Fig. 1. $Q(D^{\mathrm{res}})$-refutation of $\mathcal{F}$ from Example 7

The formula $\mathcal{F}$ is satisfiable, but Figure 4 shows a $Q(D^{\mathrm{res}})$-refutation of $\mathcal{F}$. It is straightforward to verify that the pair (x, y) is not in $D^{\mathrm{res}}(\mathcal{F})$ since every resolution path from x or $\neg x$ to y goes through y. As a consequence, one can derive

Fact 1 ([5]). *Let $\mathcal{F}$ be a QCNF formula. There exists a Q-resolution refutation of $\mathcal{F}$ if and only if $\mathcal{F}$ is unsatisfiable.*

Let $\mathcal{F}$ be a QCNF formula, let a be a universal literal, and let b be an existential literal. We say that b *blocks* a (relative to $\mathcal{F}$) if $var(a) <_{\mathcal{F}} var(b)$. We extend this to clauses C and say that C *blocks* a if there is a literal $b \in C$ such that b blocks a. If $T = S\|_a$ is a $Q(D)$-derivation from $\mathcal{F}$ and C blocks a we say that C *blocks* T. In $Q(D)$-derivations, forall-reduction can be applied even in the presence of blocking literals. We refer to such occurrences of forall-reduction as *D-reductions*.

Definition 3 (*D-reduction*). *Let D be a proto-dependency scheme, let $\mathcal{F}$ be a QCNF formula, and let T be a $Q(D)$-derivation from $\mathcal{F}$. Let $T[\pi] = S\|_u$ be a derivation of clause C. If C blocks $S\|_a$ then $S\|_a$ is a D-reduction (of T). If, in addition, there is no D-reduction $R\|_b$ of T such that $var(b) <_{\mathcal{F}} var(a)$, then $T[\pi]$ is an* outermost *D-reduction (of T).*

A $Q(D)$-derivation that does not contain D-reductions is already a Q-resolution derivation.

4   Dependency Schemes and $Q(D)$-Resolution

In the literature, there are two definitions of *Dependency Scheme* that refine our abstract notion of a proto-dependency scheme:

1. Dependency schemes and so-called *cumulative dependency schemes* that characterize truth-value preserving permutations of a formula's prefix [15].
2. Dependency schemes based on a semantic notion of *independence* [14].

It turns out these notions are too weak to characterize soundness of $Q(D)$-resolution. For dependency schemes of type 2, this can be shown using the following formula:

$$\mathcal{F} = \forall x \forall y \exists z \, \{\{x, y, \neg z\}, \{\neg x, \neg y, z\}\}$$

The formula $\mathcal{F}$ is true if the value assigned to z matches the value assigned to x or the value assigned to y. Accordingly, both $f_z(x, y) = x$ and $f'_z(x, y) = y$ are *models* [6] of $\mathcal{F}$. This implies that z is both independent of x and independent of y in the sense of [14]. Let D be the proto-dependency scheme such that $D(\mathcal{F}) = \emptyset$ and $D(\mathcal{G}) = D^{\text{trv}}(\mathcal{G})$ for every QCNF formula $\mathcal{G} \neq \mathcal{F}$. Then D is a dependency scheme of type 2, but $Q(D)$-resolution is unsound: since $(x, z) \notin D(\mathcal{F})$ and $(y, z) \notin D(\mathcal{F})$, forall-reduction derives the clauses $\{x\}$ and $\{\neg x\}$ which yield a $Q(D)$-resolution refutation of $\mathcal{F}$.

The dependency scheme D in the above example is constructed with the counterexample in mind. For dependency schemes of type 1, we do not have to come up with artificial proto-dependency schemes. We will now show that $Q(D^{\text{res}})$-resolution is unsound, where D^{res} is the *resolution-path dependency scheme* [16,18] (we give a simplified but equivalent version of the definition in [16]).

the clause $\{u, \neg y\}$ from $\{x, u, \neg y\}$, and the clause $\{\neg u, \neg y\}$ from $\{\neg x, \neg u, \neg y\}$ by forall-reduction in $Q(D^{\text{res}})$-resolution.

5 The Reflexive Resolution-Path Dependency Scheme

Motivated by Example 7, we define the following variant of D^{res} for which resolution paths inducing an (x, y)-dependency may also go through y.

Definition 8 (Reflexive Resolution-Path Dependency Scheme). *The reflexive resolution-path dependency scheme D^{rrs} maps each QCNF formula $\mathcal{F}$ to the relation $D^{\text{rrs}}_{\mathcal{F}} = \{\, (x, y) \in R_{\mathcal{F}} : (x, y) \text{ is a resolution-path dependency pair of } \mathcal{F} \text{ with respect to } R_{\mathcal{F}}(x) \cap var_{\exists}(\mathcal{F}) \,\}.$*

Proto-dependency schemes can be partially ordered by a pointwise comparison of the relations they associate with QCNF formulas: we say that a proto-dependency scheme D_1 is *at least as general as* a proto-dependency scheme D_2 if $D_1(\mathcal{F}) \subseteq D_2(\mathcal{F})$ for every QCNF formula $\mathcal{F}$. If this inclusion is strict for some formulas we say D_1 is *strictly more general* than D_2. One can show that D^{rrs} is strictly more general than the *Standard Dependency Scheme D^{std}* used in Dep-QBF (the following definition is a streamlined version of the definition of D^{std} in [15]).

Definition 9 (Primal Graph). *Let $\mathcal{F}$ be a QCNF formula with matrix F. The* primal graph *of $\mathcal{F}$ is the undirected graph with vertex set $var(\mathcal{F})$ and edge set $\{\, xy : x, y \in var(\mathcal{F}), x \neq y, \text{ and there is a clause } C \in F \text{ such that } x, y \in var(C) \,\}.$*

Definition 10 (Standard Dependency Pair). *Let $\mathcal{F}$ be a QCNF formula and let $X \subseteq var(\mathcal{F})$. We call $(x, y) \in D^{\text{trv}}_{\mathcal{F}}$ a standard dependency pair of $\mathcal{F}$ with respect to X if there is a walk from x to y in $G[X \cup \{x, y\}]$, where G denotes the primal graph of $\mathcal{F}$.*

Definition 11 (Standard Dependency Scheme). *The* standard dependency scheme D^{std} *maps every QCNF formula $\mathcal{F}$ to the relation $D^{\text{std}}_{\mathcal{F}} = \{\, (x, y) \in R_{\mathcal{F}} : (x, y) \text{ is a standard dependency pair of } \mathcal{F} \text{ with respect to } R_{\mathcal{F}}(x) \cap var_{\exists}(\mathcal{F}) \,\}.$*

Proposition 12. *D^{res} is strictly more general than D^{rrs}, and D^{rrs} is strictly more general than D^{std}.*

It is not difficult to see that if D_1 is at least as general as D_2 and $Q(D_1)$-resolution is sound then $Q(D_2)$-resolution is sound as well. Thus soundness of $Q(D^{\text{rrs}})$-resolution (proved in Section 6) carries over to $Q(D^{\text{std}})$-resolution.

We now give an alternative characterization of resolution paths in terms of walks in the *implication graph* of a formula (also known as the formula's *associated graph* [11].)

Definition 13 (Implication graph). *Let $\mathcal{F} = QF$ be a QCNF formula. The* implication graph *of $\mathcal{F}$ is the directed graph with vertex set $var(\mathcal{F}) \cup \overline{var(\mathcal{F})}$ and edge set $\{\, (\overline{a}, b) : \text{ there is a } C \in F \text{ such that } a, b \in C \text{ and } a \neq b \,\}.$*

Lemma 14. *Let $\mathcal{F}$ be a QCNF formula and let $a, b \in var(\mathcal{F}) \cup \overline{var(\mathcal{F})}$ be distinct literals. Let $X \subseteq var(\mathcal{F})$ and let G denote the implication graph of $\mathcal{F}$. The following statements are equivalent.*

1. *There is a resolution path from a to b via X.*
2. *There is a walk from $\overline{a}$ to b in $G[X \cup \overline{X} \cup \{\overline{a}, b\}]$.*

Proof. Let $p = a_1 \ldots a_{2k}$ be a resolution path from a to b via X. The sequence $p' = \overline{a_1} * \langle a_{2i} \rangle_{i=1}^{k}$ is a walk in the implication graph of $\mathcal{F}$. We have $\{a_2, \ldots, a_{2k-1}\} \subseteq X \cup \overline{X}$ by Definition 4, so p is even a walk in $G[X \cup \overline{X} \cup \{\overline{a}, b\}]$. For the converse, let $p = a_1 \ldots a_k$ be a shortest walk from $\overline{a}$ to b in $G[X \cup \overline{X} \cup \{\overline{a}, b\}]$. Then the sequence $p' = \overline{a_1} * \langle a_i \overline{a_i} \rangle_{i=2}^{k-1} * a_k$ is a resolution path in $\mathcal{F}$. Since p is a shortest walk we have $\overline{a}, b \notin \{a_2, \ldots, a_{k-1}\}$. This implies $\{a_2, \ldots, a_{k-1}\} \subseteq X \cup \overline{X}$ as well as $\overline{\{a_2, \ldots, a_{k-1}\}} \subseteq X \cup \overline{X}$. So p' is a resolution path from $\overline{a_1} = a$ to b via X.   □

The implication graph of a formula can be constructed in time quadratic in the size of $\mathcal{F}$ and directed reachability can be decided in linear time, so we obtain the following result.

Proposition 15. *Both D^{res} and D^{rrs} are tractable.*

For practical purposes the explicit construction of the implication graph can be avoided. Moreover, the following result shows that an overapproximation of D^{rrs} can be represented in terms of the strongly connected components of the implication graph.

Proposition 16. *Let $\mathcal{F}$ be a QCNF formula and let $(x, y) \in D^{trv}(\mathcal{F})$. Let $X \subseteq var(\mathcal{F})$ and let G denote the implication graph of $\mathcal{F}$. If (x, y) is a resolution-path dependency pair with respect to X there is strongly connected component $\mathcal{C}$ of $G[X \cup \overline{X} \cup \{x, y, \neg y\}]$ such that $x, y \in \mathcal{C}$ or $x, \neg y \in \mathcal{C}$.*

Proof. Let (x, y) be a resolution-path dependency pair with respect to X. Assume without loss of generality that there are resolution paths p_1 and p_2 such that p_1 is a resolution path from $\neg x$ to y via X and p_2 is a resolution path from $\neg y$ to x via X. By Lemma 14 there are walks p'_1 from x to y and p'_2 from y to x in $G[X \cup \overline{X} \cup \{x, y\}]$, so x and y are in the same strongly connected component of $G[X \cup \overline{X} \cup \{x, y, \neg y\}]$.   □

6 Soundness of $Q(D^{rrs})$-Resolution

This section is devoted to proving our main result, stated below.

Theorem 17. *For every QCNF formula $\mathcal{F}$ there is a $Q(D^{rrs})$-resolution refutation of $\mathcal{F}$ if and only if $\mathcal{F}$ is unsatisfiable.*

In fact, we are going to prove the following, stronger statement.

Proposition 18. *Given a QCNF formula $\mathcal{F}$ and a $Q(D^{rrs})$-refutation T of $\mathcal{F}$, one can compute a Q-resolution refutation of $\mathcal{F}$ of size at most $3^{|T|}$.*

We prove this proposition by demonstrating correctness and termination of an algorithm (Algorithm 3) that turns $Q(D^{rrs})$-refutations into Q-resolution refutations in Lemma 31. An outline of this algorithm is given below.

Algorithm outline. Let a be the universal literal removed by an outermost D-reduction of the input $Q(D)$-derivation. There are two cases.

1. If there is no clause containing $\bar{a}$ on the path from the D-reduction to the root of the derivation, we simply skip the D-reduction and add it at the root. If the clause derived at the root does not contain any literals blocking a, the D-reduction is turned into an ordinary forall-reduction. (This condition is satisfied by a refutation, and we can ensure that it holds for subderivations and their outermost D-reductions as well.)
2. Otherwise, the derivation must contain a resolution step on a variable x such that $x <_{\mathcal{F}} var(a)$ (see Lemma 30). We drop the lowermost such resolution step to the root of the derivation. This may introduce x-literals to the clauses on the path from the resolution step to the root. But since the D-reduction picked in the first step is outermost these literals will not interfere with D-reductions. Moreover, because the resolution step is lowermost, every clause on the path contains an existential variable y such that $var(a) <_{\mathcal{F}} y$, so introducing x to these clauses will not turn a forall-reduction into a D-reduction.

In this way, we obtain a derivation whose immediate subderivations are (a) strictly smaller than the original derivation and which (b) do not contain new D-reductions. We run the algorithm on these subderivations to rewrite them into Q-resolution derivations and add a final resolution or forall-reduction step.

Example 19. Consider the QCNF formula $\mathcal{F} = \exists e_1 \forall u \exists e_2 \exists e_3 \; \{\{u, e_2\}, \{\neg u, e_3\},$ $\{\neg e_3, e_1\}, \{\neg e_1, \neg e_2\}\}$. Derivation T_A of Figure 2 shows a $Q(D^{\mathrm{rrs}})$-refutation of $\mathcal{F}$ with D-reductions at positions e_2 and $\neg e_2$. At the root of the derivation case 1 applies, so we skip the D-reduction $T_A[e_2]$ and add it at the root of the derivation, which leads to derivation T_B. We continue with the subderivation $T_B[u]$. Since u occurs on the path from the D-reduction $T_B[u \neg e_2]$ to the root, case 2 applies and we drop the resolution step $T_B[u \neg e_2 \neg u \neg e_3]$ on $e_1 <_{\mathcal{F}} u$ to the root, resulting in the Q-resolution derivation T_C.

Example 19 also illustrates that known rewrite strategies for removing long-distance resolution steps from Q-resolution proofs [1,8] cannot be applied to remove D-reductions from $Q(D)$-resolution refutations. If a long-distance resolution step leads to a clause containing a universal variable u in both polarities, one can assume that the variable resolved on does not block u. In a refutation, the literals blocking u have to be resolved out eventually, so one can remove the long-distance resolution step by successively lowering it [1] or by (recursively) resolving out blocking literals using clauses resolved closer to the root of the derivation [8]. In refutation T_A, resolving $\{u, e_2\}$ and $\{\neg u, \neg e_2\}$ would amount to a long-distance resolution step. But e_2 is both the variable resolved on and the variable blocking u in the premises, and we cannot further lower this resolution step.

We now turn to a formal proof of Proposition 18. In order to state Algorithm 3 and prove its correctness and termination, we are going to define and characterize the following two operations:

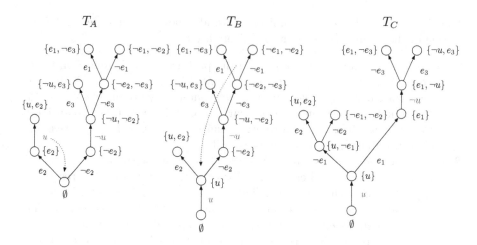

Fig. 2. Rewriting a $Q(D^{\mathrm{rrs}})$-resolution refutation (Example 19)

1. Substitution (Definition 22 and Lemma 24).
2. Dropping a resolution step (Algorithm 2 and Lemma 29).

The second operation can in turn be represented by a successive "lowering" of a resolution step (Algorithm 1, Lemmas 26 and 27). This lowering operation essentially corresponds to the rewrite rules presented in [1] for turning long-distance resolution proofs into ordinary Q-resolution proofs.[2]

For the remainder of this section, let D be an arbitrary but fixed proto-dependency scheme, and let $\mathcal{F}$ be an arbitrary but fixed QCNF formula. In order to make formal statements to the effect that operations "do not create new D-reductions", we introduce the notion of *narrowing*.

Definition 20 (Narrowing). *Let $R_1 = S_1\|_a$ and $R_2 = S_2\|_a$ be $Q(D)$-derivations from $\mathcal{F}$ with conclusions C_1 and C_2, respectively. We write $R_1 \preceq R_2$ if every literal in C_1 that blocks a is contained in C_2. For $Q(D)$-derivations T_1 and T_2 we say that T_1 narrows T_2 if, for every forall-reduction step $T_1[\pi_1]$, there is a forall-reduction step $T_2[\pi_2]$ such that $T_1[\pi_1] \preceq T_2[\pi_2]$.*

The narrowing relation defines a preorder on derivations. Throughout the rewriting process, we make sure that intermediate derivations narrow earlier ones, so as to not introduce "new" or "more complicated" D-reductions.

Rewriting a derivation may cause literals to disappear from the derivation's conclusion, so that subsequent resolution or forall-reduction steps may become inapplicable. To suppress explicit case distinctions needed for situations of this kind we define "lazy" versions of resolution and forall-reduction as follows (cf. [9]).

---

[2] The cases covered in lines 5-8 of Algorithm 1 are not explicitly dealt with in [1].

Definition 21. *Let T_1, T_2, and T be $Q(D)$-derivations from $\mathcal{F}$ of clauses C_1, C_2, and C.*

$$T_1 \odot_a^L T_2 = \begin{cases} T_1 \odot_a T_2 & \text{if } a \in C_1 \text{ and } \bar{a} \in C_2, \\ T_1 & \text{if } a \notin C_1, \\ T_2 & \text{if } \bar{a} \notin C_2. \end{cases}$$

$$T\|_a^L = \begin{cases} T\|_a & \text{if } a \in C, \\ T & \text{otherwise.} \end{cases}$$

Definition 22 (Substitution). *Let T and S be $Q(D)$-derivations from $\mathcal{F}$. For a position π of T we define $T[\pi \leftarrow S]$ recursively as follows.*

$$T[\pi \leftarrow S] = \begin{cases} S & \text{if } \pi = \varepsilon, \\ T_1[\sigma \leftarrow S] \odot_a^L T_2 & \text{if } T = T_1 \odot_a T_2 \text{ and } \pi = a * \sigma, \\ T'[\sigma \leftarrow S]\|_a^L & \text{if } T = T'\|_a \text{ and } \pi = a * \sigma. \end{cases}$$

Definition 23. *Let T be a $Q(D)$-derivation from $\mathcal{F}$, let π is a position of T, and let a be a literal. We say that T does not contain a below π if, for every proper prefix σ of π, the conclusion of $T[\sigma]$ does not contain a.*

Lemma 24. *Let T be a $Q(D)$-derivation from $\mathcal{F}$ of a clause C such that $T[\pi] = S\|_a$. If T does not contain $\bar{a}$ below π then $T[\pi \leftarrow S]$ is a $Q(D)$-derivation from $\mathcal{F}$ of a clause $C' \subseteq C \cup a$ and $T[\pi \leftarrow S]$ narrows T.*

Proof. The derivation $T[\pi \leftarrow S]$ simply omits the forall-reduction step on a, introducing a to clauses on the path from π to the root of the derivation (not necessarily all the way to the root, since there may be another forall-reduction step on a). By assumption, T does not contain $\bar{a}$ below π, so the result will be a $Q(D)$-derivation. □

Lemma 25. *Let T and S be $Q(D)$-derivations from $\mathcal{F}$. Let π be a position such that S subsumes and narrows $T[\pi]$. Then $T[\pi \leftarrow S]$ subsumes and narrows T.*

Proof. One can prove that $T[\pi \leftarrow S]$ subsumes T by an induction on the length of π. For the narrowing part, observe that every forall-reduction step $R\|_a$ of $T[\pi \leftarrow S]$ not already in T occurs in S or on the path from S to the root of $T[\pi \leftarrow S]$. If $R\|_a = S[\sigma]$ there is position ρ of $T[\pi]$ such that $S[\sigma] \preceq T[\pi][\rho]$ since S narrows $T[\pi]$. Otherwise, R derives a clause subsuming the clause to which the corresponding reduction is applied in T. □

For the lowering operation, we distinguish two cases based on whether the resolution step is lowered past a forall-reduction step (Lemma 26) or another resolution step (Lemma 27).

Lemma 26. *Let $T = (T_1 \odot_a T_2)\|_b$ be a $Q(D)$-derivation from $\mathcal{F}$ such that a does not block b. Then $\mathtt{lower}(T, b)$ is a $Q(D)$-derivation from $\mathcal{F}$ that subsumes and narrows T.*

```
1 Function lower(T, b)
      input: A Q(D)-derivation T and a literal b.
2     if T = (T₁ ⊙ₐ T₂) ⊙_b T₃ then
3         let Cᵢ be the conclusion of Tᵢ for i ∈ {1, 2, 3}
4         if a ∉ C₃ and ā ∉ C₃ then
5             return (T₁ ⊙_b^L T₃) ⊙ₐ (T₂ ⊙_b^L T₃)
6         else if a ∈ C₃ then
7             return T₁ ⊙_b^L T₃
8         else
9             return T₂ ⊙_b^L T₃
10    else if T = (T₁ ⊙ₐ T₂)‖_b then
11        return T₁‖_b^L ⊙ₐ T₂‖_b^L
12    else
13        return T
```

Algorithm 1: Lowering a resolution step

Proof. It is readily verified that $\texttt{lower}(T, b)$ is a $Q(D)$-derivation that subsumes T. The derivation $\texttt{lower}(T, b)$ may contain new forall-reduction steps $T_1\|_b$ and $T_2\|_b$, but since a does not block b we have $T_1\|_b \preceq (T_1 \odot_a T_2)\|_b$ and $T_2\|_b \preceq (T_1 \odot_a T_2)\|_b$, so $\texttt{lower}(T, b)$ narrows T. □

Lemma 27. *Let $T = (T_1 \odot_a T_2) \odot_b T_3$ be a $Q(D)$-derivation of from $\mathcal{F}$. Then $\texttt{lower}(T, b)$ is a $Q(D)$-derivation of from $\mathcal{F}$ that subsumes and narrows T.*

Proof. The case distinction in lines 2-8 of Algorithm 1 is exhaustive, and for each case the derivation returned subsumes the original derivation. Every forall-reduction step of the resulting derivation is already present in T, so $\texttt{lower}(T, b)$ narrows T. □

Definition 28. *Let T be a $Q(D)$-resolution derivation from $\mathcal{F}$, let a be an existential literal, and let π be a position of T. We say that a does not block in T below π if, for every prefix ρ of π, whenever $T[\rho] = S\|_b$ then a does not block b.*

Lemma 29. *Let T be a $Q(D)$-derivation from $\mathcal{F}$, and let $T[\pi]$ be a resolution step on a such that a does not block in T below π. Then $T' = \texttt{drop}(T, \pi, a)$ is a $Q(D)$-derivation from $\mathcal{F}$ that subsumes and narrows T, and at least one of the following conditions[3] holds.*

---

[3] The proof of Lemma 31 is by induction on the size of the input derivation. These conditions ensure that we can apply the induction hypothesis after dropping a resolution step.

```
 1  Function drop(T, π, a)
       input: A Q(D)-derivation T, a position π of T, and a literal a.
 2     if π = ε then
 3      │  return T
 4     else if π = b * ρ then
 5      │  R := drop(T[b], ρ, a)
 6      │  S := T[b ← R]
 7      │  if R = S₁ ⊙ₐ S₂ then
 8      │   │  return lower(S, b)
 9      │  else
10      │   │  return S
```

Algorithm 2: "Dropping" a resolution step

1. $T' = T_1 \odot_a T_2$, and $|T_1| < |T|$ as well as $|T_2| < |T|$
2. $|T'| < |T|$

Proof. We proceed by induction on the length of π. The base case is trivial. For the inductive case, we use the induction hypothesis in line 5 and Lemma 25 in line 6 to conclude that $S = T[b \leftarrow R]$ subsumes and narrows $T[b]$. If $R \neq S_1 \odot_a S_2$ then $|R| < |T[b]|$ by induction hypothesis and $|\text{drop}(T, \pi, a)| = |T[b \leftarrow R]| < |T|$. Otherwise, $|S_1| < |T[b]|$ and $|S_2| < |T[b]|$ by induction hypothesis. Suppose S is a forall-reduction step. Since a does not block in T below π it follows from Lemma 26 that $\text{lower}(S, b) = S_1\|_b^L \odot_a S_2\|_b^L$ is a $Q(D)$-resolution derivation that subsumes and narrows T, and $|S_1\|_b^L| < |T|$ as well as $|S_2\|_b^L| < |T|$. Now suppose S is a resolution step. Then $\text{lower}(S, b)$ is a $Q(D)$-resolution derivation that subsumes and narrows T by Lemma 27, and it is straightforward to verify that the derivation satisfies one of the above conditions. □

Lemma 30. *Let T be a $Q(D^{\text{rrs}})$-derivation from $\mathcal{F}$ with conclusion C. If $T[\pi]$ is a forall-reduction step on literal a and $\bar{a} \in C$ then there is a position π of T such that $T[\sigma]$ is a resolution step on a variable x with $x <_{\mathcal{F}} var(a)$.*

Proof (Sketch). The proof is by an induction on the length of π, using the following claim (cf. [16,18]).

Claim. Let S be a $Q(D)$-derivation from $\mathcal{F}$ with conclusion E, and let $resvar(S)$ denote the set of variables resolved on in S. If $a, b \in E$ are distinct literals, there is a resolution path from a to b via $resvar(S)$.

Suppose $|\pi| = 1$ and let $\pi = b$. Then $T = T_1 \odot_b T_2$ is a resolution step since the conclusion of $T[\pi * a]$ is non-tautological. Let C_1 be the clause derived by

```
 1  Function normalize(ℱ, T)
        input: A QCNF formula ℱ and a Q(D^rrs)-derivation T from ℱ.
 2      if T does not contain a D-reduction then
 3      │   return T
 4      else if T = S‖_a then
 5      │   return normalize(ℱ, S)‖_a^L
 6      else if T = T_1 ⊙_a T_2 then
 7      │   let T[π] = T'‖_b be an outermost D-reduction of T
 8      │   if T does not contain b̄ below π then
 9      │   │   S := T[π ← T']
10      │   │   return normalize(ℱ, S)‖_b^L
11      │   else
12      │   │   let σ be a shortest position such that T[σ] = R_1 ⊙_c R_2 and
        │   │   var(c) <_ℱ var(b)
13      │   │   S := drop(T, σ, c)
14      │   │   if S ≠ S_1 ⊙_c S_2 then
15      │   │   │   return normalize(ℱ, S)
16      │   │   else
17      │   │   │   return normalize(ℱ, S_1) ⊙_c^L normalize(ℱ, S_2)
```

Algorithm 3: Converting $Q(D^{rrs})$-derivations to Q-resolution derivations

---

$T[\pi * a] - T[ba]$ and let C_2 be the clause derived by T_2. Then $a, b \in C_1$ and $\bar{a}, \bar{b} \in C_2$. It follows from the above claim that there are resolution paths from a to b and from $\bar{a}$ to $\bar{b}$ via $resvar(T)$. That is, $(var(a), var(b))$ is a resolution-path dependency pair of $\mathcal{F}$ with respect to $resvar(T)$. If $resvar(T) \subseteq R_{\mathcal{F}}(var(a))$ then $(var(a), var(b)) \in D^{rrs}(\mathcal{F})$, a contradiction. Thus there must be an $x \in resvar(T)$ such that $x <_{\mathcal{F}} var(a)$. The inductive case is proved by a straightforward generalization of this argument. □

Lemma 31. *Let T be a $Q(D^{rrs})$-derivation of C from $\mathcal{F}$ such that C does not block a D-reduction of T. Then* normalize$(\mathcal{F}, T)$ *returns a Q-resolution derivation of size at most $3^{|T|}$ that subsumes T.*

Proof. By induction on the size of T. If T consists of a single node it does not contain a D-reduction, and the algorithm simply returns T (line 2). Suppose the lemma holds for derivations of size strictly less than $|T|$. If T does not contain a D-reduction it is already a Q-resolution derivation (line 2). Otherwise, if T ends with a forall-reduction step the induction hypothesis implies that the derivation returned in line 5 is a Q-resolution derivation from $\mathcal{F}$ that subsumes T. Suppose T ends with a resolution step (line 6). There must be some outermost D-reduction $T[\pi] = T'\|_b$ of T (line 7). There are two cases. (a) If T does not contain $\bar{b}$

below π (line 8) then by Lemma 24 the derivation $S = T[\pi \leftarrow T']$ narrows T and derives a clause $C' \subseteq C \cup b$. Since by assumption C does not block a D-reduction of T the clause C' does not block a D-reduction of S. Moreover $|S| < |T|$, so we can apply the induction hypothesis and conclude that $\texttt{normalize}(\mathcal{F}, S)$ is a Q-resolution derivation that subsumes S. It follows that $\texttt{normalize}(\mathcal{F}, S)\|_b^L$ is a Q-resolution derivation that subsumes T (line 10). (b) If T contains $\bar{b}$ below π then by Lemma 30 there must be a (shortest) position σ of T such that $T[\sigma]$ is a resolution step on $var(c)$ for some literal c, and $var(c) <_{\mathcal{F}} var(b)$ (line 12). We claim that c does not block in T below σ. Towards a contradiction assume that there is a proper prefix ψ of σ such that $T[\psi] = R\|_d$ and $var(d) <_{\mathcal{F}} var(c)$. By choice of σ the conclusion of R must contain an existential literal e such that $var(b) <_{\mathcal{F}} var(e)$. Thus $var(d) <_{\mathcal{F}} var(c) <_{\mathcal{F}} var(b) <_{\mathcal{F}} var(e)$ and e blocks d, so $R\|_d$ must be a D-reduction. But $T[\pi] = T'\|_b$ is an outermost D-reduction of T and $var(d) <_{\mathcal{F}} var(b)$, a contradiction. So c does not block in T below σ. Thus by Lemma 29 the derivation $S = \texttt{drop}(T, \sigma, c)$ subsumes and narrows T (line 13). Since $T'\|_b$ is an outermost D-reduction of T and $c <_{\mathcal{F}} var(b)$, neither c nor $\neg c$ block a D-reduction of S because S narrows T. If $S \neq S_1 \odot_c S_2$ then $|S| < |T|$ by Lemma 29 and by induction hypothesis $\texttt{normalize}(\mathcal{F}, S)$ is a Q-resolution derivation that subsumes T (line 15). Otherwise $|S_1| < |T|$ and $|S_2| < |T|$ by Lemma 29 and the conclusions of S_1 and S_2 do not block D-reductions of S by choice of c. By induction hypothesis $\texttt{normalize}(\mathcal{F}, S_1)$ and $\texttt{normalize}(\mathcal{F}, S_2)$ are Q-resolution derivations subsuming S_1 and S_2, respectively, so $\texttt{normalize}(\mathcal{F}, S_1) \odot_c^L \texttt{normalize}(\mathcal{F}, S_2)$ is a Q-resolution derivation that subsumes T (line 17). It is easy to verify that the output of $\texttt{normalize}(\mathcal{F}, T)$ satisfies the size bound claimed in the statement of the lemma. □

Proof (of Proposition 18). Immediate from Lemma 31 and the observation that the empty clause cannot not block a D-reduction. □

7   Conclusion

We proposed and studied $Q(D)$-resolution, a generalization of Q-resolution, to capture the certificates generated by DepQBF. We introduced the reflexive resolution-path dependency scheme D^{rrs}, proved soundness of $Q(D^{\mathrm{rrs}})$-resolution, and provided an alternative characterization of resolution paths that lends itself to an efficient implementation. In this manner, we hope to have created a solid theoretical basis for future incarnations of DepQBF using dependency schemes more general than D^{std}.

Expansion-based solvers can also benefit from an analysis of variable dependencies [4,12,14,15]. Recently, Janota et al. [10] introduced a proof system to capture the behavior of these solvers. We plan to study how such systems can be augmented with dependency schemes as part of future work. Another intriguing topic for further research is the relative complexity of Q-resolution and $Q(D)$-resolution.

References

1. Balabanov, V., Jiang, J.H.R.: Unified QBF certification and its applications. Formal Methods in System Design 41(1), 45–65 (2012)
2. Lonsing, F., Biere, A.: Integrating dependency schemes in search-based QBF solvers. In: Strichman, O., Szeider, S. (eds.) SAT 2010. LNCS, vol. 6175, pp. 158–171. Springer, Heidelberg (2010)
3. Boudou, J., Woltzenlogel Paleo, B.: Compression of propositional resolution proofs by lowering subproofs. In: Galmiche, D., Larchey-Wendling, D. (eds.) TABLEAUX 2013. LNCS, vol. 8123, pp. 59–73. Springer, Heidelberg (2013)
4. Bubeck, U.: Model-based transformations for quantified Boolean formulas. Ph.D. thesis. University of Paderborn (2010)
5. Büning, H.K., Karpinski, M., Flögel, A.: Resolution for quantified Boolean formulas. Information and Computation 117(1), 12–18 (1995)
6. Büning, H.K., Subramani, K., Zhao, X.: Boolean functions as models for quantified Boolean formulas. Journal of Automated Reasoning 39(1), 49–75 (2007)
7. Cadoli, M., Schaerf, M., Giovanardi, A., Giovanardi, M.: An algorithm to evaluate Quantified Boolean Formulae and its experimental evaluation. Journal of Automated Reasoning 28(2) (2002)
8. Giunchiglia, E., Narizzano, M., Tacchella, A.: Clause/term resolution and learning in the evaluation of Quantified Boolean Formulas. J. Artif. Intell. Res. 26, 371–416 (2006)
9. Goultiaeva, A., Gelder, A.V., Bacchus, F.: A uniform approach for generating proofs and strategies for both true and false QBF formulas. In: Walsh, T. (ed.) Proceedings of IJCAI 2011, pp. 546–553. IJCAI/AAAI (2011)
10. Janota, M., Marques-Silva, J.: On propositional QBF expansions and Q-resolution. In: Järvisalo, M., Van Gelder, A. (eds.) SAT 2013. LNCS, vol. 7962, pp. 67–82. Springer, Heidelberg (2013)
11. Kleine Büning, H., Lettman, T.: Propositional logic: Deduction and algorithms. Cambridge University Press, Cambridge (1999)
12. Lonsing, F.: Dependency Schemes and Search-Based QBF Solving: Theory and Practice. Ph.D. thesis. Johannes Kepler University, Linz, Austria (April 2012)
13. Niemetz, A., Preiner, M., Lonsing, F., Seidl, M., Biere, A.: Resolution-based certificate extraction for QBF. In: Cimatti, A., Sebastiani, R. (eds.) SAT 2012. LNCS, vol. 7317, pp. 430–435. Springer, Heidelberg (2012)
14. Samer, M.: Variable dependencies of quantified CSPs. In: Cervesato, I., Veith, H., Voronkov, A. (eds.) LPAR 2008. LNCS (LNAI), vol. 5330, pp. 512–527. Springer, Heidelberg (2008)
15. Samer, M., Szeider, S.: Backdoor sets of quantified Boolean formulas. Journal of Automated Reasoning 42(1), 77–97 (2009)
16. Slivovsky, F., Szeider, S.: Computing resolution-path dependencies in linear time. In: Cimatti, A., Sebastiani, R. (eds.) SAT 2012. LNCS, vol. 7317, pp. 58–71. Springer, Heidelberg (2012)
17. Stockmeyer, L.J., Meyer, A.R.: Word problems requiring exponential time. In: Proc. Theory of Computing, pp. 1–9. ACM (1973)
18. Van Gelder, A.: Variable independence and resolution paths for quantified Boolean formulas. In: Lee, J. (ed.) CP 2011. LNCS, vol. 6876, pp. 789–803. Springer, Heidelberg (2011)

Detecting Cardinality Constraints in CNF

Armin Biere[1], Daniel Le Berre[2], Emmanuel Lonca[2], and Norbert Manthey[3]

[1] Johannes Kepler University
[2] CNRS Université d'Artois
[3] Technische Universität Dresden

Abstract. We present novel approaches to detect cardinality constraints expressed in CNF. The first approach is based on a syntactic analysis of specific data structures used in SAT solvers to represent binary and ternary clauses, whereas the second approach is based on a semantic analysis by unit propagation. The syntactic approach computes an approximation of the cardinality constraints AtMost-1 and AtMost-2 constraints very fast, whereas the semantic approach has the property to be generic, i.e. it can detect cardinality constraints AtMost-k for any k, at a higher computation cost. Our experimental results suggest that both approaches are efficient at recovering AtMost-1 and AtMost-2 cardinality constraints.

1 Introduction

Current benchmarks in CNF contain various Boolean functions encoded with clauses [30,15]. Among them, cardinality constraints $\sum_{i=1}^{n} l_i \otimes k$ with $\otimes \in \{<, \leq, =, \geq, >\}$ are Boolean functions whose satisfiability is determined by counting the satisfied literals on the left hand side and compare them to the right hand side (the *threshold*). For instance, $x_1 + x_2 + \neg x_3 + \neg x_4 \leq 2$ is satisfied iff at most 2 of its literals are satisfied. A wide use case of those constraints is to encode that a domain variable v takes one value of the discrete set $\{o_1, o_2, \ldots, o_n\}$, which is represented by the n Boolean variables v_{o_i} and the cardinality constraint $\sum v_{o_i} = 1$.

Since cardinality constraints are Boolean functions, they can be expressed by an equivalent CNF. The "theoretical" approach, i.e. the one found in [12] for instance, translates a cardinality constraint $\sum_{i=1}^{n} l_i \leq k$ using $\binom{n}{k+1}$ negative clauses of size $k + 1$. Such encoding is called *binomial* because of the number of generated clauses. In practice, introducing new variables to reduce the number of clauses in the CNF usually results in a better performance. Various encodings have been proposed in the last decade (see for instance [14] for a survey). We discuss commonly used encodings in next section.

Pseudo-Boolean solvers use a proof system like *generalized resolution* [21], which is a specific form of the *cutting planes* proof system [12] that *p-simulates* resolution. This way, these solvers are able to solve instances of the Pigeon Hole Principle [19] when they are given cardinality constraints but not when they are given the same problem expressed with clauses. The reason of that behavior is

C. Sinz and U. Egly (Eds.): SAT 2014, LNCS 8561, pp. 285–301, 2014.

that applying generalized resolution on clauses is equivalent to resolution [21], while on cardinality constraints generalized resolution is a specific form of cutting planes [12]. Retrieving cardinality constraints from clauses in the cutting planes proof system requires a very specific procedure. Take for instance the cardinality constraint

$$x_1 + x_2 + x_3 + x_4 \leq 1$$

which is equivalent to

$$\overline{x_1} + \overline{x_2} + \overline{x_3} + \overline{x_4} \geq 3$$

This cardinality constraint is represented in CNF using the following clauses:

$$\neg x_1 \vee \neg x_2, \quad \neg x_1 \vee \neg x_3, \quad \neg x_1 \vee \neg x_4, \quad \neg x_2 \vee \neg x_3, \quad \neg x_2 \vee \neg x_4, \quad \neg x_3 \vee \neg x_4$$

These clauses can be represented as binary cardinality constraints:

$$x_1 + x_2 \leq 1, \quad x_1 + x_3 \leq 1, \quad x_1 + x_4 \leq 1, \quad x_2 + x_3 \leq 1, \quad x_2 + x_4 \leq 1, \quad x_3 + x_4 \leq 1$$

Retrieving the original cardinality from the clauses represented by cardinalities ≤ 1 requires to derive intermediate constraints as shown below (from [12]):

$x_1 + x_2 \leq 1$	$x_1 + x_2 \leq 1$	$x_1 + x_3 \leq 1$	$x_2 + x_3 \leq 1$
$x_1 + x_3 \leq 1$	$x_1 + x_4 \leq 1$	$x_1 + x_4 \leq 1$	$x_2 + x_4 \leq 1$
$x_2 + x_3 \leq 1$	$x_2 + x_4 \leq 1$	$x_3 + x_4 \leq 1$	$x_3 + x_4 \leq 1$
$x_1 + x_2 + x_3 \leq 1$	$x_1 + x_2 + x_4 \leq 1$	$x_1 + x_3 + x_4 \leq 1$	$x_2 + x_3 + x_4 \leq 1$

For the first column, summing the three cardinality constraints leads to $2x_1 + 2x_2 + 2x_3 \leq 3$, which can be reduced to $x_1 + x_2 + x_3 \leq 1$ by dividing the inequality by 2 and rounding down the threshold. The same process can be applied to derive the other cardinality constraints in the last line. Finally, summing up these four cardinality constraints of 3 literals results in a cardinality constraint of 4 literals: $3x_1 + 3x_2 + 3x_3 + 3x_4 \leq 4$. The expected cardinality constraint $x_1 + x_2 + x_3 + x_4 \leq \lfloor \frac{4}{3} \rfloor$ is obtained after division by 3 and rounding.

The described process is tedious and not easy to integrate in a solver. Thus, the idea is to find a way to detect those cardinality constraints in a preprocessing step, independent from the original proof system of the solver.

The motivation for this work is to allow solvers to take advantage of those cardinality constraints, at least for space efficiency (support of native cardinality constraints) or because of a better proof system (e.g. Generalized Resolution [21] or Cutting Planes [12]). Detecting cardinality constraints is also an interesting idea for pure SAT solvers, namely for constraints reencoding, e.g. to encode cardinality constraints back to CNF with an alternative and hopefully more efficient encoding [27,26]. This is especially useful in practice to replace the commonly used *pairwise* encoding of ≤ 1 constraints with a more efficient encoding.

2 Short Review of Known Encodings

Before we discuss how to find encoded cardinality constraints, a few common encodings for widely used constraints are introduced. For the AtMost-1 constraint

$\sum_{i=1}^{n} x_i \leq 1$ the *naïve encoding*, also known as *pairwise encoding*, is to exclude each pair of satisfied literals explicitly: $\bigwedge_{i=1}^{n} \bigwedge_{j>i}^{n} (\neg x_i \vee \neg x_j)$. This way, a constraint with n variables requires $\frac{n(n-1)}{2}$ clauses. This encoding is also referred to as *direct encoding* in the CP community [33].

The *nested encoding* uses auxiliary variables to reduce the number of generated clauses from a quadratic number can to a linear number, by (recursively) splitting the constraint into two constraints:

$$\sum_{i=1}^{n} x_i \leq 1 = [y + (\sum_{i=1}^{\lfloor \frac{n}{2} \rfloor - 1} x_i) \leq 1] \wedge [\neg y + (\sum_{i=\lfloor \frac{n}{2} \rfloor - 1}^{n} x_i) \leq 1].$$

For $n = 4$, the naïve encoding requires six clauses, and the nested encoding requires six clauses as well, but has more variables. Hence, as soon as the number of variables for an AtMost-1 constraint is at most four, no more recursions are applied. This way, the nested encoding requires $3n - 6$ clauses.

The currently best known asymptotic (starting from $n > 47$ [26]) encoding for the AtMost-1 constraint is the *two product encoding* [11]. For n variables in the constraint, two integers $p = \lfloor \sqrt{n} \rfloor$ and $q = \lceil \frac{n}{p} \rceil$ are used to create two more AtMost-1 constraints: $\sum_{i=1}^{p} r_i \leq 1$ and $\sum_{i=1}^{q} c_i \leq 1$. These two constraints are used as selector for a row and a column. The variables x_i are placed in a matrix, such that each variable x_i is assigned exactly to one row selector r_s and to one column selector c_t with the clauses $(\neg x_i \vee r_s)$ and $(\neg x_i \vee c_t)$, where $s = \lfloor \frac{i-1}{q} \rfloor + 1$ and $t = ((i - 1) \mod q) + 1$. An illustration for 10 variables is given in Fig. 1.

Further proposed encodings for the AtMost-1 constraint are the log encoding [33], the ladder encoding [18,17] also defined independently in [3], the commander encoding [23], generalizations of the log encoding and the two-product encoding [14], the bimander encoding [20], as well as generalizations of the bimander encoding [7].

For cardinality constraints $\sum_{i=1}^{n} x_i \leq k$ with a higher threshold $k > 1$, many encodings have been presented. Well known and sophisticated encodings are the partial sum encoding [1], totalizer encoding [6], the sequential counter encoding [31], BDDs [13] or sorting networks [13], cardinality networks [5], as well as the perfect hashing encoding [9]. As shown in [26], these specialized encodings produce much smaller CNF formulas compared to the binomial encoding. However, it is not clear whether smaller is better in all contexts. A recent survey on practical efficiency of those encodings in the context of MaxSAT solving is available in [28].

3   Static Detection of AtMost-1 and AtMost-2 Constraints

The naïve encoding of the AtMost-1 constraint can be detected by a syntactic analysis of the formula, namely by finding cliques in the *NAND graph* (NAG) of the formula, which is the undirected graph connecting literals that occur negated in the same binary clause. In [4,2], the authors modified the solvers zChaff and Satz to recognize those constraints using unit propagation and local search. A specific data structure for binary clauses is often found in modern

SAT solvers to reduce the memory consumption of the solver. From such a graph AtMost-1 constraints can be extracted by syntactic analysis. The naïve encoding of the AtMost-2 constraint can be recognized by exploring ternary clauses. The tools 3MCARD [24], LINGELING [10] and SBSAT [34] can recover cardinality constraints based on a syntactic analysis, and hence their methods are presented below additionally to the new extraction method. Both 3MCARD and SBSAT do not restrict their search on clauses of special size, but consider the whole formula: SBSAT constructs BDDs based on clauses that share the same variables. By merging and analyzing these BDDs cardinality constraints can be detected [34]. The tool 3MCARD builds a graph based on the binary clauses, and increases the current constraint while collecting more clauses [24]. Only LINGELING has special methods to extract AtMost-1 constraints, and AtMost-2 constraints.

3.1   Detecting the Pairwise Encoding

The structure of the pairwise encoding on the NAG is quite simple: if a clique is present in that graph, then the literals of the corresponding nodes form an AtMost-1 constraint. Since finding a clique of size k in a graph is NP-complete [22], a preprocessing step should not perform a full clique search. The algorithm for greedily finding cliques as implemented in LINGELING goes over all literals n which have not been included in an AtMost-1 constraint yet. For each n the set S of candidate literals is initialized with n. Then all literals l which occur negated in binary clauses $\bar{n} \vee \bar{l}$ together with n, e.g. l and n are connected in the NAG, are considered, in an arbitrary order, and greedily added to S after checking that for each previously added $k \in S$ a binary clause $\bar{l} \vee \bar{k}$ is also present in the formula, e.g. l and k have an edge in the NAG too. As an optimization, literals k which already occur in previously extracted AtMost-1 constraints are skipped. The final set S of nodes forms a clique in the graph. If $|S| > 2$ the clique is non-trivial and the AtMost-1 constraint $\sum_{l \in S} \leq 1$ is added [10].

3.2   Detecting the Nested Encoding

Consider the nested encoding of the AtMost-1 constraint $x_1 + x_2 + x_4 + x_5 \leq 1$, where the constraint is divided into the cardinality constraints $x_1 + x_2 + x_3 \leq 1$ and $\neg x_3 + x_4 + x_5 \leq 1$. They are represented in CNF by the six clauses $(\neg x_1 \vee \neg x_2), (\neg x_1 \vee \neg x_3), (\neg x_2 \vee \neg x_3), (x_3 \vee \neg x_4), (x_3 \vee \neg x_5), (\neg x_4 \vee \neg x_5)$. Since there is no binary clause $(\neg x_1 \vee \neg x_4)$, the above method cannot find this encoding. Here, we present another method that recognizes this encoding. The two smaller constraints can be recognized with the above method (their literals form two cliques in the NAG). Then, there is an AtMost-1 constraint for the literal x_3, as well as for the literal $\neg x_3$. By resolving the two constraints, the original constraint can be obtained. Algorithm 1 searches for exactly this encoding by combining pairs of constraints. For each variable v, all AtMost-1 constraints with different polarity are added and simplified. As a simplification it is checked, whether duplicate literals occur, or whether complementary literals

---

Algorithm 1: Merge AtMost-1

Input: A set of "at most 1" cardinality constraints S, the set of variables V
Output: An extended set of "at most 1" cardinality constraints

1 **foreach** $v \in \mathsf{V}$ **do**
2 | **foreach** $\mathsf{A} \in \mathsf{S}_v$ **do**
3 | | **foreach** $\mathsf{B} \in \mathsf{S}_{\neg v}$ **do**
4 | | | $S := S \cup \mathtt{simplify}(\mathsf{A} + \mathsf{B})$
5 | | **end**
6 | **end**
7 **end**
8 **return** S ;

---

occur. In the former case, the duplicated literal has to be assigned false, because that literal has now a weight of two in that constraint, while the threshold is 1. In the latter case, all literals of the constraint $(A+B)$, except the complementary literal, has to be falsified (because $x + \overline{x} = 1$, so the threshold is reduced by one to zero). The simplified constraint is added to the set of AtMost-1 constraint, which is finally returned by the algorithm.

Since the nested encoding can be encoded recursively, the algorithm can be called multiple times to find these recursive encodings. To not resolve the same constraints multiple times, for each variable the already *seen* constraints can be memorized, so that in a new iteration only resolutions with new constraints are performed. In practice, our implementation loops over the variables in ascending order exactly once. This seems to be sufficient, because the recursive encoding of constraints requires that the "fresh" variable is not present yet, so that the ascending order in the variable finds this encoding nicely.

3.3   Detecting the Two-Product Encoding

The two product encoding has a similar recursive structure as the nested encoding, however, its structure is more complex. Hence, this encoding is discussed in more details. The constraint in Fig. 1 illustrates an AtMost-1 constraint that is encoded with the two-product encoding.

For all concerned literals, in the example x_1 to x_{10}, two implications are added to set the column and row selectors. For example, as x_7 is on the second row and the fourth column, the constraints $x_7 \rightarrow r_2$ and $x_7 \rightarrow c_3$ are added. In order to prevent two rows or two columns selectors to be set simultaneously, we also add AtMost-1 cardinality constraints on the c_i and on the r_i literals. Those new cardinality constraints are encoded using the pairwise encoding if their size is low, or using the two product encoding. As the product encoding of AtMost-1 constraints may generate other AtMost-1 constraints to be encoding in the same way, the algorithm may be written in a recursive way.

In the given constraint, the following implications to select a column and a row for x_7 are entailed by the encoding: $x_7 \rightarrow c_3$ and $x_7 \rightarrow r_2$. Additionally, the implications $c_3 \rightarrow \neg c_2$ and $\neg c_2 \rightarrow (\neg x_2 \wedge \neg x_6)$ by transitivity show, that

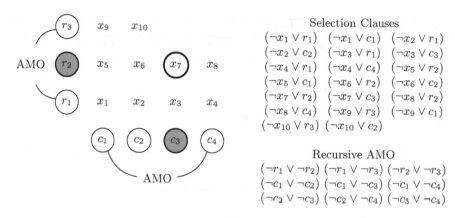

Fig. 1. Encoding the AtMost-1 constraint $\sum_{i=1}^{i\leq 10} x_i \leq 1$ with the two product encoding, and two auxiliary AtMost-1 constraints $r_1 + r_2 + r_3 \leq 1$ and $c_1 + c_2 + c_3 + c_4 \leq 1$

$x_7 \rightarrow (\neg x_6 \wedge \neg x_2)$. Since all implications are build on binary clauses, the reverse direction also holds: $x_6 \rightarrow \neg x_7$ and $x_2 \rightarrow \neg x_7$. Hence, the constraints $x_6 + x_7 \leq 1$ and $x_2 + x_7 \leq 1$ can be deduced. However, the constraint $x_2 + x_6 \leq 1$ cannot be deduced via the columns and their literals c_2 and c_3. This constraint can still be found via rows, namely with the literals r_1 and r_2. The same reasoning as for columns applies also to rows.

More generally, given an AtMost-1 constraint R, where the complement of a literal $r_i \in R$ implies some literal $\neg x_i$ ($\neg r_i \rightarrow \neg x_i$), and furthermore, this literal $\neg x_i$ implies a literal b_i, which belongs to another AtMost-1 constraint C, $\neg b_i \in C$, then by using R as row constraint, and C as column constraint, an AtMost-1 constraint that includes x_i can be constructed by searching for the remaining literals x_j. Per literal a_i in the row constraint R, literals x_i implied by $\neg r_i$ can be collected as candidates to form a row in the two-product representation. Only literals x_i that imply a different literal c_i of the column constraint C are considered, so that the literal inside each row matches exactly one column in the matrix. The literals for one row already form an AtMost-1 constraint. For the next row r_{i+1}, more literals x_i are collected in the same way, and added to the AtMost-1 constraint. This addition is sound based on the construction of the encoding: if one of the elements in the new AtMost-1 constraint is assigned to true, then this assignment implies its row and column variable to be satisfied as well. Since there is an AtMost-1 constraint enforced for both the rows and the columns, all other row and column variables are assigned false. Due to the implications in the Two-Product encoding, these falsified selector variables also falsify all variables (except the currently satisfied one) in the new AtMost-1 constraint, and hence only the initially satisfied variable remains satisfied.

To the best of our knowledge, no existing system is able to detect AtMost-1 constraints which are encoded in this way. We now present algorithm 2, that is able to find an approximation of the set of those constraints. Constructing new AtMost-1 constraints based on the idea of the two-product encoding is done by

---

Algorithm 2: Extract AtMost-1 Constraints Two Product Encodings

---

Input: A set of "at most 1" cardinality constraints S, the NAG of the formula
Output: An extended set of "at most 1" cardinality constraints

```
1  foreach R ∈ AMO do
2  |   r = min(R) ;                          // smallest literal only
3  |   l = min(NAG(r)) ;                      // r is row-selector for l
4  |   foreach c ∈ NAG(l) ;                   // c is column-selector for l
5  |   do
6  |   |   foreach C ∈ AMOₑ do
7  |   |   |   newAMO = ∅ ;          // construct a new AMO based on R and C
8  |   |   |   foreach k ∈ C do
9  |   |   |   hitSet = R ;                   // to hit each literal once
10 |   |   |   foreach hitLit ∈ NAG(k), hitLit ∉ newAMO do
11 |   |   |   |   foreach targetLit ∈ NAG(hitLit) do
12 |   |   |   |   |   if targetLit ∈ hitSet ;         // found selector pair
13 |   |   |   |   |   then
14 |   |   |   |   |   |   hitSet := hitSet \ {targetLit};    // update hit set
15 |   |   |   |   |   |   newAMO := newAMO ∪ {hitLit} ;    // update AMO
16 |   |   |   end
17 |   |   |   end
18 |   |   end
19 |   |   AMO := AMO ∪ newAMO ;          // store new AMO constraint
20 |   end
21 end
22 end
23 return AMO
```

---

first finding two AtMost-1 constraints R and C, which contain a literal r and c, which are used by some literal l as row selector and column selector (lines 1–6). Therefore, all AtMost-1 constraints R are considered, and a literal r is considered as row-selector variable. Next, the literal l is chosen to be part in the new two product AtMost-1. To reduce the computational work, the literal r is assumed to be the smallest literal in R, and the literal l is the smallest literal, such that $\neg r \to \neg l$ holds (lines 2–3). Finally, another AtMost-1 constraint C is selected, which contains the column selector literal c.

For each pair of AtMost-1s R and C, a new AtMost-1 can be constructed (line 7), by collecting all literals x_i. The literals x_i are called hitLit in the algorithm, because each such literal needs to imply a unique pair of row and column selector literals. This condition can be ensured by searching for literals that are implied by the complement of the column selector literal c: $\neg c \to \neg$hitLit. Furthermore, a literal hitLit has to imply a row selector variable $r \in R$ (lines 8–10). To ensure the second condition, an auxiliary set of literals hitSet is used, which stores all the literals of the row selector AtMost-1 constraint R during the analysis of each column. If for the current column selector c and the current literal hitLit a *new* selector targetLit ∈ hitSet is found (line 12), then the set hitSet of hit literals is updated by removing the current hit literal `targetLit`, and furthermore, the current hitting literal `hitLit` is added to the currently constructed AtMost-1 constraint (lines 14–15). Finally, the new AtMost-1 constraint is added to the set of constraints (line 19).

3.4 Detecting AtMost-2 Constraints

For a small number of literals x_i, and small thresholds k, for example $k = 2$, the naïve binomial encoding is competitive. Therefore, a method for extracting this constraint is proposed as well. Similarly to the syntactic extraction of AtMost-1 constraints, the structure of ternary clauses is analyzed by a greedy algorithm. Starting with a seed literal n which does not occur in an extracted AtMost-2 constraint yet all ternary clauses with $\bar{n}$ are considered and the set of candidate literals is initialized by all literals which occur negated at least twice in these clauses. If the candidate set at one point contains less than 4 literals the algorithm moves on to the next seed literal n. Otherwise each triple of literals in S is tested to have a corresponding ternary clause in the formula. If this test fails the set of candidates is reduced by removing from S one of the literals in a triple without a matching clause. If $|S| \geq 4$ and all triples can be matched with a clause, then the AtMost-2 constraint $\sum_{l \in S} l \leq 2$ is added.

4 Semantic Detection of AtMost-k Constraints

Another approach to detect cardinality constraints is to use unit propagation in the spirit of [25]. Using a more *semantic* approach instead of a pure *syntactic* approach allows to detect some nested cardinality constraints without requiring a specific procedure at the expense of performing unit propagation in a solver instead of traversing a NAG. The main advantage of the more semantic detection is that we may detect cardinality constraints as long as the encoding preserves arc-consistency by unit propagation. This allow us to propose an algorithm for all known encodings, that is also able to detect constraints that would not have been explicitly known at problem encoding time. However, our approach may not detect all cardinality constraints, since additional variables used in some encodings may interfere with the actual constraint variables, and make our algorithm produce truncated versions of the constraints to detect.

Basically our approach starts with a cardinality constraint $\sum_{i=1}^{n} l_i \leq k$ and tries to extend it with new literals m such that $(\sum_{i=1}^{n} l_i) + m \leq k$.

Our contribution is an algorithm to detect cardinality constraints in CNF using unit propagation, such that these constraints contain as much literals as it is possible to detect using unit propagation.

4.1 Expanding a Cardinality Constraint with One Literal

The idea of the algorithm is as follows: Given a clause $cl = l_1 \vee l_2 \vee .. \vee l_n$, we want to check if it belongs to a cardinality constraint $cc = \sum_{i=1}^{n} \bar{l_i} + \sum_j m_j \leq n - 1$.

Indeed, we know that $cl = l_1 \vee l_2 \vee .. \vee l_n \equiv \sum_{i=1}^{n} l_i \geq 1 \equiv \sum_{i=1}^{n} \overline{l_i} \leq n - 1 = cc'$.
We are thus looking for literals m_j which extend cc'.

Going back to our nested encoding example based on a CNF $\alpha = \neg x_1 \vee \neg x_2$, $\neg x_1 \vee \neg x_3, \neg x_2 \vee \neg x_3, x_3 \vee \neg x_4, x_3 \vee \neg x_5, \neg x_4 \vee \neg x_5$. $\neg x_1 \vee \neg x_2$ does represent the cardinality constraint $x_1 + x_2 \leq 1$. If we assign both x_1 or x_2 in α, we notice that the literals $\neg x_3, \neg x_4, \neg x_5$ are derived by unit propagation in both cases. hence, we can extend $x_1 + x_2 \leq 1$ by either x_3, x_4 or x_5, i.e. that the cardinality constraints $x_1 + x_2 + x_3 \leq 1$, $x_1 + x_2 + x_4 \leq 1$, $x_1 + x_2 + x_5 \leq 1$ are derivable from α. More generally: if all valid maximal combinations of the literals in cc' imply a literal $\neg m$, then m can be added to cc'. We exploit the following property.

Proposition 1. *Let α be a CNF. Let $\alpha(S)$ be the conjunction of the literals propagated in α under the set of assumptions S. Let $cc = \sum_{i=1}^{n} l_i \leq k$. Let $L = \{l_i \mid 1 \leq i \leq n\}$ and $L_k = \{S \mid [S \subseteq L] \wedge [|S| = k]\}$. If $\alpha \models cc$ and $\forall S \in L_k, \alpha(S) \models \neg m$ then $\alpha \models (\sum_{i=1}^{n} l_i) + m \leq k$.*

Proof. Let us suppose that ω is a model of α, $\alpha \models \sum_{i=1}^{n} l_i \leq k$ and $\forall S \in L_k, \alpha(S) \models \neg m$. Let us suppose that ω is not a model of $\alpha \wedge (\sum_{i=1}^{n} l_i) + m \leq k$. This implies that at least $k + 1$ literals in $\{l_1, ..., l_n\}$ are set to true. As $\alpha \models \sum_{i=1}^{n} l_i \leq k$, m must be set to true, which is inconsistent with the fact that $\forall S \in L_k, \alpha(S) \models \neg m$. □

If several of those literals exist, it is not valid to add them at once to cc'. In our running example, x_3, x_4 and x_5 are candidates to extend $x_1 + x_2 \leq 1$ but extending cc' with all literals leads to the cardinality constraint $x_1 + x_2 + x_3 + x_4 + x_5 \leq 1$, which is not derivable from α. We also need to pay attention to unit clauses in the original formula and literals implied by unit propagation. Those literals are by definition candidates to the cardinality constraint expansion. Adding those literals may results in a case where the only literals that will be able to expand the constraint through our algorithm are known to be falsified.

Consider the formula $\neg x_1 \vee \neg x_2, \neg x_1 \vee \neg x_3, \neg x_3 \vee \neg x_2, \neg x_4$, and suppose you treat $\neg x_1 \vee \neg x_2$ as the cardinality constraint $x_1 + x_2 \leq 1$. Two literals are candidates to the clause expansion: x_3 and x_4. If we choose x_4, then the generated cardinality constraint is not tight because we know that x_4 must be falsified. Note that if unit propagation leads to unsatisfiability ($\bot$ is detected), we do not have to filter out the candidates, because all literals are implied by a falsified formula. In this case, we must pay attention that the candidate we choose for the expansion is not the complement of a literal that is already present in the initial cardinality constraint. This check is done in line 1 of Algorithm 3.

Algorithm 3 exploits Proposition 1 to find the complement of a literal that may expand a cardinality constraint. The set candidates keeps all the remaining candidates (the literals whose negation may expand the constraint). For each unit propagation phase, only literals that are propagated are kept, that is eliminating the ones for which there exists a subset L_k that does not propagate them (this is, in fact, preserve arc-consistency). This procedure implies that for each literal $\neg m$ in the set candidates, the literal m may expand the current cardinality constraint.

Algorithm 3: findExpandingLiterals

Input: a CNF formula α, a cardinality constraint $\sum_{i=1}^{n} l_i \leq k$
Output: a set of literals m such that $(\sum_{i=1}^{n} l_i) + \overline{m} \leq k$

1 candidates $\leftarrow \{v_i | v_i \in VARS(\alpha)\} \cup \{\overline{v_i} | v_i \in VARS(\alpha)\} \setminus \{\overline{l_i}\}$;
2 **foreach** S $\subseteq \{l_i\}$ *such that* $|S| = k$ **do**
3 $\quad$ propagated $\leftarrow$ unitProp(α, S) ;
4 $\quad$ **if** $\bot \notin$ propagated **then**
5 $\quad\quad$ candidates $\leftarrow$ candidates $\cap$ propagated ;
6 $\quad\quad$ **if** candidates $= \emptyset$ **then**
7 $\quad\quad\quad$ **return** $\emptyset$;
8 **end**
9 **return** candidates ;

Lemma 1. *Let α be a CNF. Let $cc = \sum_{i=1}^{n} l_i \leq k$ such that $\alpha \models cc$. $\forall m \in findExpandingLiterals(\alpha, cc)$, $\alpha \models \sum_{i=1}^{n} l_i + \overline{m} \leq k$.*

4.2 Maximal Cardinality Constraint Expansion

In practice, we are not going to learn any arbitrary cardinality constraint, but only the ones which cannot be extended further. Moreover, if a cardinality constraint corresponding to a clause cannot be extended at all, we will keep it in its clausal form. Algorithm 3 computes all the literals that are propagated through unit propagation by all sets L_k. Once this set is empty, we are not able to find a literal that extends this constraint using the unit propagation. As long as there exists such literals, they may be added as proved by Proposition 1, as written in the following lemma.

Lemma 2. *Let α be a CNF. $\forall c \in \alpha$, $\alpha \models expandCardFromClause(\alpha, c)$.*

We iteratively find a new expanding literal, add the literal to the constraint, then search a new literal, and repeat these steps until there are no more expansion candidates. It is not necessary to compute all sets L_k when the second iteration is reached. In fact, to find the nth literal of a constraint, we have computed $\binom{n-1}{k}$ of the $\binom{n}{k}$ propagations that are required by the current call to Algorithm 3. The only sets L_k that are not analyzed yet are the sets containing the literal that was added to the constraint in the most recent step. This procedure is shown in Algorithm 4. With this insight, we build an efficient algorithm to compute maximum cardinality constraints in Algorithm 5.

The computation of the $\binom{n}{k}$ unit propagations is the costly part of the algorithm. We assume that the unit propagation cost is bounded by the number of literals to produce the following lemma.

Lemma 3. *Let α be a CNF with n variables and l literals. Let c be a clause of α of size $|c| = k + 1$. expandCardFromClause(α, c) has a complexity in $O(\binom{n}{k} \times l)$.*

Algorithm 4: refineExpandingLiterals

Input: a CNF formula α, a cardinality constraint $\sum_{i=1}^{n} l_i + l_{new} \leq k$, a set of literals L

Output: a set of literals m such that $(\sum_{i=1}^{n} l_i) + l_{new} + \overline{m} \leq k$

1 candidates $\leftarrow L$;
2 **foreach** $S' = S \cup \{l_{new}\}$ *such that* $S \subseteq \{l_i\}$ *and* $|S| = k - 1$ **do**
3 $\quad$ propagated $\leftarrow$ unitProp(α, S') ;
4 $\quad$ **if** $\perp \notin$ propagated **then**
5 $\quad\quad$ candidates $\leftarrow$ candidates $\cap$ propagated ;
6 $\quad\quad$ **if** candidates $= \emptyset$ **then**
7 $\quad\quad\quad$ **return** $\emptyset$;
8 **end**
9 **return** candidates ;

4.3 Replacing Clauses by Cardinality Constraints

The last step in our approach is to detect clauses that are entailed by the cardinality constraints found so far. This step is important, because it allows to avoid considering clauses that would lead to already revealed cardinality constraints. Furthermore, we need to keep the clauses not covered by any cardinality constraint to build a mixed formula of cardinality constraints and clauses, which is logically equivalent to the original formula.

We use the rule described by Barth in [8] and used in 3MCard [24] to determine if a clause (written as an at-most-k constraint) is dominated by a revealed cardinality constraint. This rule states that $L \geq d$ dominates $L' \geq d'$ iff $|L \setminus L'| \leq d - d'$. So, before considering a clause for cardinality constraint expansion, we check using this rule if the clause is dominated by one of our new constraints. In this case, we do not search any expansion, and remove this clause from the problem. We also remove the clauses that have been expanded to cardinality constraints, as they are trivially dominated by the new constraint.

Algorithm 5: expandCardFromClause

Input: a CNF formula α, a clause c

Output: a cardinality constraint cc or c

1 cc $\leftarrow \sum_{l \in c} \overline{l} \leq |c| - 1$;
2 candidates $\leftarrow findExpandingLiterals(\alpha, \mathsf{cc})$;
3 **while** candidates $\neq \emptyset$ **do**
4 $\quad$ select m in candidates;
5 $\quad$ cc $\leftarrow \sum_{l_i \in cc} l_i + \overline{m} \leq |c| - 1$;
6 $\quad$ candidates $\leftarrow refineExpandingLiteral(\alpha, \mathsf{cc}, \text{candidates} \setminus \{\mathsf{m}\})$;
7 **end**
8 **if** $|cc| > |c|$ **then return** cc;
9 **else return** c ;

Algorithm 6: revealCardsInCNF

Input: a CNF formula α and a bound k
Output: a formula $\phi \equiv \alpha$ containing cardinality constraints with threshold $\leq k$
   and clauses

1 $\phi \leftarrow \emptyset$;
2 **foreach** *clause* $c \in \alpha$ *of increasing size —c— such that* $|c| \leq k+1$ **do**
3 | **if** *there is no cardinality constraint* $cc \in \phi$ *which dominates* c **then**
4 | | $\phi \leftarrow \phi \cup expandCardFromClause(\alpha, c)$;
5 | **end**
6 **end**
7 **return** ϕ ;

To avoid redundancy it is important to first consider the smallest clauses as candidates for the expansion. In fact, while considering the smallest clauses first, we find the cardinality constraints with the lowest threshold first. Consider that $l_1 + ... + l_n \leq k$ has been discovered, and that we take a look at a constraint where the sum part sums a subset of $\{l_1, ..., l_n\}$ and where the threshold is $k + d$ ($d > 0$). In this case, the latter constraint is always dominated by the former cardinality constraint, so there is no need to expand it.

As the new cardinality constraints are consequences of the formula and the removed clauses are consequences of the cardinality constraints, we ensure that the new formula is equivalent to the original one, as written in the following theorem.

Theorem 1. *Let α be a CNF. Let k an arbitrary integer.*

$$\alpha \equiv revealCardsInCNF(\alpha, k)$$

In our nested encoding example, our approach will work as follows. We first try to extend $x_1 + x_2 \leq 1$. Our approach will find either $x_1 + x_2 + x_3 \leq 1$ or $x_1 + x_2 + x_4 + x_5 \leq 1$. Suppose it finds the longest one. The CNF is reduced from clauses dominated by that cardinality constraint: $\neg x_1 \vee \neg x_2, \neg x_4 \vee \neg x_5$. The next clause to consider is $\neg x_1 \vee \neg x_3$. We try to extend $x_1 + \neg x_3 \leq 1$. We can extend it to $x_1 + x_2 + x_3 \leq 1$. The clause $\neg x_2 \vee \neg x_3$, is removed from the CNF, because this clause is dominated by that new cardinality constraint. The next cardinality to extend is $\neg x_3 + x_4 \leq 1$. The cardinality constraint $\neg x_3 + x_4 + x_5 \leq 1$ is found. The remaining clauses are dominated by the cardinality constraints, so they are removed from the CNF. The procedure stops, since no more clauses have to be considered. Note that if the first cardinality constraint found is $x_1 + x_2 + x_3 \leq 1$, the procedure will be unable to reveal $x_1 + x_2 + x_4 + x_5 \leq 1$, because all clauses containing $\neg x_1$ would be removed from the CNF.

In terms of complexity, the worst case will be reached if we try to expand all clauses; implying the following complexity bound.

Lemma 4. *Let α be a CNF with n variables, m clauses and l literals. Let k an integer such that $0 < k \leq n$ and $m_k \leq m$ the number of clauses of size $\leq k + 1$ in α. revealCardsInCNF(α, k) has a complexity in $O(m_k \times \binom{n}{k} \times l)$.*

Preprocessor Solver	#inst.	Lingeling Lingeling	Synt.(Riss) Sat4jCP	Sem.(Riss) Sat4jCP	no SBSAT	no Sat4jCP
Pairwise	14	**14 (3s)**	13 (244s)	14 (583s)	6 (0s)	1 (196s)
Binary	14	3 (398s)	2 (554s)	**7 (6s)**	6 (7s)	2 (645s)
Sequential	14	0 (0s)	**14 (50s)**	14 (40s)	10 (6s)	1 (37s)
Product	14	0 (0s)	**14 (544s)**	11 (69s)	6 (25s)	2 (346s)
Commander	14	1 (3s)	7 (0s)	**14 (40s)**	9 (187s)	1 (684s)
Ladder	14	0 (0s)	11 (505s)	11 (1229s)	**12 (26s)**	1 (36s)

Fig. 2. Six encodings of the pigeon hole instances: number of solved instances and sum of run time for solved instances per solver configuration per encoding

5   Experimental Results

The experimental results show that the proposed methods detect a significant amount of cardinality constraints in CNFs. For this analysis we use academic benchmarks, like Sudoku puzzles and the pigeon hole problem, for which we know how many cardinality constraints are present in the CNF, and which are easy to solve using Generalized Resolution when the constraints are expressed using cardinality constraints. All the benchmarks were launched on Intel Xeon X5550 processors (@2.66GHz) with 32GB RAM and a 900s timeout.

The static approach is implemented in the latest release of Lingeling. The static approach plus the specific handling of the two product encoding are implemented on Riss (so called Syntactic). The semantic detection of cardinality constraints is implemented on Riss. As Riss does not take advantage of those cardinality constraints for solving the benchmarks, we use it as a fast preprocessor to feed Sat4j which uses Generalized Resolution to solve the new benchmark with a mix of clauses and cardinality constraints. It allows us to check if the cardinality constraints found by the incomplete approaches are sufficient to solve those benchmarks. We compare the proposed approaches against SBSAT[1].

5.1   Pigeon Hole Principle

These famous benchmark are known to be extremely hard for resolution based solvers [19]. For $n + 1$ pigeons and n holes, the problem is to assign each pigeon in a hole while not having more that one pigeon per hole. Each Boolean variable $x_{i,j}$ represents pigeon i is assigned hole j. The problem is expressed by $n + 1$ clauses $\bigvee_{j=1}^{n} x_{i,j}$ and n cardinality constraints $\sum_{j=1}^{n+1} x_{i,j} \leq 1$. Those benchmarks are generated for n from 10 to 15, and n from 25 to 200 by steps of 25, using six different encodings: binomial, product, binary, ladder, commander and sequential. The results are presented in Table 2.

As expected, without revealing the cardinality constraint, Sat4j can only solve one or two problems. The semantic approach can detect many cardinality con-

---

[1] We tried to run 3MCard on those benchmarks but the solver was not able to read or solve most of the benchmarks.

straints and let Sat4j solve more instances than the other solvers — it is particularly efficient for the pairwise, sequential and commander encodings. Note that the instance using the pairwise encoding for $n = 200$ has 402000 variables and 4020201 clauses, which shows that our approach scales. The static analysis (with specific reasoning for the two product encoding) is also very efficient on most of the encodings, and is the best on the product encoding, as intended. The commander and binary encodings are most difficult to reveal using our techniques. SBSAT is not as efficient as our approaches, even if it got the best results for the ladder encoding. Lingeling is very efficient for the pairwise encoding, but not for the others, as intended too.

5.2 Small Hard Combinatorial Benchmarks

Benchmarks of unsatisfiable balanced block designs are described in [32,16]. They contain the cardinality constraints AtMost-2. We use the benchmarks submitted to the SAT09 competition (called sgen) which were the smallest hard unsatisfiable formulas as well as benchmarks provided by Jakob Nordström and Mladen Miksa from KTH [29]. Both the syntactic and the semantic detection are able to recover quickly all cardinality constraints of these benchmarks and let the solver use them to prove unsatisfiability, as presented in Table 3. The solvers Lingeling and Sat4jCP, which do not use cardinality detection or generalized resolution, emphasize the fact these benchmarks are too difficult for existing solver.

5.3 Sudoku Benchmarks

Sudoku puzzles contain only $= 1$ cardinality constraints represented by a clause and an AtMost-1 constraint. The puzzles are trivial to solve but interesting from a preprocessing point of view because the cardinality constraints share a lot of variables which may mislead our approximation methods. We use two instances representing empty $n \times n$ for $n = 9$ and $n = 16$. The grids contain $n^2 \times 4$ AtMost-1 constraints, for $n = 9(16)$ there are 324(1024) constraints. All constraints are encoded with the pairwise encoding. The AtMost-1 constraints that can be found by the syntactic approach all contain 9 (resp. 16) literals, as expected. However, some constraints are missing: 300 constraints revealed out

Preprocessor Solver	#inst.	Lingeling Lingeling	Synt.(Riss) Sat4jCP	Sem.(Riss) Sat4jCP	no SBSAT	no Sat4jCP
Sgen unsat	13	0 (0s)	**13 (0s)**	**13 (0s)**	9 (614s)	4 (126s)
Fixed bandwidth	23	2 (341s)	**23 (0s)**	**23 (0s)**	**23 (1s)**	13 (1800s)
Rand. orderings	168	16 (897s)	**168 (7s)**	**168 (8s)**	99 (2798s)	69 (3541s)
Rand. 4-reg.	126	6 (1626s)	**126 (4s)**	**126 (5s)**	84 (2172s)	49 (3754s)

Fig. 3. Various hard combinatorial benchmarks families: number of solved instances and sum of run time for solved instances per solver configuration per family

of 324 and 980 revealed out of 1024. The semantic preprocessor finds all the cardinality constraints for those benchmarks because each cardinality constraint has a binary clause which belongs to only this constraint, and hence is not discarded when another constraint is found. This specific clause is used to retrieve the cardinality constraint by addition of literals.

6   Conclusion

We presented two approaches to derive cardinality constraints from CNF. The first approach is based on the analysis of a NAND graph, the data structure used in some SAT solvers to handle binary clauses efficiently, to retrieve AtMost-1 or AtMost-2 constraints. The second approach, based on using unit propagation on the original CNF, is a generic way to derive AtMost-k constraints. We show that both approaches are able to retrieve cardinality constraints in known hard combinatorial problems in CNF. Our experiments suggest that the syntactic approach is particularly useful to derive AtMost-1 and AtMost-2 constraints while the semantic is more robust because this method is also able to detect cardinality constraints with an arbitrary threshold. Our approaches are useful for tackling the smallest unsolved UNSAT problems of the SAT 2009 competition (sgen) which are all solved by Sat4j using generalized resolution within seconds once the AtMost-2 constraints have been revealed (this fact was already observed in [34]). The difference between the syntactic and semantic approaches is visible with the challenge benchmark from [16] that no solver could solve in one day. It is solved in a second after deriving 22 AtMost-2 and 20 AtMost-3 constraints when revealing the constraints using the semantic approach. The syntactic approach is not able to reveal those AtMost-3 constraints.

We have been able to reveal cardinality constraints on large application benchmarks. However, using that information to improve the run time of the solvers is future work. Checking if those constraints result from the original problem specification or are "hidden" constraints, i.e. constraints not explicitly known in the specification, is an open problem. An interesting research question, out of the scope of this paper, is to study the proof system of the combination of our semantic preprocessing step (extension rule and domination rule) plus generalized resolution.

Acknowledgement. The authors would like to thank Jakob Nordström and his group from KTH for providing us the hard combinatorial benchmarks, pointing out the challenge benchmark, and the fruitful discussions about Sat4j proof system which motivated this work. The authors would like to thank the Banff International Research Station which hosted the Theoretical Foundations of Applied SAT Solving workshop (14w5101) which essentially contributed to this joint work. The authors would like to thank the anonymous reviewers for their helpful comments to improve this paper. This work has been partially supported by ANR TUPLES.

References

1. Aloul, F.A., Ramani, A., Markov, I.L., Sakallah, K.A.: Generic ilp versus specialized 0-1 ilp: An update. In: Pileggi, L.T., Kuehlmann, A. (eds.) ICCAD, pp. 450–457. ACM (2002)
2. Ansótegui, C., Larrubia, J., Li, C.M., Manyà, F.: Exploiting multivalued knowledge in variable selection heuristics for sat solvers. Ann. Math. Artif. Intell. 49(1-4), 191–205 (2007)
3. Ansótegui, C., Manyà, F.: Mapping problems with finite-domain variables to problems with boolean variables. In: Hoos, H.H., Mitchell, D.G. (eds.) SAT 2004. LNCS, vol. 3542, pp. 1–15. Springer, Heidelberg (2005)
4. Ansótegui Gil, C.J.: Complete SAT solvers for Many-Valued CNF Formulas. Ph.D. thesis. University of Lleida (2004)
5. Asín, R., Nieuwenhuis, R., Oliveras, A., Rodríguez-Carbonell, E.: Cardinality networks and their applications. In: Kullmann, O. (ed.) SAT 2009. LNCS, vol. 5584, pp. 167–180. Springer, Heidelberg (2009)
6. Bailleux, O., Boufkhad, Y.: Efficient cnf encoding of boolean cardinality constraints. In: Rossi, F. (ed.) CP 2003. LNCS, vol. 2833, pp. 108–122. Springer, Heidelberg (2003)
7. Barahona, P., Hölldobler, S., Nguyen, V.-H.: Representative Encodings to Translate Finite CSPs into SAT. In: Simonis, H. (ed.) CPAIOR 2014. LNCS, vol. 8451, pp. 251–267. Springer, Heidelberg (2014)
8. Barth, P.: Linear 0-1 inequalities and extended clauses. In: Voronkov, A. (ed.) LPAR 1993. LNCS, vol. 698, pp. 40–51. Springer, Heidelberg (1993)
9. Ben-Haim, Y., Ivrii, A., Margalit, O., Matsliah, A.: Perfect hashing and cnf encodings of cardinality constraints. In: Cimatti, A., Sebastiani, R. (eds.) SAT 2012. LNCS, vol. 7317, pp. 397–409. Springer, Heidelberg (2012)
10. Biere, A.: Lingeling, plingeling and treengeling entering the sat competition 2013. In: Balint, A., Belov, A., Heule, M., Järvisalo, M. (eds.) Proceedings of SAT Competition 2013. Solver and Benchmark Descriptions, vol. B-2013-1, pp. 51–52. University of Helsinki, Department of Computer Science Series of Publications B (2013)
11. Chen, J.C.: A new sat encoding of the at-most-one constraint. In: Proc. of the Tenth Int. Workshop of Constraint Modelling and Reformulation (2010)
12. Cook, W., Coullard, C., Turán, G.: On the complexity of cutting-plane proofs. Discrete Applied Mathematics 18(1), 25–38 (1987)
13. Eén, N., Sörensson, N.: Translating pseudo-boolean constraints into sat. JSAT 2(1-4), 1–26 (2006)
14. Frisch, A., Giannaros, P.: Sat encodings of the at-most-k constraint: Some old, some new, some fast, some slow. In: Proceedings of the 9th International Workshop on Constraint Modelling and Reformulation, ModRef 2010 (2010)
15. Fu, Z., Malik, S.: Extracting logic circuit structure from conjunctive normal form descriptions. In: VLSI Design, pp. 37–42. IEEE Computer Society (2007)
16. Van Gelder, A., Spence, I.: Zero-one designs produce small hard sat instances. In: Strichman, O., Szeider, S. (eds.) SAT 2010. LNCS, vol. 6175, pp. 388–397. Springer, Heidelberg (2010)
17. Gent, I.P., Nightingale, P.: A new encoding of alldifferent into sat. In: Proc. 3rd International Workshop on Modelling and Reformulating Constraint Satisfaction Problems, pp. 95–110 (2004)
18. Gent, I., Prosser, P., Smith, B.: A 0/1 encoding of the gaclex constraint for pairs of vectors. In: ECAI 2002 workshop W9: Modelling and Solving Problems with Constraints. University of Glasgow (2002)

19. Haken, A.: The intractability of resolution. Theoretical Computer Science 39(0), 297–308 (1985)
20. Hölldobler, S., Nguyen, V.H.: On SAT-Encodings of the At-Most-One Constraint. In: Katsirelos, G., Quimper, C.G. (eds.) Proc. The Twelfth International Workshop on Constraint Modelling and Reformulation, Uppsala, Sweden, September 16-20, pp. 1–17 (2013)
21. Hooker, J.N.: Generalized resolution and cutting planes. Ann. Oper. Res. 12(1-4), 217–239 (1988)
22. Karp, R.: Reducibility among combinatorial problems. In: Miller, R., Thatcher, J., Bohlinger, J. (eds.) Complexity of Computer Computations. The IBM Research Symposia Series, pp. 85–103. Springer, US (1972)
23. Klieber, W., Kwon, G.: Efficient cnf encoding for selecting 1 from n objects. In: International Workshop on Constraints in Formal Verification (2007)
24. van Lambalgen, M.: 3MCard 3MCard A Lookahead Cardinality Solver. Master's thesis. Delft University of Technology (2006)
25. Le Berre, D.: Exploiting the real power of unit propagation lookahead. Electronic Notes in Discrete Mathematics 9, 59–80 (2001)
26. Manthey, N., Heule, M.J.H., Biere, A.: Automated reencoding of boolean formulas. In: Biere, A., Nahir, A., Vos, T. (eds.) HVC. LNCS, vol. 7857, pp. 102–117. Springer, Heidelberg (2013)
27. Manthey, N., Steinke, P.: Quadratic direct encoding vs. linear order encoding, a one-out-of-n transformation on cnf. In: Proceedings of the First International Workshop on the Cross-Fertilization Between CSP and SAT, CSPSAT11 (2011)
28. Martins, R., Manquinho, V.M., Lynce, I.: Parallel search for maximum satisfiability. AI Commun. 25(2), 75–95 (2012)
29. Mikša, M., Nordström, J.: Long proofs of (Seemingly) simple formulas. In: Sinz, C., Egly, U. (eds.) SAT 2014. LNCS, vol. 8561, pp. 122–138. Springer, Heidelberg (2014)
30. Ostrowski, R., Grégoire, É., Mazure, B., Sais, L.: Recovering and exploiting structural knowledge from cnf formulas. In: Van Hentenryck, P. (ed.) CP 2002. LNCS, vol. 2470, pp. 185–199. Springer, Heidelberg (2002)
31. Sinz, C.: Towards an optimal cnf encoding of boolean cardinality constraints. In: van Beek, P. (ed.) CP 2005. LNCS, vol. 3709, pp. 827–831. Springer, Heidelberg (2005)
32. Spence, I.: sgen1: A generator of small but difficult satisfiability benchmarks. ACM Journal of Experimental Algorithmics 15 (2010)
33. Walsh, T.: Sat v csp. In: Dechter, R. (ed.) CP 2000. LNCS, vol. 1894, pp. 441–456. Springer, Heidelberg (2000)
34. Weaver, S.: Satisfiability Advancements Enabled by State Machines. Ph.D. thesis. University of Cincinnati (2012)

Improving Implementation of SLS Solvers for SAT and New Heuristics for k-SAT with Long Clauses

Adrian Balint[1], Armin Biere[2], Andreas Fröhlich[2], and Uwe Schöning[1]

[1] Institute of Theoretical Computer Science, Ulm University, Ulm, Germany
{adrian.balint,uwe.schoening}@uni-ulm.de
[2] Inst. Formal Models and Verification, Johannes Kepler University, Linz, Austria
{armin.biere,andreas.froehlich}@jku.at

Abstract. Stochastic Local Search (SLS) solvers are considered one of the best solving technique for randomly generated problems and more recently also have shown great promise for several types of hard combinatorial problems. Within this work, we provide a thorough analysis of different implementation variants of SLS solvers on random and on hard combinatorial problems. By analyzing existing SLS implementations, we are able to discover new improvements inspired by CDCL solvers, which can speed up the search of all types of SLS solvers. Further, our analysis reveals that the multilevel break values of variables can be easily computed and used within the decision heuristic. By augmenting the probSAT solver with the new heuristic, we are able to reach new state-of-the-art performance on several types of SAT problems, especially on those with long clauses. We further provide a detailed analysis of the clause selection policy used in focused search SLS solvers.

1 Introduction

The Satisfiability problem (SAT) is one of the best known NP-complete problems. Besides its theoretical importance, it has also many practical applications in different domains such as software and hardware verification. Two of the best known solving approaches for the SAT problem are Conflict Driven Clause Learning (CDCL) and Stochastic Local Search (SLS). Solvers based on the CDCL solving approach have a very good performance on structured and application problems, and they are highly engineered for this type of problems. Stochastic Local Search solvers, on the other side, are good at solving randomly generated problems and several types of hard combinatorial problems.

The wide use of CDCL solvers in different applications has led to highly efficient implementations for these type of solvers. In contrast, SLS solvers do not have many practical applications, and their implementations have not been optimized as excessively as those of CDCL solvers. We think that SLS solvers have a high potential on hard combinatorial problems and that it is worth investing effort in improving algorithmic aspects of their implementation.

C. Sinz and U. Egly (Eds.): SAT 2014, LNCS 8561, pp. 302–316, 2014.

The contribution of this paper are three-fold. First we propose an improvement of the implementations techniques used in SLS solvers inspired from the CDCL solver NanoSAT. During the analysis of implementation techniques we discovered that the multilevel *break* value can be computed cheaply and can speed-up the search of SLS solvers when used in decision heuristics, this being our second contribution. Further we observed that the selection of unsatisfied clauses influences the performance of SLS solvers considerably. Our third contribution consists in the analysis of the clause selection heuristics.

1.1 Related Work

Implementation methods play a crucial role for SAT solvers and can influence their performance considerably. Fukunaga analyzed different implementations of common SLS solvers in [1]. He observed that during search the transition of clauses from having one satisfied to having two satisfied literals, or backwards occurs in structured problems very often. As a consequence, he proposed to introduce the 2-watched literals scheme for SLS solvers.

Multilevel properties of variables like the $make_l$ and $break_l$ values were first considered in [2], where the level one and two make value were combined as a linear function in order to break ties in WalkSAT. They were able to show that the performance of the WalkSAT solver can be considerably increased with this new tie breaking mechanism on 5-SAT and 7-SAT problems. The possible use of the $break_2$ value was also mentioned, but was not practically applied, probably due to the involved implementation complexity. In [3] the second level score value is being used ($score_2 = make_2 - break_2$) in the clause weighting solver CScoreSAT. The use of higher levels is considered but was not analyzed.

Within a caching implementation, the $make_2$ value can be computed with little overhead, while the $break_2$ value needs additional data structures. As one result of our implementation analysis, we have figured out that the $break_l$ value can be computed very easily within the non-caching implementations and consequently analyzed its role in the probSAT solver. The authors of [2] also mention that the role of *make* should not be neglected as was shown in [4]. This is of course true for their findings, but as the *make* and *break* value are complements of each other, it is sufficient to consider only the break value.

The importance of clause selection is strongly related to the class of GSAT solvers [5] and to those of weighted solvers [6,7,8]. Though, we are not aware of any analysis performed for focused random walk solvers like WalkSAT or probSAT. The pseudo breadth first search (PBFS) scheme was proposed in [9, p. 93] and only analyzed on a small set of randomly generated 3-SAT problems.

2   Implementations of SLS Solvers

SLS solvers work on complete assignments as opposed to partial assignments used by CDCL solvers. Starting from a random generated assignment, an SLS solver selects a variable according to some heuristic ($pickVar$) and then changes its value (a *flip*). This process is repeated until a satisfying assignment has been found or some limits have been reached.

The input to the SLS solver is a formula F in conjunctive normal form (CNF), i.e., a conjunction of clauses. A clause is a disjunction of literals, which are defined as variables or negation of these. An assignment α is called satisfying for formula F if every clause contains at least one satisfying literal. A literal l within a clause is satisfying if the value within the assignment α corresponds to its polarity (i.e. false for negated and true for positive literals).

The complexity of an SLS solver is completely dominated by two operations: $pickVar()$ and $flip(var)$. The complexity of the $pickVar$ method depends on the heuristic used by the SLS solver, while the complexity of the $flip(var)$ method depends on the computation of the information that has to be updated.

For the moment, we will restrict our analysis to simple SLS solvers like Walk-SAT [5] and probSAT [4], and later extend it to other more complex solvers like Sparrow [10]. The information needed by WalkSAT and probSAT is identical. Both need the set of unsatisfied clauses under the current assignment and the *break* value of variables from those clauses. The $break(x)$ value is the number of clauses that become unsatisfied after flipping x. Within their $pickVar$ method, both solvers randomly select an unsatisfied clause and then pick a variable from this clause according to their particular heuristic.

For each variable x in the selected clause, the probSAT solver uses a flip probability proportional to $cb^{-break(x)}$ or, alternatively, to $(break(x) + \epsilon)^{-cb}$. The WalkSAT solver randomly selects a variable x with $break(x) = 0$ if such a variable exists in the selected clause. Otherwise, it picks a variable randomly from selected clause with probability p and the best variable with respect to the *break* value with probability $1 - p$.

In general, SLS solvers keep track of the transition of clauses from different satisfaction states. By applying an assignment α to the formula F, denoted by $F|\alpha$, we can categorize clauses according to their satisfaction status. If a clause has t true literals, we say that the clause is t-satisfied.

To maintain the set of unsatisfied clauses $falseClause$, an SLS solver keeps track of the clause transitions from 0-satisfied to 1-satisfied and back. For a variable x, the value of $break(x)$ is equal to the cardinality of the set of 1-satisfied clauses that contain x as a satisfying literal, because by flipping x these clauses will get 0-satisfied (unsatisfied). Further, $break(x)$ changes if and only if any of these clauses become 0-satisfied or 2-satisfied.

The *break* value of variables can be either maintained incrementally (also called caching) or computed in every iteration (also called non-caching). The original implementation of WalkSAT and of probSAT uses the caching scheme while the WalkSAT implementation within the UBCSAT framework [11] is using the non-caching implementation.

To monitor the satisfaction status of clauses, SLS solvers use an array *trueLit* that stores for each clause the number of true literals and maintains it incrementally. Besides the before mentioned data structures, an SLS solver additionally needs the occurrence list for each literal, which is the set of clauses where a literal occurs. Further, for each 1-satisfied clause, also the satisfying variable is stored (this is also known as the critical variable $critVar$, or as the $watch1$ scheme).

Having introduced these data structures, we can now describe the standard implementation [1] of SLS solvers that use caching of the $break$ value and of $falseClause$. The $flip(var)$ method using caching is described in Algorithm 1.

Algorithm 1. Variable flip including caching (see also [1]).

Input: Variable to flip v

```
1  α[v] = ¬α[v];                                    /* change variable value */
2  satisfyingLiteral = α[v] ? v : ¬v;
3  falsifyingLiteral = α[v] ? ¬v : v;
4  for clause in occurrenceList[satisfyingLiteral] do
5  │   if numTrueLit[clause] == 0 then               /* transition 0 → 1 */
6  │   │   remove clause from falseClause ;
7  │   │   break[v] + +;
8  │   │   critVar[clause] = v
9  │   else
10 │   │   if numTrueLit[clause] == 1 then           /* transition 1 → 2 */
11 │   │   └   break[critVar[clause]]- -;
12 │   numTrueLit[clause]++;
13 for clause in occurrenceList[falsifyingLiteral] do
14 │   if numTrueLit[clause] == 1 then               /* transition 0 ← 1 */
15 │   │   add clause to falseClause ;
16 │   │   break[v]- -;
17 │   │   critVar[clause] = v
18 │   else
19 │   │   if numTrueLit[clause] == 2 then           /* transition 1 ← 2 */
20 │   │   │   for var in clause do            /* find the critical variable */
21 │   │   │   │   if var is satisying literal in clause then
22 │   │   │   │   │   critVar[clause]=var;
23 │   │   │   │   └   break[var]++;
24 │   numTrueLit[clause]- -;
```

Except for transition $1 \leftarrow 2$, all others have constant complexity. Whenever the number of satisfied literals decreases from two to one, we have to search for the critical variable in the clause, thus rendering the complexity of this transition to be $O(len(C))$, where $len(C)$ denotes the length of the clause. This is the previous state-of-the-art and was first toroughly analyzed in [1].

With a simple trick, first introduced in the CDCL solver NanoSAT [12], we can reduce the complexity of this step to $O(1)$. Instead of storing the critical variable for each clause, we store the XOR concatenation of all satisfying variables $trueVarX$. As alternative one can maintain the sum of the literals, using addition and subtraction during updates, while for the XOR scheme only the XOR operation is needed.

If there is only one satisfying variable per clause, $trueVarX(clause)$ will contain this variable. If there are two satisfying variables in a clause C, i.e. $trueVarX[C] = x_i \oplus x_j$, then we can obtain the second variable in constant many steps if we know the first variable. This is the case in the transition $1 \leftarrow 2$. Consequently, we can obtain x_j by removing x_i from $trueVarX[C]$, i.e., $x_j = x_i \oplus trueVarX[C]$. This new flip method is described in Algorithm 2

Algorithm 2. Variable flip with XOR caching

Input: Variable to flip v

```
1  α[v] = ¬α[v];                               /* change variable value */
2  satisfyingLiteral = α[v] ? v : ¬v;
3  falsifyingLiteral = α[v] ? ¬v : v;
4  for clause in occurrenceList[satisfyingLiteral] do
5      if numTrueLit[clause] == 0 then         /* transition 0 → 1 */
6          remove clause from falseClause ;
7          break[v] + +;
8          trueVarX[clause] = 0
9      else
10         if numTrueLit[clause] == 1 then     /* transition 1 → 2 */
11             break[trueVarX[clause]]- -;
12     numTrueLit[clause]++;
13     trueVarX[clause] ⊕= v;
14 for clause in occurrenceList[falsifyingLiteral] do
15     trueVarX[clause] ⊕= v;
16     if numTrueLit[clause] == 1 then         /* transition 0 ← 1 */
17         add clause to falseClause ;
18         break[v]- -;
19     else
20         if numTrueLit[clause] == 2 then     /* transition 1 ← 2 */
21             break[trueVarX[clause]]++;
22     numTrueLit[clause]- -;
```

The XOR scheme can be used in all types of solvers which need the break value and compute it incrementally. This is the case for almost all SLS solvers. Note, in CDCL solvers lazy data structures [13,14] have proven to be superior to schemes based on counting assigned literals. Counting, as well as the XOR scheme, both need full occurrence lists and their traversal for every assignment is too costly for those long clauses learned in CDCL [15].

To show the practical relevance of this implementation improvement, we have implemented the XOR scheme in probSAT and also in the way more complex SLS solver Sparrow.

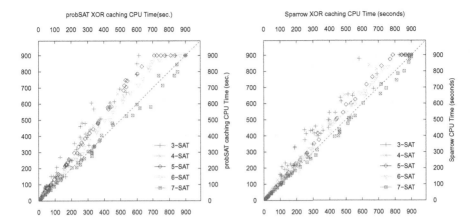

Fig. 1. Scatter plot between the caching implementation of probSAT (left side) and Sparrow (right side) and their respective XOR caching implementation on the set of random problems from the SAT Challenge 2012

We have evaluated the standard and the XOR implementation on the random and hard combinatorial benchmarks of the SAT Challenge 2012 [1]. We have opted to use this benchmark set instead of the latest, because it contains more benchmarks and is designed for lower run times of 900 seconds. All solvers have been started with the same seed, i.e. their implementation variants are semantically identical and produce the exact same search trace. The only difference is the complexity of a search step, which will cause the runtime to vary.

The results of our evaluations on the random instances can be seen in Fig. 1. The solvers were evaluated on the same hardware as the SAT Challenge and we used a cutoff of 900 seconds. In most cases, the XOR implementation is faster than the standard implementation. The quantity of improvement of the XOR implementation over the standard implementation seems to be in inverse proportional to k (the clause length of the problem in uniform randomly generated problems), i.e. the best improvement could be achieved for 3-SAT problems, while the XOR implementation is actually slower for 7-SAT problems.

This phenomenon can be explained by analyzing the number of clause transitions from 2-satisfied to 1-satisfied. The more transitions of this kind take place during search, the better the XOR implementation, because this is the step where the XOR implementation reduces the complexity. During search on 3-SAT problems, almost 30% of the transitions are of the type $2 \to 1$. On 7-SAT problems, only 5% are of this kind. Further, the occurrence list of a variable in 7-SAT problems contains on average 600 clauses, which means that we have to perform as many XOR operations to update the $trueVarX$ values. The overhead introduced by the XOR operations cannot be compensated by constant complexity of the $2 \to 1$ transition, which occurs rarely.

---

[1] http://baldur.iti.kit.edu/SAT-Challenge-2012/

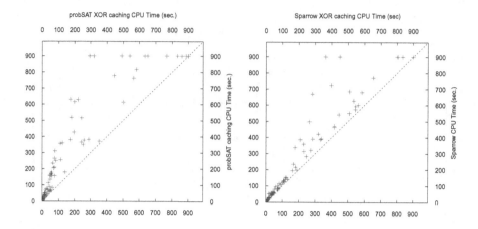

Fig. 2. Scatter plot between the caching implementation of probSAT (left side) and Sparrow (right side) and their respective XOR caching implementation on the set of hard combinatorial problems from the SAT Challenge 2012

Hard combinatorial problems and structured problems are known to contain many $2 \to 1$ transitions [1]. We evaluated the different implementations on the hard combinatorial satisfiable problems from the SAT Challenge 2012. The XOR implementation of probSAT and of Sparrow is always faster than their standard implementation as can be seen from Fig. 2. Moreover, the XOR implementation of probSAT and Sparrow are able to solve four and six additional instances, respectively. The speed-up with respect to solving time was on some instances more than a factor of two. During the search on these instances, almost 50% of the steps were $2 \to 1$ transitions.

3   Incorporating Multilevel Break

Most modern SLS solvers use the *make* and *break* value of variables or the combination of both, e.g. given by the *score*. For a given variable x, the *score* is defined by $score(x) := make(x) - break(x)$. Solvers like probSAT or WalkSAT only use the *break* value within their decision heuristic. Within the non-caching implementations of theses solvers, the *break* value is computed from scratch. If we want to compute the *break* value of a literal l, occuring within a non-satisfied clause, we have to traverse the occurrence list of $\neg l$ (the clauses that contain l as a satisfying literal) and count the number of clauses with $numTrueLit = 1$. The *break* value is defined as the number of clauses that change their satisfiability status from 1-satisfied to 0-satisfied. In the following, we will also refer to this *break* value as the $break_1$ value. Accordingly, we can define the $break_l$ value of a variable x as the number of clauses that are l-satisfied and will become $l - 1$ satisfied when x is flipped. The $make_l$ value can be defined similarly. Within the non-caching implementations, we can compute the $break_l$ value with little overhead. Instead of counting only the number of 1-satisfied clauses, we also count the number of l-satisfied clauses.

The question that arises now, is how to integrate these *break* values into the heuristic of the solvers. In [2], the $make_1$ and $make_2$ values were used in the form $lmake = w_1 \cdot make_1 + w_2 \cdot make_2$. The $lmake$ value was used as a tie breaker in the WalkSATlm solver. In [3] the $score_2 = make_2 - break_2$ value is being used in the solver CScoreSAT within the score function $score = score_1 + \lfloor score_2 \rfloor$.

Within the probSAT solver, new variable properties can be included quite easily by incorporating the property in the probability distribution. We propose the following inclusion of higher level *break* values into the probability distribution function of probSAT:

$$p(x) = cb^{-break(x)} \quad \longrightarrow \quad p(x) = \prod_l cb_l^{-break_l(x)}$$

$$p(x) = (1 + break(x))^{-cb} \quad \longrightarrow \quad p(x) = \prod_l (1 + break_l(x))^{-cb_l}$$

The constants cb_l specify the influence of the $break_l$ value within the probability distribution. To figure out which values to take for the cb_l variables, we used an automated algorithm configurator. More specifically, we used a parallel version of the SMAC configurator [16], which is implemented in the EDACC [17] configuration framework. The instances of interest are the random k-SAT problems ($k \in \{3, 5, 7\}$), resulting in three scenarios. For each scenario, we have performed two types of configuration experiments. First, we allowed the configuration of the cb_1 and cb_2 parameters (i.e. only include the $break_1$ and $break_2$ value). In the second experiment, we allowed the configuration of all cb_l. For the configuration instances, we have used the train sets of the benchmark sets used in [18] and [4], which are two independent set of instances. Each set contains 250 instances of 3-SAT ($n = 10.000$, $r = 4.2$), 5-SAT ($n = 500$, $r = 20$) and 7-SAT ($n = 90$, $r = 85$) problems. Each configuration scenario was allowed to use up to $5 \cdot 10^5$ seconds and we optimized the PAR10 statistics which measures the average runtime and counts unsuccessful runs with ten times the time limit.

3.1 Configuration of $break_l$

For 3-SAT problems, the configurator reported a cb_2 value of one, meaning that it could not improve upon the default configuration, which is not using the $break_2$ value. For the 5-SAT problems, the best configuration had $cb_1 = 3.7$, $cb_2 = 1.125$, and for 7-SAT problems it was $cb_1 = 5.4$, $cb_2 = 1.117$. The low values of cb_2, when compared to those of cb_1 shows that the $break_2$ values have a lower importance than $break_1$ or that the $break_2$ values are considerably larger than those of $break_1$. This motivates the analysis of the $break_2$ values within the search of the solver.

Similar as in the configuration of $break_2$, for 3-SAT problems the configurator could not find better configurations than the default configuration (that ignores the $break_2$ and $break_3$ values). The best configuration found for 5-SAT problem had the parameters:

$$cb_1 = 3.729 \quad cb_2 = 1.124 \quad cb_3 = 1.021 \quad cb_4 = 0.990 \quad cb_5 = 1.099$$

The best configuration for 7-SAT had the parameters:

$$cb_1 = 4.596 \quad cb_2 = 1.107 \quad cb_3 = 0.991 \quad cb_4 = 1.005 \quad cb_5 = 1.0 \quad cb_6 = 1.0 \quad cb_7 = 1.0$$

The value of the constants seems to be very low, but the high $break_l$ values that occur during search probably require such low cb_l values.

3.2   Results

We compare the $break_2$ and the $break_l$ implementations with the XOR implementation of probSAT and with the solvers WalkSATlm [2] and CScoreSAT [3], two solvers that established the latest state-of-the-art results in solving 5-SAT and 7-SAT problems. The binaries of the latter two solvers were the ones submitted to the SAT Competition 2013. We evaluate the solvers on the Competition benchmarks from 2011, 2012 and 1013, which all together represent a very heterogeneous set of instances, as they were generated with different parameters. An additional benchmark set of randomly generated 5-SAT problems with $n = 4000$ and a clause to variable ratio of $r = 20$ shall show where the limits of our new approach lies. The results of our evaluation can be seen in Tab. 1.

Table 1. The evaluation results on different 5-SAT and 7-SAT benchmarks sets. For each solver the number of solved instances (#sol.) and the average run time (time) is reported (unsuccessful runs are also counted in the run time). Bold values represents the best achieved results for that particular instance class.

	probSAT$_x$		probSAT$_2$		probSAT$_l$		WalkSATlm		CScoreSAT	
	#sol.	time	#sol.	time	#sol.	time	#sol.	time	#sol.	time
SC11-5-SAT	32	333s	40	34s	**40**	**19s**	39	108s	39	118s
SC12-5-SAT	67	578s	103	246s	**107**	**207s**	82	427s	77	465s
SC13-5-SAT	7	809s	5	836s	**7**	**808s**	6	817s	5	824s
5sat4000	0	900s	36	470s	**41**	**382s**	8	827s	0	900s
SC11-7-SAT	8	721s	10	521s	12	495s	11	571s	**14**	**437s**
SC12-7-SAT	49	633s	67	548s	69	488s	66	520s	**73**	**462s**
SC13-7-SAT	19	666s	17	671s	**20**	652s	16	673s	18	652s

By only incorporating the $break_2$ values in the probability distribution improves the performance of the probSAT solver considerably. On the SC12-5-SAT instances probSAT$_2$ is almost twice as fast as probSAT in terms of solved instanced and also in terms of runtime. It also dominates the WalkSATlm solver, which was shown to be the state-of-the-art solver for 5-SAT problems in [2].

Table 2. The theoretical (th.) expected number of l-satisfied clauses that contain an arbitrary variable x in a randomly generated k-SAT formula and the average $break_l$ values (pr.) encountered during search for variables within a randomly picked unsatisfied clause (value in brackets)

k-SAT	$break_1$ th.	pr.	$break_2$ th.	pr.	$break_3$ th.	pr.	$break_4$ th.	pr.	$break_5$ th.	pr.	$break_6$ th.	pr.	$break_7$ th.	pr.
3-SAT	1.6	1.7	3.2	3.7	1.6	1.6								
4-SAT	2.4	2.2	7.1	7.4	7.1	7.2	2.4	2.3						
5-SAT	3.1	2.8	12.5	13.1	18.6	19.5	12.5	12.7	3.1	3.2				
6-SAT	3.9	3.6	19.7	19.4	39.4	38.9	39.4	38.9	19.7	19.5	3.9	3.9		
7-SAT	4.6	4.5	27.9	28.8	69.7	72.2	93.0	96.7	69.7	72.9	27.9	29.2	4.6	4.9

The probSAT$_2$ solver achieves better results also on the 7-SAT instances. One exception are the SC13 problems, that have been generated exactly on the threshold. For these problems probSAT$_2$ is worse than probSAT.

By allowing also higher level break values to influence the probability distribution further improvements can be achieved. The probSAT$_l$ solver achieves the best results on all 5-SAT and on SC13-7-SAT problems. On the remaining 7-SAT problems it is only slightly worse than the CScoreSAT solver but still better than the WalkSATlm solver.

It is also worth mentioning that our non-caching implementation is on average about 30% slower than the caching implementation of the WalkSATlm and CScoreSAT solver in terms of flips per second performed by the solver.

3.3 Theoretical Distribution of $break_l$ Values

Given a randomly generated formula with constant clause length k and $m = r_k n$ many clauses (where the ratios $r_3 = 4.2$, $r_4 = 9.5$, $r_5 = 20$, $r_6 = 42$, and $r_7 = 85$ have been used) we are interested in the distribution and the expectation of the $break_l$ value for a given variable. The probability that a random clause contains this variable (in the correct polarization) is $k/(2n)$. Furthermore, for having l satisfied literals in such a random clause we need that among the remaining $k - 1$ literals exactly $l - 1$ of them are satisfied. This happens with probability $\binom{k-1}{l-1}/2^{k-1}$. Therefore, the $break_l$ value is binomially distributed as $Bin(m, p)$ where $m = r_k n$ and $p = \frac{k}{2n} \cdot \frac{\binom{l-1}{k-1}}{2^{k-1}}$. The expectation of the $break_l$ value is

$$r_k n \cdot \frac{k}{2n} \cdot \frac{\binom{l-1}{k-1}}{2^{k-1}} = \frac{r_k \cdot k \cdot \binom{k-1}{l-1}}{2^k}$$

Table 2 lists these values for common k-SAT problems.

Comparing these two values (the theoretical $break_l$ values and the actually observed values) there is a very good agreement. Interestingly SLS solver take into consideration only the transitions $0 \leftarrow 1$ (break) and $0 \rightarrow 1$ (make).

SLS solvers are optimizers that try to minimize the number of 0-satisfied clauses to zero, and thus only these two transitions play a role.

As we can see from Tab. 2 by far the largest $break_l$ values occur in the middle range, $l \approx k/2$. Therefore, the cb_l values even when they are relatively close to 1.0 should not be disregarded since they are raised to the $break_l$-th power, and so play an important role for the success of the algorithm. These seems to open up opportunities for further algorithmic improvements.

4   Clause Selection

While there has been lots of work on different variable picking schemes, comparatively few work has considered clause selection so far. In a certain way, clause weighting schemes [6,7,8] in combination with GSAT algorithms [5] relate to this because they influence the likelihood of a clause to become satisfied in the current step. However, this effect is rather indirect, and when it comes to WalkSAT algorithms, most implementations simply select a clause randomly.

We propose to question this selection and analyze several alternative schemes. For this, it is important to be aware of the data structures and algorithms that are used to save possible candidate clauses (i.e. the unsatisfied ones in a *falseClause* container). In most implementations, all unsatisfied clauses are stored in a list which is then updated in each iteration. The update procedure consists of removing newly satisfied clauses (the *make-step*) and adding newly unsatisfied clauses (the *break-step*).

The list itself is usually implemented as an array with a non-fixed length m'. If a new clause is added in the break-step, m' is increased and the clause is simply put at the last position. Whenever a clause is removed in the make-step, the element from the last position is used to replace it and m' is decreased. This is an easy but very efficient implementation used in most SLS solvers. To select a random clause during the clause selection phase, one can use a random number $r \in \{0, \ldots, m'\}$ and choose the clause at index r. We will call this approach RS, which is short for *random selection*.

One alternative way to select a clause is by using the current flip count j instead of choosing a clause at a random index. This was first proposed in [9, p. 93] and was the standard implementation of probSAT in the SAT Competition 2013. This version of probSAT selects the clause at index $j \bmod m'$ in each step. Interestingly, this version of probSAT, denoted with *pseudo breadth first search* (PBFS), performed much better than the original one from [4] (c.f. Fig. 3) which used random selection. Note, that under the (unrealistic) assumption of using real random numbers, this almost trivial change actually renders the original state-less implementation of probSAT to rely on the search history now, more precisely on the number of flipped clauses sofar.

Considering the success of DLS solvers compared to GSAT, it is not surprising that clause selection policies can have a big influence on the performance of WalkSAT solvers as well. Nevertheless, it is not fully clear why the particular approach of using the flip count provides this remarkable increase in performance. We therefore analyze this heuristic in more detail.

It is easy to see, that the candidate clauses are traversed in the same order in which they are contained in the array. As already described, new clauses are added to the end of the list in the break-step. This approach is somehow similar to the behavior of a queue. On the other hand, it is obviously not a real queue because the flip count is used for clause selection. Also, a certain shuffling takes place whenever clauses are removed in the make-step.

As long as a clause is selected randomly, the order of the candidate list does not matter. For PBFS, the situation is quite different. Once we start selecting clauses according to their position in the array, where exactly they are put becomes important. For example, there is a difference in whether the make-step or the break-step is performed first. Assume the list of unsatisfied clauses includes $[C_0, C_1, C_2]$ in the given order in iteration 0. C_0 will be selected. Further, assume that by flipping a variable in C_0, both C_0 and C_1 will get satisfied while previously satisfied clauses C_3, C_4, and C_5, will become unsatisfied. If the make-step is performed first, the list will change to $[C_2]$ and then to $[C_2, C_3, C_4, C_5]$ after the break-step. Therefore, C_3 will be selected in iteration 1 while C_2 is "skipped". If the break-step is performed first, the list will contain $[C_0, C_1, C_2, C_3, C_4, C_5]$ and finally become $[C_5, C_4, C_2, C_3]$ after the make-step. This leads to C_4 being selected in iteration 1. Obviously, this will also have an influence on the selection process in all following iterations.

To distinguish between the different order of make-step and break-step, we will use indices mf and bf to denote *make-first* and *break-first*, respectively. If the index is omitted, we refer to the make-first implementation. In Fig. 3, the performance of $PBFS_{mf}$ and $PBFS_{bf}$ is compared. We can see that $PBFS_{mf}$ outperforms $PBFS_{bf}$ significantly. Actually, $PBFS_{bf}$ is even slower than RS.

Again, it is not clear why this is the case. Adding clauses to the end in combination with the use of the modulo operator makes it hard to predict the effect on the clause selection order. The closer the index is to the end of the array, the sooner a newly added clause will be selected. Shuffling clauses in the break-step obfuscates the order of the clauses even more. However, better understanding the cause for this difference in performance might help us to further improve clause selection heuristics.

One of our conjectures was that maybe $PBFS_{mf}$ resembles a true *breadth first search* (BFS) more closely, leading to the increased performance. We therefore implemented a real queue in our solver. The result of this BFS approach is also shown in Fig. 3. Interestingly, BFS performs much worse than $PBFS_{mf}$ and even worse than RS and $PBFS_{bf}$. Apparently, the conjecture does not hold. Just for reference, we also implemented a stack for the clauses in order to simulate a *depth first search* (DFS). As expected, DFS performs very poorly and worse than all other approaches. This confirms that, although pure BFS does not work out, some kind of breadth first search seems to be beneficial.

Since a true BFS is too strict, we decided to use a modification inspired by the example given in the context of comparing $PBFS_{mf}$ and $PBFS_{bf}$. As pointed out in connection with the (better-performing) make-first version, clauses in this kind of implementation sometimes are "skipped", i.e. they will only be visited in

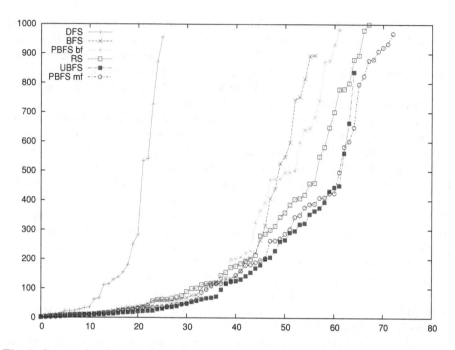

Fig. 3. Cactus plot showing the performance of a faithful reimplementation of probSAT in our internal Yals SAT solver with various different clause selection heuristics on the satisfiable instances from the SAT Competition 2013 hard combinatorial track (cutoff time ist 1000 seconds). The X-Axis and Y-Axis represent the number of solved instances and the runtime (in seconds), respectively.

the next full cycle. We simulated this in an implementation called *unfair breadth first search* (UBFS) in the following way. The list of clauses is still saved in a queue. When the first element is touched, we select it with probability p_u. With probability $1 - p_u$, it is moved to the end of the queue and the second clause is picked instead. For our experiments, we set $p_u = \frac{1}{2}$. In Fig. 3, we can see that this version performs much better than the previous queue implementations and already comes close to the results of PBFS.

Finally, an additional remark considering the efficiency of the different data structures. While it is easy to implement the array structure for RS and PBFS efficiently, more care has to be taken when implementing a queue for the other approaches. In our implementation we used a dedicated memory allocator and moving garbage collector for defragmentation. This achieves, roughly the same time efficiency as using the original array structure, considering average flips per second but needs slightly more memory.

Our results show that clause selection can have a large impact on the performance of WalkSAT solvers. There are still plenty of open questions. For example, it is not clear yet which is the optimal value for p_u. Also, it might be interesting to combine UBFS-like heuristics with clause weights, assigning different probabilities to clauses depending on different properties, such as its variable score distribution, the number of flips since the last time it was touched, the number of times it has been satisfied, or its length for non-uniform problems.

5 Summary and Future Work

We started with a detailed analysis of standard SLS implementations. This analysis revealed that there is room for improvement for the caching variants and that the $break_l$ property can be computed cheaply in the non-caching variants.

For caching implementations, we have shown that we can speed up the flip procedure of SLS solvers by using an XOR implementation as done in NanoSAT. This approach provides a great flexibility because it can be applied to all SLS solvers that use the *break* value within their heuristics and compute it in an incremental way. Our experimental results showed that the XOR implementation should be used for random k-SAT problems (except for 7-SAT problems) and especially for structured problems. For future work, it will be interesting to look at non-uniform problems more closely. As we have seen from random SAT, the clause length is the crucial factor for determining whether an XOR implementation helps to increase performance. In non-uniform formulas, it might be beneficial to combine XOR implementations with common ones and benefit from both advantages. For short clauses the XOR scheme should be used, while for long clauses the classical ones.

Within the non-caching implementations, it turned out that the multilevel $break_l$ value can be computed cheaply. By incorporating $break_l$ into the probSAT solver, we were able to discover better heuristics for 5-SAT and 7-SAT problems that establish new state-of-the-art results, especially on 5-SAT problems. We further extended our practical results by giving a detailed theoretical analysis regarding the distribution of different $break_l$ values in random formulas, providing a better understanding of their individual impact.

While most WalkSAT implementations simply select a random clause, we proposed several alternative ways. Our experimental results clearly show that selection heuristics influence the performance and picking a clause randomly is not the best choice. For structured problems the PBFS scheme is currently the best one and can be implemented very easily. Analyzing the impact of clause selection heuristics in more detail is a another promising research direction for future work. For example, clause weights might be one possible way to further improve the quality of clause selection heuristics.

Acknowledgments. We would like to thank the BWGrid [19] project for computational resources. This work was supported by the Deutsche Forschungsgemeinschaft (DFG) under the number SCHO 302/9-1, and by the Austrian Science Fund (FWF) through the national research network RiSE (S11408-N23).

References

1. Fukunaga, A.S.: Efficient implementations of SAT local search. In: Proc. SAT 2004 (2004)
2. Cai, S., Su, K., Luo, C.: Improving WalkSAT for random k-satisfiability problem with k > 3. In: Proc. AAAI 2013 (2013)
3. Cai, S., Su, K.: Comprehensive score: Towards efficient local search for SAT with long clauses. In: Proc. IJCAI 2013 (2013)
4. Balint, A., Schöning, U.: Choosing probability distributions for stochastic local search and the role of make versus break. In: Cimatti, A., Sebastiani, R. (eds.) SAT 2012. LNCS, vol. 7317, pp. 16–29. Springer, Heidelberg (2012)
5. McAllester, D.A., Selman, B., Kautz, H.A.: Evidence for invariants in local search. In: Proc. AAAI/IAAI 1997, pp. 321–326 (1997)
6. Wu, Z., Wah, B.W.: An efficient global-search strategy in discrete lagrangian methods for solving hard satisfiability problems. In: Proc. AAAI/IAAI 2000, pp. 310–315 (2000)
7. Hutter, F., Tompkins, D.A.D., Hoos, H.H.: Scaling and probabilistic smoothing: Efficient dynamic local search for SAT. In: Van Hentenryck, P. (ed.) CP 2002. LNCS, vol. 2470, pp. 233–248. Springer, Heidelberg (2002)
8. Thornton, J., Pham, D.N., Bain, S., Ferreira. Jr., V.: Additive versus multiplicative clause weighting for SAT. In: Proc. AAAI 2004, pp. 191–196 (2004)
9. Balint, A.: Engineering stochastic local search for the satisfiability problem. PhD thesis. Ulm University (2013)
10. Balint, A., Fröhlich, A.: Improving stochastic local search for SAT with a new probability distribution. In: Strichman, O., Szeider, S. (eds.) SAT 2010. LNCS, vol. 6175, pp. 10–15. Springer, Heidelberg (2010)
11. Tompkins, D.A.D., Hoos, H.H.: UBCSAT: An implementation and experimentation environment for SLS algorithms for SAT & MAX-SAT. In: Hoos, H.H., Mitchell, D.G. (eds.) SAT 2004. LNCS, vol. 3542, pp. 306–320. Springer, Heidelberg (2005)
12. Biere, A.: The evolution from limmat to NanoSAT. Technical Report 444, Dept. of Computer Science, ETH Zürich (2004)
13. Zhang, H.: Sato: An efficient propositional prover. In: McCune, W. (ed.) CADE 1997. LNCS, vol. 1249, pp. 272–275. Springer, Heidelberg (1997)
14. Moskewicz, M.W., Madigan, C.F., Zhao, Y., Zhang, L., Malik, S.: Chaff: Engineering an efficient SAT solver. In: Proc. DAC 2001, pp. 530–535 (2001)
15. Biere, A.: PicoSAT essentials. JSAT 4(2-4), 75–97 (2008)
16. Hutter, F., Hoos, H.H., Leyton-Brown, K.: Sequential model-based optimization for general algorithm configuration. In: Coello, C.A.C. (ed.) LION 5. LNCS, vol. 6683, pp. 507–523. Springer, Heidelberg (2011)
17. Balint, A., Diepold, D., Gall, D., Gerber, S., Kapler, G., Retz, R.: EDACC - an advanced platform for the experiment design, administration and analysis of empirical algorithms. In: Coello, C.A.C. (ed.) LION 5. LNCS, vol. 6683, pp. 586–599. Springer, Heidelberg (2011)
18. Tompkins, D.A.D., Balint, A., Hoos, H.H.: Captain jack: New variable selection heuristics in local search for SAT. In: Sakallah, K.A., Simon, L. (eds.) SAT 2011. LNCS, vol. 6695, pp. 302–316. Springer, Heidelberg (2011)
19. BWGrid: Member of the German D-Grid initiative, funded by the Ministry for Education and Research (Bundesministerium für Bildung und Forschung) and the Ministry for Science, Research and Arts Baden-Württemberg (Ministerium für Wissenschaft, Forschung und Kunst Baden-Württemberg), http://www.bw-grid.de

Everything You Always Wanted to Know about Blocked Sets (But Were Afraid to Ask)

Tomáš Balyo[1], Andreas Fröhlich[2], Marijn J.H. Heule[3], and Armin Biere[2,*]

[1] Charles University, Prague, Czech Republic
Faculty of Mathematics and Physics
[2] Johannes Kepler University, Linz, Austria
Institute for Formal Models and Verification
[3] The University of Texas at Austin
Department of Computer Science

Abstract. Blocked clause elimination is a powerful technique in SAT solving. In recent work, it has been shown that it is possible to decompose any propositional formula into two subsets (blocked sets) such that both can be solved by blocked clause elimination. We extend this work in several ways. First, we prove new theoretical properties of blocked sets. We then present additional and improved ways to efficiently solve blocked sets. Further, we propose novel decomposition algorithms for faster decomposition or which produce blocked sets with desirable attributes. We use decompositions to reencode CNF formulas and to obtain circuits, such as AIGs, which can then be simplified by algorithms from circuit synthesis and encoded back to CNF. Our experiments demonstrate that these techniques can increase the performance of the SAT solver Lingeling on hard to solve application benchmarks.

1 Introduction

Boolean satisfiability (SAT) solvers have seen lots of progress in the last two decades. They have become a core component in many different areas connected to the analysis of systems, both in hardware and software. Also, the performance of state-of-the-art satisfiability modulo theories (SMT) solvers often heavily relies on an integrated SAT solver or, as for example when bit-blasting bit-vectors, just encodes the SMT problem into SAT and then uses the SAT solver as reasoning engine. This gives motivation to develop even more efficient SAT techniques.

One crucial factor for the performance of state-of-the-art SAT solvers are sophisticated preprocessing and inprocessing techniques [1]. Conjunctive normal form (CNF) level simplification techniques, such as blocked clause elimination (BCE) [2,3,4], play an important role in the solving process. A novel use of blocked clauses was proposed in [5], which showed, that it is possible to decompose any propositional formula into two so called blocked sets, both of which can

---

* This work is partially supported by FWF, NFN Grant S11408-N23 (RiSE), the Grant Agency of Charles University under contract no. 600112 and by the SVV project number 260 104, and DARPA contract number N66001-10-2-4087.

C. Sinz and U. Egly (Eds.): SAT 2014, LNCS 8561, pp. 317–332, 2014.

be solved solely by blocked clause elimination. This blocked clause decomposition was then used to efficiently find backbone variables [6] and implied binary equivalences through SAT sweeping. Using this technique the performance of the state-of-the-art SAT solver Lingeling [7] could be improved on hard application benchmarks from the last SAT Competition 2013.

The success of these techniques gives reason to investigate blocked clauses in more detail. In this paper, we revisit topics from previous work [5] and present several new results. The paper is structured as follows. In Sect. 2, we first give definitions used throughout the rest of the paper. We introduce the notion of *blocked sets*, and, in Sect. 3, present new properties of blocked sets. We then discuss several new and improved ways to efficiently solve blocked sets in Sect. 4. In Sect. 5, we revisit *blocked clause decomposition* (BCD), as earlier proposed in [5], and suggest extensions to increase performance of decomposition algorithms and to generate blocked sets with certain useful attributes. We apply these techniques in Sect. 6 to extract compact circuit descriptions in AIG format from arbitrary CNF inputs, allowing the use of sophisticated circuit simplification techniques as implemented in state-of-the-art synthesis tools and model checkers like ABC [8,9]. As described in Sect. 7, this extraction mechanism can also be used to reencode the original formula into a different CNF which can sometimes be solved more efficiently by existing SAT solvers. All our experimental results are discussed in Sect. 8. We conclude our paper in Sect. 9.

2   Preliminaries

In this section, we give the necessary background and definitions for existing concepts used throughout our paper.

CNF.   Let F be a Boolean formula. Given a Boolean variable x, we use x and $\overline{x}$ (alternatively $\neg x$) to denote the corresponding positive and negative literal, respectively. A clause $C := (x_1 \vee \cdots \vee x_k)$ is a disjunction of literals. We say F is in conjunctive normal form (CNF) if $F := C_1 \wedge \cdots \wedge C_m$ is a conjunction of clauses. Clauses can also be seen as a set of literals. A formula in CNF can be seen as a set of clauses. A clause is called a unit clause if it contains exactly one literal. A clause is a tautology if it contains both x and $\overline{x}$ for some variable x. The sets of variables and literals occurring in a formula F are denoted by $vars(F)$ and $lits(F)$, respectively. A literal l is pure within a formula F if and only if $\overline{l} \notin lits(F)$.

AIGs.   An *and-inverter-graph* (AIG) [10] is a directed acyclic graph that represents a structural implementation of a circuit. Each node corresponds to a logical and-gate and each edge can either be positive or negative, representing whether the gate is negated. Since $\{\neg, \wedge\}$ is a functionally complete set of Boolean operators, obviously all Boolean formulas can be represented by AIGs.

(Partial) Assignments.   An *assignment* for a formula F is a function α that maps all variables in F to a value $v \in \{1, 0\}$. We extend α to literals, clauses and formulas by using the common semantics of propositional logic. Therefore $\alpha(F)$ corresponds to the truth value of F under the assignment α. Similarly, a *partial assignment* is a function β that maps only some variables of F to $v \in \{1, 0\}$. The value of the remaining variables in F is undefined and we write $\beta(x) = *$. Again, we extend β to literals, clauses and formulas by using the common semantics and simplification rules of propositional logic. Therefore $\beta(F)$ denotes the resulting formula under the partial assignment β. If we only want to assign one specific variable, we also use $F_{x=v}$ to represent the simplified formula.

Resolution. The resolution rule states that, given two clauses $C_1 = (l \vee a_1 \vee \cdots \vee a_{k_1})$ and $C_2 = (\bar{l} \vee b_1 \vee \cdots \vee b_{k_2})$, the clause $C = (a_1 \vee \cdots \vee a_{k_1} \vee b_1 \vee \cdots \vee b_{k_2})$, called the *resolvent* of C_1 and C_2, can be inferred by resolving on the literal l. This is denoted by $C = C_1 \otimes_l C_2$. A special case of resolution where C_1 or C_2 is a unit clause is called *unit resolution*.

Unit Propagation. If a unit clause $C = (x)$ or $C = (\bar{x})$ is part of a formula F, then F is equivalent to $F_{x=1}$ or $F_{x=0}$, respectively, and can be replaced by it. This process is called *unit propagation*. By $\mathrm{UP}(F)$ we denote the fixpoint obtained by iteratively performing unit propagation until no more unit clauses are part of the formula.

Blocked Clauses. Given a CNF formula F, a clause C, and a literal $l \in C$, l blocks C w.r.t. F if (i) for each clause $C' \in F$ with $\bar{l} \in C', C \otimes_l C'$ is a tautology, or (ii) $\bar{l} \in C$, i.e., C is itself a tautology. A clause C is blocked w.r.t. a given formula F if there is a literal that blocks C w.r.t. F. Removal of such blocked clauses preserves satisfiability [2]. For a CNF formula F, *blocked clause elimination* (BCE) repeats the following until fixpoint: If there is a blocked clause $C \in F$ w.r.t. F, let $F := F \setminus \{C\}$. BCE is confluent and in general does not preserve logical equivalence [3,4]. The CNF formula resulting from applying BCE on F is denoted by $\mathrm{BCE}(F)$. We say that BCE can solve a formula F if and only if $\mathrm{BCE}(F) = \emptyset$. Such an F is called a *blocked set*. We define $\mathcal{BS} := \{F \mid \mathrm{BCE}(F) = \emptyset\}$. A pure *literal* in F is a literal which occurs only positively. It blocks the clauses in which it occurs. As special case of blocked clause elimination, eliminating pure literals removes clauses with pure literals until fixpoint.

3 Properties

In this section, we revisit two monotonicity properties of blocked sets from previous work [5,4] and then prove several new properties. One basic property of blocked sets is the following monotonicity property: If $G \subset F$ and C is blocked w.r.t. F, then C is blocked w.r.t. G [5,4]. A direct implication is given by the following second version: If $F \in \mathcal{BS}$ and $G \subseteq F$ then $G \in \mathcal{BS}$.

Resolution does not affect the set of solutions of a Boolean formula. In CDCL solvers resolution is often used to learn additional information to help in the solving process. Therefore, it is interesting to see, that adding resolvents to a blocked set can destroy this property, which might thus make the formula much harder to solve:

Proposition 1. *Blocked sets are not closed under resolution, not even unit resolution.*

Proof. We give a counter-example:

$$F = (x_1 \vee x_2 \vee x_3) \wedge (\overline{x}_1 \vee \overline{x}_3) \wedge (\overline{x}_2 \vee x_3) \wedge (\overline{x}_3)$$

We can check that $\text{BCE}(F) = \emptyset$ by removing the clauses in the order in which they appear in F using the first literal of each clause as blocking literal. However, $F \wedge (x_1 \vee x_2)$ (with $(x_1 \vee x_2)$ obtained by unit resolution from the first and the last clause) is not a blocked set anymore.   □

As being one of the most important techniques in SAT solvers, it is of interest whether unit propagation can be combined with blocked clause elimination without destroying blockedness. In contrast to (unit) resolution, it is possible to apply unit propagation to blocked sets without the risk of obtaining a set of clauses that is not blocked anymore:

Proposition 2. *Blocked sets are closed under unit propagation.*

Proof. Since $F \in \mathcal{BS}$, $\text{BCE}(F) = \emptyset$. Let $C_1, \ldots, C_m$ be the clauses of F in the order in which they are removed by blocked clause elimination. Consider the formula F' obtained from F by unit propagation of a single unit clause $C_i = \{x\}$ and the sequence of formulas $F_i = \{C_i, \ldots, C_m\}$. We remove all clauses from F' by BCE in the same order as the corresponding clauses in F were removed by BCE. For each C_j, we have to consider three different cases. Case 1: $\overline{x}$ was the blocking literal of C_j. This is however actually not possible for $j < i$. Note that C_j is not blocked in F_j since C_i is still in F_i and the resolvent of C_j with C_i on $\overline{x}$ is not a tautology. Similarly observe for $j > i$ that the clause C_i is not blocked in F_i. Finally for $j = i$, the literal $\overline{x}$ is not part of C_i. Case 2: x was the blocking literal of C_j. In this case the clause is not part of F' anymore and therefore does not need to be removed. Case 3: $l \notin \{x, \overline{x}\}$ was the blocking literal of C_j. In order to remove C_j, we have to show that all resolvents of C_j with other clauses $C_k \in F_i$ on l with $j \leq k \leq m$ are still tautologies. Since unit propagation did not remove any literals other than $\overline{x}$ from clauses in the formula, only resolvents being tautologies due to containing both x and $\overline{x}$ can be affected. However, in any case, either $x \in C_j$ or $x \in C_k$, and one of the two clauses has been removed from F due to unit propagation and thus this resolution does not have to be considered anymore.   □

Given a set of clauses $F \in \mathcal{BS}$, we know that F has at least one satisfying assignment. We could try to find this satisfying assignment by choosing a random unassigned variable x, setting the variable to both possible values and checking which of the reduced formulas $F_{x=0}$ and $F_{x=1}$ is still satisfiable. One could hope that the satisfiability of $F_{x=v}$ could again be proved by showing that $\text{BCE}(F_{x=v}) = \emptyset$. However, this does not hold:

Proposition 3. *Blocked sets are not closed under partially assigning variables, furthermore, a blocked set may become non-blocked even in the subspace where this formula remains satisfiable.*

Proof. Consider the following example:

$$F = (\overline{x}_3 \vee \overline{x}_1 \vee x_4) \wedge (x_3 \vee x_2 \vee \overline{x}_4) \wedge (\overline{x}_2 \vee \overline{x}_1) \wedge (x_1 \vee x_4) \wedge (x_1 \vee x_5) \wedge (\overline{x}_5 \vee \overline{x}_4)$$

It is easy to verify that $\mathrm{BCE}(F) = \emptyset$ by removing the clauses in the order in which they appear in F with each clause being blocked on its first literal. If we now assign a value to x_3, we get the following two formulas:

$$F_{x_3=0} = (x_2 \vee \overline{x}_4) \wedge (\overline{x}_2 \vee \overline{x}_1) \wedge (x_1 \vee x_4) \wedge (x_1 \vee x_5) \wedge (\overline{x}_5 \vee \overline{x}_4)$$
$$F_{x_3=1} = (\overline{x}_1 \vee x_4) \wedge (\overline{x}_2 \vee \overline{x}_1) \wedge (x_1 \vee x_4) \wedge (x_1 \vee x_5) \wedge (\overline{x}_5 \vee \overline{x}_4)$$

Neither of the resulting formulas is blocked and both are satisfiable.   □

The difference between unit propagation and applying partial assignment is that in the former we assume the unit clause to be part of the blocked set. In this case the result after unit propagation or just applying the corresponding assignment can not destroy blockedness, while adding an arbitrary unit clause to a blocked set does not have this property.

4   Solving Blocked Sets

Blocked sets are always satisfiable and it is easy to find a satisfying assignment for them [4]. However, as shown in [5] it is also possible to efficiently find multiple satisfying assignments in a bit-parallel fashion by generalizing the original solution extraction algorithm [4].

In the original algorithm, we start from a random full truth assignment α for a blocked set B. We then traverse the clauses $C \in B$ in the reverse order of their elimination, e.g. the clause eliminated first will be considered last. For each clause C, we check if $\alpha(C) = 1$ and, if it is not, we flip the truth value of the blocking literal variable in α.

Our new generalized version of the reconstruction algorithm presented next, uses ternary logic consisting of values $v \in \{1, 0, *\}$. Instead of starting from a random (full) assignment, we start from a partial assignment β with $\beta(x) := *$ for all $x \in vars(B)$. We then traverse the clauses $C \in B$ in the same order as the original algorithm. If the clause is satisfied under the current assignment, we continue with the next clause. Otherwise, we do the following:

1. if there is a literal $l \in C$ with variable x and $\beta(x) = *$,
   then we set the variable to 1 or 0 which satisfies the clause.
2. otherwise, we flip the assignment of the blocking literal variable.

This algorithm is presented in Fig. 1. It will terminate with a partial satisfying assignment (some variables might have the value $*$). The values of the $*$ variables

```
      Solve (Blocked set B)
S1      β := [*, *, . . . , *]
S2      for Clause C ∈ reverse(eliminationStack) do
S3        if C is satisfied under β then continue
S4        V := getUnassignedVars(C)
S5        if V = ∅ then flip the blocking literal of C in β
S6        else
S7          select x ∈ V
S8          set x in β to a value that satisfies C
S9      return β
```

Fig. 1. Pseudo-code of the generalized blocked set solution reconstruction algorithm. The clauses are traversed in the reverse order of their elimination.

can be chosen arbitrarily. In this way, the algorithm finds several solutions at the same time. If the number of $*$ variables is k then the algorithm found 2^k solutions. Note, that any unassigned variable in a partial satisfying assignment can not be in the backbone nor part of an implied equivalence.

Making different choices on line S7 will give us different solutions. In fact, the non-determinism allows the algorithm to find all satisfying assignments of a blocked set. We show this in the following proposition.

Proposition 4. *The generalized reconstruction algorithm can find all the solutions of a blocked set.*

Proof. Let $β$ be an arbitrary satisfying assignment of a blocked set B. When the reconstruction algorithm encounters an unsatisfied clause C, it can choose to satisfy C using the same variable value pair as in $β$. The case that the blocking literal needs to be flipped will never occur, since unsatisfied clauses with no $*$ variable will not be encountered. This is due to the fact that we set all variables to their proper values when first changing the value from $*$. The algorithm will terminate with the solution $β$ (possibly some $*$ values remain). □

The next proposition demonstrates the correctness and time complexity of the generalized reconstruction algorithm.

Proposition 5. *The generalized reconstruction algorithm always terminates in linear time with a satisfying assignment of the blocked set.*

Proof. The algorithm visits each clause exactly once and performs a constant number of operations for each of its literals. Therefore the algorithm runs in linear time in the size of the blocked set.

We will prove that at the end of each for-cycle iteration all the clauses that have been traversed so far are satisfied by the assignment $β$. From this, the correctness of the returned solution follows. We need to examine three cases. If the clause is already satisfied by $β$ or it is satisfied by setting a value of a $*$ variable, then none of the traversed clauses can become unsatisfied. Otherwise, the clause becomes satisfied by flipping the blocked literal. In this case, we know from the definition of the blocked set that all previous clauses which contain the negation of the blocking literal are already satisfied by another literal. Namely, by the literal whose negation appears in the current clause and is false here.   □

We have seen that one can easily find solutions of blocked sets using a simple algorithm which in a way resembles a local search SAT solver (i.e. by "flipping" assignments to certain variables). This gives rise to the question of how state-of-the-art stochastic local search (SLS) SAT solvers perform on blocked sets. We conducted experiments using several well known SLS solvers and added our own specialized local search solver which always flips the blocking literal. From the results in Fig. 2, we can conclude that standard local search solvers struggle with solving blocked sets. However, when only the blocking literal is flipped, the problems are easily solved. This suggests that the crucial point in solving blocked sets is to know which is the blocked literal of a clause and the order in which the clauses are addressed is less important.

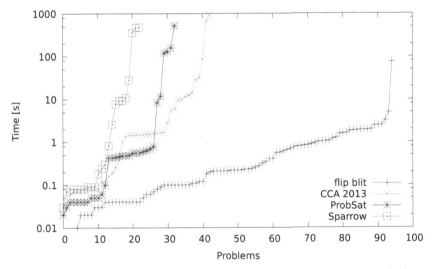

Fig. 2. Performance of the stochastic local search SAT solvers ProbSat [11], Sparrow [12], CCA2013 [13], and a modified ProbSat which always flips the blocking literal. The used instances are the 95 benchmarks described in Section 5.1 (instances from the 2013 SAT competition where unit decomposition is successful). The experiments were run on a PC with Intel Core i7-2600 3.4GHz CPU and 16GB of memory. The time limit for each instance was 1000 seconds.

5   Blocked Clause Decomposition

Blocked clause decomposition (BCD) [5] is the process of splitting a clause set of a CNF formula F into two sets such that at least one of them is a blocked set. If both sets are blocked, the decomposition is called symmetric.

Usually, we want to decompose the formula in a way that one of the blocked sets is large (i.e. contains many clauses) while the other is small. The motivation is that at least one of the blocked sets (the large one) should resemble the original formula as much as possible. Therefore, we will use the size difference of the two sets as a measure of the *quality of the decomposition*. We denote the larger blocked set by L (large set) and the smaller one by R (rest of the clauses). Symmetric BCD can be defined by the following formula:

$$\text{BCD}(F) = (L, R); \ F = L \cup R; \ L, R \in \mathcal{BS}; \ |L| \geq |R|$$

The simplest way of doing BCD is *pure decomposition* (described in [5]). First, we start with two empty clause sets L and R. Then, for each variable $x \in vars(F)$, we do the following. Let F_x be the set of clauses of F where x occurs positively and $F_{\bar{x}}$ where it occurs negatively. We add the larger of F_x and $F_{\bar{x}}$ to L and the smaller to R. We remove F_x and $F_{\bar{x}}$ from F and continue with the next variable. This algorithm produces two blocked sets which can be solved by pure literal elimination, hence the name pure decomposition.

Pure decomposition can be easily implemented to run in linear time and, therefore, is very fast on all formulas. A drawback is given by the fact that the difference in the sizes of L and R is usually not very big. This disadvantage can be addressed by post-processing, i.e., moving clauses from R to L as long as L (and R) remain blocked. An obvious way of post-processing (already described in [5]) is to move blocked (with respect to the current L) clauses from R to L. We extend this method by also moving so called *blockable* clauses.

Definition 1. *A clause C is* blockable *w.r.t. a blocked set L if each literal $l \in C$ potentially blocks C where a literal $l \in C$ potentially blocks C if each clause $C' \in L$ containing $\bar{l}$ has a different blocking literal (its blocking literal is not $\bar{l}$).*

It is easy to observe that a blockable clause will not prevent any other clause in the blocked set from being eliminated by blocked clause elimination and, therefore, adding it to the blocked set will not destroy its blocked status. A blockable clause can be easily detected if we maintain a literal occurrence list for the clauses in the blocked set. Using this data structure, and remembering for each clause which is its blocking literal, allows an efficient checking of the blockable property.

Moving blocked and blockable clauses can be considered to be lightweight post-processing methods since they are fast but often cannot move too many clauses. Another kind of post-processing algorithm is based on the following idea. If $\text{BCE}(L \cup S) = \emptyset$ for some $S \subset R$, all clauses in S can be moved from R to L. We refer to S as a candidate set. This kind of post-processing is more time consuming but also much more powerful. Different strategies can be employed for the selection of S such as *QuickDecompose* [5], which continues to move clauses from R to L until no more clauses can be moved. L is then called a maximal blocking set. Since *QuickDecompose* requires a lot of time for many instances, we decided to use a heuristic approach called *EagerMover*, which is described in Fig. 3. The algorithm tries to move $1/4$ of all clauses in R until none of the 4 parts can be moved. Trying to move smaller parts than $1/4$ in each step makes the algorithm slower but possibly more clauses can be moved. Experiments revealed that $1/4$ is a good compromise.

EagerMover (Blocked set L, R)

```
EM1      moved := True
EM2      while moved do
EM3        moved := False
EM4        for i := 0 to 3 do
EM5          S := R : {i · 0.25 · |R|, (i + 1) · 0.25 · |R|}
EM6          if BCE(L ∪ S) = ∅ then
EM7            L := L ∪ S; R := R \ S; moved := True
```

Fig. 3. Pseudo-code of the *EagerMover* post-processing algorithm, $R : \{a, b\}$ is the subsequence of R starting with the a-th element until the b-th element

5.1 Unit Decomposition

To achieve good quality decomposition, the following heuristic (called *unit decomposition* in [5]) was introduced: remove unit clauses from the original formula and test if the remaining clauses are a blocked set. If they are a blocked set, put the unit clauses into R and we are done. If the formula is an encoding of a circuit SAT problem, then this approach will always succeed [5]. This heuristic works on 77 of the 300 instances of the application track of the 2013 SAT Competition.

This heuristic can be generalized, by running unit propagation on the input clauses and removing satisfied clauses. Clauses are not simplified by removing false literals since those might be used as blocking literals. Next, we test if the clause set is blocked and, if it is, we put the unit clauses into R and we are done. Our improved heuristic succeeds on 95 instances, 18 more than the original. In the case that unit decomposition is not successful, we use pure decomposition.

After unit decomposition succeeds, there still might be some unit clauses which can be moved from R to L. Therefore, it is useful to run some of the mentioned post-processing algorithms to improve the quality of the decomposition.

5.2 Solitaire Decomposition

In this section, we define a special type of blocked clause decomposition called *solitaire blocked clause decomposition* (SBCD). In solitaire decomposition, we require that the small set R contains only a single unit clause. We will use this concept to translate a SAT problem into a circuit SAT problem (see Sect. 6). Solitaire decomposition cannot be achieved for every formula unless we allow an additional variable. A formal definition of solitaire decomposition follows.

Definition 2. *Let F be CNF formula. SBCD(F) = $(L, \{l\})$ where $L \in \mathcal{BS}$ and l is a literal, F and $L \cup \{l\}$ have the same set of satisfying assignments on the variables of F.*

A trivial solitaire decomposition of an arbitrary CNF is obtained by adding a new fresh variable x to each clause of the formula and then using $l := \overline{x}$ as the only literal in the small set:

$$\text{SBCD}(C_1 \wedge C_2 \wedge \cdots \wedge C_m) = (\{x \vee C_1\} \wedge \{x \vee C_2\} \wedge \cdots \wedge \{x \vee C_m\}, \{\overline{x}\})$$

It is easy to see that $\{x \vee C_1\} \wedge \{x \vee C_2\} \wedge \cdots \wedge \{x \vee C_m\}$ is a blocked set with x being the blocking literal for each clause. As improvement, perform blocked clause decomposition and then add a new fresh variable only to the clauses in the small set R. These clauses can then be moved to L and the new R will contain only the negation of the new variable.

6   Extracting Circuits

Using the reconstruction algorithm discussed in Sect. 4, it is possible to construct a circuit representation of a blocked set. Let F be a blocked set and let $C_1, \ldots, C_m$ be the clauses of F in the order in which they are removed by blocked clause elimination. During the reconstruction algorithm, we iterate through all the clauses of F starting from C_m to C_1 and flip the blocking literal of a clause if and only if that clause is not satisfied, i.e., all its literals are set to 0.

For each original variable $x_1, \ldots, x_n$ of F, we will have several new variables called *versions*. By $x_{i,k}$, we denote the k-th version of x_i ($x_{i,0}$ is x_i) and by $x_{i,\$}$, the latest version of x_i ($x_{i,\$} = x_{i,k}$ if k is the largest integer such that $x_{i,k}$ is defined). The notation is extended to literals in the same way.

Starting with C_m, we traverse the clauses in the reverse order of their elimination. For each clause $C = (x_i \vee y_{j_1} \vee \cdots \vee y_{j_k})$, where x_i is the blocking literal, we create a new version of x_i. We do this to represent the fact that the literal might have been flipped and can have a different value in the following iterations. The value of the new version is given by the following definition.

$$x_{i,\$+1} := x_{i,\$} \vee (\overline{y}_{j_1,\$} \wedge \cdots \wedge \overline{y}_{j_k,\$})$$

The following example demonstrates the definitions obtained from a blocked set.

Example 1. Let $F = (\overline{x}_1 \vee x_3 \vee x_2) \wedge (x_3 \vee x_4 \vee \overline{x}_1) \wedge (x_1 \vee \overline{x}_2 \vee \overline{x}_3)$. Using BCE, the clauses can be eliminated in the order of their appearance with the first literal being the blocking literal. We proceed in the reverse order and obtain the following definitions:

$$x_{1,1} := x_{1,0} \vee (x_{2,0} \wedge x_{3,0})$$
$$x_{3,1} := x_{3,0} \vee (\overline{x}_{4,0} \wedge x_{1,1})$$
$$\overline{x}_{1,2} := \overline{x}_{1,1} \vee (x_{3,1} \wedge x_{2,0})$$

These equations can be directly implemented as a circuit or, w.l.o.g., as an AIG. The first versions of the variables ($x_{i,0}$) are the inputs of the circuit and the higher versions are defined by the definitions. Using this construction, together with the solitaire decomposition as defined in Sect. 5, we can translate an arbitrary CNF (i.e. $F \notin \mathcal{BS}$) into an instance of the circuit SAT problem in the form of an AIG as follows: The large set L of the decomposition (with $L \in \mathcal{BS}$) is first encoded into an AIG G as already described. Then the output is defined to be the conjunction of G with the latest version of the unit literal corresponding to the small set. By doing this, one can potentially apply simplification techniques known from circuit synthesis and model checking (e.g. use ABC [9] to rewrite the circuit) and potentially profit from the similarity of blocked sets to circuits.

7   CNF Reencodings

In this section, we describe several ways of reencoding SAT problems using the result of blocked clause decomposition. Our input as well as our output is a CNF formula. By reencoding, we hope to increase the speed of SAT solving. The idea is similar to the one of circuit extraction described in Sect. 6. In comparison to the AIG encoding, our CNF encoding is more complex and versatile.

First, we describe how to encode the progression of the solution reconstruction for one blocked set $C_1 \wedge \cdots \wedge C_m$. For each variable x_i, we will have several versions as already described in Sect. 6, with $x_{i,\$}$ being the latest version of x_i. As we did previously, we traverse the clauses in the order $C_m, \ldots, C_1$ and introduce a new version for the blocking literal for each clause using the definition

$$x_{i,\$+1} := x_{i,\$} \vee (\overline{y}_{j_1,\$} \wedge \cdots \wedge \overline{y}_{j_k,\$})$$

which can be expressed by the following $k + 2$ clauses:

$$(\overline{x}_{i,\$} \vee x_{i,\$+1}) \wedge (x_{i,\$+1} \vee y_{j_1,\$} \vee \cdots \vee y_{j_k,\$}) \wedge \bigwedge_{l=1}^{k} (\overline{y}_{j_l,\$} \vee \overline{x}_{i,\$+1} \vee x_{i,\$}) \quad (1)$$

This formula represents the main step of the reconstruction algorithm, which states that the blocking literal is flipped if and only if the clause is not satisfied.

While symbolically encoding the progression of the reconstruction algorithm, we might decide not to have new versions for some of the variables. During reconstruction this corresponds to disallow that their values are flipped. Thus these variables need to have the right truth value from the beginning. Specifying that x_i should not have versions, means that $x_{i,\$+1} = x_{i,\$} = x_i$ which turns $k+1$ of the $k+2$ clauses of (1) into a tautology. The remaining clause is just a copy of the original clause using the latest versions of the variables. If we decided that none of the variables should have new versions, then the result of the reencoding would simply be the original formula.

In the rest of this section we propose several options to reencode a pair of blocked sets. The simplest way is to reencode the large blocked set L and append the clauses of R with variables renamed according to the last versions of variables from the reencoding of L. This can be done even for asymmetric decompositions since we have no requirement on R. Another way is to reencode both blocked sets and then make the last versions of the corresponding variables equal. This can be done by renaming the last versions from one blocked set to match the last versions in the other. As mentioned earlier, we can decide not to have versions for some of the variables. There are several heuristics which can be used to select these variables. In our experiments we found the following two heuristics useful.

1. Have versions only for variables that occur in both sets.
2. Have versions only for variables that occur as a blocking literal in both sets.

The entire reencoding process can be done in several ways. First, we need to choose a decomposition and a post-processing algorithm to obtain the blocked sets. Next, we need to decide which variables should have versions and whether to reencode both blocked sets. The choice of the decomposition and post-processing algorithms strongly influences the runtime but also the quality of the reencoding. As we will see from the experiments, better decompositions usually result in reencoded formulas that are easier to solve. The number of variables which have versions has a strong impact on the size of the reencoded formula which also influences the runtime of SAT solvers. Unfortunately, there is no clear choice since different combinations work best for different benchmark formulas.

8   Experiments

The algorithms described above were implemented in a Java tool which takes CNF formulas as input and produces an AIG or a reencoded CNF formula as output. The tool, its source code, and log files of experiments described in this section are available at `http://ktiml.mff.cuni.cz/~balyo/bcd/`.

We evaluated the proposed methods on the 300 instances from the application track of the SAT Competition 2013. All experiments were run on a 32-node cluster with Intel Core 2 Quad (Q9550 @2.83GHz) processors and 8GB of memory. We used a time limit of 5000 seconds and a memory limit of 7000MB, a similar set-up as in the SAT Competition 2013.

To solve the reencoded instances, we used the SAT solver Lingeling [7], the winner of the application track of the SAT Competition 2013. We used the synthesis and verification system ABC [9] to simplify AIG circuits and the AIGER library to convert between different AIG formats and converting AIG to CNF.

Figure 4 shows the performance of various decomposition methods. The best decomposition can be obtained by using unit decomposition followed by the lightweight and eager post-processing methods. Further, we observe that the eager post-processing method can significantly increase the quality after pure decomposition and often helps the unit decomposition. On the other hand, the eager post-processing can increase the time of decomposition up to 1000 seconds, which takes away a lot of time from the SAT solver.

In Fig. 5 the time required to solve the instances before and after reencoding are shown, as described in Sect. 7. For the experiments, we used different decomposition methods (see Fig. 5) followed by the simple reencoding method (only reencode the large set and rename the small set). Versions were only introduced for variables that occur as a blocking literal in both sets. For most of the problems, Lingeling with the original formula dominates the other methods and it is only for the hard problems (solved after 3500 seconds) that the reencoding starts to pay off. Overall, Lingeling without reencoding solved 232 instances, with Unit+Blockable decomposition it solved 237, and with Pure+Eager 240.

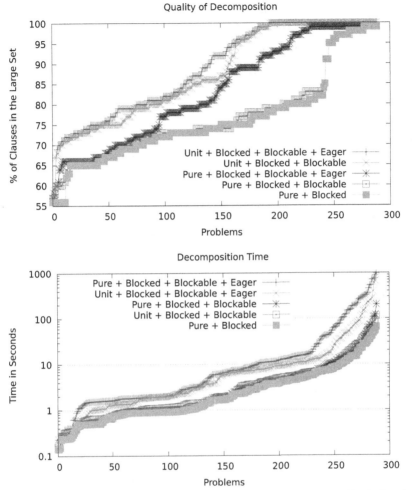

Fig. 4. The quality (percentage of the number of clauses in the large blocked set) and runtime of several decomposition and post-processing algorithm combinations.

Figure 6 demonstrates the usefulness of the circuit extraction and simplification approach. Similarly to the reencoding approach, this is also only useful for harder instances (solved after 3000 seconds). For the hardest 20 problems, our approach clearly takes over and ends up solving 7 more instances. These include two instances which were not solved by any sequential solver in the SAT competition 2013: *kummling_grosmann−pesp/ctl_4291_567_8_unsat_pre.cnf* and *kummling_grosmann−pesp/ctl_4291_567_8_unsat.cnf*. For this experiment, we only used those 135 instances where we can obtain a decomposition with a certain quality, namely 90% of the clauses must be in the large set. To obtain the blocked sets we used unit decomposition with eager post-processing.

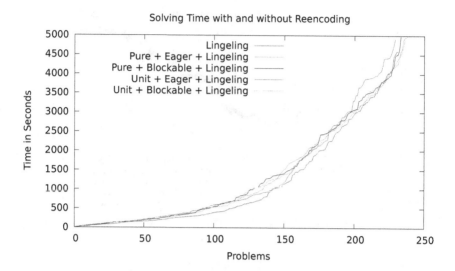

Fig. 5. The time required to solve the benchmark formulas which can be solved in 5000 seconds. The time limit for the reencoded instances was decreased by the time required for the reencoding process.

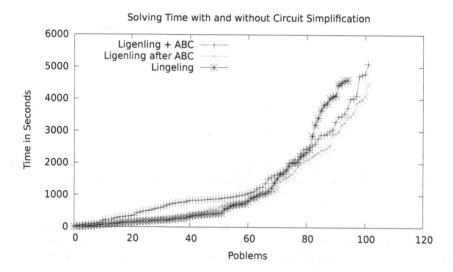

Fig. 6. The time required to solve the benchmark formulas, which can be decomposed with a quality of at least 90% (135 of 300 instances). The plot shows Lingeling on the original formula and the formula obtained by AIG circuit extraction followed by circuit simplification using the dc2 method of ABC (with a 500 seconds time limit) and finally reencoding the simplified circuit back to CNF. The data labeled "Lingeling + ABC" represents the total time required by the solving, reencodings and simplification. Lingeling alone can solve 95 instances and 102 after reencoding and simplification.

9   Conclusion

Simplification techniques based on blocked clauses can have a big influence on SAT solver performance. In this paper, we looked at blocked clauses in detail once again. We showed that blocked sets are closed under unit propagation while they are not closed under resolution or partial assignments. We proposed new ways for finding solutions of blocked sets and modified existing reconstruction algorithms to work with partial assignments. This enables us to find multiple solutions in one reconstruction run. This allows to rule out backbones and implied binary equivalences faster. Further, we analyzed the performance of local search on blocked sets.

While existing local search solvers performed rather poorly a simple extension improves the efficiency of local search on blocked sets significantly. We revisited blocked clause decomposition and described several new decomposition techniques as well as improved versions of existing ones. In particular, we showed how unit decomposition heuristics can be extended to be successful on more problems. We defined solitaire decomposition and described how it can be used to translate SAT to circuit SAT. We proposed various reencoding techniques to obtain different CNF representations. In the experimental section, we evaluated the performance of the state-of-the-art SAT solver Lingeling on the reencoded benchmarks. Our results showed that Lingeling can benefit from the reencodings being able to solve more formulas in a given time limit. The performance of solving is increased mainly for harder instances.

As future work, we want to optimize our circuit extraction techniques. If parts of a CNF were obtained from a circuit SAT problem, the original representation might help to solve the problem. It is unclear if or how this original structure can best be extracted from a CNF. However, there is a close connection between blocked clause elimination and operations on the circuit structure of a formula [3]. Therefore a better understanding of blocked sets might be an important step into this direction.

From a theoretical point of view, it is still not clear whether or how all solutions of a blocked set of clauses can be enumerated in polynomial time. Although this was conjectured to be possible in [5], it still has not been proven. While this conjecture is interesting from the theoretical point of view, it also would have important implications in practice since it guarantees the efficiency of algorithms on formulas containing blocked subsets with few solutions.

Finally, blocked clause decomposition and solving blocked sets needs a non-negligible portion of time in the solving process.Therefore, further improvements for decomposition techniques or the solving process of blocked sets will also directly affect the number of formulas that can be solved.

References

1. Järvisalo, M., Heule, M.J.H., Biere, A.: Inprocessing rules. In: Gramlich, B., Miller, D., Sattler, U. (eds.) IJCAR 2012. LNCS (LNAI), vol. 7364, pp. 355–370. Springer, Heidelberg (2012)
2. Kullmann, O.: On a generalization of extended resolution. Discrete Applied Mathematics 96–97, 149–176 (1999)

3. Järvisalo, M., Biere, A., Heule, M.J.H.: Blocked clause elimination. In: Esparza, J., Majumdar, R. (eds.) TACAS 2010. LNCS, vol. 6015, pp. 129–144. Springer, Heidelberg (2010)
4. Järvisalo, M., Biere, A., Heule, M.J.H.: Simulating circuit-level simplifications on cnf. Journal of Automated Reasoning 49(4), 583–619 (2012)
5. Heule, M.J.H., Biere, A.: Blocked clause decomposition. In: McMillan, K., Middeldorp, A., Voronkov, A. (eds.) LPAR-19 2013. LNCS, vol. 8312, pp. 423–438. Springer, Heidelberg (2013)
6. Parkes, A.J.: Clustering at the phase transition. In: Proc. of the 14th Nat. Conf. on AI, pp. 340–345. AAAI Press / The MIT Press (1997)
7. Biere, A.: Lingeling, plingeling and treengeling entering the sat competition 2013. In: Balint, A., Belov, M.J.H., Heule, M. (eds.) Proceedings of SAT Competition 2013. Department of Computer Science Series of Publications B, vol. B-2013-1, pp. 51–52. University of Helsinki (2013)
8. Mishchenko, A., Chatterjee, S., Brayton, R.K.: DAG-aware AIG rewriting: A fresh look at combinational logic synthesis. In: Sentovich, E. (ed.) Proceedings of the 43rd Design Automation Conference (DAC 2006), pp. 532–535. ACM (2006)
9. Brayton, R., Mishchenko, A.: Abc: An academic industrial-strength verification tool. In: Touili, T., Cook, B., Jackson, P. (eds.) CAV 2010. LNCS, vol. 6174, pp. 24–40. Springer, Heidelberg (2010)
10. Kuehlmann, A., Paruthi, V., Krohm, F., Ganai, M.K.: Robust boolean reasoning for equivalence checking and functional property verification. IEEE Transactions on Computer-Aided Design of Integrated Circuits and Systems 21(12) (2002)
11. Balint, A., Schöning, U.: Choosing probability distributions for stochastic local search and the role of make versus break. In: Cimatti, A., Sebastiani, R. (eds.) SAT 2012. LNCS, vol. 7317, pp. 16–29. Springer, Heidelberg (2012)
12. Balint, A., Fröhlich, A.: Improving stochastic local search for sat with a new probability distribution. In: Strichman, O., Szeider, S. (eds.) SAT 2010. LNCS, vol. 6175, pp. 10–15. Springer, Heidelberg (2010)
13. Li, C., Fan, Y.: Cca2013. In: Proceedings of SAT Competition 2013: Solver and Benchmark Descriptions (2013)

Minimal-Model-Guided Approaches to Solving Polynomial Constraints and Extensions*

Daniel Larraz, Albert Oliveras, Enric Rodríguez-Carbonell, and Albert Rubio

Universitat Politècnica de Catalunya, Barcelona, Spain

Abstract. In this paper we present new methods for deciding the satisfiability of formulas involving integer polynomial constraints. In previous work we proposed to solve SMT(NIA) problems by reducing them to SMT(LIA): non-linear monomials are linearized by abstracting them with fresh variables and by performing case splitting on integer variables with finite domain. When variables do not have finite domains, artificial ones can be introduced by imposing a lower and an upper bound, and made iteratively larger until a solution is found (or the procedure times out). For the approach to be practical, unsatisfiable cores are used to guide which domains have to be relaxed (i.e., enlarged) from one iteration to the following one. However, it is not clear then how large they have to be made, which is critical.

Here we propose to guide the domain relaxation step by analyzing minimal models produced by the SMT(LIA) solver. Namely, we consider two different cost functions: the number of violated artificial domain bounds, and the distance with respect to the artificial domains. We compare these approaches with other techniques on benchmarks coming from constraint-based program analysis and show the potential of the method. Finally, we describe how one of these minimal-model-guided techniques can be smoothly adapted to deal with the extension Max-SMT of SMT(NIA) and then applied to program termination proving.

1 Introduction

Polynomial constraints are ubiquitous. They arise naturally in many contexts, ranging from the analysis, verification and synthesis of software and cyber-physical systems [17,43,44,42,14] to, e.g., game theory [6]. In all these cases, it is critical to have efficient automatic solvers that, given a formula involving polynomial constraints with integer or real variables, either return a solution or report that the formula is unsatisfiable.

However, solving this kind of formulas has been a challenging problem since the early beginnings of mathematics. A landmark result is due to Tarski [48], who constructively proved that the problem is decidable for the first-order theory of real closed fields, in particular if variables are reals. Still, the algorithm in the proof has no use in practice as it has non-elementary complexity. More feasible procedures for solving polynomial constraints on the reals are based on cylindrical algebraic decomposition

---

* This work has been supported by the Spanish Ministry MICINN/MINECO under the project SweetLogics-UPC (TIN2010-21062-C02-01) and the FPI grant (Daniel Larraz) BES-2011-044621.

C. Sinz and U. Egly (Eds.): SAT 2014, LNCS 8561, pp. 333–350, 2014.

(CAD) [16,2]. However, their applicability is limited, as their complexity is still doubly exponential.

With the breakthrough of SAT and SMT solving [7,39], numerous techniques and tools have been developed which exploit the efficiency and automaticity of this technology. Many of these approaches for solving polynomial constraints on the reals are numerically-driven. E.g., in [25] interval constraint propagation is integrated with SMT(LRA) solving. In [23], non-linear formulas are pre-processed and then fed to an off-the-shelf SMT(LRA) solver. Other works for instance integrate interval-based arithmetic constraint solving in the SAT engine [21], combine interval arithmetic and testing [29], or focus on particular kinds of constraints like convex constraints [40]. In order to address the ever-present concern that numerical errors can result in incorrect answers in these methods, it has been proposed to relax constraints and consider δ-complete decision procedures [24,26]. As opposed to numerically-driven approaches, recently symbolic CAD-based techniques have been successfully integrated in a model-constructing DPLL(T)-style procedure [28,36], and several libraries and toolboxes have been made publicly available for the development of symbolically-driven solvers [19,37].

On the other hand, when variables must take integer values, even the problem of solving a single polynomial equation is undecidable (Hilbert's 10th problem, [18]). In spite of this theoretical limitation, and similarly to the real case, several methods that take advantage of the advancements in SAT and SMT solving have been proposed for solving integer polynomial constraints. The common idea of these methods is to reduce instances of integer non-linear arithmetic into problems of a simpler language that can be directly handled by existing SAT/SMT tools, e.g., propositional logic [22], linear bit-vector arithmetic [49], or linear integer arithmetic [11]. All these approaches are satisfiability-oriented, which makes them more convenient in contexts in which finding solutions is more relevant than proving that none exists (e.g., in invariant generation [31]).

In this paper we build upon our previous method [11] for deciding the satisfiability of formulas involving integer polynomial constraints. In that work, non-linear monomials are linearized by abstracting them with fresh variables and by performing case splitting on integer variables with finite domain. In the case in which variables do not have finite domains, *artificial* ones are introduced by imposing a lower and an upper bound, and made iteratively larger until a solution is found (or the procedure times out). For the approach to be useful in practice, unsatisfiable cores are employed to guide which bounds have to be relaxed (i.e., enlarged) from one iteration to the following one. However, one of the shortcomings of the approach is that unsatisfiable cores provide no hint on how large the new bounds have to be made. This is critical, since the size of the new formula (and hence the time required to determine its satisfiability) can increase significantly depending on the number of new cases that must be added.

The contributions of this paper are twofold:

1. We propose heuristics for guiding the domain relaxation step by means of the analysis of minimal models [4,5,47] generated by the SMT(LIA) solver. More specifically, we consider two different cost functions: first, the number of violated artificial domain bounds, which leads to *Maximum Satisfiability Modulo Theories* (Max-SMT, [38,15]) problems; and second, the distance with respect to the artificial domains, which boils down to *Optimization Modulo Theories* (OMT, [46,41])

problems. The results of comparing these approaches with other techniques show
the potential of the method.

2. We extend the first of these approaches to handle problems in Max-SMT(NIA).

This paper is structured as follows. Section 2 reviews basic background on SMT,
Max-SMT and OMT, and also on our previous approach in [11]. In Section 3 two
different heuristics for guiding the domain relaxation step are proposed, together with
an experimental evaluation. Then Section 4 presents the extension of the technique from
SMT(NIA) to Max-SMT(NIA). Finally, Section 5 summarizes the conclusions of this
work and sketches lines for future research.

2 Preliminaries

2.1 SMT, Max-SMT and OMT

Let $\mathcal{P}$ be a fixed finite set of *propositional variables*. If $p \in \mathcal{P}$, then p and $\neg p$ are *literals*.
The *negation* of a literal l, written $\neg l$, denotes $\neg p$ if l is p, and p if l is $\neg p$. A *clause*
is a disjunction of literals $l_1 \vee \cdots \vee l_n$. A (CNF) *propositional formula* is a conjunction
of clauses $C_1 \wedge \cdots \wedge C_n$. The problem of *propositional satisfiability* (abbreviated SAT)
consists in, given a propositional formula, to determine whether it is *satisfiable*, i.e., if
it has a *model*: an assignment of Boolean values to variables that satisfies the formula.

A generalization of SAT is the *satisfiability modulo theories (SMT)* problem: to decide the satisfiability of a given quantifier-free first-order formula with respect to a
background theory. In this setting, a model (which we may also refer to as a *solution*)
is an assignment of values from the theory to variables that satisfies the formula. Here
we will focus on integer variables and the theories of *linear integer arithmetic (LIA)*,
where literals are linear inequalities, and the more general theory of *non-linear integer
arithmetic (NIA)*, where literals are polynomial inequalities.[1]

Another generalization of SAT is *Max-SAT* [32,1,34], which extends the problem by
asking for more information when the formula turns out to be unsatisfiable: namely,
the Max-SAT problem consists in, given a formula $\mathcal{F}$, to find an assignment such that
the number of satisfied clauses in $\mathcal{F}$ is maximized, or equivalently, that the number of
falsified clauses is minimized. This problem can in turn be generalized in a number
of ways. For example, in *weighted Max-SAT* each clause C_i of $\mathcal{F}$ has a *weight* ω_i (a
positive natural or real number), and then the goal is to find the assignment such that
the *cost*, i.e., the sum of the weights of the falsified clauses, is minimized. Yet a further
extension of Max-SAT is the *partial weighted Max-SAT* problem, where clauses in
$\mathcal{F}$ are either weighted clauses as explained above, called *soft clauses* in this setting,
or clauses without weights, called *hard clauses*. In this case, the problem consists in
finding the model of the hard clauses such that the sum of the weights of the falsified
soft clauses is minimized. Equivalently, hard clauses can also be seen as soft clauses
with infinite weight.

The problem of *Max-SMT* merges Max-SAT and SMT, and is defined from SMT
analogously to how Max-SAT is derived from SAT. Namely, the *Max-SMT* problem

---

[1] In some classes of formulas of practical interest, real variables can also be handled by our
methods. See Section 2.2 for details.

consists in, given a set of pairs $\{[C_1, \omega_1], \ldots, [C_m, \omega_m]\}$, where each C_i is a clause and ω_i is its weight (a positive number or infinity), to find a model that minimizes the sum of the weights of the falsified clauses in the background theory.

Finally, the problem of *Optimization Modulo Theories (OMT)* is similar to Max-SMT in that they are both optimization problems, rather than decision problems. It consists in, given a formula $\mathcal{F}$ involving a particular variable called *cost*, to find the model of $\mathcal{F}$ such that the value assigned to *cost* is minimized. Note that this framework allows one to express a wide variety of optimization problems (maximization, piecewise linear functions, etc.).

2.2   Solving SMT(NIA) with Unsatisfiable Cores

In [11], we proposed a method for solving SMT(NIA) problems based on encoding them into SMT(LIA). The basic idea is to linearize each non-linear monomial in the formula by applying a case analysis on the possible values of some of its variables. For example, if the monomial x^2yz appears in the input SMT(NIA) formula and x must satisfy $0 \leq x \leq 2$, we can introduce a fresh variable v_{x^2yz}, replace the occurrences of x^2yz by v_{x^2yz} and add to the clause set the following three *case splitting clauses*: $x = 0 \rightarrow v_{x^2yz} = 0$, $x = 1 \rightarrow v_{x^2yz} = yz$ and $x = 2 \rightarrow v_{x^2yz} = 4yz$. In turn, new non-linear monomials may appear, e.g., yz in this example. All non-linear monomials are handled in the same way until a formula in SMT(LIA) is obtained, for which efficient decision procedures exist [20].

Note that, in order to linearize a non-linear monomial, there must be at least one variable in it which is both lower and upper bounded. When this property does not hold, new *artificial* bounds can be introduced for the variables that require them. In principle, this implies that the procedure is no longer complete, since a linearized formula with artificial bounds may be unsatisfiable while the original SMT(NIA) formula is actually satisfiable. A way to overcome this problem is to proceed iteratively: variables start with bounds that make the size of their domains small, and then the domains are enlarged on demand if necessary, i.e., if the formula turns out to be unsatisfiable. The decision of which bounds are to be relaxed is heuristically taken based on the analysis of an *unsatisfiable core* (an unsatisfiable subset of the clause set) that is obtained when the solver reports unsatisfiability (e.g. by writing a trace on disk one can extract a resolution refutation, whose leaves form a core [50]). Note that the method tells *which* bounds should be enlarged, but does not provide any guidance in regard to *how large* the new bounds should be. This is critical, as the size of the formula in the next iteration (and so the time needed to determine its satisfiability) can grow significantly depending on the number of new case splitting clauses that have to be added.

Altogether, the overall algorithm in [11] for solving a given formula in SMT(NIA) is as follows (see Figure 1). First, the needed artificial bounds are added (procedure *initial_bounds*) and the linearized formula (procedure *linearize*) is passed to an SMT(LIA) solver (procedure *solve_LIA*). If the solver returns SAT, we are done. If the solver returns UNSAT, then an unsatisfiable core is computed. If this core does not contain any of the artificial bounds, then the original non-linear formula must be unsatisfiable, and again we are done. Otherwise, at least one of the artificial bounds appearing in the core must be chosen for relaxation (procedure *relax_domains*). Once the domains are enlarged and the formula is updated (procedure *update*), the new linearized formula is tested for satisfiability, and the process is repeated (typically, while a prefixed time limit is not exceeded). We refer the reader to [11] for a more formal description.

```
status  solve_NIA(Formula F₀) {
    b = initial_bounds (F₀); // enough artificial bounds to linearize F₀
    F = linearize(F₀, b);
    while (not timed_out ()) {
        ⟨st, core⟩ = solve_LIA(F); // core computed here to ease presentation
        if (st == SAT) return SAT;
        else if (b ∩ core == ∅) return UNSAT;
        else {
            b = relax_domains(b, core); // at least one in the intersection is relaxed
            F = update(F, b); // add new bounds and case splitting clauses
        } }
    return UNKNOWN;
}
```

Fig. 1. Algorithm in [11] based on unsatisfiable cores

Finally, notice that the assumption that all variables should have integer type can be weakened, since it suffices that there are *enough* finite domain variables to perform the linearization. For example, this can be exploited in our SMT problems coming from constraint-based program analysis [31,30]. Those formulas are produced by applying Farkas' Lemma [45], and therefore only quadratic monomials of the form $\lambda \cdot u$ appear. Although in principle both λ and u are real unknowns, in the context of invariant and ranking function generation it is reasonable to assume that u should be integer. Hence, by case splitting on u one can linearize the monomial and does not need to force λ to take integer values. Moreover, when analyzing programs with integer variables, one often needs to be able to reason taking into account the integrality of the variables. In this situation integer versions of Farkas' Lemma [9] can be used, which when applied in the context of, e.g., invariant generation, require again the unknowns u to be in $\mathbb{Z}$.

3   Domain Relaxation with Minimal Models

Taking into account the limitations of the method based on cores when domains have to be enlarged, in this section we propose a model-guided approach to perform this step. The idea is to replace the satisfiability check in linear arithmetic with an optimization call, so that the best model found by the linear solver can be used as a reference for relaxing bounds (e.g., by extending the domains up to the value in that best model for those bounds that have to be relaxed).

Thus, the high-level algorithm we propose for solving a given formula in SMT(NIA) is shown in Figure 2 (cf. Figure 1). Here the SMT(LIA) black box does not just decide satisfiability, but finds the minimum model of the formula according to a prefixed non-negative cost function (procedure *optimize_LIA*). This function must have the property

```
status  solve_NIA (Formula F₀) {
    b = initial_bounds (F₀); // enough artificial bounds to linearize F₀
    F = linearize(F₀, b);
    while (not timed_out ()) {
        ⟨st, model⟩ = optimize_LIA(F);
        if (st == UNSAT) return UNSAT;
        else if (cost(model) == 0) return SAT;
        else {
            b = relax_domains(b, model);
            F = update(F, b); // add new bounds and case splitting clauses
        } }
    return UNKNOWN;
}
```

Fig. 2. Algorithm for solving SMT(NIA) based on minimal models

that the models of the linearized formula with cost 0 are true models of the original non-linear formula, and that if the linearization is unsatisfiable then so is the original formula. In addition to procedure *optimize_LIA*, the concrete implementations of procedures *linearize*, *relax_domains* and *update* also depend on the cost function.

Below we suggest two such cost functions: the number of violated artificial bounds (Section 3.1), and the distance with respect to the artificial domains (Section 3.2).

3.1   A Max-SMT(LIA) Approach

A possibility is to define the cost of an assignment as the number of violated artificial domain bounds. A natural way of implementing this is to transform the original non-linear formula into a linearized weighted formula and use a Max-SMT(LIA) tool. In this setting, *linearize* works as in the core-based algorithm, with the following difference: the clauses of the original formula (after being linearized by replacing non-linear monomials with fresh variables) together with the case splitting clauses are considered to be hard, while the artificial bounds are soft (with weight 1). Following the same construction, procedure *update* updates the soft clauses with the relaxed bounds, and adds the new case splitting clauses as hard clauses.

As regards the optimization step, procedure *optimize_LIA* boils down to making a call to a Max-SMT(LIA) solver on the linearized formula (which, e.g., can be implemented on top of an SMT(LIA) solver with branch-and-bound). In this case, the status *st* in Figure 2 corresponds to the satisfiability of the hard clauses. It is clear that if this status is UNSAT, then the original non-linear clause set is also unsatisfiable, given that the models of the original formula are a subset of the models of the hard clauses of the linearized formula. Another important property is that, if a model of the linearization has cost 0, then it is a true model of the non-linear formula.

Finally, procedure *relax_domains* determines the bounds to be relaxed by inspecting the soft clauses that are falsified. Moreover, as outlined above, the bounds are enlarged as follows. Let us assume that $x \leq u$ is an artificial bound that is falsified in the minimal model. If x is assigned value U in that model (and, hence, $u < U$), then $x \leq U$ becomes the new upper bound of x. A similar construction applies for lower bounds.

Regarding the weights of the soft clauses, in general it is not necessary to have unit weights. One may use different values, provided they are positive, and then the cost function corresponds to a weighted sum. Moreover, note that weights can be different from one iteration of the loop of *solve_NIA* to the next one.

Example 1. Let us consider the formula $tx + wy \geq 4 \wedge t^2 + x^2 + w^2 + y^2 \leq 12$, where variables t, x, w, y are integer. Let us also assume that we add the following artificial bounds in order to linearize: $-1 \leq t, x, w, y \leq 1$. Then we obtain the following linearized weighted formula:

$$v_{tx} + v_{wy} \geq 4 \wedge v_{t^2} + v_{x^2} + v_{w^2} + v_{y^2} \leq 12 \wedge$$

$$
\left.
\begin{array}{llll}
(t = -1 \rightarrow v_{tx} = -x) & \wedge & (w = -1 \rightarrow v_{wy} = -y) & \wedge \\
(t = 0 \quad \rightarrow v_{tx} = 0) & \wedge & (w = 0 \quad \rightarrow v_{wy} = 0) & \wedge \\
(t = 1 \quad \rightarrow v_{tx} = x) & \wedge & (w = 1 \quad \rightarrow v_{wy} = y) & \wedge \\
\\
(t = -1 \rightarrow v_{t^2} = 1) & \wedge & (w = -1 \rightarrow v_{w^2} = 1) & \wedge \\
(t = 0 \quad \rightarrow v_{t^2} = 0) & \wedge & (w = 0 \quad \rightarrow v_{w^2} = 0) & \wedge \\
(t = 1 \quad \rightarrow v_{t^2} = 1) & \wedge & (w = 1 \quad \rightarrow v_{w^2} = 1) & \wedge \\
\\
(x = -1 \rightarrow v_{x^2} = 1) & \wedge & (y = -1 \rightarrow v_{y^2} = 1) & \wedge \\
(x = 0 \quad \rightarrow v_{x^2} = 0) & \wedge & (y = 0 \quad \rightarrow v_{y^2} = 0) & \wedge \\
(x = 1 \quad \rightarrow v_{x^2} = 1) & \wedge & (y = 1 \quad \rightarrow v_{y^2} = 1) & \wedge
\end{array}
\right\} (\star)
$$

$$[-1 \leq t, 1] \wedge [-1 \leq x, 1] \wedge [-1 \leq w, 1] \wedge [-1 \leq y, 1] \wedge$$
$$[t \leq 1, 1] \wedge \quad [x \leq 1, 1] \wedge \quad [w \leq 1, 1] \wedge [y \leq 1, 1] ,$$

where $v_{tx}, v_{wy}, v_{t^2}, v_{x^2}, v_{w^2}, v_{y^2}$ are integer fresh variables standing for non-linear monomials. Soft clauses are written $[C, \omega]$, while clauses without weight are hard clauses.

In this case minimal solutions have cost 1, since at least one of the artificial bounds has to be violated so as to satisfy $v_{tx} + v_{wy} \geq 4$. For instance, the Max-SMT(LIA) solver could return the assignment: $t = 1$, $x = 4$, $v_{tx} = 4$, $w = y = v_{wy} = v_{w^2} = v_{y^2} = 0$, $v_{t^2} = 1$ and $v_{x^2} = 0$, where the only soft clause that is violated is $[x \leq 1, 1]$. Note that, as $x = 4$ is not covered by the case splitting clauses for v_{x^2}, the values of v_{x^2} and x are unrelated. Now the new upper bound for x would be $x \leq 4$ (so the soft clause $[x \leq 1, 1]$ would be replaced by $[x \leq 4, 1]$), and the following hard clauses would be added: $x = 2 \rightarrow v_{x^2} = 4$, $x = 3 \rightarrow v_{x^2} = 9$ and $x = 4 \rightarrow v_{x^2} = 16$. In the next iteration there are solutions with cost 0, e.g., $t = 1$, $x = 3$, $v_{tx} = 3$, $w = y = v_{wy} = v_{w^2} = v_{y^2} = 1$, $v_{t^2} = 1$ and $v_{x^2} = 9$. ∎

One of the disadvantages of this approach is that potentially the Max-SAT(LIA) solver could return models with numerical values much larger than necessary. Since the model is used for extending the domains, it could be the case that a prohibitive number of case splitting clauses are added, and at the next iteration the Max-SAT(LIA) solver is not able to handle the formula with a reasonable amount of resources. For instance, in Example 1, it could have been the case that the Max-SAT(LIA) solver returned $u = y =$

0, $t = 1$, $x = 10^5$, $v_{x^2} = 0$, etc. However, as far as we have been able to experiment, this kind of behaviour is rarely observed in our implementation; see Section 3.3 for more details. On the other hand, the cost function in Section 3.2 below does not suffer from this drawback.

3.2 An OMT(LIA) Approach

Another possibility is to define the cost of an assignment as the distance with respect to the artificial domains. This can be cast as a problem in OMT(LIA) as follows.

First of all, given a non-linear formula $\mathcal{F}_0$, the linearization $\mathcal{F}$ (procedure *linearize*) is computed like in the algorithm based on cores, except for the fact that artificial bounds are not included in the linearization: $\mathcal{F}$ consists only of the clauses of $\mathcal{F}_0$ (after being linearized), and of the case splitting clauses (together with other constraints to express the cost function, to be described below).

Now, let S be the set of variables x for which an artificial domain $[\lambda_x, v_x]$ is added in the linearization. Formally, the cost function is $\sum_{x \in S} \delta(x, [\lambda_x, v_x])$, where $\delta(z, [\lambda, v])$ is the *distance* of z with respect to $[\lambda, v]$:

$$\delta(z, [\lambda, v]) = \begin{cases} \lambda - z & \text{if } z < \lambda \\ 0 & \text{if } \lambda \leq z \leq v \\ z - v & \text{if } z > v \end{cases}$$

Note that, in the definition of the cost function, one could also include true original bounds: the contribution to the cost of these is null, since they are part of the formula and therefore must be respected.

In procedure *optimize_LIA*, the OMT(LIA) solver (which can be implemented by adding a phase II [45] to the consistency checks of an SMT(LIA) solver) minimizes this function, expressed in the following way. Let *cost* be the variable that the solver minimizes. For each variable $x \in S$ with domain $[\lambda_x, v_x]$, let us introduce once and for all two extra integer variables l_x and u_x (meaning the distance with respect to the lower and to the upper bound of the domain of x, respectively) and the *auxiliary constraints* $l_x \geq 0$, $l_x \geq \lambda_x - x$, $u_x \geq 0$, $u_x \geq x - v_x$. Then the cost function is determined by the equation $cost = \sum_{x \in S}(l_x + u_x)$, which is added to the linearization together with the auxiliary constraints listed above.

Note that a model of the linearization that has cost 0 must assign values within the bounds for all variables. Therefore the variables standing for non-linear monomials must be assigned consistent values with their semantics, by virtue of the case splitting clauses. Thus, models of the linearization with null cost are models of the original non-linear formula. Moreover, if the linearized formula is unsatisfiable, then the original formula must be unsatisfiable too, since the models of the original formula are included in the models of the linearized formula.

As regards domain relaxation, procedure *relax_domains* determines the bounds to be enlarged by identifying the variables l_x, u_x that are assigned a non-null value. Further, again the bounds are enlarged by taking the optimal model as a reference: similarly as in Section 3.1, if $x \leq u$ is an artificial bound to be relaxed and x is assigned value U in the best model, then $x \leq U$ becomes the new upper bound. Then procedure *update* updates the auxiliary constraints (e.g., $u_x \geq x - u$ is replaced by $u_x \geq x - U$), and adds the new case splitting clauses (for the $U - u$ cases $x = u + 1, ..., x = U$, etc.). Note that precisely the value of u_x in the optimal model is $U - u > 0$. Hence, intuitively the cost function corresponds to the *number of new cases* that will have to be taken into account for the next iteration of the loop of *solve_NIA*.

It is also possible to consider a slightly different cost function, which corresponds to the *number of new clauses* that will have to be added for the next iteration. For that purpose, it is only necessary to multiply variables l_x, u_x in the equation that defines *cost* by the number of monomials whose value is determined by case splitting on x. In general, similarly to Section 3.1, one may have a generic cost function of the form $cost = \sum_{x \in S} (\alpha_x l_x + \beta_x u_x)$, where $\alpha_x, \beta_x > 0$ for all $x \in S$. Further, again these coefficients may be changed from one iteration to the next one.

Example 2. Let us consider again the same non-linear formula from Example 1: $tx + wy \geq 4 \wedge t^2 + x^2 + w^2 + y^2 \leq 12$, where variables t, x, w, y are integer. Let us also assume that we add the following artificial bounds in order to linearize: $-1 \leq t, x, w, y \leq 1$. Then we obtain the following OMT(LIA) problem:

$$\text{min } cost \quad \text{subject to}$$

$$\text{constraints } (\star) \text{ from Example 1 } \wedge$$
$$cost = l_t + u_t + l_x + u_x + l_w + u_w + l_y + u_y \wedge$$

$$l_t \geq 0 \ \wedge \ l_t \geq -1 - t \ \wedge \ u_t \geq 0 \ \wedge \ u_t \geq t - 1 \ \wedge$$
$$l_x \geq 0 \ \wedge \ l_x \geq -1 - x \ \wedge \ u_x \geq 0 \ \wedge \ u_x \geq x - 1 \ \wedge$$
$$l_w \geq 0 \ \wedge \ l_w \geq -1 - w \ \wedge \ u_w \geq 0 \ \wedge \ u_w \geq w - 1 \ \wedge$$
$$l_y \geq 0 \ \wedge \ l_y \geq -1 - y \ \wedge \ u_y \geq 0 \ \wedge \ u_y \geq y - 1$$

In this case, it can be seen that minimal solutions have cost 1. For example, the OMT(LIA) solver could return the assignment: $x = 1$, $v_{x^2} = 1$, $t = 2$, $v_{tx} = 4$, $v_{t^2} = 0$ and $w = y = v_{wy} = v_{w^2} = v_{y^2} = 0$. Note that, as $t = 2$ is not covered by the case splitting clauses, the values of v_{tx} and v_{t^2} are unrelated to t. Now the new upper bound for t is $t \leq 2$, constraint $u_t \geq t - 1$ is replaced by $u_t \geq t - 2$, and clauses $t = 2 \rightarrow v_{tx} = 2x$ and $t = 2 \rightarrow v_{t^2} = 4$ are added.

At the next iteration there is still no solution with cost 0, and at least another further iteration is necessary before a true model of the non-linear formula can be found.   ∎

One of the drawbacks of this approach is that, as the previous example suggests, domains may be enlarged very slowly. This implies that, in cases where solutions have large numbers, many iterations are needed before one of them is discovered. See Section 3.3 below for more details on the performance of this method in practice.

3.3 Experiments

In this section we evaluate experimentally the performance of the two minimal-model-guided approaches proposed above, and compare them with other competing non-linear solvers. Namely, we consider the following tools[2] :

---

[2] We also experimented with other tools, namely dReal [26], SMT-RAT [19] and MiniSMT [49]. It turned out that the kind of instances we are considering here are not well-suited for these solvers, and many timeouts were obtained.

- bcl-maxsmt, our Max-SMT-based algorithm from Section 3.1;
- bcl-omt, our OMT-based algorithm from Section 3.2;
- bcl-cores, our core-based algorithm [11];
- Z3 version 4.3.1 [35].

The experiments were carried out on an Intel Core i7 with 3.40GHz clock speed and 16 GB of RAM. We set a timeout of 60 seconds.

All bcl-* solvers[3] share essentially the same underlying SAT engine and LIA theory solver. Moreover, some strategies are also common:

- procedure *initial_bounds* uses a greedy algorithm to approximate the minimal set of variables that have to be introduced in the linearization [11]. For each of them, we force the domain $[-1, 1]$, even if variables have true bounds (for ease of presentation, we will assume here that true bounds always contain $[-1, 1]$). This turns out to be useful in practice, as in some cases formulas have solutions with small coefficients. By forcing the domain $[-1, 1]$, unnecessary case splitting clauses are avoided and the size of the linearized formula is reduced.
- the first time a bound has been chosen to be enlarged is handled specially. Let us assume it is the first time that a lower bound (respectively, an upper bound) of x has to be enlarged. By virtue of the remark above, the bound must be of the form $x \geq -1$ (respectively, $x \leq 1$). Now, if x has a true bound of the form $x \geq l$ (respectively, $x \leq u$), then the new bound is the true bound. Otherwise, if x does not have a true lower bound (respectively, upper bound), then the lower bound is decreased by one (respectively, the upper bound is increased by one). Again, this is useful to capture the cases in which there are solutions with small coefficients.
- from the second time a bound has to be enlarged onwards, domain relaxation of bcl-maxsmt and bcl-omt follows basically what is described in Section 3, except for a correction factor aimed at instances where solutions have some large values. Namely, if $x \leq u$ has to be enlarged and in the minimal model x is assigned value U, then the new upper bound is $U + \alpha \cdot \min(\beta, \frac{n}{m})$, where α and β are parameters, n is the number of times the upper bound of x has been relaxed, and m is the number of occurrences of x in the original formula. As regards bcl-cores, a similar expression is used in which the current bound u is used instead of U, since there is no notion of "best model". The analogous strategy is applied for lower bounds.

In this evaluation we considered two different sets of benchmarks. The first benchmark suite consists of 1934 instances generated by our constraint-based termination prover [30]. As pointed out in Section 2.2, in these problems non-linear monomials are quadratic. Moreover, since it makes sense in our application, for each benchmark we have run Z3 (which cannot solve any of our non-linear integer instances) on versions of the instances where all variables are reals. This has been done in order to perform a fairer comparison, since unlike our approaches, Z3 is targeted to the real case. Results can be seen in Table 1, where columns represent systems and rows possible outcomes (SAT, UNSAT, UNKNOWN and TIMEOUT). Each cell contains the number of problems with that outcome obtained with the corresponding system, or the total time to process them.

Table 1. Experiments with benchmarks from Termination prover

	z3		bcl-cores		bcl-maxsmt		bcl-omt	
	#prob	secs	#prob	secs	#prob	secs	#prob	secs
SAT	1136	2578	1838	5464	1852	3198	1798	7896
UNSAT	0	0	0	0	4	0	62	112
UNKNOWN	11	2	0	0	0	0	0	0
TIMEOUT	787	47220	96	5760	78	4680	74	4440

Table 2. Experiments with benchmarks from model checking

	z3		bcl-cores		bcl-maxsmt		bcl-omt	
	#prob	secs	#prob	secs	#prob	secs	#prob	secs
SAT	30	2	35	55	35	72	34	263
UNSAT	1	0	1	0	1	0	1	0
UNKNOWN	0	0	0	0	0	0	0	0
TIMEOUT	5	300	0	0	0	0	1	60

The second benchmark suite consists of 36 examples of SMT(NIA) generated by the QArmc-Hsf(c) tool [27], a predicate-abstraction-based model checker with a special focus on liveness properties. In these problems all variables are integer, and monomials beyond quadratic appear. Results are in Table 2 and follow the same format as in Table 1.

As we can see in the tables, bcl-cores and bcl-maxsmt are the most efficient systems on satisfiable instances. While bcl-omt is doing slightly worse, Z3 is clearly outperformed, even when variables have real type. After inspecting the traces, we have seen that bcl-omt enlarges the domains too slowly, which is hindering the search.

Regarding unsatisfiable instances, it can also be observed that bcl-cores performs worse than the model-guided approaches, and that in particular bcl-omt is surprisingly effective. The reason is that, while the latter will always identify when the linear abstraction of the formula is unsatisfiable, this may not be the case with the former, which depending on the computed core may detect or not the unsatisfiability. In fact, for the sake of efficiency, bcl-cores does not guarantee that cores are minimal with respect to subset inclusion: computing *minimial unsatisfiable sets* [3] to eliminate irrelevant clauses implies an overhead that in our experience does not pay off.

Finally, as a side note, it is worth mentioning that we also experimented with a mixed version of the Max-SMT and OMT approaches. This version works as follows. Once the Max-SMT(LIA) finds a propositional model of the (propositional skeleton of the) linearization that minimizes the number of violations of the artificial bounds (this is the Max-SMT part), instead of taking any of the solutions that satisfy this propositional model, one finds a solution among those that minimizes the distance with respect to the artificial domains (this is the OMT part). This hybridization did not perform

---

[3] Available at www.lsi.upc.edu/~albert/sat14.tgz

significantly better than the Max-SMT approach, because most often the solution computed by default by the Max-SMT(LIA) solver turns out to be already optimal with respect to the distance cost function, and in general the gain obtained with this final optimization does not compensate the overhead it incurs in the total execution time.

4   Extension to Max-SMT(NIA)

As we showed in previous work [30], the framework of Max-SMT(NIA) is particularly appropriate for constraint-based termination proving. Other applications of Max-SMT(NIA) in program analysis can be envisioned given the enormous expressive power of its language. For the feasibility of this kind of applications, it is of paramount importance that efficient solvers are available. For this reason, this section will be devoted to the extension of our techniques for SMT(NIA) to Max-SMT(NIA).

More specifically, the experiments in Section 3.3 indicate that, when applied to satisfiable instances of SMT(NIA), the Max-SMT(LIA) approach behaves better than the OMT(LIA) one, and similarly to the core-based one, although on the instances coming from our program analysis applications it tends to perform better. Because of this, in Section 4.1 the Max-SMT(LIA) approach will be taken as a basis upon which a new algorithm for Max-SMT(NIA) will be proposed, which is more simple and natural than what a Max-SMT(NIA) system built on top of a core-based SMT(NIA) solver would be. Finally, in Section 4.2 we will report on the application of an implementation of this algorithm to program termination.

4.1   Algorithm

We will represent the input $\mathcal{F}_0$ of a Max-SMT(NIA) instance as a conjunction of a set of hard clauses $\mathcal{H}_0 = \{C_1, \cdots, C_n\}$ and a set of soft clauses $\mathcal{S}_0 = \{[D_1, \omega_1], \cdots, [D_m, \omega_m]\}$. The aim is to decide whether there exist assignments α such that $\alpha \models \mathcal{H}_0$, and if so, to find one such that $\sum_{[D,\omega] \in \mathcal{S}_0 \mid \alpha \not\models D} \omega$ is minimized.

The algorithm for solving Max-SMT(NIA) is shown in Figure 3. In its first step, as usual the initial artificial bounds are chosen (procedure *initial_bounds*) and the input formula $\mathcal{F}_0 \equiv \mathcal{H}_0 \wedge \mathcal{S}_0$ is linearized (procedure *linearize*). As a result, a weighted linear formula $\mathcal{F}$ is obtained with hard clauses $\mathcal{H} \wedge C$ and soft clauses $\mathcal{S} \wedge \mathcal{B}$, where:

- $\mathcal{H}$ and $\mathcal{S}$ are the result of replacing the non-linear monomials in $\mathcal{H}_0$ and $\mathcal{S}_0$ by their corresponding fresh variables, respectively;
- C are the case splitting clauses;
- $\mathcal{B}$ is the set of artificial bounds of the form $[x \geq l, \Omega], [x \leq u, \Omega']$, where the weights Ω, Ω' are positive numbers that are introduced in the linearization.

Now notice that there are two kinds of weights: those from the original soft clauses, and those produced by the linearization. As they have different meanings, it is convenient to consider them separately. Thus, given an assignment α, we define its *(total) cost* as $cost(\alpha) = (cost_{\mathcal{B}}(\alpha), cost_{\mathcal{S}}(\alpha))$, where $cost_{\mathcal{B}}(\alpha) = \sum_{[B,\Omega] \in \mathcal{B} \mid \alpha \not\models B} \Omega$ is the *bound cost*, i.e., the contribution to the total cost due to artificial bounds, and $cost_{\mathcal{S}}(\alpha) =$

```
<Status, Model> solve_Max_SMT_NIA(Formula F₀) {
    b = initial_bounds (F₀);
    F = linearize(F₀, b);
    best_so_far = ⊥;
    max_soft_cost = ∞
    while (not timed_out ()) {
        ⟨st, model⟩ = solve_Max_SMT_LIA(F, max_soft_cost);
        if (st == UNSAT)
            if (best_so_far == ⊥) return < UNSAT, ⊥ >;
            else                  return < SAT, best_so_far >;
        else if (cost_B(model) == 0) {
            best_so_far = model ;
            max_soft_cost = cost_S(model) − 1;
        }
        else {
            b = relax_domains(b, model);
            F = update(F, b);
        } }
    return < UNKNOWN, ⊥ >;
}
```

Fig. 3. Algorithm for solving Max-SMT(NIA) based on Max-SMT(LIA)

$\sum_{[D,\omega]\in S \mid \alpha\not\models D} \omega$ is the *soft cost*, corresponding to the original soft clauses. Equivalently, if weights are written as pairs, so that artificial bound clauses become of the form $[C, (\Omega, 0)]$ and soft clauses become of the form $[C, (0, \omega)]$, we can write $cost(\alpha) = \sum_{[C,(\Omega,\omega)]\in B\cup S \mid \alpha\not\models C}(\Omega, \omega)$, where the sum of the pairs is component-wise.

In what follows, pairs $(cost_B(\alpha), cost_S(\alpha))$ will be lexicographically compared, so that the bound cost (i.e., to be consistent in NIA) is more relevant than the soft cost. Hence, by taking this cost function and this ordering we have a Max-SMT(LIA) instance in which weights are not natural or non-negative real numbers, but pairs of them.

In the next step of the algorithm, procedure *solve_Max_SMT_LIA* calls a Max-SMT(LIA) tool to solve this instance. A difference with the usual setting is that the Max-SMT(LIA) solver admits a parameter *max_soft_cost* that restrains the models of the hard clauses we are considering: only assignments α such that $cost_S(\alpha) \leq max\_soft\_cost$ are taken into account. Thus, this adapted Max-SMT(LIA) solver computes, among the models α of the hard clauses such that $cost_S(\alpha) \leq max\_soft\_cost$ (if any), one that minimizes $cost(\alpha)$. This allows one to prune the search lying under the Max-SMT(LIA) solver when it is detected that the best soft cost found so far cannot be improved. This is not difficult to implement if the Max-SAT solver follows a branch-and-bound scheme, as it is our case.

Now the algorithm examines the result of the call to the Max-SMT(LIA) solver. If it is **UNSAT**, then there are no models of the hard clauses with soft cost at most *max_soft_cost*. Therefore, the algorithm can stop and report the best solution found so far, if any. Otherwise, *model* satisfies the hard clauses and has soft cost at most *max_soft_cost*. If it has null bound cost, i.e., it is a true model of the hard clauses of the original formula, then the best solution found so far and *max_soft_cost* are updated, in order to search for a solution with better soft cost. Finally, if the bound cost is not null, then domains are relaxed as described in Section 3.1, in order to widen the search space. In any case, the algorithm jumps back to a new call to *solve_Max_SMT_LIA*.

4.2 Application

As far as we know, none of the competing non-linear solvers is providing native support for Max-SMT, and hence no fair comparison is possible. For this reason, in order to give empirical evidence of the usefulness of the algorithm described in Section 4.1, here we opt for giving a brief summary of the application of Max-SMT to program termination [30] and, most importantly, highlighting the impact of our Max-SMT solver on the efficacy of the termination prover built on top of it.

Termination proving requires the generation of ranking functions as well as support- ing invariants. Previous work [12] formulated invariant and ranking function synthesis as constraint problems, thus yielding SMT instances. In [30], Max-SMT is proposed as a more convenient framework. The crucial observation is that, albeit the goal is to show that program transitions cannot be executed infinitely by finding a ranking function or an invariant that disables them, if we only discover an invariant, or an invariant and a *quasi-ranking function* that almost fulfills all needed properties for well-foundedness, we have made some progress: either we can remove part of a transition and/or we have improved our knowledge on the behavior of the program. A natural way to implement this idea is by considering that some of the constraints are hard (the ones guaranteeing invariance) and others are soft (those guaranteeing well-foundedness).

Thus, efficient Max-SMT solvers open the door to more refined analyses of termina- tion, which in turn allows one to prove more programs terminating. In order to support this claim, we carried out the experiment reported in Table 3, where we considered two termination provers:

- The tool (SMT) implements the generation of invariants and ranking functions us- ing a translation to SMT(NIA), where all constraints are hard.
- The tool (Max-SMT) is based on the same infrastructure, but expresses the synthesis of invariants and ranking functions as Max-SMT(NIA) problems. As outline above, this allows performing more refined analyses.

Table 3 presents the number of instances (#ins.) in each benchmark suite we con- sidered (from [13]) and the number of those that respectively each system proved ter- minating (with a timeout of 300 seconds). As can be seen in the results, there is a

Table 3. Comparison of SMT-based and Max-SMT-based termination provers

	#ins.	SMT	Max-SMT
Set1	449	212	228
Set2	472	245	262

non-negligible improvement in the number of programs proved terminating thanks to the adoption of the Max-SMT approach and the efficiency of our Max-SMT(NIA) solver.

5 Conclusions and Future Work

In this paper we have proposed two strategies to guide domain relaxation in the instantiation-based approach for solving SMT(NIA) [11]. Both are based on computing minimal models with respect to a cost function, respectively, the number of violated artificial domain bounds, and the distance with respect to the artificial domains. The results of comparing them with other techniques show their potential. Moreover, we have developed an algorithm for Max-SMT(NIA) building upon the first of these approaches, and have shown its impact on the application of Max-SMT(NIA) to program termination.

As for future work, several directions for further research can be considered. Regarding the algorithmics, it would be interesting to look into different cost functions following the minimal-model-guided framework proposed here, as well as alternative ways for computing those minimal models (e.g., by means of *minimal correction subsets* [33]). On the other hand, one of the shortcomings of our instantiation-based approach for solving Max-SMT/SMT(NIA) is that unsatisfiable instances that require non-trivial non-linear reasoning cannot be captured. In this context, the integration of real-goaled CAD techniques adapted to SMT [28] appears to be a promising line of work.

Another direction for future research concerns applications. So far we have applied Max-SMT(NIA) to array invariant generation [31] and termination proving [30]. Other problems in program analysis where we envision the Max-SMT(NIA) framework could help in improving the state-of-the-art are, e.g., the analysis of worst-case execution time and the analysis of non-termination. Also, so far we have only considered sequential programs. The extension of Max-SMT-based techniques to concurrent programs is a promising line of work with a potentially high impact in the industry.

Acknowledgments. We thank C. Popeea and A. Rybalchenko for their benchmarks.

References

1. Ansótegui, C., Bonet, M.L., Levy, J.: SAT-based MaxSAT algorithms. Artif. Intell. 196, 77–105 (2013)
2. Basu, S., Pollack, R., Roy, M.F.: Algorithms in Real Algebraic Geometry. Springer, Berlin (2003)
3. Belov, A., Lynce, I., Marques-Silva, J.: Towards efficient MUS extraction. AI Commun. 25(2), 97–116 (2012)
4. Ben-Eliyahu, R., Dechter, R.: On computing minimal models. Ann. Math. Artif. Intell. 18(1), 3–27 (1996)
5. Ben-Eliyahu-Zohary, R.: An incremental algorithm for generating all minimal models. Artif. Intell. 169(1), 1–22 (2005)

6. Beyene, T., Chaudhuri, S., Popeea, C., Rybalchenko, A.: A constraint-based approach to solving games on infinite graphs. In: Proceedings of the 41st ACM SIGPLAN-SIGACT Symposium on Principles of Programming Languages, POPL 2014, pp. 221–233. ACM, New York (2014)

7. Biere, A., Heule, M.J.H., van Maaren, H., Walsh, T. (eds.): Handbook of Satisfiability. Frontiers in Artificial Intelligence and Applications, vol. 185. IOS Press (February 2009)

8. Bloem, R., Sharygina, N. (eds.): Proceedings of 10th International Conference on Formal Methods in Computer-Aided Design, FMCAD 2010, Lugano, Switzerland, October 20-23. IEEE (2010)

9. Bockmayr, A., Weispfenning, V.: Solving numerical constraints. In: Robinson, J.A., Voronkov, A. (eds.) Handbook of Automated Reasoning, pp. 751–842. Elsevier and MIT Press (2001)

10. Bonacina, M.P. (ed.): CADE 2013. LNCS, vol. 7898. Springer, Heidelberg (2013)

11. Borralleras, C., Lucas, S., Oliveras, A., Rodríguez-Carbonell, E., Rubio, A.: SAT Modulo Linear Arithmetic for Solving Polynomial Constraints. J. Autom. Reasoning 48(1), 107–131 (2012)

12. Bradley, A.R., Manna, Z., Sipma, H.B.: Linear ranking with reachability. In: Etessami, K., Rajamani, S.K. (eds.) CAV 2005. LNCS, vol. 3576, pp. 491–504. Springer, Heidelberg (2005)

13. Brockschmidt, M., Cook, B., Fuhs, C.: Better termination proving through cooperation. In: Sharygina, N., Veith, H. (eds.) CAV 2013. LNCS, vol. 8044, pp. 413–429. Springer, Heidelberg (2013)

14. Cheng, C.H., Shankar, N., Ruess, H., Bensalem, S.: EFSMT: A Logical Framework for Cyber-Physical Systems, coRR abs/1306.3456 (2013)

15. Cimatti, A., Franzén, A., Griggio, A., Sebastiani, R., Stenico, C.: Satisfiability Modulo the Theory of Costs: Foundations and Applications. In: Esparza, J., Majumdar, R. (eds.) TACAS 2010. LNCS, vol. 6015, pp. 99–113. Springer, Heidelberg (2010)

16. Collins, G.E.: Hauptvortrag: Quantifier elimination for real closed fields by cylindrical algebraic decomposition. In: Brakhage, H. (ed.) GI-Fachtagung 1975. LNCS, vol. 33, pp. 134–183. Springer, Heidelberg (1975)

17. Colón, M.A., Sankaranarayanan, S., Sipma, H.B.: Linear Invariant Generation Using Nonlinear Constraint Solving. In: Hunt Jr., W.A., Somenzi, F. (eds.) CAV 2003. LNCS, vol. 2725, pp. 420–432. Springer, Heidelberg (2003)

18. Cooper, S.B.: Computability Theory. Chapman Hall/CRC Mathematics Series (2004)

19. Corzilius, F., Loup, U., Junges, S., Ábrahám, E.: SMT-RAT: An SMT-Compliant Nonlinear Real Arithmetic Toolbox - (Tool Presentation). In: Cimatti, A., Sebastiani, R. (eds.) SAT 2012. LNCS, vol. 7317, pp. 442–448. Springer, Heidelberg (2012)

20. Dutertre, B., de Moura, L.: A Fast Linear-Arithmetic Solver for DPLL(T). In: Ball, T., Jones, R.B. (eds.) CAV 2006. LNCS, vol. 4144, pp. 81–94. Springer, Heidelberg (2006)

21. Fränzle, M., Herde, C., Teige, T., Ratschan, S., Schubert, T.: Efficient solving of large nonlinear arithmetic constraint systems with complex boolean structure. JSAT 1(3-4), 209–236 (2007)

22. Fuhs, C., Giesl, J., Middeldorp, A., Schneider-Kamp, P., Thiemann, R., Zankl, H.: SAT Solving for Termination Analysis with Polynomial Interpretations. In: Marques-Silva, J., Sakallah, K.A. (eds.) SAT 2007. LNCS, vol. 4501, pp. 340–354. Springer, Heidelberg (2007)

23. Ganai, M.K., Ivancic, F.: Efficient decision procedure for non-linear arithmetic constraints using CORDIC. In: FMCAD, pp. 61–68. IEEE (2009)

24. Gao, S., Avigad, J., Clarke, E.M.: δ-complete decision procedures for satisfiability over the reals. In: Gramlich, B., Miller, D., Sattler, U. (eds.) IJCAR 2012. LNCS (LNAI), vol. 7364, pp. 286–300. Springer, Heidelberg (2012)

25. Gao, S., Ganai, M.K., Ivancic, F., Gupta, A., Sankaranarayanan, S., Clarke, E.M.: Integrating ICP and LRA solvers for deciding nonlinear real arithmetic problems. In: Bloem and Sharygina [8], pp. 81–89

26. Gao, S., Kong, S., Clarke, E.M.: dReal: An SMT Solver for Nonlinear Theories over the Reals. In: Bonacina [10], pp. 208–214

27. Grebenshchikov, S., Gupta, A., Lopes, N.P., Popeea, C., Rybalchenko, A.: HSF(C): A Software Verifier Based on Horn Clauses - (Competition Contribution). In: Flanagan, C., König, B. (eds.) TACAS 2012. LNCS, vol. 7214, pp. 549–551. Springer, Heidelberg (2012)

28. Jovanović, D., de Moura, L.: Solving non-linear arithmetic. In: Gramlich, B., Miller, D., Sattler, U. (eds.) IJCAR 2012. LNCS (LNAI), vol. 7364, pp. 339–354. Springer, Heidelberg (2012)

29. Khanh, T.V., Ogawa, M.: SMT for Polynomial Constraints on Real Numbers. Electr. Notes Theor. Comput. Sci. 289, 27–40 (2012)

30. Larraz, D., Oliveras, A., Rodríguez-Carbonell, E., Rubio, A.: Proving termination of imperative programs using Max-SMT. In: FMCAD, pp. 218–225. IEEE (2013)

31. Larraz, D., Rodríguez-Carbonell, E., Rubio, A.: SMT-Based Array Invariant Generation. In: Giacobazzi, R., Berdine, J., Mastroeni, I. (eds.) VMCAI 2013. LNCS, vol. 7737, pp. 169–188. Springer, Heidelberg (2013)

32. Li, C.M., Manyà, F.: MaxSAT, Hard and Soft Constraints. In: Biere, A., Heule, M., van Maaren, H., Walsh, T. (eds.) Handbook of Satisfiability. Frontiers in Artificial Intelligence and Applications, vol. 185, pp. 613–631. IOS Press (2009)

33. Marques-Silva, J., Heras, F., Janota, M., Previti, A., Belov, A.: On computing minimal correction subsets. In: Rossi, F. (ed.) IJCAI. IJCAI/AAAI (2013)

34. Morgado, A., Heras, F., Liffiton, M.H., Planes, J., Marques-Silva, J.: Iterative and coreguided MaxSAT solving: A survey and assessment. Constraints 18(4), 478–534 (2013)

35. de Moura, L., Bjørner, N.S.: Z3: An Efficient SMT Solver. In: Ramakrishnan, C.R., Rehof, J. (eds.) TACAS 2008. LNCS, vol. 4963, pp. 337–340. Springer, Heidelberg (2008)

36. de Moura, L., Jovanović, D.: A model-constructing satisfiability calculus. In: Giacobazzi, R., Berdine, J., Mastroeni, I. (eds.) VMCAI 2013. LNCS, vol. 7737, pp. 1–12. Springer, Heidelberg (2013)

37. Moura, L.M.d., Passmore, G.O.: Computation in real closed infinitesimal and transcendental extensions of the rationals. In: Bonacina [10], pp. 178–192

38. Nieuwenhuis, R., Oliveras, A.: On SAT Modulo Theories and Optimization Problems. In: Biere, A., Gomes, C.P. (eds.) SAT 2006. LNCS, vol. 4121, pp. 156–169. Springer, Heidelberg (2006)

39. Nieuwenhuis, R., Oliveras, A., Tinelli, C.: Solving SAT and SAT Modulo Theories: From an abstract Davis–Putnam–Logemann–Loveland procedure to DPLL(T). J. ACM 53(6), 937–977 (2006)

40. Nuzzo, P., Puggelli, A., Seshia, S.A., Sangiovanni-Vincentelli, A.L.: CalCS: SMT solving for non-linear convex constraints. In: Bloem and Sharygina [8], pp. 71–79

41. Oliver, R.: Optimization Modulo Theories. Master's thesis. Universitat Politècnica de Catalunya, Spain (January 2012)

42. Platzer, A., Quesel, J.-D., Rümmer, P.: Real world verification. In: Schmidt, R.A. (ed.) CADE 2009. LNCS (LNAI), vol. 5663, pp. 485–501. Springer, Heidelberg (2009)

43. Sankaranarayanan, S., Sipma, H., Manna, Z.: Non-linear loop invariant generation using Gröbner bases. In: Jones, N.D., Leroy, X. (eds.) POPL, pp. 318–329. ACM (2004)

44. Sankaranarayanan, S., Sipma, H.B., Manna, Z.: Constructing invariants for hybrid systems. Formal Methods in System Design 32(1), 25–55 (2008)

45. Schrijver, A.: Theory of Linear and Integer Programming. Wiley (June 1998)

46. Sebastiani, R., Tomasi, S.: Optimization in SMT with $\mathcal{LA}(\mathbb{Q})$ Cost Functions. In: Gramlich, B., Miller, D., Sattler, U. (eds.) IJCAR 2012. LNCS (LNAI), vol. 7364, pp. 484–498. Springer, Heidelberg (2012)

47. Soh, T., Inoue, K.: Identifying necessary reactions in metabolic pathways by minimal model generation. In: Coelho, H., Studer, R., Wooldridge, M. (eds.) ECAI. Frontiers in Artificial Intelligence and Applications, vol. 215, pp. 277–282. IOS Press (2010)

48. Tarski, A.: A decision method for elementary algebra and geometry. Bulletin of the American Mathematical Society 59 (1951)

49. Zankl, H., Middeldorp, A.: Satisfiability of non-linear (ir)rational arithmetic. In: Clarke, E.M., Voronkov, A. (eds.) LPAR-16 2010. LNCS (LNAI), vol. 6355, pp. 481–500. Springer, Heidelberg (2010)

50. Zhang, L., Malik, S.: Validating SAT Solvers Using an Independent Resolution-Based Checker: Practical Implementations and Other Applications. In: 2008 Design, Automation and Test in Europe, vol. 1, p. 10880 (2003)

Simplifying Pseudo-Boolean Constraints in Residual Number Systems*

Yoav Fekete and Michael Codish

Department of Computer Science, Ben-Gurion University, Israel

Abstract. We present an encoding of pseudo-Boolean constraints based on decomposition with respect to a residual number system. We illustrate that careful selection of the base for the residual number system, and when bit-blasting modulo arithmetic, results in a powerful approach when solving hard pseudo-Boolean constraints. We demonstrate, using a range of pseudo-Boolean constraint solvers, that the obtained constraints are often substantially easier to solve.

1 Introduction

Pseudo-Boolean constraints take the form $a_1 x_1 + a_2 x_2 + \cdots + a_n x_n \# k$, where $a_1, \ldots, a_n$ are integer coefficients, $x_1, \ldots, x_n$ are Boolean literals (i.e., Boolean variables or their negation), $\#$ is an arithmetic relation ($\leq$, $<$, $=$, etc.), and k is an integer. Pseudo-Boolean constraints are well studied and arise in many different contexts, for example in verification [9] and in operations research [7]. Typically we are interested in the satisfiability of a conjunction of Pseudo-Boolean constraints, often in conjunction with additional CNF clauses involving (among others also) Boolean variables from the Pseudo-Boolean constraints. Since 2005 there is a series of Pseudo-Boolean Evaluations [16] which aim to assess the state of the art in the field of Pseudo-Boolean constraint solvers.

There are variety of types of pseudo-Boolean constraint solvers. Most of these either encode constraints directly to SAT and then use a SAT solver to solve them, or else adapt methods applied in SAT solvers, such as conflict analysis and constraint learning, to deal directly with pseudo-Boolean constraints.

From the encoding based solvers, many are designed in terms of a BDD (Binary Decision Diagram) interpretation. These are typically the fastest when applicable. However, in some cases BDD based encodings are exponential in size (this is unavoidable unless NP = co-NP) and lead to memory problems in the encoding phase (even before solving) [2]. MINISAT$^+$ [14] provides solvers based on three different encoding techniques. In addition to the BDD based encoding, two alternatives are provided and applied when the BDD encoding is too large: one based on binary adders and the other on a mixed radix representation and unary sorting networks. The approach based on adders results in the most concise encodings, but it has the weakest propagation properties and often leads to high SAT solving times. Sat4j [6] provides two different categories of direct pseudo-Boolean solvers: one based on resolution, and the other on cutting planes.

---

* Supported by the Israel Science Foundation, grant 182/13.

C. Sinz and U. Egly (Eds.): SAT 2014, LNCS 8561, pp. 351–366, 2014.
© Springer International Publishing Switzerland 2014

This paper introduces a new encoding technique for solving pseudo-Boolean constraints. It is based on the well-studied residual number system and its correctness is justified by the Chinese Remainder Theorem [12].[1] Our approach is based on a transformation of pseudo-Boolean constraints to equi-satisfiable conjunctions of pseudo-Boolean constraints with smaller coefficients. As such, it can be coupled with any solver for pseudo-Boolean constraints. Each constraint in the transformed system concerns the same Boolean variables as the original, and possibly a few additional. To evaluate our approach we illustrate the impact of applying the transformation, comparing for a given solver, performance with and without the proposed transformation. We consider both types of solvers: based on SAT encodings as implemented in MiniSat$^+$, and based on a direct approach as implemented in Sat4j [6].

In a residual number system, given a (finite) base ρ of co-prime natural values, all natural numbers $n < \pi_\rho$, where π_ρ denotes the multiplication of the elements in ρ, are uniquely represented in terms of their residuals with respect to the elements of ρ. In our approach, given a pseudo-Boolean constraint C, we select a suitable residual base $\rho = \{m_1, \ldots, m_k\}$, and transform c to an equi-satisfiable conjunction of pseudo-Boolean constraints $\{C_1, \ldots, C_k\}$ in which the coefficients are smaller than those in in C. The choice of the residual base ρ for the given constraint C is an important factor in the quality of the encoding and an important attribute of our contribution.

Another major theme in our approach is "bit-blasting". This is a term referring to the encoding of finite-precision arithmetic to a CNF formula based on its bit-level representation (see for example [10]). The traditional notion of bit-blasting assumes a binary representation of integers. Recent research, e.g. [18,17], demonstrates the potential when considering unary representations for bit-blasting. In this paper we propose to consider also mixed radix representations.

The mixed radix number system is also at the heart of the sorting network encodings applied in MiniSat$^+$. In brief, the sum on the left side of the pseudo-Boolean constraint C is represented in terms of a (finite) mixed radix base $\mu = \{r_1, \ldots, r_k\}$, and a network of $k+1$ sorting networks is constructed, so that there is one sorting network to represent the value of each digit in the representation. For details, see [14]. However these are not essential for our presentation. What is important, is that the quality of this encoding is governed by the choice of the mixed radix base μ. The objective is to find a base which minimizes the size of the sorting networks. Roughly this means, to minimize the sum of the digits of the coefficients in C when represented in the number system determined by μ. For experimental evaluation, when considering pseudo-Boolean constraint solvers based on sorting networks, we apply an improved version of MiniSat$^+$, which we call MiniSat^{++}, that is configured to encode the constraints to SAT using an optimal mixed radix base. We note that the standard MiniSat$^+$ tool does not always apply an optimal base and ours is based on the application of the optimal base algorithm described in [11].

---

[1] The Chinese Remainder Theorem is introduced in a fifth century book by Chinese mathematician Sun Tzu. We cite the more accessible textbook presentation.

Our technique builds on recent work by Aavani *et al.*[1] where the authors also propose to apply a residual number system to encode pseudo-Boolean constraints. Similar to their work, we first apply the Chinese Remainder Theorem to derive a conjunction of "pseudo-Boolean modulo" constraints and then consider how to encode these constraints. Our work differs in two main points. First, we define new criteria on the selection of the residual base, and make an optimal choice with respect to these criteria. Second, we bit-blast pseudo-Boolean modulo constraints to standard pseudo-Boolean constraints instead of encoding them to SAT. We consider unary, binary and mixed-radix bit-blasts. The resulting constraints can be solved using any pseudo-Boolean constraint solver. We illustrate the impact of these contributions when solving hard pseudo-Boolean constraints.

We assume that pseudo-Boolean constraints are in Pseudo-Boolean normal form [5] and follow [1] (Proposition 1) to further assume that constraints are also normalized so that the relation between the sum on the left hand side and the constant on the right hand side is an equality.

2   Preliminaries

Underlying the techniques presented in this paper are concepts concerning the residual- and the mixed-radix- number systems. In this section we provide the basic definitions for these number systems and state the corresponding optimal base problems: the problem of selecting an "optimal" base for a given encoding task. Both tasks concern finding a suitable representation for the multiset of coefficients occurring on the left side in a given pseudo-Boolean constraint. The following introduces two pseudo-Boolean constraints as running examples.

Example 1. We demonstrate two pseudo-Boolean constraints:

$$621x_1 + 459x_2 + 323x_3 + 7429x_4 = 7888 \tag{1}$$
$$20x_1 + 493x_2 + 561x_3 + 1071x_4 = 1942 \tag{2}$$

Constraint (1) involves the coefficients $s_1 = \{323, 459, 621, 7429\}$ and is satisfiable as illustrated by the assignment $\{x_1 = 0, x_2 = 1, x_3 = 0, x_4 = 1\}$. Constraint (2) involves the coefficients $s_2 = \{20, 493, 561, 1071\}$ and is not satisfiable.

Mixed Radix Base: A finite mixed radix base, $\mu = \langle r_0, r_1, \ldots, r_{k-1} \rangle$, is a sequence of k integer values, called *radices*, where for each radix, $r_i > 1$. A number in base μ has $k + 1$ digits (possibly zero padded) and takes the form $\mathbf{d} = \langle d_0, \ldots, d_k \rangle$. For each digit, except the last, $0 \leq d_i < r$ $(i < k)$. There is no bound on the value of the most significant digit, $d_k \geq 0$. The integer value associated with $\mathbf{d}$ is $v = d_0 w_0 + d_1 w_1 + d_2 w_2 + \cdots + d_k w_k$ where $\langle w_0, \ldots, w_k \rangle$, called the *weights* for μ, are defined by $w_0 = 1$ and for $0 \geq i < k$, $w_{i+1} = w_i r_i$.

Example 2. Consider Constraint (1) with coefficients $s_1 = \{323, 459, 621, 7429\}$ represented in bases $\mu_1 = \langle 2, 2, 2, 2, 2, 2, 2, 2, 2, 2, 2, 2 \rangle$ (a binary base) and $\mu_2 = \langle 17, 3, 3, 2, 2, 2, 3 \rangle$. The columns in the two tables (below) correspond to the digits in the corresponding mixed radix representations.

$$\mu_1 = \langle 2, 2, 2, 2, 2, 2, 2, 2, 2, 2, 2, 2 \rangle$$

s_1 \ d	d_0	d_1	d_2	d_3	d_4	d_5	d_6	d_7	d_8	d_9	d_{10}	d_{11}	d_{12}
323	1	1	0	0	0	0	1	0	1	0	0	0	0
459	1	1	0	1	0	0	1	1	1	0	0	0	0
621	1	0	1	1	0	1	1	0	0	1	0	0	0
7429	1	0	1	0	0	0	0	0	1	0	1	1	1

$$\mu_2 = \langle 17, 3, 3, 2, 2, 2, 2, 3 \rangle$$

s_1 \ d	d_0	d_1	d_2	d_3	d_4	d_5	d_6	d_7	d_8
323	0	1	0	0	1	0	0	0	0
459	0	0	0	1	1	0	0	0	0
621	9	0	0	0	0	1	0	0	0
7429	0	2	1	0	0	0	0	0	1

Let $\mu = \langle r_0, r_1, \ldots, r_{k-1} \rangle$ be a mixed radix base. We denote by $v_{(\mu)} = \langle d_0, d_1, \ldots, d_k \rangle$ the representation of a natural number v in base μ. We let $ms(\mathbb{N})$ denote the set of finite non-empty multisets of natural numbers. For $S \in ms(\mathbb{N})$ with n elements we denote the $n \times (k+1)$ matrix of digits of elements from S in base μ as $S_{(\mu)}$. So, the ith row in $S_{(\mu)}$ is the vector $S(i)_{(\mu)}$. We denote the sum of the digits in $S_{(\mu)}$ by $|S|_\mu^{sod}$. When clear from the context we omit the superscript. Observe in the context of Example 2 that $|s_1|_{\mu_1}^{sod} = 22$ and $|s_1|_{\mu_2}^{sod} = 18$. These are the sums of the mixed radix digits in the corresponding tables.

The following optimal base problem is introduced in [11]. That work introduces an algorithm to find an optimal base and demonstrates that when solving pseudo-Boolean constraints encoded as networks of sorting networks it is beneficial to prefer a construction defined in terms of an optimal base.

Definition 1. *Let $S \in ms(\mathbb{N})$. We say that a mixed radix base μ is an optimal base for S, if for all bases μ', $|S|_\mu^{sod} \leq |S|_{\mu'}^{sod}$. The corresponding optimal base problem is to find an optimal base μ for S.*

Residual Number Systems: An RNS base is a sequence, $\rho = \langle m_1, \ldots, m_k \rangle$, of co-prime integer values, called *moduli*, where for each modulo, $m_i > 1$, and we denote $\pi_\rho = m_1 \times \cdots \times m_k$. Given the base ρ every natural number $n < \pi_\rho$ is uniquely represented as $n_{(\rho)} = \langle (n \bmod m_1), \ldots, (n \bmod m_k) \rangle$. That this is a unique representation is a direct consequence of the Chinese reminder theorem.

Let $\rho = \langle m_1, \ldots, m_k \rangle$ be an RNS base and consider natural numbers a, b and $c = a \text{ op } b$ where $\text{op} \in \{+, *\}$ and such that $a, b, c < \pi_\rho$. From the Chinese remainder theorem, $c_{(\rho)} = \langle (a_1 \text{ op } b_1) \bmod m_1, \ldots, (a_k \text{ op } b_k) \bmod m_K \rangle$. This implies that one can implement operations of addition and multiplication on RNS representations with no carry between the RNS digits. In particular, this implies that corresponding CNF encodings based on RNS representations result in a formula with small depth. The downside of the RNS representation is that encoding other arithmetic operations as well as encoding the comparison of RNS representations (except equality) is difficult.

The following two examples illustrate the basic idea underlying the application of an RNS to the encoding of pseudo-Boolean constraints. These examples also illustrate the first phase in the approach suggested in [1].

Example 3. Consider the RNS bases $\rho_1 = \langle 2, 3, 5, 7, 11, 13 \rangle$ and $\rho_2 = \langle 17, 3, 19, 23 \rangle$ and again Constraint (1). Base ρ_1, consists of the first 6 prime numbers and ρ_2 consists of carefully selected prime numbers as explained in the sequel. From the Chinese remainder theorem we know that solving Constraint (1) is the same as solving the system of equivalences that derive from its residuals modulo the given base elements. For ρ_1 this means solving the system of 6 modulo-equivalences on the left, and for ρ_2, the 4 modulo-equivalences on the right:

$$
\begin{aligned}
1x_1 + 1x_2 + 1x_3 + 1x_4 &\equiv 0 \quad \mod 2 \\
2x_3 + 1x_4 &\equiv 1 \quad \mod 3 \\
1x_1 + 4x_2 + 3x_3 + 4x_4 &\equiv 3 \quad \mod 5 \\
5x_1 + 4x_2 + 1x_3 + 2x_4 &\equiv 6 \quad \mod 7 \\
5x_1 + 8x_2 + 4x_3 + 4x_4 &\equiv 1 \quad \mod 11 \\
10x_1 + 4x_2 + 11x_3 + 6x_4 &\equiv 10 \quad \mod 13
\end{aligned}
$$

$$
\begin{aligned}
9x_1 &\equiv 0 \quad \mod 17 \\
2x_3 + 1x_4 &\equiv 1 \quad \mod 3 \\
13x_1 + 3x_2 &\equiv 3 \quad \mod 19 \\
22x_2 + 1x_3 &\equiv 22 \quad \mod 23
\end{aligned}
$$

In both cases the coefficients in the modulo-equivalences are smaller than those in the original constraint. In the sequel we will see how to encode modulo-equivalences in terms of pseudo-Boolean constraints. Meanwhile, observe that the system on the right is extremely easy to solve. The first equivalence implies that $x_1 = 0$, then the third that $x_2 = 1$, then the fourth that $x_3 = 0$, and finally the second that $x_4 = 1$.

Example 4. Consider again Constraint (2) and the RNS bases $\rho_1 = \langle 2, 3, 5, 7, 11 \rangle$ and $\rho_2 = \langle 17, 3, 2, 5, 7 \rangle$. Base, ρ_1 consists of the first 5 prime numbers and ρ_2 consists of carefully selected prime numbers as explained in the sequel. Just as in Example 3, the given constraint is equivalent to the system of modulo-equivalences obtained with respect to either of the two given bases. The residuals with respect to ρ_1 are given on the left and with respect to ρ_2, on the right:

$$
\begin{aligned}
1x_2 + 1x_3 + 1x_4 &\equiv 0 \quad \mod 2 \\
2x_1 + 1x_2 &\equiv 1 \quad \mod 3 \\
3x_2 + 1x_3 + 1x_4 &\equiv 2 \quad \mod 5 \\
6x_1 + 3x_2 + 1x_3 &\equiv 3 \quad \mod 7 \\
9x_1 + 9x_2 + 4x_4 &\equiv 6 \quad \mod 11
\end{aligned}
$$

$$
\begin{aligned}
3x_1 &\equiv 4 \quad \mod 17 \\
2x_1 + 1x_2 &\equiv 1 \quad \mod 3 \\
1x_2 + 1x_3 + 1x_4 &\equiv 0 \quad \mod 2 \\
3x_2 + 1x_3 + 1x_4 &\equiv 2 \quad \mod 5 \\
6x_1 + 3x_2 + 1x_3 &\equiv 3 \quad \mod 7
\end{aligned}
$$

Observe that the system on the right is easy to solve as the first equivalence (modulo 17) is not satisfiable, and hence also the original constraint.

The next example illustrates that RNS decomposition is not arc consistent.

Example 5. Consider the constraint $7x_1 + 3x_2 + 1x_3 = 5$ and the partial assignment $\alpha = \{x_1 = true\}$ which renders the constraint false. Now consider the RNS decomposition given the base $B = \langle 3, 5 \rangle$ resulting in the two modulo constraints $1x_1 + 1x_3 \equiv 2 \mod 3$ and $2x_1 + 3x_2 + 1x_3 \equiv 0 \mod 5$. One cannot deduce the conflict with partial assignment α on the two modulo constraints.

We conclude this section with a statement of two questions that this paper addresses: (1) Given a pseudo-Boolean constraint, how to best choose an RNS base for its encoding? In each of Examples 3 and 4, one of the bases considered results in a system of modulo-equivalences that is easy to solve. But how to select such a base? (2) Given a system of pseudo-Boolean modulo-equivalences, such as those derived in Examples 3 and 4, how to solve or encode them?

To answer these questions Section 3 proposes an encoding of pseudo-Boolean modulo-equivalences to pseudo-Boolean constraints. As such, we can apply any pseudo-Boolean constraint solver to solve the pseudo-Boolean modulo equivalences. Section 4 proposes two new measures on RNS bases, introduces corresponding optimal base problems, and provides simple algorithms to select optimal RNS bases with respect to these measures for a given pseudo-Boolean constraint.

3   The RNS Pseudo-Boolean Transformation

Let $C = (a_1 x_1 + a_2 x_2 + \cdots + a_n x_n = c)$ be a pseudo-Boolean constraint with $S = \langle a_1, \ldots, a_n \rangle$ the multiset of its coefficients. We say that ρ is an RNS base for C if it is an RNS base and $\pi_\rho > \Sigma S$. In this case, we also say that ρ is a base for (sums from) S. We say that ρ is non-redundant for C if for every $p \in \rho$, $p < max(S)$; and no prefix of ρ is also a base for C. Not all pseudo-Boolean constraints have a non-redundant RNS base, as is the case when the sum of the elements in S is larger than the multiplication of the primes smaller than $max(S)$. For example, cardinality constraints do not have a non-redundant RNS base. We assume that C has a non-redundant base, and we describe an encoding of C to an equivalent system of "smaller" pseudo-Boolean constraints derived from ρ. In the following we denote $\rho = \langle p_1, \ldots, p_m \rangle$, $a_{ij} = a_i \bmod p_j$, and $c_j = c \bmod p_j$ for $1 \le i \le n$ and $1 \le j \le m$. We detail the encoding in three phases.

The first phase, demonstrated in Examples 3 and 4, is formalized as Equation (3). The pseudo-Boolean constraint on the left is equivalent to the conjunction of pseudo-Boolean modulo-equivalences, on the right. Stated as a transition: to encode the left side, it is sufficient to encode the right side.

$$\sum_{i=1}^{n} a_i x_i = c \qquad \Leftrightarrow \qquad \bigwedge_{j=1}^{m} \left(\sum_{i=1}^{n} a_{ij} x_i \equiv c_j \mod p_j \right) \tag{3}$$

In the second phase, we apply the definition of integer division modulo p_j to replace the equivalence modulo p_j on the left side with an equality, on the right side, introducing integer variables t_j and constants $c_j < p_j$ for $1 \le j \le m$.

$$\bigwedge_{j=1}^{m} \left(\sum_{i=1}^{n} a_{ij} x_i \equiv c_j \mod p_j \right) \qquad \Leftrightarrow \qquad \bigwedge_{j-1}^{m} \left(\sum_{i=1}^{n} a_{ij} x_i = t_j p_j + c_j \right) \tag{4}$$

The right side of Equation (4) is not yet a system of pseudo-Boolean constraints due to the integer variables t_j ($1 \le j \le m$). However, these variables take values from a finite domain, as we have, recalling that $a_{ij} < p_j$ and $c_j < p_j$:

$$0 \le t_j = \left\lfloor \frac{(\sum a_{ij} x_i) - c_j}{p_j} \right\rfloor \le \left\lfloor \frac{(\sum a_{ij}) - c_j}{p_j} \right\rfloor \le n \tag{5}$$

In the third phase we bit-blast the integer variables in the right side of Equation (4) to obtain a proper pseudo-Boolean constraint. We consider three alternative bit-blasting techniques: binary, unary, and mixed radix and obtain Equation (6) where $[\![t_j]\!]_{bb}$ indicates the corresponding bit-blast of integer variable t_j with bb equal to b (binary), u (unary), or μ (mixed radix). Once substituting $[\![t_j]\!]_{bb}$, the right side of Equation (6) is a pseudo-Boolean constraint.

$$\bigwedge_{j=1}^{m} \left(\sum_{i=1}^{n} a_{ij} x_i = p_j t_j + c_j \right) \qquad \Leftrightarrow \qquad \bigwedge_{j=1}^{m} \left(\sum_{i=1}^{n} a_{ij} x_i = p_j [\![t_j]\!]_{bb} + c_j \right) \tag{6}$$

For all three bit-blasting alternatives, assume that t is an integer variable taking values in the range $[0, \ldots, n]$.

$$[t]_b = \sum_{i=0}^{k} 2^i y_i \qquad [t]_u = [t]_u = \sum_{i=1}^{n} y_i \qquad [t]_\mu = \sum_{i=0}^{k} w_i \times \Sigma y_i$$
$$\text{(a) binary} \qquad\qquad \text{(b) unary} \qquad\qquad \text{(c) mixed-radix}$$

Fig. 1. Bit-blasting the integer variable t

Binary Bit-Blast: Let $k = \lfloor \log_2 n \rfloor$ and let $y_0, \ldots, y_k$ be fresh Boolean variables corresponding to the binary representation of integer variable t. The equation in Figure 1(a) specifies the binary bit-blast of t and substituting in Equation 6 results in a system of pseudo-Boolean constraints introducing an order of $m \log n$ fresh Boolean variables with integer coefficients of value up to $\max(\rho) \cdot n$.

Unary Bit-Blast: In the order encoding [4,3,13], an integer variable $0 \leq t \leq n$ is represented in n bits, $[y_1, \ldots, y_n]$ such that $y_1 \geq y_2 \geq \cdots \geq y_n$. For example, if $n = 5$ and $t = 3$, then the representation is $[1, 1, 1, 0, 0]$. Let $y_1, \ldots, y_n$ be fresh Boolean variables corresponding to the unary, order encoding, representation of the integer variable t. Note that with the unary encoding we add also binary clauses to keep the order between the bits ($y_1 \geq y_2 \geq \cdots \geq y_n$). The equation in Figure 1(b) specifies the unary bit-blast of integer variable t. Substituting in Equation (6) results in a system of pseudo-Boolean constraints introducing an order of $m \cdot n$ fresh Boolean variables with integer coefficients of value up to $\max(\rho)$. It also introduces an order of $m \cdot n$ binary CNF clauses.

Mixed-Radix Bit-Blast: Let $\mu = \langle r_0, r_1, \ldots, r_{k-1} \rangle$ be a mixed radix base, let $\langle w_0, \ldots, w_k \rangle$ be the weights for μ, and let $y_0, \ldots, y_k$ be a sequence of bit-vectors representing, in the order encoding, the values of the digits of the mixed radix bit-blast of t. For $0 \leq i < k$, the vector y_i consists of r_i fresh Boolean variables, and y_k consists of $\lfloor n \div w_k \rfloor$ fresh Boolean variables. Denoting by Σy_i the sum of the Boolean variables in bit-vector y_i, the equation in Figure 1(c) specifies the mixed radix bit-blast of integer variable t. Substituting in Equation 6 results in a system of pseudo-Boolean constraints introducing at most $m \cdot n$ fresh Boolean variables with integer coefficients. It also introduces at most $m \cdot n$ binary CNF clauses to encode the order between the bits in the corresponding bit-vectors y_i. For a given pseudo-Boolean modulo-constraint the choice of a mixed radix base to bit-blast it to a pseudo-Boolean constraint is left to a brute-force search algorithm which aims to minimize the size of the representation of the resulting pseudo-Boolean constraint. This works well in practice as the number of potential bases is not large.

The following example demonstrates the transformation with unary bit-blasting. It also illustrates that in practice, the domain of the integer variables t_j of Equation 5 is tighter than the prescribed $0 \leq t_j \leq n$.

Example 6. Consider the bases, $\rho_1 = \langle 2, 3, 5, 7, 11, 13 \rangle$ and $\rho_2 = \langle 17, 3, 19, 23 \rangle$ and again Constraint (1). Recall from Example 3 the derived pseudo-Boolean equivalences with respect to the two bases. To detail the RNS decomposition of Constraint (1) to pseudo-Boolean constraints we illustrate the (tighter) domains of the integer variables t_j of Equation (5) where $0 \leq t_j \leq \left\lfloor \frac{(\sum a_{ij}) - c_j}{p_j} \right\rfloor$.

Substituting in Equation (5), we have the following bounds for t_j with respect to ρ_1, on the left and with respect to ρ_2, on the right.

$$0 \le t_1 \le \lfloor (1+1+1+1-0)/2 \rfloor = 2 \qquad 0 \le t_1 \le \lfloor (9-0)/17 \rfloor = 0$$
$$0 \le t_2 \le \lfloor (2+1-1)/3 \rfloor = 0 \qquad 0 \le t_2 \le \lfloor (2+1-1)/3 \rfloor = 0$$
$$0 \le t_3 \le \lfloor (1+4+2+4-3)/5 \rfloor = 0 \qquad 0 \le t_3 \le \lfloor (13+3-3)/19 \rfloor = 0$$
$$0 \le t_4 \le \lfloor (5+4+1+2-6)/7 \rfloor = 0 \qquad 0 \le t_4 \le \lfloor (22+1-22)/23 \rfloor = 0$$
$$0 \le t_5 \le \lfloor (5+8+4+4-1)/11 \rfloor = 1$$
$$0 \le t_6 \le \lfloor (10+4+11+6-10)/13 \rfloor = 1$$

Based on these bounds we obtain the following encodings to pseudo-Boolean constraints using unary bit blasting with respect to ρ_1, on the left and with respect to ρ_2, on the right. Here, $[y_{11}, y_{12}]$ represents t_1, $[y_{31}]$ represents t_3, $[y_{51}]$ represents t_5, and $[y_{61}]$ represents t_6, all in the order encoding.

$$1x_1 + 1x_2 + 1x_3 + 1x_4 - 2y_{11} - 2y_{12} = 0 \qquad\qquad 9x_1 = 0$$
$$2x_3 + 1x_4 = 1 \qquad\qquad 2x_3 + 1x_4 = 1$$
$$1x_1 + 4x_2 + 3x_3 + 4x_4 - 5y_{31} = 3 \qquad\qquad 13x_1 + 3x_2 = 3$$
$$5x_1 + 4x_2 + 1x_3 + 2x_4 = 6 \qquad\qquad 22x_2 + 1x_3 = 22$$
$$5x_1 + 8x_2 + 4x_3 + 4x_4 - 11y_{51} = 1$$
$$10x_1 + 4x_2 + 11x_3 + 6x_4 - 13y_{61} = 10$$

4 Choosing an RNS Base

In [1], the authors consider two different options for an RNS base when encoding a given pseudo-Boolean constraint.

Definition 2 (primes and prime powers RNS bases [1]). *Let $S \in ms(\mathbb{N})$, and for any k let $p_1, p_2, \ldots, p_k$ denote the sequence of the first k prime numbers. The **primes** RNS base is $\rho_p = \{p_1, p_2, \ldots, p_k\}$ where k is the smallest integer such that $\pi_{\rho_p} > \Sigma S$. The **prime powers** RNS base is $\rho_{pp} = \{p_1^{n_1}, p_2^{n_2}, \ldots, p_k^{n_k}\}$ where for each i, n_i is the smallest integer such that $\log_2(\Sigma S) \le p_i^{n_i}$ and k is the smallest integer such that $\pi_{\rho_{pp}} > \Sigma S$.*

Definition 2 determines an RNS base focusing only on the maximal number that needs to be represented in the base (the sum of the coefficients). As already hinted at in Examples 3 and 4, we maintain that the choice of the RNS base should be geared towards the specific coefficients occurring in the constraint. We consider two additional options to select an RNS base. The first option focuses on the number of variables in the derived pseudo-Boolean modulo constraints, and the second on the number of clauses in their resulting CNF encoding.

For a given pseudo-Boolean constraint, the following definition specifies an RNS base which minimizes the variable occurrences in the pseudo-Boolean modulo constraints resulting from the RNS transformation specified in Equation (3). So we want as many as possible of the base elements to divide as many as possible of the coefficients in the constraint. Let $f_S(p)$ denote the number of elements from $S \in ms(\mathbb{N})$ that are divided by the (prime) number p and for $\rho = \langle p_1, \ldots, p_k \rangle$ denote $f_S(\rho) = \langle f_S(p_1), \ldots, f_S(p_k) \rangle$.

Definition 3 (optV RNS base). *RNS base ρ for $S \in ms(\mathbb{N})$ is **optV**, if ρ is a non-redundant RNS base for S; and the tuple $f_S(\rho)$ is maximal (in the lexicographic order) over all of the non-redundant RNS bases for S.*

Example 7. Consider the multiset $S = \langle 621, 459, 323, 7429 \rangle$ of coefficients from Constraint (1). The following tables illustrate the representations of the elements of S in the three RNS bases defined above, highlighting in the bottom line(s) the number of elements divided by each base element.

ρ_P \ S	2	3	5	7	11	13
323	1	2	3	1	4	11
459	1	0	4	4	8	4
621	1	0	1	5	5	10
7429	1	1	4	2	4	6
#divided	0	2	0	0	0	0

ρ_{PP} \ S	16	27	25
323	3	26	23
459	11	0	9
621	13	0	21
7429	5	4	4
#divided	0	2	0

ρ_{optV} \ S	17	3	19	23
323	0	2	0	1
459	0	0	3	22
621	9	0	13	0
7429	0	1	0	0
#divided	3	2	2	2

The following theorem defines an algorithm to construct an **optV** RNS base.

Theorem 1 (finding an optV RNS base). *Let $S \in ms(\mathbb{N})$ be such that it has a non-redundant RNS base and let $P = \langle p_1, \ldots, p_n \rangle$ be the sequence of distinct primes smaller than $max(S)$, sorted so that for all $0 < i < n$, $\langle f_S(p_i), p_i \rangle \succ_{lex} \langle f_S(p_{i+1}), p_{i+1} \rangle$. Let $\rho = \langle p_1, \ldots, p_k \rangle$ ($k \leq n$) be the shortest prefix of P such that $\pi_\rho > \Sigma S$. Then ρ is an **optV** RNS base for S.*

Example 8. Consider again the multiset S from Example 7. In the context of the statement of Theorem 1 we have $P = \langle 17, 3, 19, 23, 2, 5, 7, 11, 13, \ldots, 7417 \rangle$ and $\rho = \langle 17, 3, 19, 23 \rangle$.

Proof. (of Theorem 1) First note that ρ as constructed in the statement of the theorem is well-defined: That S has a non-redundant RNS base implies that $\pi_P > \Sigma S$ and hence there exists $k \leq n$ such that $\rho = \langle p_1, \ldots, p_k \rangle$ is a non-redundant RNS base for S. Assume falsely that there exists another non-redundant RNS base $\rho' = \langle q_1, \ldots, q_l \rangle$ such that $f_S(\rho) \prec_{lex} f_S(\rho')$. It follows that neither ρ nor ρ' are a prefix of the other (otherwise contradicting that both are non-redundant) which implies that there exists an index $i \leq min(k, l)$ such that $p_i \neq q_i$ and for $0 < j < i$, $p_j = q_j$. Namely, ρ and ρ' have a common prefix of length $i - 1$. Assume, without loss of generality, that there is no choice of ρ' which has a longer common prefix with ρ. Otherwise, take for ρ' the choice with a longer common prefix (observe that the value of i is bound from above by k). From the construction of ρ we have that $f_S(p_i) \geq f_S(q_i)$ (otherwise we would have selected q_i in the construction of ρ). From the selection of ρ' such that $f(\rho) \prec_{lex} f(\rho')$ we have that $f_S(p_i) \leq f_S(q_i)$. It follows that $f_S(p_i) = f_S(q_i)$. Again from the construction of ρ (this time because $f_S(p_i) = f_S(q_i)$) we have that $p_i < q_i$. Now consider $\rho'' = \langle p_1, \ldots, p_i, q_{i+1}, \ldots q_l \rangle$ and observe that $\pi_{\rho''} < \pi_{\rho'}$ because all elements are the same except at position i where $p_i < q_i$. If ρ'' is a base for S, then it is non-redundant and we have a contradiction because $f_S(\rho) \prec f_S(\rho') = f_S(\rho'')$ and ρ and ρ'' have a common prefix of length i. If ρ'' is not a base for S then it follows that $\rho''' = \langle p_1, \ldots, p_i, q_{i+1}, \ldots q_l, q_i \rangle$ is a non-redundant base for S, and we have a contradiction because $f_S(\rho) \prec f_S(\rho') \preceq_{lex} f_S(\rho''')$ and ρ and ρ''' have a common prefix of length i.   □

We now consider an RNS base which aims to minimize the number of CNF clauses resulting from the eventual encoding of the modulo-constraints specified in Equation (3). The size of the CNF encoding is determined not only by the RNS base but also by the bit-lasting technique and the subsequent encoding technique for the resulting pseudo-Boolean constraints of Equation (6).

Given a given pseudo-Boolean constraint C with coefficients $S \in ms(\mathbb{N})$, for each prime number $p < max(S)$, let $h_C^\tau(p)$ denote the number of CNF clauses when encoding C modulo p given the encoding technique τ. Here, τ specifies the bit-blast alternative adopted in Equation (6) and the subsequent encoding to CNF technique applied to the resulting pseudo-Boolean constraints. For a specific choice of τ, an **optC** RNS base is one that minimizes the sum of $h_C^\tau(p)$ (over the elements of ρ).

Definition 4 (optC RNS base). *RNS base ρ for pseudo-Boolean constraint C with coefficients $S \in ms(\mathbb{N})$ using encoding technique τ is **optC**, if ρ is a non-redundant RNS base for S and the value of $\sum_{p \in \rho} h_C^\tau(p)$ is minimal.*

In our implementation the search for an **optC** RNS base is brute-force. Given a pseudo-Boolean constraint with coefficients S and an encoding technique τ, for each prime $p < max(S)$ we encode the derived bit-blast of C mod p to obtain the value of $h_C^\tau(p)$. We then seek a set of primes with sufficiently large multiplication and minimal sum of $h_C^\tau(p)$ values. Practice shows that the search time is negligible. However, finding an efficient way to perform this search is an interesting research problem on its own.

5   Experiments

Our objective is to illustrate the benefit of our proposed RNS based transformation comparing, for a variety of different "off-the-shelf" solvers, the time to solve an instance using or not using the proposed RNS decomposition. We emphasize that the intention is not to compare between the solvers, nor to present results with full range of other available solvers.

We have selected 4 basic solvers: the Sat4j solver based on resolution [6] (the Sat4j solver based on cutting planes was consistently slower for our benchmarks); and three solvers based on MINISAT$^+$ [14], which apply encoding techniques based on: BDDs, binary adders, and sorting networks. For the encodings based on sorting networks, where MINISAT$^+$ employs a mixed radix representation, we apply an improved version of the tool which we call MINISAT^{++}. Improvements include: the application of an optimal mixed radix base as described in [11] (the base selected by MINISAT$^+$ is often not optimal); and fixing several bugs in MINISAT$^+$ such as the one reported in [1] (in the section on experiments).

Table 1. PB12 with various solvers — solving times (seconds) with 1800 sec. timeout

instance (s/u)	without RNS				optV RNS			ρ$_p$ RNS	
	Adders	Sorters	BDDs	S4J.R	U×Sorters	B×BDDs	MR×Adders	PBM	B×BDDs
BI_cuww4 (s)	∞	∞	0.5	∞	**0.1**	14.9	98.3	∞	∞
BI_prob10 (s)	326.5	351.5	**5.9**	∞	412.3	26.6	996.5	∞	147.0
ML_cuww1 (s)	**0.0**	0.2	0.2	0.1	7.6	∞	1.1	∞	∞
ML_cuww2 (s)	∞	∞	**0.8**	∞	245.8	∞	∞	∞	∞
ML_cuww3 (s)	∞	106.9	0.5	0.5	3.9	1.87	**0.1**	∞	∞
ML_cuww5 (s)	27.8	∞	**1.1**	∞	∞	∞	∞	∞	∞
ML_prob1 (s)	4.5	**0.4**	4.3	303.0	13.9	100.5	154.3	∞	127.2
ML_prob2 (s)	1.8	∞	4.9	49.5	**0.5**	∞	∞	∞	∞
ML_prob3 (u)	∞	∞	**0.2**	∞	∞	∞	∞	∞	∞
ML_prob4 (s)	602.3	14.4	**1.7**	133.2	4.4	21.3	9.6	∞	∞
ML_prob5 (s)	44.8	17.14	6.8	∞	**3.8**	60.9	863.2	∞	388.5
ML_prob6 (s)	18.7	∞	**3.0**	508.1	100.2	∞	113.2	∞	∞
ML_prob7 (s)	∞	163.7	**8.1**	109.1	17.3	27.4	118.9	∞	∞
ML_prob8 (s)	∞	13.7	**8.0**	440.9	494.2	∞	751.5	∞	∞

In addition to the four basic solvers (without RNS), we consider the full range of RNS based solvers based on these which are parametrized by three choices: The selection of the RNS base, the selection of technique to bit-blast modulo arithmetic, and the selection of the underlying solver (from the four above mentioned) applied to the result of the RNS decomposition. We also consider the RNS based solver described in [1] which we call PBM. All solvers run on the same machine and apply the same underlying MINISAT solver (where relevant).

We consider two benchmark suites. The first suite, which we call PB12, consists of the 28 instances from the DEC-INT-LIN category of the Pseudo-Boolean Competition 2012 [15] (we show results for the 14 largest instances). Each instance consists of a single pseudo-Boolean constraint (with equality) and is categorized depending on the size of its coefficients: BI (big integers), MI (medium integers) or SI (small integers). The benchmark suite consists of both satisfiable and unsatisfiable instances. The largest constraint involves 106 coefficients, and the largest coefficient is 106,925,262. The instance name indicates to which of the three categories the instance belongs to (see for example Table 1).

For the second suite, which we call RNP, we follow [1] and consider a collection of randomly generated number partition problems. Here, an (n, L)-instance is a set of integer values $S = \{a_1, \ldots, a_n\}$ such that each element is selected randomly in the range $0 \leq a_i < 2^L$. The objective is to determine a partition of S into two parts that sum to the same value. Each such instance is straightforward to express in terms of a single pseudo-Boolean constraint: $\sum a_i x_i = \sum a_i / 2$. We have selected instances randomly generated for $3 \leq n \leq 30$ and $3 \leq L \leq 2n$. The problem is known to have a phase transition when $L/n = 1$ [8], and so we focus on these instances. Here, the instance name indicates the values of the parameters n and L (see for example Table 2). For example, the instance 25-25-1 is instance number 1 with 25 coefficients, each involving a 25 bit random value.

All of the following tables report solving times, indicated in seconds, with a timeout of 1800 seconds. The symbol ∞ indicates that the timeout was encountered, and the symbol − indicates that memory was insufficient to construct the (BDD-based) encoding. The tables detail solving times for the 14 larger

Table 2. RNP with various solvers — solving times (seconds) with 1800 sec. timeout

instance (s/u)	without RNS				optC RNS			ρ_p RNS	
	Adders	Sorters	BDDs	S4J.R	U×BDDs	B×BDDs	MR×Sorters	PBM	B×BDDs
25-25-1 (s)	192.7	18.1	**0.2**	20.9	7.9	2.2	13.6	20.2	29.1
25-25-2 (u)	181.1	159.2	**0.4**	69.4	15.9	9.6	37.9	15.5	45.1
25-25-3 (u)	63.8	8.2	**0.1**	6.2	0.2	3.8	28.6	11.0	26.8
25-25-4 (u)	58.0	55.6	**0.2**	36.3	1.1	5.2	41.8	47.6	70.9
25-25-5 (u)	108.9	34.0	**0.1**	9.4	0.9	25.9	108.7	18.8	22.7
25-25-6 (u)	41.9	42.5	**0.2**	26.5	0.5	11.4	22.6	18.5	30.5
25-25-7 (u)	99.9	22.4	**0.1**	7.3	21.9	8.6	34.2	68.8	39.5
25-25-8 (u)	119.5	51.5	**0.4**	63.3	24.1	15.9	59.6	25.8	43.4
25-25-9 (s)	26.2	25.7	**0.4**	0.4	2.6	0.6	5.8	1.9	1.8
25-25-10 (s)	21.9	16.9	**0.3**	17.0	4.4	2.8	1.3	0.7	9.7
30-30-1 (u)	∞	∞	**9.0**	∞	14.3	∞	∞	1696.7	∞
30-30-2 (s)	∞	34.9	**4.6**	∞	867.8	994.7	∞	1006.2	653.9
30-30-3 (u)	∞	∞	**8.7**	∞	∞	1751.4	∞	∞	∞
30-30-4 (s)	844.8	418.6	-	∞	269.0	**66.89**	211.6	757.0	689.2
30-30-5 (u)	∞	∞	-	∞	**52.8**	∞	∞	∞	∞
30-30-6 (u)	∞	1762.3	**7.8**	∞	139.1	∞	∞	∞	∞
30-30-7 (u)	∞	∞	**9.9**	∞	1416.3	∞	∞	∞	∞
30-30-8 (u)	∞	1448.2	**7.6**	∞	397.3	497.9	1362.6	1346.6	∞
30-30-9 (u)	∞	∞	-	∞	982.3	721.3	1143.8	**345.7**	1129.1
30-30-10 (u)	∞	∞	**9.8**	∞	∞	1394.1	∞	∞	∞
35-35-1 (?)	∞	∞	-	∞	∞	∞	∞	∞	∞
35-35-2 (s)	∞	∞	-	∞	∞	∞	1602.9	∞	∞
35-35-3 (s)	∞	∞	-	∞	∞	∞	1152.9	∞	∞
35-35-4 (s)	∞	∞	-	∞	**669.4**	∞	∞	∞	∞
35-35-5 (s)	∞	333.2	-	∞	∞	∞	∞	∞	∞
35-35-6 (?)	∞	∞	-	∞	∞	∞	∞	∞	∞
35-35-7 (s)	∞	∞	-	∞	∞	∞	∞	∞	1788.1
35-35-8 (s)	∞	∞	-	∞	∞	807.5	∞	∞	∞
35-35-9 (s)	∞	∞	-	∞	**427.6**	∞	∞	∞	∞
35-35-10 (s)	∞	∞	-	104.0	**48.9**	1143.5	∞	∞	∞

instances of the PB12 benchmarks and for 30 instances from the RNP benchmark. For each instance we indicate if it is satisfiable or not (s/u). Experiments run on a Quad-core Intel i5-2400 at 3.1 GHz with 4 GB RAM with at most 3GB allocated to the representation BDDs.

Tables 1 and 2 present results for the benchmark instances using a series of solvers. The first four are the basic solvers: "without RNS". The next three are "with RNS". In each of the two tables we portray results for one solver for each of the three types of bit-blasting combined with one of the four basic solvers. In Table 1 we consider an optV RNS base and in Table 2, an optC RNS base. Both tables indicate a comparison (in the two right columns) of our approach with the PBM solver which is also based on RNS decomposition. Here we apply their dynamic programming based transformation with the ρ_p RNS base (their best configuration for the given benchmarks) and our RNS decomposition with binary bit-blasting and an underlying BDD solver (with the same RNS base).

Tables 1 and 2 indicate the general utility of the RNS based approach. In Table 1 the BDD based encodings (without RNS) perform best. But, the RNS solvers come in second. Except for the case of BDDs, the combination of a base solver with RNS solves more instances than without. Table 2 illustrates the case where the BDD based approach (without RNS) suffers from the size of the encodings as it fails to encode most of the large instances with the allocated

Table 3. PB12 benchmarks; Solvers based on RNS decomposition and Sorters; Impact of RNS base selection and bit-blast technique. Solving times (seconds) with 1800 sec. timeout.

instance (s/u)	with RNS:U×Sorters				with RNS:B×Sorters				with RNS:MR×Sorters			
	$\rho_{\mathbf{p}}$	$\rho_{\mathbf{pp}}$	ρ_{optV}	ρ_{optC}	$\rho_{\mathbf{p}}$	$\rho_{\mathbf{pp}}$	ρ_{optV}	ρ_{optC}	$\rho_{\mathbf{p}}$	$\rho_{\mathbf{pp}}$	ρ_{optV}	ρ_{optC}
BI_cuww4 (s)	∞	∞	**0.1**	∞	∞	∞	0.4	∞	∞	∞	0.5	∞
BI_prob10 (s)	∞	∞	412.3	278.9	∞	∞	**6.1**	55.7	949.2	∞	55.1	1523.3
MI_cuww1 (s)	1134.4	∞	7.6	3.6	∞	∞	0.6	**0.2**	∞	∞	6.9	3.1
MI_cuww2 (s)	∞	∞	**245.9**	∞	∞	∞	∞	∞	∞	∞	∞	∞
MI_cuww3 (s)	∞	∞	3.9	3.9	∞	∞	**0.6**	∞	∞	∞	1.2	522.4
MI_cuww5 (s)	∞	∞	∞	∞	∞	∞	∞	∞	∞	∞	∞	∞
MI_prob1 (s)	57.5	1423.8	13.9	11.0	241.7	∞	**9.8**	226.2	784.2	685.9	28.7	174.2
MI_prob2 (s)	∞	∞	**0.5**	11.2	∞	∞	1.5	284.1	∞	406.6	∞	41.5
MI_prob3 (u)	∞	∞	∞	∞	∞	∞	∞	∞	∞	∞	∞	∞
MI_prob4 (s)	∞	∞	**4.4**	1504.2	∞	∞	16.4	32.5	∞	∞	∞	394.7
MI_prob5 (s)	330.9	∞	3.4	3.3	1769.5	630.5	1.4	3.3	44.9	138.3	2.5	**1.2**
MI_prob6 (s)	137.0	26.9	100.2	∞	1772.9	∞	∞	∞	∞	∞	**9.0**	∞
MI_prob7 (s)	∞	593.7	**17.2**	∞	∞	∞	∞	∞	∞	∞	17.7	1019.8
MI_prob8 (s)	537.2	∞	494.2	**18.0**	∞	∞	816.6	∞	∞	∞	∞	∞

3GB of RAM. For the second benchmark, overall, the RNS based solvers are superior. The comparison with the PBM solver indicates that our approach based on bit-blasting to pseudo-Boolean constraints is superior to encoding the pseudo-Boolean modulo constraints directly to CNF.

Tables 3 and 4 illustrate the impact of the choice of the RNS base and bit-blasting technique on our approach for the PB12 and RNP benchmarks respectively. In Table 3 we focus on solvers based on RNS decomposition using Sorters. In Table 4 we focus on solvers based on RNS decomposition using BDDs. For each instance we consider its RNS transformations using the four choices of RNS base: $\rho_{\mathbf{p}}$ (primes), $\rho_{\mathbf{pp}}$ (prime powers), ρ_{optV} (optimal with respect to number of variables), and ρ_{optC} (optimal with respect to number of clauses) and the three choices of bit-blasting techniques. We can clearly see that the optimal bases are superior. As for the bit-blasting no conclusion can be made.

Table 5 presents the "bottom-line" advantage of applying RNS decomposition. For each of the benchmark instances we indicate the best result "without" and "with" RNS decomposition. For each such result we also indicate the solver (configuration) with which this result is obtained. For RNS that means specifying the base solver, the choice of RNS base, and the choice of bit-blast for modulo arithmetic. Table 5(a), for the PB12 benchmark, indicates that (except for one instance) the RNS based techniques perform better, always with one of the optimal RNS bases introduced in Section 4. Table 5(b), for the RNP benchmark, indicates that when the BDD encoding without RNS applies, it is typically fastest. However, when the encoding with BDDs exhausts memory the RNS based approach is typically faster. Especially note that for all of the instances where the "No RNS" column does not apply BDDs (including those where there

Table 4. RNP benchmarks; Solvers based on RNS decomposition and BDDs; Impact of RNS base selection and bit-blast technique. Solving times (seconds) with 1800 sec. timeout.

instance (s/u)	with RNS: U×BDDs				with RNS: B×BDDs				with RNS: MR×BDDs			
	ρ_p	ρ_{pp}	ρ_{optV}	ρ_{optC}	ρ_p	ρ_{pp}	ρ_{optV}	ρ_{optC}	ρ_p	ρ_{pp}	ρ_{optV}	ρ_{optC}
25-25-1 (s)	34.0	62.0	0.7	7.9	29.1	50.2	1.2	1.6	26.8	68.7	**0.4**	14.0
25-25-2 (u)	36.9	52.0	10.5	15.9	45.1	53.1	18.5	**6.6**	42.4	56.4	17.0	12.7
25-25-3 (u)	33.5	45.1	7.7	**0.2**	26.8	66.9	10.5	2.5	28.6	48.6	15.0	4.9
25-25-4 (u)	63.5	43.9	4.2	**1.1**	70.9	46.3	6.9	2.3	59.7	49.2	7.1	6.5
25-25-5 (u)	26.9	81.3	39.7	**0.9**	22.7	77.0	41.0	13.8	38.6	74.2	41.6	19.7
25-25-6 (u)	36.5	70.4	12.1	**0.5**	30.5	70.2	12.7	8.9	35.5	60.3	12.9	18.5
25-25-7 (u)	38.1	53.6	9.28	21.9	39.5	45.7	8.8	**6.4**	38.6	45.4	9.4	12.9
25-25-8 (u)	44.5	73.3	28.36	24.1	43.4	81.9	34.0	**11.4**	49.3	73.6	19.9	21.7
25-25-9 (s)	6.6	27.8	1.5	2.6	1.8	15.8	0.4	0.5	2.6	7.7	**0.3**	2.7
25-25-10 (s)	3.3	12.4	4.6	4.5	9.7	5.0	4.2	2.1	10.1	17.9	**1.5**	2.2
30-30-1 (u)	∞	∞	∞	**14.3**	∞	∞	∞	∞	∞	∞	∞	∞
30-30-2 (s)	1543.0	∞	583.6	867.8	653.9	∞	**192.1**	1197.9	681.2	∞	1126.9	885.3
30-30-3 (u)	∞	∞	∞	∞	∞	∞	∞	∞	∞	∞	∞	∞
30-30-4 (s)	760.6	∞	112.6	269.0	689.2	∞	89.6	**84.8**	710.2	∞	98.0	423.7
30-30-5 (u)	∞	∞	∞	**52.8**	∞	∞	1751.7	∞	∞	∞	1362.8	∞
30-30-6 (u)	∞	∞	∞	139.1	∞	∞	∞	136.3	∞	∞	∞	**107.9**
30-30-7 (u)	∞	∞	1113.3	1416.3	∞	∞	1130.4	∞	∞	∞	1139.3	∞
30-30-8 (u)	∞	∞	1120.9	**397.3**	∞	∞	915.1	597.8	∞	∞	801.1	1191.1
30-30-9 (u)	1287.3	∞	604.3	982.3	1129.1	∞	623.4	843.3	1224.5	∞	580.77	**476.9**
30-30-10 (u)	∞	∞	∞	∞	∞	∞	1719.0	**1603.2**	∞	∞	1603.6	∞
35-35-1 (?)	∞	∞	∞	∞	∞	∞	∞	∞	∞	∞	∞	∞
35-35-2 (s)	∞	∞	**1250.9**	∞	∞	∞	∞	∞	∞	∞	∞	∞
35-35-3 (s)	∞	∞	∞	∞	∞	∞	∞	∞	∞	∞	∞	∞
35-35-4 (s)	∞	∞	∞	**669.4**	∞	∞	∞	∞	∞	∞	∞	∞
35-35-5 (s)	∞	∞	∞	∞	∞	∞	∞	∞	**1397.1**	∞	∞	∞
35-35-6 (?)	∞	∞	∞	∞	∞	∞	∞	∞	∞	∞	∞	∞
35-35-7 (s)	∞	∞	170.3	∞	1788.1	∞	630.7	∞	∞	∞	590.5	∞
35-35-8 (s)	∞	∞	∞	∞	∞	∞	∞	**831.1**	∞	∞	∞	∞
35-35-9 (s)	∞	∞	∞	**427.6**	∞	∞	∞	∞	∞	∞	∞	∞
35-35-10 (s)	∞	∞	∞	**48.9**	∞	∞	∞	1242.6	∞	∞	∞	∞

is a timeout because of the other base solvers), it is the case that the BDD technique exhausts memory. For most of these instances the fastest RNS solver is the one that applies BDDs and in this case there is no memory problem.

In a final experiment we observe that the average size of the conflict clauses reported by MINISAT^{++} (using sorters) is smaller with RNS decomposition. We compute the average size of the conflict clauses considering the first k clauses generated with each technique (where k is as large as possible but required to be the same per instance for all techniques compared). We consider the average over all relevant combinations of RNS base and bit blasting options. For the PB12 benchmark, where the number of variables is much larger, the average learned clause size is 281 without decomposition, and 188 with. For the RNP benchmark the averages are 68 (without) and 62 (with). The RNS decomposition replaces an original constraint by a conjunction of "smaller" constraints, each one typically involving less of the original constraint problem variables. Conflicts derived from a single constraint will then generate smaller learned clauses and support better propagation during search.

Table 5. Best solving times (seconds) with and without RNS (1800 sec. timeout)

(a) PB12 benchmarks

instance	No RNS		With RNS	
BI_cuww4	0.5	bdd	0.1	sort.ρ_{optV} (u)
BI_prob10	4.9	sort	0.5	add.ρ_{optV} (b)
MI_cuww1	0.1	add	0.1	add.ρ_{optV} (b)
MI_cuww2	0.8	bdd	0.6	add.ρ_{optV} (b)
MI_cuww3	0.5	bdd	0.1	add.ρ_{optV} (mr)
MI_cuww5	1.1	bdd	0.4	sort.ρ_{optC} (mr)
MI_prob1	**0.4**	sort	0.5	sort.ρ_{optV} (b)
MI_prob2	1.8	add	0.3	sort.ρ_{optC} (u)
MI_prob3	**0.2**	bdd	∞	
MI_prob4	1.7	bdd	0.4	sort.ρ_{optV} (u)
MI_prob5	5.9	sort	0.3	S4J.ρ_{optV} (b)
MI_prob6	3.0	bdd	0.6	add.ρ_{optV} (b)
MI_prob7	8.1	bdd	2.4	S4J.ρ_{optV} (u)
MI_prob8	8.0	bdd	1.6	sort.ρ_{optC} (mr)

(b) RNP benchmarks

instance	No RNS		With RNS	
25-25-1	**0.2**	bdd	0.3	sort.ρ_{optC} (b)
25-25-2	**0.4**	bdd	6.6	bdd.ρ_{optC} (b)
25-25-3	**0.1**	bdd	0.2	bdd.ρ_{optC} (u)
25-25-4	**0.2**	bdd	0.8	bdd.ρ_{optC} (u)
25-25-5	**0.1**	bdd	0.7	bdd.ρ_{optC} (u)
25-25-6	**0.2**	bdd	0.5	bdd.ρ_{optC} (u)
25-25-7	**0.1**	bdd	6.5	bdd.ρ_{optC} (b)
25-25-8	**0.4**	bdd	11.4	bdd.ρ_{optC} (b)
25-25-9	**0.4**	bdd	0.2	sort.ρ_{optC} (b)
25-25-10	**0.4**	bdd	0.3	sort.ρ_{optV} (mr)
30-30-1	**9.0**	bdd	10.0	bdd.ρ_{optC} (u)
30-30-2	**4.6**	bdd	151.1	bdd.ρ_{optV} (b)
30-30-3	**8.7**	bdd	1588.8	sort.ρ_{optC} (mr)
30-30-4	418.6	sort	**66.9**	bdd.ρ_{optC} (b)
30-30-5	∞		**44.9**	bdd.ρ_{optC} (u)
30-30-6	**7.8**	bdd	95.0	bdd.ρ_{optC} (mr)
30-30-7	**9.9**	bdd	1069.8	bdd.ρ_{optV} (b)
30-30-8	**7.6**	bdd	345.4	bdd.ρ_{optC} (u)
30-30-9	∞		**449.5**	bdd.ρ_{optC} (mr)
30-30-10	**9.8**	bdd	1201.9	sort.ρ_{optC} (b)
35-35-1	∞		∞	
35-35-2	598.7	sort	**539.7**	sort.ρ_{optV} (b)
35-35-3	∞		**868.7**	sort.ρ_{optC} (b)
35-35-4	∞		**590.2**	bdd.ρ_{optC} (u)
35-35-5	**333.2**	sort	362.7	sort.ρ_{optC} (mr)
35-35-6	∞		∞	
35-35-7	44.6	sort	**43.7**	sort.ρ_{optC} (b)
35-35-8	∞		**807.5**	bdd.ρ_{optC} (b)
35-35-9	∞		**422.9**	bdd.ρ_{optC} (u)
35-35-10	∞		**42.1**	bdd.ρ_{optC} (u)

6   Conclusion

We introduce a new encoding technique for pseudo-Boolean constraints based on RNS decomposition. The new approach, defined as a transformation from pseudo-Boolean constraints to conjunctions of pseudo-Boolean constraints, can be combined with any pseudo-Boolean constraint solver. We demonstrate that careful selection of the RNS base, fine-tuned for each specific pseudo-Boolean instance, is key to the success of the approach. We show that when applying our approach we can solve instances that we cannot solve without. In particular, in cases where the BDD encoding is out of scope of current solvers due to the size of the encoding, we can apply and solve the constraints using BDDs after RNS decomposition. We compare our approach with previous work which applies RNS decomposition but encodes pseudo-Boolean modulo constraints directly to SAT. The comparison indicates the advantage of our approach where pseudo-Boolean modulo constraints are bit-blasted to standard pseudo-Boolean constraints. A repository of benchmark instances (before and after applying our transformation) as well as a more extensive presentation of the experimental results, are available at http://www.cs.bgu.ac.il/~mcodish/Benchmarks/SAT2014.

Acknowledgment. We thank the authors of [1] for making their PBM solver available, as well as their generator for RNP instances.

References

1. Aavani, A., Mitchell, D.G., Ternovska, E.: New encoding for translating pseudo-Boolean constraints into SAT. In: Frisch, A.M., Gregory, P. (eds.) SARA. AAAI (2013)
2. Abío, I., Nieuwenhuis, R., Oliveras, A., Rodríguez-Carbonell, E., Mayer-Eichberger, V.: A new look at BDDs for pseudo-Boolean constraints. J. Artif. Intell. Res. (JAIR) 45, 443–480 (2012)
3. Ansótegui, C., Manyà, F.: Mapping problems with finite-domain variables into problems with Boolean variables. In: Hoos, H.H., Mitchell, D.G. (eds.) SAT 2004. LNCS, vol. 3542, pp. 1–15. Springer, Heidelberg (2005)
4. Bailleux, O., Boufkhad, Y.: Efficient CNF encoding of Boolean cardinality constraints. In: Rossi, F. (ed.) CP 2003. LNCS, vol. 2833, pp. 108–122. Springer, Heidelberg (2003)
5. Barth, P.: Logic-based 0-1 constraint programming. Kluwer Academic Publishers, Norwell (1996)
6. Berre, D.L., Parrain, A.: The Sat4j library, rel. 2.2. JSAT 7(2-3), 6–59 (2010)
7. Bixby, R.E., Boyd, E.A., Indovina, R.R.: MIPLIB: A test set of mixed integer programming problems. SIAM News 25, 16 (1992)
8. Borgs, C., Chayes, J.T., Pittel, B.: Phase transition and finite-size scaling for the integer partitioning problem. Random Struct. Algorithms 19(3-4), 247–288 (2001)
9. Bryant, R.E., Lahiri, S.K., Seshia, S.A.: Deciding CLU logic formulas via Boolean and pseudo-Boolean encodings. In: Proc. Intl. Workshop on Constraints in Formal Verification, CFV 2002 (2002)
10. Bryant, R.E., Kroening, D., Ouaknine, J., Seshia, S.A., Strichman, O., Brady, B.A.: Deciding bit-vector arithmetic with abstraction. In: Grumberg, O., Huth, M. (eds.) TACAS 2007. LNCS, vol. 4424, pp. 358–372. Springer, Heidelberg (2007)
11. Codish, M., Fekete, Y., Fuhs, C., Schneider-Kamp, P.: Optimal base encodings for pseudo-Boolean constraints. In: Abdulla, P.A., Leino, K.R.M. (eds.) TACAS 2011. LNCS, vol. 6605, pp. 189–204. Springer, Heidelberg (2011)
12. Cormen, T.H., Leiserson, C.E., Rivest, R.L., Stein, C.: Introduction to Algorithms, 3rd edn. MIT Press (2009)
13. Crawford, J.M., Baker, A.B.: Experimental results on the application of satisfiability algorithms to scheduling problems. In: Hayes-Roth, B., Korf, R.E. (eds.) AAAI, vol. 2, pp. 1092–1097. AAAI Press / The MIT Press, Seattle (1994)
14. Eén, N., Sörensson, N.: Translating pseudo-Boolean constraints into SAT. JSAT 2(1-4), 1–26 (2006)
15. Manquinho, V., Roussel, O.: Pseudo-Boolean competition (2012), http://www.cril.univ-artois.fr/PB12/
16. Manquinho, V.M., Roussel, O.: The first evaluation of Pseudo-Boolean solvers (PB 2005). Journal on Satisfiability, Boolean Modeling and Computation (JSAT) 2(1-4), 103–143 (2006)
17. Metodi, A., Codish, M., Stuckey, P.J.: Boolean equi-propagation for concise and efficient SAT encodings of combinatorial problems. J. Artif. Intell. Res. (JAIR) 46, 303–341 (2013)
18. Tamura, N., Taga, A., Kitagawa, S., Banbara, M.: Compiling finite linear CSP into SAT. Constraints 14(2), 254–272 (2009)

An Ising Model Inspired Extension of the Product-Based MP Framework for SAT

Oliver Gableske

Ulm University, Theoretical Computer Science, 89077 Ulm, Germany
oliver.gableske@uni-ulm.de
https://www.gableske.net

Abstract. Message Passing (MP) has been presented in a unified and consistent notational frame in [7]. The paper explained the product-based MP framework (PMPF) and various MP heuristics (BP, SP, and several interpolations). The paper concluded, that an increased flexibility of MP heuristics (in the form of tunable parameters) leads to a tunable MP algorithm that can be helpful to solve a wide variety of different CNFs.

Based on this work, the paper at hand makes three contributions. First, we extend the PMPF regarding flexibility based on theoretical insights from the Ising Model [13]. As an immediate result, this extended PMPF (ePMPF) provides new possibilities for parameter tuning. Second, the ePMPF will also allow us to uncover various theoretical connections between well-known variable and value ordering heuristics for SAT (Zero-first, One-first, occurrence-based heuristics). We show, that these heuristics can be understood as special cases of Message Passing. Third, we show that the ePMPF provides numerous possibilities to tightly integrate Message Passing with various aspects of the CDCL search paradigm.

1 Introduction

The language SAT (comprised of all satisfiable Boolean formulas) is one of the most studied languages in computer science. The interest in the SAT problem (deciding whether a given formula is in SAT) was once purely academic (due to its NP-completeness [4]). Even though there is little hope to find an algorithm that solves the SAT problem (a SAT solver) in an efficient way, ongoing research has constantly improved the performance of SAT solvers during the past decades.

In practice, SAT solvers search for an assignment to the Boolean variables of a given formula under which the formula evaluates to true (a solution). Major advances have been made w.r.t. the design of efficient data-structures for problem representation, the design of paradigms to conduct the search, as well as the design of powerful variable and value ordering heuristics to advance the search.

Our work focuses on Message Passing (MP) algorithms [7], which help to determine variable and value orderings to advance the search. The task of an MP algorithm is to provide biases for the variables of a formula. Roughly speaking, a bias estimates the marginal assignment to a variable in all solutions of a formula.

C. Sinz and U. Egly (Eds.): SAT 2014, LNCS 8561, pp. 367–383, 2014.
© Springer International Publishing Switzerland 2014

Such biases can be used for both variable ordering (implied by the absolute value of the biases), and value ordering (implied by the signs of the biases) [2,7,8].

Numerous MP algorithms have been proposed so far. The first was Belief Propagation (BP) [14], which is based on graphical models (e.g. Bayesian networks) in order to perform inference. Later, Survey Propagation (SP) was proposed [2], which is based on the Ising Model [13] from physics. Since BP and SP might fail to provide biases (by failing to converge), their expectation maximization (EM) variants have been proposed [9,10,11,12]. They are derived with the EM algorithm [5] and are guaranteed to provide biases (by guaranteed convergence).

An MP *algorithm* is comprised of two parts: an MP *framework* and an MP *heuristc*. We suggest reading [7] in case the reader is not familiar with MP.

The MP *framework* formalizes how Message Passing works in general. It clarifies that biases are the result of an interaction of variables and clauses facilitated by an exchange of messages (δ and ω equations). This exchange might converge into an equilibrium in which all the messages do not change notably anymore. The framework clarifies that the equilibrium ω-messages are used in order to compute variable freedom values ($\mathcal{T}$ and $\mathcal{F}$ equations). These values are then used in magnetization functions (μ equations) in order to compute biases (β equations). Note, that the *MP framework* explicitly defines all but the δ and μ equations. Therefore, *it governs the overall process to compute biases.*

The MP *heuristic* formalizes how the disrespect- and magnetization-values are computed. Put differently, the MP heuristic provides the definitions for the δ and μ equations. Therefore, *it influences the overall process to compute biases.*

All together, an MP algorithm performs Message Passing according to an MP framework applying an MP heuristic in order to compute biases. As shown in [7], all of the above MP algorithms follow the same product-based MP framework (PMPF), and differ only in the applied MP heuristic. Furthermore [7] showed, that all these heuristics can be combined into a single MP heuristic called $\rho\sigma\text{PMP}^i$. The BP, SP, EMBPG, and EMSPG *heuristics* can all be represented by $\rho\sigma\text{PMP}^i$. Hence, applying $\rho\sigma\text{PMP}^i$ in the PMPF allows us to represent the BP, SP, EMBPG, and EMSPG *algorithms*. This insight will be needed later.

The "helpfulness" of biases provided by a specific MP algorithm depends on the *type of formula* (random 3-SAT, battleship, edge-matching, etc.) that the SAT solver must solve. For example, we might consider the biases computed by the SP algorithm (using the SP heuristic in the PMPF). They can be very helpful to solve large uniform random 3-SAT formulas. In contrast, they are not helpful when trying to solve hard combinatorial formulas [2,9,8].

Given a specific type of formula, there are three main approaches to improve the helpfulness of biases in order to increase an MP-based SAT solver's performance. First, we can design a new MP *heuristic* that *influences* the bias computations to increase their helpfulness. Second, we can design a new MP *framework* that *alters* the bias computations to increase their helpfulness. Third, we can do *both simultaneously* by providing a new MP heuristic *and* a new MP framework.

The first approach can be realized in two ways. First, analyze the formulas in question and derive strong theoretical insights on how the MP heuristic should look like. This is basically what was done in the design of SP [2]. Here, the Ising Model was used in order to acquire theoretical insight strong enough to design an MP heuristic for random k-SAT. However, this approach is not feasible in general. One cannot possibly analyze all types of formulas and provide a "custom made" MP heuristic for each type. This leads us to the second possibility. One might try to design an MP heuristic with increased flexibility, realized by parameters that can be tuned. Parameter tuning on different types of formulas will then lead to suitable parameter settings for the MP heuristic. This approach was already covered in [7,8]. The fundamental insight here is that an increased *flexibility of the MP heuristic* provides a meaningful possibility to conduct parameter tuning. This allows us to influence the bias computations on the given type of formula in order to increase an MP-based SAT solver's performance.

In analogy to the above, the second approach can be realized in two ways as well. First, try to design a specific MP framework based on a theoretical analysis of the formula type. Again, this was done in the design of SP. The Ising Model was used to design the MP framework to conduct Message Passing in SP. But again, following this purely theoretical approach to design "custom made" MP frameworks for the various formula types is not feasible in practice. In analogy to the above, one might try to increase the *flexibility of the MP framework* by introducing tunable parameters. Parameter tuning w.r.t. the current PMPF is impossible since it has no parameters. Hence, this approach has never been tried.

The third approach is basically a combination of the first two, i.e. designing a flexible MP heuristic (using parameters) and a flexible MP framework (using additional parameters). Tuning all parameters simultaneously for a given type of formula should provide the best performance achievable with the means of MP. As stated above, the MP heuristic $\rho\sigma\text{PMP}^i$ can represent all product-based MP heuristics currently available for SAT. We do not see how to improve this any further. We might, however, try to increase the flexibility of the MP *framework*.

The first contribution of this paper is to provide an extended PMPF (ePMPF) with increased flexibility. Recall, that the SP algorithm is based on the Ising Model, which provides concepts to design an MP framework. However, some of these concepts were *ignored* in the design of SP. We will explain these concepts and show, how they can be integrated into the PMPF (Sections 4 and 5).

The ePMPF will provide new possibilities for parameter tuning. However, we will show that the ePMPF also allows us to draw additional theoretical conclusions. More precisely, we will uncover various connections between well-known heuristics like Zero-first, One-first, as well as the conflict avoiding and conflict seeking literal-occurrence based heuristics. The second contribution of this paper is to show, that all these heuristics can be understood as special cases of Message Passing using the ePMPF (Section 6).

Additionally, the ePMPF allows us to tightly integrate MP with various aspects of the CDCL search paradigm (e.g. feedback for the variable ordering based on conflict analysis, clause database maintenance, and restarts). Furthermore, we explain how such a CDCL-based MP-inspired decimation (MID) solver [7] can use a-priori knowledge about solutions of a formula to increase performance.

The third contribution of this paper is to elaborate on the possibilities of integrating CDCL and MP, based on the theoretical insights mentioned above (Section 6).

2   Notational Details and Definitions

Let $\mathcal{V} = \{v_1, \ldots, v_n\}$ be a set of n Boolean *variables*, and let $\mathcal{L} = \{v, \bar{v} | v \in \mathcal{V}\}$ be the set of *literals*. A clause c is a set of literals, interpreted as a disjunction. A *CNF formula* F is a set of m *clauses*, interpreted as a conjunction. We call $|c|$ the *size* of clause c. A clause c is called *unit*, iff $|c| = 1$. A clause c is called *tautological*, iff $\exists l \in c : \bar{l} \in c$. A literal l is called *pure*, iff $\nexists c \in F : \bar{l} \in c$. We call F a *k-CNF formula*, iff $\max_{c \in F} |c| = k$. Additionally, the sets C_v^+, C_v^-, and C_v comprise the *sets of clauses* containing v as a positive, negative, or arbitrary literal, respectively. Furthermore, we denote with $L(v, c)$ the *literal* as which v occurs in c, and we denote with $V(l, c)$ the *variable* that corresponds to l in c.

For convenience, we assume that F does not contain tautological clauses or units, and that every variable occurs at least twice as positive literal and at least twice as negative literal. We will use this in Section 5 to ensure that for all $v \in \mathcal{V}$ and for all $l \in \mathcal{L}$ it is $s(l), u(l), t(v), f(v) \geq 2$ (see Definition 7 in [7]).

An *assignment* $\alpha : \mathcal{V} \to \{0, 1\}$ is called *satisfying* (or a solution) for F, iff $\alpha(F)$ evaluates to true (abbr. with $\alpha(F) = 1$). We call an assignment *total*, iff it assigns all variables in $\mathcal{V}$. In case there is at least one satisfying assignment for F, we call F *satisfiable*, and *unsatisfiable* otherwise.

The *language k-SAT* is defined as the set of all satisfiable k-CNF formulas. For a k-CNF F we write $F \in k\text{-SAT}$, if F is satisfiable, and $F \notin k\text{-SAT}$ otherwise.

Let $b \in \{0, 1\}$. Let $A(v, b, F)$ be the set of all total satisfying assignments for F that assign variable v to value b, i.e. $A(v, b, F) = \{\alpha | \alpha(F) = 1, \alpha(v) = b, \alpha \text{ total}\}$. We define the *solution marginal* of variable v for a formula F as the real number $\text{mar}(v, F) = (|A(v, 1, F)| - |A(v, 0, F)|) / (|A(v, 1, F)| + |A(v, 0, F)|) \in [-1, 1]$.

A *factor graph* [14] of a CNF F is an undirected and bipartite graph that contains two types of nodes and two types of edges. The nodes are partitioned into *variable nodes* (one for each $v \in \mathcal{V}$) and *clause nodes* (one for each $c \in F$). The edges are partitioned into *positive and negative edges*. A positive (negative) edge between a variable and a clause node exists, iff the variable appears as positive (negative) literal in the clause. The next section briefly covers the connections between the factor graph, the PMPF, marginals, and biases.

3   The PMPF – From Warnings to Biases

An MP *algorithm* performs Message Passing according to an MP *framework* applying an MP *heuristic* in order to compute biases. The biases are supposed to estimate the solution marginals for the variables. As explained in [7], the biases are the result of an exchange of messages between variable nodes and clause nodes in the factor graph. The PMPF governs this exchange based on *disrespect messages* δ (send from variable nodes to clause nodes) and *warning messages* ω (send from clause nodes to variable nodes). We cannot repeat all the details from [7] here, but it is important to understand that an MP *heuristic* H *defines* the equation for the disrespect message $\delta_H \in [0,1]$. The MP *framework* (PMPF) then *uses* δ_H to define the warning message $\omega_H \in [0,1]$ as well as the literal cavity freedom values $S_H, U_H \in [0,1]$ as follows.

$$\substack{y\\z}\omega_H(c_i, v) = \prod_{l \in c_i \setminus \{L(v, c_i)\}} \substack{y\\z}\delta_H(l, c_i) \tag{1}$$

$$\substack{y\\z}S_H(l, c_i) = \prod_{d \in C_{\bar{l}}} [1 - \substack{y\\z}\omega_H(d, V(l, c_i))] \quad \substack{y\\z}U_H(l, c_i) = \prod_{d \in C_l \setminus \{c_i\}} [1 - \substack{y\\z}\omega_H(d, V(l, c_i))] \tag{2}$$

In summary, an MP algorithm performs a cycle y of iterations z until the following abort condition holds: $\forall c \in F : \forall v \in c : \left| \substack{y\\z}\omega_H(c, v) - \substack{y\\z-1}\omega_H(c, v) \right| < \omega_{max}$ (assume $\omega_{max} = 0.01$). The messages of the iteration in which the convergence was reached are called *equilibrium messages*, denoted $\substack{y\\*}\delta_H(l, c)$ and $\substack{y\\*}\omega_H(c, v)$. Based on $\substack{y\\*}\omega_H(c, v)$, the MP framework defines the *variable freedom* of a variable to be assigned to 1 ($\mathcal{T}$) or 0 ($\mathcal{F}$) as follows.

$$\substack{y}\mathcal{T}_H(v) = \prod_{c \in C_v^-} [1 - \substack{y\\*}\omega_H(c, v)] \quad \substack{y}\mathcal{F}_H(v) = \prod_{c \in C_v^+} [1 - \substack{y\\*}\omega_H(c, v)] \tag{3}$$

The PMPF defines how to use these values to estimate the solution marginal of v, i.e. it defines how to compute the bias $\substack{y}\beta_H(v)$ with $\substack{y}\mathcal{T}_H(v)$ and $\substack{y}\mathcal{F}_H(v)$.

Eqn. 1 will be at the heart of our modifications to the PMPF. Note, that altering this equation will have an impact on the iteration process and will therefore have an impact on the resulting values from Eqn. 3. This in turn will impact the resulting biases. The ultimate purpose of altering Eqn. 1 is to provide promising possibilities for integrating MP with the search paradigm (i.e. CDCL) of the MP-based SAT solver in order to be able to increase its performance.

Arguing about *how well the modifications serve that purpose* is, however, impossible without a theoretical context. As mentioned before, the Ising Model has served as the theoretical basis in the design of the SP framework. However, various concepts from the Ising Model (*forms of agitation*) have been *ignored* in the design of SP, even though these concepts provide strong theoretical insights. The Ising Model furthermore *dictates* the modifications to Eqn. 1 in case we want to harness these insights. This will be discussed further in the next section.

4   The Ising Model and Forms of Agitation

In principle, physicists try to discover fundamental laws that describe our world. One aspect of their work is to derive models that explain properties of condensed matter. Roughly speaking, a condensed-matter object consists of a large number of *sites*. These sites have local properties that can induce a global property of the object. The Ising Model is used to describe and study the interconnections between local site-properties and global object-properties.

For example, we might use a block of iron (the object) and study the iron-atoms' electrons (the sites) w.r.t. a local property (electron spin). A *spin* can be understood as an *angular momentum* (think of a direction that the electron points to in 3D space) [18]. If the spins have a specific "alignment" (local property) the block of iron induces a magnetic field (global property).

Site agitation. The type of matter (iron) implies a connection of the sites (electrons). These connections are called *couplings*. Couplings allow sites to influence each other in their local properties (the spins of the electrons influence each other). This influence is called *site agitation*. One of the fundamental questions that physicists address when studying condensed matter is as follows. *Given a random configuration of the spins and a fixed set of couplings (implied by the matter), can the spins re-arrange (via agitation), such that a stable configuration is reached (in which no further re-arrangement takes place)?* The difficulty of answering this question (and the quality of the answer w.r.t. the real-world) depends on how sophisticated the model is in which we formalize the question.

Intriguingly, this question from physics has a strong connection to the question whether $F \in k\text{-SAT}$. The connection between these two questions can be formalized using the Ising Model [13]. Physics has produced a vast body of literature to answer the above question [2,18]. The Ising Model allows us to translate results from physics into the domain of theoretical computer science, providing us with new techniques to answer whether a given k-CNF formula is in k-SAT.

In the Ising Model, we understand the structure of the matter to be a d-dimensional lattice of n *sites* $x \in \{1, \ldots, n\}$. Furthermore, the *spin* s_x of site x is $s_x \in \{+1, -1\}$ (it points either up or down). The *spin configuration* of all sites is denoted $s = (s_1, \ldots, s_n) \in \{+1, -1\}^n$. Finally, site agitation is confined to neighboring sites, represented by the edges (x, y) in the lattice. The strength of the coupling associated with the edge (x, y) is given as a *constant* $J_{x,y} \in \mathbb{R}$.

In that sense, the condensed-matter object relates to a k-CNF formula F, the lattice structure relates to the factor graph, the sites relate to Boolean variables, the spins to variable assignments, the spin configurations to total assignments, and the couplings relate to literal-occurrences in clauses. In addition, asking for a maximally stable spin configuration w.r.t. the condensed-matter object can be understood as asking for an assignment that satisfies the most clauses of F.

Before we can explain any results from physics (and answer the above question), we must define what "stability of a configuration" means in the Ising Model. In order to do this, physicists use a *Hamiltonian Operator*. The Hamiltonian Operator fulfills two tasks. First, it governs the site agitation, thereby describing how spins influence each other via the couplings. Second, it measures the stability of a configuration under a given set of couplings. A smaller value computed by the Hamiltonian refers to a more stable configuration. Various Hamiltonian Operators have been proposed in the literature. We will, however, focus on the *Edwards-Anderson Hamiltonian* (EAH) given as follows [18].

$$\text{EAH}(s, J, \Phi) = - \left[\sum_{(x,y)} J_{x,y} s_x \cdot s_y + \Phi \sum_x s_x \right]$$

The first sum is over all pairs of nearest neighbors in the lattice. The second sum represents the influence by a uniform external magnetic field in which the matter resides, where $\Phi \in \mathbb{R}$ gives the strength and direction of the magnetic field ($\Phi > 0$ means it "points up" and $\Phi < 0$ means it "points down"). The case $\Phi = 0$ is called the zero-field case. Keep in mind, that the $J_{x,y}$ are constants.

The stability of a spin configuration s is implied by $\mathrm{EAH}(s, J, \Phi)$. We call s more stable than s', iff $\mathrm{EAH}(s, J, \Phi) < \mathrm{EAH}(s', J, \Phi)$. In case $\Phi = 0$, the second sum is zero, and the value computed by the EAH is implied by s alone. During the design of SP, the couplings were restricted to $J_{x,y} \in \{+1, -1\}$. Under this restriction, an increased number of satisfied couplings implies an increased stability. This makes sense when relating the term of "stability" of a spin configuration to the amount of satisfied clauses of a k-CNF formula under a given assignment. However, this restriction is *not required*. Allowing $J_{x,y} \in \mathbb{R}$ as provided by the Ising Model can, as we will later see, be understood as allowing *weighted literal-occurrences* in an MP algorithm. The first extension of the PMPF will introduce literal-weights based on arbitrary coupling-strengths (see the next section).

Magnetic agitation. Note, that a strong positive (negative) magnetic field $\Phi \gg 0$ ($\Phi \ll 0$) will result in a smaller EAH-value as soon as more spins point up (down). A strong magnetic field can be understood to "overrule" the couplings. It forces the spins into the direction of the magnetic field in order for them to realize a stable configuration. We call this influence *magnetic agitation*. In relation to k-SAT, this influence relates to the ability to *prefer* a specific 0-1-ratio in an assignment. The second extension of the PMPF will introduce such 0-1-ratio preferences based on magnetic agitation (see the next section).

Thermal agitation. Given a condensed-matter object in the Ising Model (defined by the couplings J), we might ask the following question. *What happens if a spin configuration does not satisfy all couplings?* Roughly speaking, any object in the real world will try to acquire a state of maximum stability (e.g. liquids cool off, apples fall down, etc.). This, in conjunction with the site agitation governed by the EAH, implies that the spins re-arrange themselves in order to increase stability. In relation to k-SAT, this process of re-arrangement is what Message Passing facilitates by an exchange of messages – the convergence into an equilibrium relates to the re-arrangement of spins into a configuration with maximum stability. The Ising Model allows us to analyze this process in more detail. In order to do so we need to introduce a fundamental equation from physics, called the *partition function* [3].

$$Z(\Upsilon, J, \Phi, n) = \sum_{s \in \{+1, -1\}^n} e^{-T \cdot \mathrm{EAH}(s, J, \Phi)} \text{ , with } T = \frac{1}{b \cdot \Upsilon}$$

Here, b is Boltzmann's constant, and $\Upsilon \in (0, \infty)$ is the temperature of the matter. The partition function allows us to define the *probability function* P, that gives the probability for a specific spin configuration s to be realized by agitation [3].

$$P(s, \Upsilon, J, \Phi, n) = \frac{e^{-T \cdot \mathrm{EAH}(s, J, \Phi)}}{Z(\Upsilon, J, \Phi, n)}$$

Note, that $\lim_{\Upsilon \to \infty} P(s, \Upsilon, J, \Phi, n) = \frac{1}{2^n}$, i.e. all configurations have an equal probability to be realized by agitation under high temperature. Let us now investigate $\lim_{\Upsilon \to 0} P(s, \Upsilon, J, \Phi, n)$. Given the spin configuration s, we want to know how probable its realization by agitation is at zero temperature. We partition the configuration space $\{+1, -1\}^n$ w.r.t. the stability of s as follows.

$M^+(s) = \{s' \in \{+1, -1\}^n | \, \mathrm{EAH}(s, J, \Phi) > \mathrm{EAH}(s', J, \Phi)\}$ (s' is more stable)

$M^-(s) = \{s' \in \{+1, -1\}^n | \, \mathrm{EAH}(s, J, \Phi) < \mathrm{EAH}(s', J, \Phi)\}$ (s' is less stable)

$M^=(s) = \{s' \in \{+1, -1\}^n | \, \mathrm{EAH}(s, J, \Phi) = \mathrm{EAH}(s', J, \Phi)\}$ (s' is equally stable)

Furthermore, and for reasons of simplicity, instead of calculating the limit for P, we will calculate the limit for the (analytically simplified) reciprocal of P, i.e.

$$\lim_{\Upsilon \to 0} P(s, \Upsilon, J, \Phi, n)^{-1} = \sum_{s' \in \{+1, -1\}^n} \lim_{\Upsilon \to 0} e^{[\mathrm{EAH}(s, J, \Phi) - \mathrm{EAH}(s', J, \Phi)]/[b \cdot \Upsilon]}$$

Since $\{+1, -1\}^n$ is partitioned into $M^+(s) \cup M^-(s) \cup M^=(s)$, we can separate the above sum into three disjoint parts for $s' \in M^+(s), s' \in M^-(s)$, or $s' \in M^=(s)$. The three separate limits are then calculated as follows.

$$G^+(s) := \sum_{s' \in M^+(s)} \lim_{\Upsilon \to 0} e^{[H(s, J, \Phi) - H(s', J, \Phi)]/[b \cdot \Upsilon]} = \begin{cases} \infty, & \text{if } M^+(s) \neq \emptyset \\ 0, & \text{otherwise.} \end{cases}$$

$$G^-(s) := \sum_{s' \in M^-(s)} \lim_{\Upsilon \to 0} e^{[\mathrm{EAH}(s, J, \Phi) - \mathrm{EAH}(s', J, \Phi)]/[b \cdot \Upsilon]} = 0 \quad (\text{even if } M^-(s) = \emptyset)$$

$$G^=(s) := \sum_{s' \in M^=(s)} \lim_{\Upsilon \to 0} e^{[\mathrm{EAH}(s, J, \Phi) - \mathrm{EAH}(s', J, \Phi)]/[b \cdot \Upsilon]} = \sum_{s' \in M^=(s)} \lim_{\Upsilon \to 0} e^0 = |M^=(s)|$$

Note, that $s \in M^=(s)$, so $M^=(s) \neq \emptyset$. In summary, for the original limit we get

$$\lim_{\Upsilon \to 0} P(s, \Upsilon, J, \Phi, n) = \frac{1}{G^+(s) + G^-(s) + G^=(s)} \overset{G^-(s)=0}{=} \frac{1}{G^+(s) + |M^=(s)|}.$$

First, assume there is a configuration s' that is more stable than s. Here, $G^+(s) = \infty$ and $\lim_{\Upsilon \to 0} P(s, \Upsilon, J, \Phi, n) = 0$, i.e. the probability for s to be realized by agitation at zero temperature is zero, as soon as a more stable configuration s' exists. Second, assume there is no configuration more stable than s. Here, $G^+(s) = 0$ and $\lim_{\Upsilon \to 0} P(s, \Upsilon, J, \Phi, n) = (|M^=(s)|)^{-1}$, i.e. all most stable configurations are equally likely to be realized by agitation at zero temperature.

Obviously, temperature plays an important role in the realization of spin configurations. It makes perfect sense to assume $\Upsilon \to 0$ from a physics perspective, because this will, *in theory*, result in a spin configuration with maximum stability [16]. The problem is, that the *process* that leads to a spin configuration with maximum stability is not specified. All that P does is to provide a "landscape of probabilities" for the spin configurations, which can be *altered* by Υ.

In practice, we must, however, specify the process that *leads to* a satisfying assignment in the end – the SAT solver's algorithm. Just instantiating a total assignment using the biases will most likely not yield a solution, because biases are estimations of solution marginals (for single variables), not solutions (total assignments). Therefore, a SAT solver that wants to *make use of these biases* must employ some form of search that leads towards a satisfying assignment.

Currently, the process that is supposed to lead to a satisfying assignment in MP-based SAT solvers is Message Passing Inspired Decimation (MID) [7,2]. Here, the MP-based SAT solver uses an MP algorithm to compute biases. Then, it assigns the variable with strongest bias to the implied value. After that, it will have the MP algorithm refine the biases for the remaining variables. This MID is repeated until a satisfying assignment is found (or a conflict occurred). In any case, we can understand MID as a gradient-based approach. The biases reflect the gradient, and the search space landscape is implied by the used MP heuristic in conjunction with the PMPF.

The influence of Υ on the "landscape of probabilities" implied by P (the thermal agitation) is fundamental – the temperature can drastically change the landscape. In contrast, the landscape implied by the MP heuristic in conjunction with the PMPF, which provides the "bias gradient", cannot be influenced at all. Put differently, the concept of thermal agitation is missing. The current PMPF assumes $\Upsilon \to 0$, which, as stated above, makes perfect sense from a physics perspective. In practice, altering the bias-gradient by altering the landscape implied by the application of an MP heuristic in the PMPF *might* improve the performance of an MP-based SAT solver. However, it is currently impossible to study this – due to the lack of thermal agitation in the PMPF. Therefore, the third extension of the PMPF will introduce thermal agitation.

In summary, the Ising Model applies three forms of agitation to describe how spin configurations are realized for a given set of couplings. First, site agitation (governed by the EAH) captures the ability of neighboring sites to influence each other. Second, magnetic agitation (governed by Φ) influences all sites simultaneously and might force their spins into a specific direction. Third, thermal agitation (governed by Υ) influences the probability for a specific configuration to be realized. Currently, the PMPF only uses site agitation with couplings in $\{+1, -1\}$ – it assumes the zero-field case at zero temperature. In what follows, we extended the PMPF s.t. it can apply all forms of agitation described above.

5 Introducing New Forms of Agitation into the PMPF

As stated above, our goal is to extend the PMPF in such a way that it realizes site agitation with arbitrary coupling-strengths, magnetic agitation, as well as thermal agitation. It is, however, important to understand the following fact.

Agitation as described above influences *the stability of configurations*. That is, agitation must influence how the MP algorithm iterates (using δ, ω, U, and S [7])

in order to *influence the equilibrium*. Since the δ equation is defined by an MP heuristic, we cannot use this equation to extend the PMPF. Furthermore, the S and U equations are used as "helpers" to define the δ equations recursively, so we cannot use them either. The only remaining option to introduce the new forms of agitation is a modification of Eqn. 1. All other equations defined by the PMPF take effect *after* convergence and cannot influence the equilibrium.

Site agitation with arbitrary coupling-strength. In principle, the coupling-strength dictates how much influence a coupling between two given sites has on the configuration stability computed by EAH. The larger $J_{x,y}$, the larger the influence. With respect to CNF formulas we can understand the sites as variables and the couplings as literal-occurrences. A coupling-strength in the Ising Model can therefore be understood as a literal-weight in the context of SAT. We introduce the *set of weights* $\Psi = \{\psi_{i,l} | \psi_{i,l} \in [0, \infty), c_i \in F, l \in c_i\}$, i.e. weight $\psi_{i,l}$ is associated with literal l in clause c_i. Altering Eqn. 1 to

$$\omega_H(c_i, v, \Psi) = \prod_{l \in c_i \setminus \{L(v,c_i)\}} {}^{\psi_{i,l}}\!\!\sqrt{\delta_H(l, c_i, \Psi)} \qquad (4)$$

introduces a weight for every literal-occurrence. Recall, that $\delta_H \in [0,1]$. In case $\psi_{i,l} > 1$ ($\psi_{i,l} < 1$), the root will increase (decrease) the factor for the product, thereby increasing (decreasing) the resulting warning. This influences the values computed in Eqn. 3, and ultimately, the biases. In case we set $\psi_{i,l} = 1$ for all literal-occurrences, Eqn. 4 is similar to Eqn. 1. How literal-weights influence an MP algorithm is discussed in the next section.

Magnetic agitation. As mentioned in the previous section, the external magnetic field (represented by Φ) has the ability to *force* spins into the direction of the field. In case $\Phi \gg 0$ (or $\Phi \ll 0$), the spin configuration will stabilize towards $s = (1, \ldots, 1)$ (or $s = (-1, \ldots, -1)$), as this will minimize EAH(s, J, Φ). A site-spin in the Ising Model can be understood as a variable assignment in the context of SAT (where $s_x = +1$ can be understood as $\alpha(v) = 1$). Introducing magnetic agitation into the PMPF should therefore influence the computation of biases, such that $\Phi \gg 0$ ($\Phi \ll 0$) results in ${}^y\beta_H(v) = 1.0$ (${}^y\beta_H(v) = -1.0$) for all $v \in V$. Put differently, a strong positive (negative) magnetic field should force the assignment of all variables to 1 (0). Instead of $\Phi \in \mathbb{R}$, we will use $\Phi \in [-1.0, 1.0]$ in the context of SAT. Furthermore, we define the *literal-sign function* as sgn $: \mathcal{L} \rightarrow \{-1, +1\}, \bar{v} \mapsto -1, v \mapsto 1$. We alter Eqn. 4 as follows.

$$\omega_H(c_i, v, \Psi, \Phi) = (1 - |\Phi|) \left[\prod_{l \in c_i \setminus \{L(v,c_i)\}} {}^{\psi_{i,l}}\!\!\sqrt{\delta_H(l, c_i, \Psi, \Phi)} \right] + \frac{|\Phi| + \text{sgn}(L(v,c_i)) \cdot \Phi}{2} \qquad (5)$$

In the zero-field case ($\Phi = 0$) this equation is similar to Eqn. 4. Any setting $\Phi \neq 0$ can either increase a warning (if the literal and Φ have the same sign), or decrease the warning (if the literal and Φ have different signs). How the magnetic field influences the MP algorithm will be discussed in the next section.

Thermal agitation. As mentioned in the previous section, the temperature Υ influences the probability landscape of all configurations implied by P. The first extremal case is $\Upsilon \to 0$, which is the case used in the original PMPF. The second extremal case is $\Upsilon \to \infty$, which results in equal probabilities for all configurations to be realized by agitation. Based on this, we will introduce thermal agitation into the PMPF. However, instead of using $\Upsilon \in (0, \infty)$, we define $\Upsilon \in [0, 1]$ in the context of SAT. We alter Eqn. 5 to $\omega_\mathrm{H}(c_i, v, \Psi, \Phi, \Upsilon) =$

$$(1 - \Upsilon)\left\{ (1 - |\Phi|) \left[\prod_{l \in c_i \setminus \{L(v, c_i)\}} \!\!\! {}^{\psi_{i,l}}\!\!\sqrt{\delta_\mathrm{H}(l, c_i, \Psi, \Phi, \Upsilon)} \right] + \frac{|\Phi| + \mathrm{sgn}(L(v, c_i)) \cdot \Phi}{2} \right\} \quad (6)$$

In the zero-temperature case ($\Upsilon = 0$) this equation is similar to Eqn. 5. Setting $\Upsilon = 1$ will result in $\omega_\mathrm{H}(c_i, v, \Psi, \Phi, 1) = 0$ for *all* warning messages. How temperature influences the MP algorithm is discussed in the next section.

Replacing Eqn. 1 in the PMPF with Eqn. 6 results in the new ePMPF.

6   Theoretical and Practical Applications of the ePMPF

In this section, we will use the new forms of agitation in the ePMPF (facilitated by Ψ, Φ, and Υ) to show, that several well-known heuristics used in SAT solving (occurrence-based heuristics, Zero-first, One-first) can be understood as special cases of Message Passing. Additionally, we will use the ePMPF to provide several possibilities to tightly integrate MP with CDCL. In summary, this corroborates that the ePMPF is a useful extension of the PMPF in both theory and practice.

The influence of Ψ on Message Passing from a theoretical perspective. We will first analyze the influence of literal-weights Ψ on the behavior of an MP algorithm with $\Phi = \Upsilon = 0$ (zero-field, zero-temperature). The extremal cases are the infinity-limit $\psi_{i,l} \to \infty$ for all occurrences (abbr. $\Psi \to \infty$), and the zero-limit $\psi_{i,l} \to 0$ for all occurrences (abbr. $\Psi \to 0$). We assume similar weights for all occurrences, i.e. for $c_i, c_j \in F$ and $l \in c_i, l' \in c_j$, it is $\psi_{i,l} = \psi_{j,l'}$ for all i, j, l, l'. For the remaining section, keep in mind that $s(l), u(l), t(v), f(v) \geq 2$ (Sec. 2).

We first investigate literal-weights in the infinity-limit. Assume, that the MP algorithm uses the MP heuristic $\rho\sigma\mathrm{PMP}^i$ [7] with constant interpolation parameters $\rho \in [0, 1]$ and $\sigma \in (0, 1]$. Additionally, we assume non-extremal initializations of the δ-messages in iteration $z = 0$, i.e. ${}^y_0\delta_{\rho\sigma\mathrm{PMP}}(l, c_i, \rho, \sigma, \Psi, 0, 0) \in (0, 1)$. Then

$$\lim_{\Psi \to \infty} {}^y_0\omega_{\rho\sigma\mathrm{PMP}}(c_i, v, \rho, \sigma, \Psi, 0, 0) = \lim_{\Psi \to \infty} \prod_{l \in c_i \setminus \{L(v, c_i)\}} \!\!\! {}^{\psi_{i,l}}\!\!\sqrt{{}^y_0\delta_{\rho\sigma\mathrm{PMP}}(l, c_i, \rho, \sigma, \Psi, 0, 0)} = 1.$$

Using Eqn. 2 gives ${}^y_0 S_{\rho\sigma\mathrm{PMP}}(l, c_i, \rho, \sigma, \Psi, 0, 0) = {}^y_0 U_{\rho\sigma\mathrm{PMP}}(l, c_i, \rho, \sigma, \Psi, 0, 0) = 0$ ($\Psi \to \infty$) for all literals and clauses in F. These values give the δ-messages for the next iteration $z = 1$. Since $s(l), u(l), t(v), f(v) \geq 2$ we get

$$\lim_{\Psi \to \infty} {}^y_1 \delta_{\rho\sigma\text{PMP}}(l, c_i, \rho, \sigma, \Psi, 0, 0) = \frac{s(l)}{s(l) + u(l)} > 0. \tag{7}$$

Therefore, $\lim_{\Psi \to \infty} {}^y_1 \omega_{\rho\sigma\text{PMP}}(c_i, v, \rho, \sigma, \Psi, 0, 0) = 1$ also holds, i.e. the MP algorithm converges in iteration $z = 1$. With Eqn. 3 we get ${}^y\mathcal{T}_{\rho\sigma\text{PMP}}(v, \rho, \sigma, \Psi, 0, 0)$ $= {}^y\mathcal{F}_{\rho\sigma\text{PMP}}(v, \rho, \sigma, \Psi, 0, 0) = 0$ ($\Psi \to \infty$). The biases in the infinity-limit are

$$\lim_{\Psi \to \infty} {}^y\beta_{\rho\sigma\text{PMP}}(v, \rho, \sigma, \Psi, 0, 0) = \frac{f(v) - t(v)}{f(v) + t(v)} = \frac{|C_v^+| - |C_v^-|}{|C_v|} \quad \rho \in [0,1], \sigma \in (0,1].$$

In case $\sigma > 0$, the MP algorithm (using $\rho\sigma\text{PMP}^i$) computes the "standard" literal occurrence heuristic in the infinity-limit for Ψ. The variable ordering prefers variables with stronger occurrence-imbalances. The value ordering prefers the value that satisfies more clauses (it acts *conflict avoiding*).

In case $\sigma = 0$, the above approach fails (Eqn. 7 does not hold). Here, the $\rho\sigma\text{PMP}^i$ heuristic collapses to ρSP^i [7] and it is unclear what the result of $\lim_{\Psi \to \infty} {}^y_1 \delta_{\rho\text{SP}}(l, c_i, \rho, \Psi, 0, 0)$ is. Calculating the behavior of ρSP^i (including BP and SP) in the infinity-limit for literal-weights remains an open problem.

In contrast, with $\Psi \to 0$ we get $\lim_{\Psi \to 0} {}^y_0 \omega_{\rho\sigma\text{PMP}}(c_i, v, \rho, \sigma, \Psi, 0, 0) = 0$. This results in ${}^y_0 S_{\rho\sigma\text{PMP}}(l, c_i, \rho, \sigma, \Psi, 0, 0) = {}^y_0 U_{\rho\sigma\text{PMP}}(l, c_i, \rho, \sigma, \Psi, 0, 0) = 1$ ($\Psi \to 0$). Since $u(l), s(l), t(v), f(v) \geq 2$, we can use arbitrary $\rho, \sigma \in [0,1]$ to show that

$$\lim_{\Psi \to 0} {}^y_1 \delta_{\rho\sigma\text{PMP}}(l, c_i, \rho, \sigma, \Psi, 0, 0) = \frac{1 - \rho + \sigma[u(l) + s(l) - 1 - \rho(s(l) - 1)]}{(1 - \sigma)(2 - \rho) + 2\sigma[u(l) + s(l)]} < 1.$$

Therefore, $\lim_{\Psi \to 0} {}^y_1 \omega_{\rho\sigma\text{PMP}}(c_i, v, \rho, \sigma, \Psi, 0, 0) = 0$ also holds, i.e. the MP algorithm converges in iteration $z = 1$. This in turn gives ${}^y\mathcal{T}_{\rho\sigma\text{PMP}}(v, \rho, \sigma, \Psi, 0, 0) = {}^y\mathcal{F}_{\rho\sigma\text{PMP}}(v, \rho, \sigma, \Psi, 0, 0) = 1$ ($\Psi \to 0$), and $\lim_{\Psi \to 0} {}^y\beta_{\rho\sigma\text{PMP}}(v, \rho, \sigma, \Psi, 0, 0) =$

$$\frac{\rho\sigma[t(v) - f(v)]}{(1 - \sigma)(2 - \rho) + 2(t(v) + f(v))} \overset{\rho=1}{\underset{\sigma=1}{=\!=}} \frac{|C_v^-| - |C_v^+|}{2|C_v|} \overset{(\star)}{\approx} \frac{|C_v^-| - |C_v^+|}{|C_v|}$$

In case $\rho = \sigma = 1$ and after re-scaling all biases by a factor of 2 ($\star$), the MP algorithm (using $\rho\sigma\text{PMP}^i$) computes a literal occurrence heuristic in the zero-limit for Ψ as well. Again, the variable ordering prefers variables with stronger occurrence-imbalances. In contrast to the infinity-limit case, the value ordering now prefers values that satisfy *less* clauses (it acts *conflict seeking*).

In summary, the extremal cases for Ψ result in biases that reflect the "standard" literal occurrence heuristics (conflict avoiding with $\sigma > 0$ and $\Psi \to \infty$, and conflict seeking with $\rho = \sigma = 1$ and $\Psi \to 0$). Note, that intermediate settings to Ψ allow for a *seamless adjustment* of the MP algorithm's behavior. Roughly speaking, Ψ allows us to seamlessly control whether we want to avoid or seek conflicts during MID in the search of an MP-based SAT solver.

The influence of Ψ in CDCL-based MID from a practical perspective.
In practice, CDCL solvers [15] use heuristics to influence how search advances, but search might also *provide feedback to the heuristic*. For example, the VSIDS heuristic [17] uses variable activities to compute a variable ordering according

to which the variables are assigned during search. However, if the CDCL runs into a conflict it performs conflict analysis which will (among other things) *alter* the variable activities. Hence, advancing the search also influences VSIDS. Such bidirectional influence is crucial for the performance of CDCL solvers. How strong the feedback is (e.g. for VSIDS) is controlled by a parameter of the solver.

In contrast, the search of a CDCL-based MID solver, that uses the original PMPF, *cannot provide feedback to the MP algorithm* – the PMPF lacks the necessary flexibility. However, the ePMPF allows the search to provide feedback by altering the literal-weights in Ψ. In what follows, we will briefly discuss how such feedback can be used to integrate MP with various aspects of the CDCL search.

Assume, that the formula F is satisfiable and that we use a CDCL-based MID solver (using the ePMPF) to solve it. Ideally, following the bias-gradient during MID (extending the decision stack) results in a satisfying assignment α. However, in practice, the CDCL will run into conflicts, i.e. $\exists c_i \in F : \alpha(c_i) = 0$ during search. Conflict analysis (CA) resolves the conflict and adds a new learned clause d to F (implying a back-jump) [19]. However, adding a single (possibly very long) clause will most likely have no influence on the biases. Arguably, the impact that c_i had on the bias computation (that led to α) was not strong enough. This implies, that we must increase the weights $\psi_{i,l}$ of literals in c_i for future computations. Additionally, we can also increase the weights of literals in clauses that act as reasons in the conflict graph (similar to the activity increase of variables in VSIDS). Note, that we can understand clause learning (adding d) as empowering unit propagation (UP) to avoid previously discovered conflicts. Similarly, we can understand weight increases as empowering MP to provide biases that avoid previously discovered conflicts. The CDCL-based MID solver can do both, which constitutes a tight integration of CA, UP, and MP via Ψ.

Additionally, CDCL solvers perform clause database maintenance (CDBM) that removes learned clauses from the formula. Various heuristics for CDBM exist, e.g. the activity-based CDBM heuristic that associates an activity value with every learned clause [6]. During conflict analysis, the activity of learned clauses is increased in case they help to resolve the conflict. Then, maintenance will remove clauses that have a relatively small activity. This in turn increases the performance of UP (less clauses must be checked). In that sense, activity-based CDBM can be understood as a tool to improve the performance of UP. Instead of using activity values for the clauses, we might rely on the $\psi_{i,l}$ values to perform the same scheme of CDBM. Here, removing learned clauses with small literal-weights can be understood as improving the performance of UP *and* MP. This constitutes a tight integration of CDBM with UP and MP via Ψ.

Altering literal-weights during search will, eventually, result in biases (using the updated Ψ) that strongly differ from the partial assignment on the decision stack of the CDCL-based MID solver. Arguably, reverting all decisions and restarting search with the new biases increases the chances for MID to find a solution. In that sense, we can also integrate restarts with MP via Ψ.

The influence of Φ on Message Passing from a theoretical perspective. We now analyze the influence of the magnetic field parameter Φ on the behavior of an MP algorithm with $\Upsilon = 0$ and $\Psi = 1$ (zero-temperature, unit-weights). The first extremal case is $\Phi = -1$. Using Eqn. 6 in iteration $z = 0$ results in two possible warnings. First, $_{0}^{y}\omega_{\rho\sigma\text{PMP}}(c_i, v, \rho, \sigma, 1, -1, 0) = 0$ iff $l = v$. Second, $_{0}^{y}\omega_{\rho\sigma\text{PMP}}(c_i, v, \rho, \sigma, 1, -1, 0) = 1$ iff $l = \bar{v}$. Note, that these results are *independent* of the δ messages. Hence, the same warnings will be computed in iteration $z = 1$ and the MP algorithm converges. Using Eqn. 3 gives us $^{y}\mathcal{T}_{\rho\sigma\text{PMP}}(v, \rho, \sigma, 1, -1, 0) = 0$ and $^{y}\mathcal{F}_{\rho\sigma\text{PMP}}(v, \rho, \sigma, 1, -1, 0) = 1$. Therefore, $^{y}\beta_{\rho\sigma\text{PMP}}(v, \rho, \sigma, 1, -1, 0) = -1.0 \, \forall v \in \mathcal{V}$. Put differently, the MP algorithm provides biases that reflect the Zero-first heuristic in case $\Phi = -1$. Similarly, we can show that the MP algorithm provides biases that reflect the One-first value ordering heuristic with $\Phi = +1$. In analogy to the Ising Model, it is possible to use $\Phi \in [-1, +1]$ in order to force the variable biases into a specific direction. This allows us to *seamlessly* influence the MP algorithm to prefer variable biases with a specific value, i.e. we can "ask" for biases that reflect a specific 0-1-ratio.

The influence of Φ in CDCL-based MID from a practical perspective.
Assume, that F is satisfiable and that we use a CDCL-based MID solver (that performs MP according to the ePMPF) to solve it. Assume furthermore, that we know the 0-1-ratio of solutions in advance.

The assumption, that we know the 0-1-ratios of solutions in advance might seem very artificial. However, there are problems that, when encoded to SAT, allow us to make such assumptions. For example, consider the Graph Isomorphism (GI) problem of two graphs with $n \geq 1$ nodes. Translating the problem to SAT can be achieved in a trivial way by representing a mapping with n^2 variables, where $v_{i,j} = 1$ iff node i is mapped to node j. The formula can be constructed s.t. it is satisfiable iff an isomorphism between the two graphs exists. In this case, the solution must assign exactly n variables to 1, and $n^2 - n$ variable to 0, respectively. Therefore, we know the 0-1-ratio of solutions in case they exist. We do, however, not know which variable must be assigned to what value – which is precisely what the CDCL-based MID solver must determine.

The setting to the field parameter Φ, that is implied by the observations above, is $\Phi = (2/n) - 1 \in [-1, +1]$. Using this will force the MP algorithm into an equilibrium that prefers biases with the desired 0-1-ratio. Arguably, these biases will be more helpful to find a satisfying assignment with the CDCL-based MID solver. Note, that $\lim_{n \to \infty}(2/n) - 1 = -1$, i.e. the Zero-first heuristic should be used in case the graphs become very large in this trivial encoding.

The influence of Υ on Message Passing from a theoretical perspective.
We will now analyze the influence of the temperature parameter Υ on the behavior of an MP algorithm with $\Phi = 0$ and $\Psi = 1$ (zero-field, unit-weights). First, note that $\Upsilon = 1$ results in $_{0}^{y}\omega_{\rho\sigma\text{PMP}}(c_i, v, \rho, \sigma, 1, 0, 1) = 0$ for all literals and clauses. Therefore, the MP algorithm converges in iteration $z = 1$ with $^{y}\mathcal{T}_{\rho\sigma\text{PMP}}(v, \rho, \sigma, 1, 0, 1) = {}^{y}\mathcal{F}_{\rho\sigma\text{PMP}}(v, \rho, \sigma, 1, 0, 1) = 1$. Finally, we get $\forall v \in \mathcal{V}$: $^{y}\beta_{\rho\sigma\text{PMP}}(v, \rho, \sigma, 1, 0, 1) = 0.0$. This means, that under high temperature, all variables are unbiased and all assignments are estimated to be satisfying with equal probability. This is what $\lim_{\Upsilon \to \infty} P(s, \Upsilon, J, \Phi, n) = \frac{1}{2^n}$ means in physics.

A setting to $\Upsilon \in (0,1]$ allows us to enforce the upper boundary $1 - \Upsilon \in [0,1)$ for the warnings. This in turn enforces that $\mathcal{T}, \mathcal{F} > 0$. Therefore, we can alter the biases, and ultimately, the bias-gradient used by MID solvers.

Let $\gamma \in (0,1]$ s.t. $\gamma_{max} = \gamma^{-1} \in \mathbb{N}$ (e.g. $\gamma = 0.1$ which gives $\gamma_{max} = 10$). We can use γ to determine a sequence of temperatures $\Gamma = (0, \gamma, 2\gamma, \ldots, 1) = (t\gamma)_{t=0}^{\gamma_{max}}$. We can perform MP sequentially using $\Upsilon \in \{t\gamma | t\gamma \in \Gamma\}$. Due to the above, we know that in general $\Upsilon = 0\gamma = 0$ gives ${}^y\beta_{\rho\sigma\mathrm{PMP}}(v, \rho, \sigma, 1, 0, 0) \neq 0 \; \forall v \in \mathcal{V}$. We also know, that $\Upsilon = \gamma_{max}\gamma = 1$ gives ${}^y\beta_{\rho\sigma\mathrm{PMP}}(v, \rho, \sigma, 1, 0, 1) = 0 \; \forall v \in \mathcal{V}$. Put differently, the biases decay when we increase the temperature. However, note that biases do not necessarily decay equally fast. Some variables will have a bias even for large Υ, while others will be unbiased. More formally, let $\epsilon, \gamma \in [0,1]$. We define the $\epsilon\gamma$-*heat resistance of variable* v as $\max t\gamma$ s.t. $|{}^y\beta_{\mathrm{H}}(v, \rho, \sigma, 1, 0, t\gamma)| > \epsilon$ when running MP. Put differently, the heat resistance is the largest temperature for v, s.t. its bias has an absolute value larger ϵ (i.e. it is still "strongly" biased w.r.t. ϵ). Arguably, the larger the heat resistance of a variable, the more stability is implied for its bias. This can be used as follows.

The influence of Υ in CDCL-based MID from a practical perspective. Empirical observations (e.g. in [8]) suggest, that the CDCL search paradigm cannot be used to solve large uniform random 3-SAT formulas with clauses-to-variable ratios above 4.2. However, [8] showed that it is possible to use a CDCL-based MID solver to do this. It was observed that, on the one hand, the solver can find a satisfying assignment without running into a single conflict. On the other hand, the solver will time-out in case it runs into a conflict. The case in which the CDCL-based MID solver resolves a conflict and finds a solution afterwards never occurred in practice. In summary we can conclude, that, when solving uniform random 3-SAT formulas with a CDCL-based MID solver, it is beneficial to use more computational time in order to make *correct decisions*, instead of using more computational time in order to *fix mistakes*.

Using Υ to compute the $\epsilon\gamma$-heat resistance for variables (which can be costly for small γ and ϵ) gives an indication which variables should be assigned first (those with largest heat resistance) and into what direction we should assign them (implied by the bias at the given temperature). Arguably, performing MID by following the most stable biases should enable the CDCL-based MID solver to avoid conflicts. In the above scenario this should help the solver to avoid the case in which it times out – even though advancing the search will be slower.

7   Conclusions and Future Work

The first contribution of this paper was to extend the product-based MP framework (PMPF) for SAT. The extensions, resulting in the ePMPF, are dictated by the Ising Model and facilitate forms of agitation motivated by condensed-matter physics. The ePMPF allows us to draw analytical conclusions and provide suggestions for the integration of MP algorithms with the CDCL search paradigm.

More precisely, we showed that several well-known heuristics used in SAT solving (e.g. Zero-First, One-first, conflict avoiding/seeking occurrence heuristics) can be understood as special cases of an MP algorithm using the ePMPF.

Additionally, the ePMPF also allows us to provide (theoretically motivated) possibilities to integrate an MP algorithm with CDCL search. More precisely, we argued that the search of a CDCL-based MID solver (using the ePMPF) can provide feedback to MP with the information gathered during search (by using literal-weights in Ψ). Additionally, it can harness a-priori knowledge about 0-1-ratio of solutions (by preferring bias directions with Φ). Even further, it can determine the most stable variable biases (by computing the $\epsilon\gamma$-heat resistance using Υ) in order to focus on making *correct decisions* instead of *fixing mistakes*.

Future work must now investigate the performance of such a CDCL-based MID solver empirically. We need to assess the feasibility of (the theoretically motivated) possibilities to integrate MP with CDCL (using Ψ), the possibilities to harness a-priori knowledge about 0-1-ratios of solutions (using Φ), as well as the option to compute the most heat-resistant variables during MP (using Υ).

References

1. Battaglia, D., Kolář, M., Zecchina, R.: Minimizing energy below the glass thresholds. Physical Review E 70, 036107 (2004)
2. Braunstein, A., Mézard, M., Zecchina, R.: Survey Propagation: An Algorithm for Satisfiability. Journal of Rand. Struct. and Algo., 201 (2005)
3. Cipra, B.A.: An Introduction to the Ising Model. The American Mathematical Monthly 94(10), 937 (1987)
4. Cook, S.: The complexity of theorem-proving procedures. In: Symposium on the Theory of Computing (STOC 1971), p. 151. ACM, New York (1971)
5. Dempster, A., Laird, N., Rubin, D.: Maximum Likelihood from Incomplete Data via the EM Algorithm. Jrn. o. t. Royal Statistical Society, Series B 39, 1 (1977)
6. Eén, N., Sörensson, N.: An extensible SAT solver. In: Giunchiglia, E., Tacchella, A. (eds.) SAT 2003. LNCS, vol. 2919, pp. 502–518. Springer, Heidelberg (2004)
7. Gableske, O.: On the Interpolation between Product-Based Message Passing Heuristics for SAT. In: Järvisalo, M., Van Gelder, A. (eds.) SAT 2013. LNCS, vol. 7962, pp. 293–308. Springer, Heidelberg (2013)
8. Gableske, O., Müelich, S., Diepold, D.: On the Performance of CDCL based Message Passing Inspired Decimation using $\rho\sigma PMP^i$. In: Pragmatics of SAT, POS 2013 (2013)
9. Hsu, E.I., McIlraith, S.A.: VARSAT: Integrating Novel Probabilistic Inference Techniques with DPLL Search. In: Kullmann, O. (ed.) SAT 2009. LNCS, vol. 5584, pp. 377–390. Springer, Heidelberg (2009)
10. Hsu, E.I., McIlraith, S.A.: Characterizing Propagation Methods for Boolean Satisfiability. In: Biere, A., Gomes, C.P. (eds.) SAT 2006. LNCS, vol. 4121, pp. 325–338. Springer, Heidelberg (2006)
11. Hsu, E.I., Muise, C.J., Beck, J.C., McIlraith, S.A.: Probabilistically Estimating Backbones and Variable Bias: Experimental Overview. In: Stuckey, P.J. (ed.) CP 2008. LNCS, vol. 5202, pp. 613–617. Springer, Heidelberg (2008)
12. Hsu, E., Kitching, M., Bacchus, F., McIlraith, S.: Using Expectation Maximization to Find Likely Assignments for Solving CSPs. In: Conference on Artificial Intelligence (AAAI 2007), p. 224. AAAI Press (2007)

13. Ising, E.: Beitrag zur Theorie des Ferromagnetismus. Zeitschrift für Physik, Nummer 31, 253 (1925)
14. Schischang, K., Frey, B., Loeliger, H.: Factor Graphs and the sum-product algorithm. In: IEEE Transactions on Information Theory (IT 2001), vol. 47, p. 498. IEEE (2002)
15. Marques-Silva, J.P., Sakallah, K.A.: GRASP: A Search Algorithm for Propositional Satisfiability. IEEE Transactions on Computers (C 1999) 48, 506 (1999)
16. Mézard, M., Parisi, G.: The cavity method at zero temperature. Journal of Statistical Physics 111(1/2) (April 2003)
17. Moskewicz, M., Madigan, C., Zhao, Y., Zhang, L., Malik, S.: Chaff: Engineering an Efficient SAT Solver. In: Design Automation Conf. (DAC 2001), p. 530. ACM (2001)
18. Stein, D., Newman, C.: Spin Glasses and Complexity. Princeton Press (2013) ISBN 978-0-691-14733-8
19. Zhang, L., Madigan, C., Moskewicz, M., Malik, S.: Efficient Conflict Driven Learning in a Boolean Satisfiability Solver. In: International Conference on Computer Aided Design (ICCAD 2001), p. 279 (2001)

Approximating Highly Satisfiable Random 2-SAT

Andrei A. Bulatov and Cong Wang

Simon Fraser University
8888 University Drive, Burnaby BC, V5A 1S6, Canada
{abulatov,cwa9}@sfu.ca

Abstract. In this paper we introduce two distributions for the Max-2-SAT problem similar to the uniform distribution of satisfiable CNFs and the planted distribution for the decision version of SAT. In both cases there is a parameter p, $0 \le p \le \frac{1}{4}d$, such that formulas chosen according to both distributions are p-satisfiable, that is, at least $(\frac{3}{4}d + p)n$ clauses can be satisfied. In the planted distribution this happens for a fixed assignment, while for the p-satisfiable distribution formulas are chosen uniformly from the set of all p-satisfiable formulas. Following Coja-Oghlan, Krivelevich, and Vilenchik (2007) we establish a connection between the probabilities of events under the two distributions. Then we consider the case when p is sufficiently large, $p = \gamma\sqrt{d \log d}$ and $\gamma > 2\sqrt{2}$. We present an algorithm that in the case of the planted distribution for any ε with high probability finds an assignment satisfying at least $(\frac{3}{4}d + p - \varepsilon)n$ clauses. For the p-satisfiable distribution for every d there is $\varepsilon(d)$ (which is a polynomial in d of degree depending on γ) such that the algorithm with high probability finds an assignment satisfying at least $(\frac{3}{4}d + p - \varepsilon(d))n$ clauses. It does not seem this algorithm can be converted into an expected polynomial time algorithm finding a p-satisfying assignment. Also we use the connection between the planted and uniform p-satisfiable distributions to evaluate the number of clauses satisfiable in a random (not p-satisfiable) 2-CNF. We find the expectation of this number, but do not improve the existing concentration results.

1 Introduction

The random SAT and Max-SAT problems have received quite a bit of attention in the last decades. In this paper we work with one of the most popular distributions used for these problems, $\Phi_k(n, dn)$, the uniform one of k-CNFs of fixed density d: Fix n and d (the number of variables, and the density, it may depend on n), and choose dn clauses uniformly at random out of $2^k \binom{n}{k}$ possible ones. It was originally suggested as a source of problems hard for SAT-solvers [18]. On the other hand, instances sampled from this distribution may, with high probability (whp), behave better than in the worst case, which opens up an interesting line of research. For instance, by [19,14] whp Random Max-3-SAT can be approximated within a factor 1.0957, a marked improvement over Håstad's worst case inapproximability bound ($\frac{22}{21} - \varepsilon$ and $\frac{8}{7} - \varepsilon$ for Max-2-SAT and Max-3-SAT, respectively, see [3,13]). However, it is not known if this bound can be further improved. Thus, in spite of large amount of research done on this model many questions remain open [9].

The phase transition phenomenon [11] also led to the study of other distributions, biased towards 'more satisfiable' instances. The most interesting of them is $\Phi_k^{sat}(n, dn)$

C. Sinz and U. Egly (Eds.): SAT 2014, LNCS 8561, pp. 384–398, 2014.
© Springer International Publishing Switzerland 2014

the uniform distribution of satisfiable formulas from $\Phi_k(n, dn)$. However, this distribution is difficult to sample, and a significant amount of attention has been paid to a related planted distribution $\Phi_k^{pl}(n, dn)$: First, choose a random assignment f of n Boolean variables, and then choose dn k-clauses satisfied by f uniformly at random from the $(2^k - 1)\binom{n}{k}$ possible ones. An important property of both satisfiable and planted distributions is that if d is large enough, they demonstrate very high concentration of satisfying assignments. All such assignments have very small distance from each other [17,6]. This property has been used to solve instances from such distributions efficiently. In [10,17] different techniques were used to solve whp planted instances provided d is sufficiently large. The technique from [17] was later generalized in [6,5], where an algorithm was suggested that solves instances from $\Phi_k^{sat}(n, dn)$ in expected polynomial time. One of the improvements made in the latter paper and the one most relevant for this paper is a link between the probability of arbitrary events in the planted and satisfiable models. This allowed to transfer the methods developed in [17] to the more general satisfiable distribution.

While the decision random SAT has been intensively studied, its optimization counterpart, random Max-SAT has received substantially less attention. In this paper we focus on the random Max-2-SAT problem. While 2-SAT is solvable in polynomial time, as we mentioned before, Max-2-SAT is hard to approximate within the factor $\frac{22}{21} - \varepsilon$ [3,13]. The uniform Max-2-SAT $\Phi_2(n, dn)$ (as from now on we only deal with Max-2-SAT, we will denote this distribution simply by $\Phi(n, dn)$) demonstrates phase transition at density $d = 1$. If density is greater than 1, whp $\varphi \in \Phi(n, dn)$ cannot be satisfied, but a random assignment satisfies the expected 3/4 of all clauses. As it is shown in [8], the number of clauses that can be satisfied whp lies within the interval $[\frac{3}{4}dn + 0.34\sqrt{dn}, \frac{3}{4}dn + 0.51\sqrt{dn}]$, provided d is sufficiently large. For MAX-k-SAT, $k > 2$, the similar interval is shown to be much narrower. Achlioptas et al. [1] proved that for any $k \geq 2$ and any $d > 2^k \log 2$, for a random k-CNF from $\Phi_k(n, dn)$ the number of clauses that can be satisfied lies in the interval $[1 - 2^{-k} + p(k, d) - \delta_k, 1 - 2^{-k} + p(k, d)]$, where $p(k, d) = 2^k \Psi\left(\frac{2^k \log 2}{d}\right)$ and $\delta_k = O(k2^{-k/2})$. Note that $d = 2^k \log 2$ is the phase transition threshold for random k-SAT [2]. Thus the length of the interval decreases exponentially as k grows; however, it gives a weaker bound for $k = 2$ than that from [8].

We first introduce two distributions of Max-2-SAT formulas similar to the satisfiable and planted distributions for the decision problem. An assignment f of variables of a formula $\varphi \in \Phi(n, dn)$ is said to be p-satisfying, $0 \leq p \leq \frac{d}{4}$, if it satisfies $\left(\frac{3}{4}d + p\right)n$ clauses. Formula φ is called p-satisfiable if it has a p'-satisfying assignment for some $p' \geq p$. By $\Phi^*(n, dn, p)$ we denote the uniform distribution of p-satisfiable formulas. Similar to the satisfiable distribution Φ^{sat}, this distribution is difficult to sample. Therefore we also introduce the planted approximation of $\Phi^*(n, dn, p)$ as follows: Choose a random assignment f to the n variables, and then choose uniformly at random $\left(\frac{3}{4}d + p\right)n$ clauses satisfied by f and $\left(\frac{1}{4}d - p\right)n$ clauses that are not satisfied by f.

Our first result is similar to the 'Exchange Rate' lemma from [6]. More precisely, if $\mathcal{P}_d^u(\mathcal{A}), \mathcal{P}_{d,p}^{pl}(\mathcal{A}), \mathcal{P}_{d,p}^{u*}(\mathcal{A})$ denote the probabilities of event $\mathcal{A}$ under the uniform distribution, the planted distribution of p-satisfiable formulas, and the uniform distribution of p-satisfiable formulas then the following holds.

Proposition 1. *There exists a function $\xi(d, p, n)$, $\xi(d, p, n) = e^{\Theta(n)}$ for all d, p, such that for any p_0, $0 \leq p_0 \leq \frac{1}{4}d$, and any event $\mathcal{A} \subseteq \Phi(n, dn)$ there is p with $p_0 \leq p \leq \frac{3}{4}d$ such that*

$$\mathcal{P}_d^u(\mathcal{A}) \leq \mathcal{P}_d^u(C_{p_0}) + \xi(d, p, n)\mathcal{P}_{d,p}^{pl}(\mathcal{A}),$$

where C_{p_0} is the set of all formulas from $\mathcal{A}$ that are not p_0-satisfiable. In particular,

$$\mathcal{P}_{d,p_0}^{u*}(\mathcal{A}) \leq \xi(d, p, n)\mathcal{P}_{d,p}^{pl}(\mathcal{A}).$$

As is shown in [6], satisfiable random 3-SAT formulas possess a very nice structure that whp allows one to find a satisfying assignment for formulas, provided the density is sufficiently large. We show that it suffices to assume sufficiently high level of satisfiability to infer a similar result. The level of satisfiability that allows our proofs to go through is $2\sqrt{2}\sqrt{d \log d}$. Throughout the paper log denotes the natural logarithm. More precisely for the planted distribution we prove the following

Theorem 2. *There is a polynomial time algorithm that for any $\varepsilon > 0$ and any $p > \gamma\sqrt{d \log d}$, $\gamma > 2\sqrt{2}$ (γ does not have to be a constant), where d is sufficiently large, given $\varphi \in \Phi^{pl}(n, dn, p)$ whp finds an at least $(p - \varepsilon)$-satisfying assignment.*

Observe that the probability that the algorithm succeeds does not necessarily go to 1 uniformly over ε. This theorem can be restated in terms of approximability: For any $\beta = \frac{3}{4}d + \gamma\sqrt{d \log d}$, $\gamma > 2\sqrt{2}$, and any $\varepsilon > 0$ there is a $(\beta - \varepsilon, \beta)$-approximation algorithm for $\Phi^{pl}(n, dn, p)$ that succeeds whp. That is, an algorithm that given an instance in which at least β-fraction of clauses are satisfiable whp returns an assignment that satisfies at least $\beta - \varepsilon$-fraction of clauses.

Using the 'Exchange Rate' proposition we then can transfer this result to the uniform distribution

Theorem 3. *There is an algorithm that for every $p > \gamma\sqrt{d \log d}$, $\gamma > 2\sqrt{2}$, given $\varphi \in \Phi^*(n, dn, p)$ finds a p'-satisfying assignment, $p' \geq p - \varepsilon + o(\varepsilon)$, where $\varepsilon = \frac{1}{2}d^{1 - \frac{4\gamma^2}{9}} + \sqrt{6}d^{\frac{1}{2} - \frac{2\gamma^2}{9}}$.*

In both cases we use the same algorithm, very similar to those from [17,6,5,7], although the analysis is somewhat different, in particular, it has to be tighter.

Finally, we make an attempt to estimate the range $[p_1, p_2]$ such that whp a formula from $\Phi(n, dn)$ is p_1-satisfiable, but not p_2-satisfiable. As was mentioned before, the best result known to date is $[0.34\sqrt{d}, 0.51\sqrt{d}]$ [8]. Although we were unable to improve upon this result, we use a careful analysis of the majority vote algorithm and the 'Exchange Rate' proposition to find the likely location of the threshold for the level of satisfiability. It turns out to be very close to $0.5\sqrt{d}$.

2 Preliminaries

2.1 Random Max-2-SAT

We denote the four possible types of 2-clauses (depending on the polarity of literals) by $(+, +), (+, -), (-, +), (-, -)$. The distributions of 2-CNFs we consider are the following:

- The uniform distribution $\Phi(n, dn)$ with n variables and dn 2-clauses, d constant. Whether or not repetitions of variables in a clause or repetitions of clauses are allowed is immaterial. A slightly different distribution is $\Phi(n, \varrho)$, in which each of the possible $\binom{n}{2}$ pairs is included with probability ϱ. It is well known that properties of $\Phi(n, dn)$ and $\Phi(n, \varrho)$ for $\varrho = \frac{2d}{n}$ are nearly identical.
- The uniform distribution of p-satisfiable 2-CNFs $\Phi^*(n, dn, p)$, $0 \le p \le \frac{d}{4}$, with n variables, dn clauses. Every $\varphi \in \Phi^*(n, dn, p)$ has a p'-satisfying assignment for $p' \ge p$.
- The planted distribution $\Phi^{pl}(n, dn, p)$ of level p. To generate a formula from $\Phi^{pl}(n, dn, p)$, we first choose a random assignment f of the n variables, and then select $\left(\frac{3}{4}d + p\right) n$ random 2-clauses that are satisfied by f and $\left(\frac{1}{4}d - p\right) n$ random 2-clauses that are not satisfied by f. Since the statistical properties of a planted formula do not depend on the planted assignment, we will always assume that the planted assignment f assigns 1 to all variables. Under this assumption planted formulas look particularly simple: they contain $\left(\frac{3}{4}d + p\right) n$ clauses of types $(+, +)$, $(+, -)$, and $(-, +)$, and $\left(\frac{1}{4}d - p\right) n$ clauses of type $(-, -)$.

The degrees of variables of formulas in $\Phi(n, dn)$ or $\Phi(n, \varrho)$ are not completely independent that makes the analysis of algorithms more difficult. To overcome this difficulty [15] suggested the *Poisson cloning model*. Under this model random formulas are generated as follows: First, take Poisson random variables $\deg(x)$ with the mean $2d$ for each of the n variables x. Then take $\deg(x)$ copies, or clones, of each variable x. If the sum of $d(x)$'s is even, then generate a uniform random matching on the set of all clones; for each edge in the matching select its type $(+, +), (+, -), (-, +), (-, -)$ at random. The clause $x^\alpha \vee y^\beta$ is in the formula $\varphi \in \Phi^{poi}(n, dn)$ if a clone of x is matched with a clone of y with an edge of type (α, β). If the sum of $d(x)$'s is odd, one of the edges must be a loop.

Let $\mathcal{P}_\varrho^u(\mathcal{A})$ and $\mathcal{P}_\varrho^{poi}(\mathcal{A})$ denote the probability of event $\mathcal{A}$ under distributions $\Phi(n, \varrho)$ and $\Phi^{poi}(n, \varrho)$. In the case of 2-SAT the result from [15] is the following

Theorem 4 ([15]). *Suppose $\varrho = \Theta(n^{-1})$. Then for any event $\mathcal{A}$*

$$c_1 \mathcal{P}_\varrho^{poi}(\mathcal{A}) \le \mathcal{P}_\varrho^u(\mathcal{A}) \le c_2 \left(\mathcal{P}_\varrho^{poi}(\mathcal{A})^{\frac{1}{2}} + e^{-n}\right),$$

where

$$c_1 = \sqrt{2} e^{\frac{\varrho(n-1)}{2} + \frac{\varrho^2 n(n-1)}{4}} + O(n^{-\frac{1}{2}}), \qquad c_2 = \sqrt{c_1} + o(1).$$

2.2   Random graph

We will need two standard results about random graphs. It is convenient for us to state them as follows. They can be proved in the standard way (see, e.g. Theorems 2.8, 2.9 of [4]).

Lemma 1. *Let $G = G(n, dn)$ (or $G = G(n, \varrho)$, $\varrho = \frac{2d}{n}$) be a random graph with n vertices and dn edges (or a random graph with n vertices and density ϱ), and let $0 < \alpha < 1$ and $\beta > \alpha^2 d + \sqrt{3d}\alpha\sqrt{\alpha(1 - \log \alpha)}$. Then whp there is no set of vertices $S \subseteq V$ such that $|S| = \alpha n$ and the subgraph $G_{|S}$ of G induced by S contains βn edges.*

More precisely, if $\beta = \alpha^2 d + \tau\sqrt{3d}\alpha\sqrt{\alpha(1 - \log\alpha)}$, $\tau > 1$, then the probability such a set exists is at most $\exp\left[-\alpha(1 - \log\alpha)(\tau^2 - 1)n\right]$.

Lemma 2. *Let $G = G(n, dn)$ (or $G = G(n, \varrho)$, $\varrho = \frac{2d}{n}$) be a random graph with n vertices and dn edges (or a random graph with n vertices and density ϱ), and let $0 < \alpha < 1$ and $\beta > 2\alpha d + \sqrt{6d}\alpha\sqrt{1 - \log\alpha}$. Then whp there is no set of variables $S \subseteq V$ such that $|S| = \alpha n$ and $\deg S = \sum_{x \in S} \deg x \geq \beta n$.*

More precisely, if $\beta = 2\alpha d + \tau\sqrt{6d}\alpha\sqrt{1 - \log\alpha}$, $\tau > 1$, then the probability such a set exists is at most $\exp\left[-\alpha(1 - \log\alpha)(\tau^2 - 1)n\right]$.

3   From Planted to Uniform

First, we establish a connection between the probabilities of events under the uniform and planted distributions. Note that the density d is NOT assumed to be large in this section. The probability of event $\mathcal{A}$ w.r.t. the uniform distribution over this set is denoted by $\mathcal{P}_d^u(\mathcal{A})$. Observe that, $\Phi^{pl}(n, dn, p)$ is uniform on the set of 2-CNFs with $(\frac{3}{4}d + p)n$ clauses of types $(+, +), (+, -), (-, +)$ and $(\frac{1}{4}d - p)n$ clauses type $(-, -)$. By $\mathcal{P}_{d,p}^{u*}(\mathcal{A})$ and $\mathcal{P}_{d,p}^{pl}(\mathcal{A})$ we denote the probability of event $\mathcal{A}$ under distributions $\Phi^*(n, dn, p)$ and $\Phi^{pl}(n, dn, p)$, respectively.

Proposition 5. *(1) Let C_{p_0} denote the event that a formula from $\Phi(n, dn)$ does not have a p-satisfying assignment for any $p \geq p_0$. For any $p_0 \leq \frac{d}{4}$ and event $\mathcal{A}$ there is $p \in [p_0, \frac{d}{4}]$ such that*

$$\mathcal{P}_d^u(\mathcal{A}) \leq \mathsf{Prob}[C_{p_0}] + \left(\frac{d}{4} - p_0\right)n \cdot \exp\left[n\log\left(1 + \exp\left[-\frac{\tau p^2}{d}\right]\right)\right]\mathcal{P}_{d,p}^{pl}(\mathcal{A}),$$

where $\tau = \frac{16}{9}$ if $p \leq \frac{d}{8}$ and $\tau = \frac{2}{3}$ if $p > \frac{d}{8}$.

(2) For any $p_0 \in [0, \frac{d}{4}]$ there is $p \in [p_0, \frac{d}{4}]$ such that

$$\mathcal{P}_{d,p_0}^{u*}(\mathcal{A}) \leq \left(\frac{d}{4} - p_0\right)n \cdot \exp\left[n\log\left(1 + \exp\left[-\frac{\tau p^2}{d}\right]\right)\right]\mathcal{P}_{d,p}^{pl}(\mathcal{A}),$$

where $\tau = \frac{16}{9}$ if $p \leq \frac{d}{8}$ and $\tau = \frac{2}{3}$ if $p > \frac{d}{8}$.

The proof follows the lines of Lemma 1 from [5] with certain modifications.

4   Approximating Max-2-SAT

In this section we present an algorithm to find an approximate solution of a Max-2-SAT instance from $\Phi^{pl}(n, dn, p)$ or $\Phi^*(n, dn, p)$ and prove Theorems 2 and 3. Some parts of the proof follow [17,6,7]. We try to highlight as many of the new proofs as the page limit allows.

4.1   Core and Majority Vote

Let φ be a 2-CNF and f its (partial) assignment. Variable x *supports* a clause C if the value of x under f satisfies C, while the value of the other variable from C exists, but does not satisfy C.

A set of variables $W \subseteq V$ is called a *core* of φ with respect to an assignment f, if every variable $x \in W$ supports at least $\frac{d}{2} - \frac{2}{3}\gamma\sqrt{d\log d}$ clauses in $\varphi(W)$. A core that contains at least $(1 - d^{-\nu})n$ variables, $\nu \geq 2$, will be called a *large core*.

We show that similar to [17] every highly satisfiable instance has a large core. Majority Vote is a one of the key ingredients in this proof. Majority Vote is a simple heuristic to solve Max-2-SAT instances. Given a 2-CNF φ it sets a variable x to 0 if $\neg x$ occurs in more clauses than x; it sets x to 1 if x occurs in more clauses than $\neg x$; and breaks the tie at random with probability $1/2$. In this section we use Majority Vote to construct an algorithm to solve highly satisfiable Random Max-2-SAT instances.

Let φ be sampled from $\Phi^{pl}(n, dn, p)$ or $\Phi^*(n, dn, p)$ and f some assignment. The following procedure defining set W will be called $\mathsf{Core}(f)$:

- let $U \subseteq V$ be the set of variables on which the majority vote assignment disagrees with f;
- let B be the set of variables that support less than $\frac{d}{2}$ clauses with respect to f;
- now W is given inductively
  - $W_0 = V - (U \cup B)$;
  - while there is $x_i \in W_i$ supporting less than $\frac{d}{2} - \frac{2}{3}\gamma\sqrt{d\log d}$ clauses in $\varphi(W_i)$ with respect to f, set $W_{i+1} = W_i - \{x_i\}$;
- set $W = W_i$.

In the rest of this subsection f is the planted assignment (recall it is assumed to be the all-ones assignment).

Lemma 3. *Let* $\varphi \in \Phi(n, dn, p)$, $p = \gamma\sqrt{d\log d}$, $\gamma > 2\sqrt{2}$, *where* d *is sufficiently large and let* W *be the result of* $\mathsf{Core}(f)$ *where* f *is the planted assignment and* $W^* = V - W$. *Then for any* $\alpha \in [d^{-\frac{\gamma^2}{4}}, d^{-1/2}]$,

$$\mathsf{Prob}[|W^*| \geq \alpha n] \leq \exp\left[-\frac{\gamma^2\log d}{48}d^{-\frac{\gamma^2}{4}}n\right].$$

Proposition 6. *Let* φ *be sampled according to* $\Phi^*(n, dn, p)$ *or* $\Phi^{pl}(n, dn, p)$, $p = \gamma\sqrt{d\log d}$, $\gamma > 2\sqrt{2}$, *where* d *is sufficiently large. Then whp there exists a large core with respect to the assignment given by Majority Vote.*

Proof. Consider first the planted distribution. Let f, f_m be the planted assignment and the assignment given by Majority Vote. By Lemma 3 with probability at least $1 - \exp\left[-\frac{\gamma^2\log d}{48}d^{-\frac{\gamma^2}{4}}n\right]$ there is a core of size at least $\left(1 - d^{-\frac{\gamma^2}{4}}\right)n$ with respect to the planted assignment, which is p-satisfying. However, since f coincides with f_m on W, it is also a large core with respect to f_m. Thus, observing that if $\gamma > 2\sqrt{2}$ then $\frac{\gamma^2}{4} > 2$, the claim of the proposition is true whp for the planted distribution.

By Proposition 5 there is a core of size $\left(1 - d^{-\frac{\gamma^2}{4}}\right) n$ with respect to the assignment given by the Majority Vote with probability at least

$$1 - \exp\left[n\log\left(1 + e^{-\frac{16}{9}p^2}\right)\right] \cdot \exp\left[-\frac{\gamma^2\log d}{48}d^{-\frac{\gamma^2}{4}}n\right] \geq 1 - \exp\left[-d^{-\frac{\gamma^2}{4}}\right].$$

4.2 Cores and Connected Components

Next we show that a large core and its complement satisfy certain strong conditions.

Proposition 7. *Let φ be sampled according to $\Phi^*(n, dn, p)$, $p = \gamma\sqrt{d\log d}$, $\gamma > 2\sqrt{2}$, where d is sufficiently large. Let W be a core of φ with respect to an at least p-satisfiable assignment f and such that $|W| \geq (1 - d^{-\nu})n$, $\nu > 2$. Then whp the largest connected component in subformula of φ induced by $V - W$ is of size at most $\log n$.*

Observe that if we choose a random set X of variables of size αn then, as the probability of each pair of variables to form a clause is $\frac{d}{n}$, by the results of [4] if $\alpha < \frac{1}{2d}$, the connected components of the formula induced by X whp are of size at most $\log n$. However, high probability in this case means $1 - n^{\log(\alpha d)}$, and we cannot claim that each of the $\binom{n}{\alpha n}$ sets of αn variables satisfies this condition. Therefore, we follow [17] and provide a fairly sophisticated proof that the complement of a core has connected components of size at most $\log n$.

Fix a set of variables $T \subseteq V$, $t = |T| = \log n$, and a collection τ of $t - 1$ clauses that induce a tree on T. We aim to bound the probability that $\tau \subseteq \varphi$. Let $J \subseteq T$ be a set of variables which appear in at most 2 clauses of τ. By a simple counting argument we have $|J| \geq \frac{1}{2}t$.

Lemma 3 amounts to saying that for $\varphi \in \Phi^{pl}(n, dn, p)$ whp there exists a p-satisfying assignment f such that $\text{Core}(f)$ produces sets U, B, W satisfying the conditions proved there. Now by Proposition 5 the same is true whp for $\varphi \in \Phi^*(n, dn, p)$. More precisely, we have the following

Lemma 4. *Let φ be sampled according to $\Phi^*(n, dn, p)$ or $\Phi^{pl}(n, dn, p)$, $p = \gamma\sqrt{d\log d}$, $\gamma > 2\sqrt{2}$, where d is sufficiently large. Then for any $\alpha \in [d^{-\frac{\gamma^2}{4}}, d^{-1/2}]$ whp there exists a p-satisfying assignment f_φ such that if U is the set of variables on which Majority Vote disagrees with f_φ and B is the set of variables that support at most $\frac{d}{2}$ clauses with respect to f_φ, then $|U \cup B| \leq \frac{\alpha}{2}n$.*

Let φ be sampled from $\Phi^{pl}(n, dn, p)$ or $\Phi^*(n, dn, p)$ and f some assignment. The following procedure, denoted $\text{Core}'(f)$, is a modification of Core from Section 4.1.

– let $U' \subseteq V$ be the set of variables on which the majority vote assignment disagrees with f or is right with advantage smaller than 4;
– let B' be the set of variables that support less than $\frac{d}{2}$ clauses with respect to the planted assignment;
– now W is given inductively
   • $W'_0 = V - (U \cup B \cup (T - J))$, where J is defined as above;

- while there is $x_i \in W_i$ supporting less than $\frac{d}{2} - \frac{2}{3}\gamma\sqrt{d\log d}$ clauses in $\varphi(W_i')$,
  set $W_{i+1}' = W_i' - \{x_i\}$;
- set $W' = W_i'$.

The following lemma is proved in the same way as Lemma 3.

Lemma 5. *Let* $\varphi \in \Phi(n, dn, p), p = \gamma\sqrt{d\log d}, \gamma > 2\sqrt{2}$, *where d is sufficiently large,* W' *be the result of* $\mathsf{Core}'(f)$ *where* f *is the planted assignment and* $W^* = V - W'$. *Then for any* $\alpha \in [d^{-\frac{\gamma^2}{4}}, \sqrt{d}]$,

$$\mathsf{Prob}[|W^*| \geq \alpha n] \leq \exp\left[-\frac{\gamma^2 \log d}{48} d^{-\frac{\gamma^2}{4}} n\right].$$

Observe also that Lemmas 3 and 5 remain true if instead of a formula $\varphi \in \Phi^*(n, dn, p)$ we apply the procedures to φ along with a small number of extra clauses. More precisely,

Lemma 6. *Let* $\varphi \in \Phi^*(n, dn, p), p = \gamma\sqrt{d\log d}, \gamma > 2\sqrt{2}$, *where d is sufficiently large, and* ψ *a set of clauses of size* $\log n$. *Also let* $W^* = V - W$ *or* $W^* = V - W'$ *where* W, W' *are the sets obtained by applying* $\mathsf{Core}(f_\varphi)$ *and* $\mathsf{Core}'(f_\varphi)$ *to* $\varphi \cup \psi$; *here* f_φ *is as found in Lemma 4. Then for any* $\alpha \in [d^{-\frac{\gamma^2}{4}}, \sqrt{d}]$,

$$\mathsf{Prob}[|W^*| \geq \alpha n] \leq \exp\left[-\frac{\gamma^2 \log d}{48} d^{-\frac{\gamma^2}{4}} n\right].$$

Lemma 7. *Let* $\varphi \in \Phi^*(n, dn, p)$ *and let* $W(\varphi \cup \tau)$ *be the value of* W *if* $\mathsf{Core}(f_\varphi)$ *is applied to* $\varphi \cup \tau$. *Let* $W'(\varphi)$ *be the value of* W' *obtained by* $\mathsf{Core}'(f_\varphi)$ *applied to* φ. *Then* $W'(\varphi) \subseteq W(\varphi \cup \tau)$.

Proof. We proceed by induction. First, show that $W_0'(\varphi) \subseteq W_0(\varphi \cup \tau)$. If $x \in W_0'(\varphi)$ then Majority Vote is right with advantage at least 3, and x supports at least $\frac{d}{2}$ clauses. Since the degree of x in τ is at most 2, Majority Vote cannot be flipped by adding τ to φ. Also, support can only increase after adding τ. Therefore $s \in W_0(\varphi \cup \tau)$.

Now, suppose $W_i'(\varphi) \subseteq W_i(\varphi \cup \tau)$. We show that the statement holds for $i + 1$. If x supports the required number of clauses with only variables from $W_i'(\varphi)$ then it also supports sufficiently many variables with respect to $W_i(\varphi \cup \tau)$. Therefore $x \in W_{i+1}'(\varphi) \subseteq W_{i+1}(\varphi \cup \tau)$. The lemma is proved.

Proposition 7 now follows from a sequence of lemmas we borrow from [17]. In all of them we assume $p = \gamma\sqrt{d\log d}, \gamma > 2\sqrt{2}, d$ is large enough.

Lemma 8. $\mathsf{Prob}[\tau \subseteq \varphi \text{ and } T \cap W = \varnothing] \leq \mathsf{Prob}[\tau \subseteq \varphi] \cdot \mathsf{Prob}[J \cap W' = \varnothing]$.

Lemma 9. *Let* $\varrho \in [\frac{t}{n}, 1]$ *be such that* $|W'| = (1 - \varrho)n$. *Then* $\mathsf{Prob}[J \cap W' = \varnothing] \leq (6\varrho)^{t/2}$.

Corollary 1. *Let* $\varrho \in [\frac{t}{n}, 1]$ *be such that* $|W'| = (1 - \varrho)n$. *Then*

$$\mathsf{Prob}[\tau \subseteq \varphi \text{ and } T \cap W = \varnothing] \leq (6\varrho)^{t/2}\left(\frac{d}{n}\right)^t.$$

Lemma 10. *The probability of a tree of size* $t = \log n$ *in the formula induced by* W^* *is at most* $\exp[-(\delta \log d - 1 - \log(\sqrt{6})) \log n]$ *for some* $\delta > 0$.

Proof. We assume that ϱ in the previous lemmas is $o(d^{-(2+2\delta)})$. By Cayley theorem there are t^{t-2} trees on a given set of t vertices. Then

$$\text{Prob[there is a tree of size} \geq \beta n] \leq \text{Prob}[\exists \tau \subseteq \varphi, |\tau| = t \text{ and } T = V(\tau) \subseteq W = \varnothing]$$

$$\leq \binom{n}{t} t^{t-2} \cdot \left(\frac{d}{n}\right)^t \cdot (6\varrho)^t \leq \left(\frac{en}{t}\right)^t t^t \cdot \left(\frac{d}{n}\right)^t \cdot (6\varrho)^{t/2}$$

$$\leq (\sqrt{6}\varrho ed)^t \leq (\sqrt{6}ed^{-\delta})^{\log n} = \exp[-(\delta \log d - 1 - \log(\sqrt{6})) \log n],$$

which completes the proof.

Proposition 7 follows.

4.3   The Algorithm

The solution algorithm follows the approach from [17]. However, the brute force search part of that algorithm is not applicable in our case, as we cannot guarantee that sufficiently few wrongly assigned variables remain assigned after the unassigning step.

Here $\varphi^h(A)$ denotes the formula φ after substituting the partial assignment h and restricted to $V - A$.

Algorithm 1. The algorithm

Require: $\varphi \in \Phi^*(n, dn, p)$
1. $f_1 \leftarrow \mathsf{MajorityVote}(\varphi)$
2. $i \leftarrow 1$
3. **while** there is x supporting less than $\frac{d}{2} - \frac{2}{3}\gamma\sqrt{d \log d}$ clauses with respect to f_i **do**
4. $f_{i+1} \leftarrow f_i$ with x unassigned
5. $i \leftarrow i + 1$
6. **end while**
7. let h be the final partial assignment
8. let A be the set of assigned variables in h
9. exhaustively search the formula $\varphi^h(A)$, component by component looking for the assignment satisfying the maximum number of clauses

---

Lemma 11. *Let f be the assignment given in line 1 of the algorithm; suppose f has a large core W. Then no variable from W is unassigned in lines 3–6 of the algorithm.*

Proof. Every $x \in W$ supports at least $\frac{d}{2} - \frac{2}{3}\gamma\sqrt{d \log d}$ clauses from $\varphi(W)$. Then the statement follows by induction.

Lemma 12. *Let $\varphi \in \Phi^{pl}(n, dn, p)$, $p = \gamma\sqrt{d \log d}$, $\gamma > 2\sqrt{2}$, where d is sufficiently large, and let f be the planted assignment. Let U denote the set of assigned variables after line 7 of the algorithm, whose value differs from that under f. Then $|U| \geq \chi n$ with probability at most $\exp[-2\chi n]$ for any $\chi < d^{-1/2}$.*

Proof. (Sketch.) Let $\varphi \in \Phi^{pl}(n, dn, p)$, and $p = \gamma\sqrt{d \log d}$. Let Z denote the set of unassigned variables, and U the set of variables x with $h(x) \neq f(x)$. In other words, $U = \{x | h(x) = 0\}$. Let $|U| = \chi n$. By Lemmas 3 and 11 and $|Z \cup U| \leq d^{-1/2}n$; in particular, $\chi < d^{-1/2}$.

We evaluate the number of clauses supported by variables from U. There are two kinds of such clauses. In clauses of the first kind both variables are from U, we call such clauses *internal*. In clauses of the second kind the second variable belongs to $V - (Z \cup U)$, we call such clauses *external*. Every internal clause has the type $(-, +)$ or $(+, -)$, every external clause has the type $(-, -)$. Using Lemma 1 we bound the number of internal clauses by $\beta_1 n < (\frac{1}{2} + o(1))(\chi^2 d + 2\chi\sqrt{3d})n$. Since every variable from U supports at least $\frac{d}{2} - \frac{2}{3}\gamma\sqrt{d \log d}$ clauses, the total number of internal and external clauses is at least $(\frac{d}{2} - \frac{2}{3}\gamma\sqrt{d \log d})\chi n$. By Lemma 2 and the bound on the number of internal variables we show that the probability that there exist a sufficient number of external clauses is bounded as in the statement of the lemma; that completes the proof.

Proposition 8. *Let $\varphi \in \Phi^{pl}(n, dn, p)$, $p = \gamma\sqrt{d \log d}$, $\gamma > 2\sqrt{2}$, where d is sufficiently large, let U be the set of assigned variables after line 7 of the algorithm, whose value differs from that under f, and let $|U| = \chi n$. The exhaustive search of the algorithm whp completes in polynomial time and returns an extension of h which is at least p'-satisfiable for $p' = p - \chi(\frac{1}{2}d - \sqrt{6d} + o(d^{-1}))$ with probability at least $1 - e^{-2\chi n}$ for any $\chi < d^{-1/2}$.*

Proof. (Sketch.) This proof is similar to the previous one. As φ has a large core with respect to f, by Lemma 11 there are at most $d^{-2}n$ unassigned variables. Therefore it follows from Proposition 7 that whp the algorithm completes in polynomial time.

To prove the second claim of the lemma observe that if all unassigned variables are assigned according to f then the only clauses that are satisfied under f but not satisfied under the assignment produced by the algorithm are the clauses of the type $(+, +)$, both of whose variables belong to U, and the clauses of the type $(+, -)$, whose first variable is from U, and the other is from $V - U$.

By Lemma 1 we bound the number of clauses of the first kind by $(1 + o(1))\left(\frac{1}{4}\chi^2 d + \chi\sqrt{3d}\right) n$. Furthermore, by Lemma 2 the number of clauses of the second kind can be bounded by $(1 + o(1))\chi\left(\frac{1}{2}d + \sqrt{6d}\right) n$. Since $\chi < 1$, the number of clauses of the second kind dominates. Thus, in the planted case the assignment produced by the algorithm has level at least $p - \chi(\frac{1}{2}d - \sqrt{6d} + o(d^{-1}))$.

Proof. (of Theorem 2). If $\gamma > 2\sqrt{2}$ then $\frac{\gamma^2}{4} > 2$. By Lemma 3 in this case the conditions of Lemma 12 hold. Therefore whp $|U| < \chi n$, where χ is such that $\chi(\frac{1}{2}d + \sqrt{6d} + o(d^{-1})) < \varepsilon$, and the result follows by Proposition 8.

Proof. (of Theorem 3). First, we compute the 'exchange rate' from Proposition 5 for $p = \gamma\sqrt{d \log d}$:

$$\exp\left[n \log\left(1 + \exp\left[-\frac{16p^2}{9d}\right]\right)\right] < \exp\left[nd^{-\frac{16\gamma^2}{9}}\right].$$

By Lemma 11 and Proposition 5, if $2\chi > d^{-\frac{16\gamma^2}{9}}$ whp the set U produced in line 7 of the algorithm has size $|U| < \chi n$. We now conclude by Proposition 8.

5   Majority vote

In this section we use Majority Vote to estimate the likely level of satisfiability of Random Max-2-SAT instances. The results of this section are true for ANY value of d.

5.1   Majority Vote and Its Performance

First, we use the simple approach from [12] to estimate the number of clauses unsatisfied by Majority Vote algorithm under the uniform distribution.

Proposition 9. *Given a random formula of density d Majority Vote whp does not satisfy $\beta^2(d)dn + o(n)$ clauses, where $\beta(d) = \mathbb{E}[\min(X,Y)]$, and X,Y are independent identical Poisson variables with the mean d.*

Remark 1. Analytically, the expected ratio of clauses unsatisfied by Majority Vote can be expressed as follows

$$\beta(d) = 1 - \frac{1}{d}\int_0^\infty t \cdot e^{-d} \frac{d^t}{\Gamma(t)} \frac{\Gamma(t+1,d)}{\Gamma(t)} dt.$$

Since there is no convenient approximation or asymptotic for function $\beta(d)$, we can only estimate it numerically. Table 1 contains some values of it.

Table 1. Minimum of two Poisson distributions. α and β denote the values $\alpha(d) = 1 - \beta(d)$ and $\beta(d)$; Unsat stands for the number of unsatisfied clauses.

d	α	β	Unsat	d	α	β	Unsat	d	α	β	Unsat
25	0.440	0.560	4.838	64	0.463	0.537	13.740	121	0.474	0.526	27.143
36	0.450	0.550	7.305	81	0.468	0.532	17.964	144	0.476	0.524	32.610
49	0.458	0.542	10.273	100	0.471	0.529	22.175				

5.2   Majority Vote on Planted Instances

Here we study how Majority Vote performs on planted instances of different level. As before, we assume that the planted assignment f is the all-ones one, and is p-satisfying, $0 \leq p \leq \frac{d}{4}$.

Proposition 10. *Let $\varphi \in \Phi^{pl}(n, dn, p)$, let f be the planted assignment and h the assignment produced by Majority Vote.*

(1) Whp the assignments f and h disagree on $\beta(d,p)n + o(n)$ variables, where $\beta(d,p) = \text{Prob}[X \geq Y]$ and X, Y are independent Poisson variables with means $\mu_1 = d - \frac{4}{3}p$ and $\mu_2 = d + \frac{4}{3}p$.

(2) Let $x(d, p, n)$ be a random variable equal to the number of clauses unsatisfied by h. Then

$$\text{num}(d, p) = \frac{1}{n} \mathbb{E}[x(d, p, n)] = \frac{1}{4}d - p + \frac{1}{3}p\beta(d, p)(2 - \beta(d, p)).$$

(3) $\text{Prob}[|x(d, p, n) - \text{num}(d, p)n| > \delta\sqrt{dn}] < e^{-16\delta^2 n}$.

Proof. Using the Poisson cloning model proofs of items (1) and (2) are straightforward.

(3) To evaluate the concentration of the number of unsatisfied clauses we again use the Poisson cloning approach as follows. For each variable v we create two sets of clones v^+ and v^- such that $|v^+|$ is a Poisson variable with the mean $d + \frac{4}{3}p$ and $|v^-|$ is a Poisson variable with the mean $d - \frac{4}{3}p$, which are the expected numbers of positive and negative occurrences of v, respectively. We now select a random matching among the generated clones.

We consider 2 cases depending on p. First, when $p = C\sqrt{d}$, C constant; we will use this case to compare against the uniform distribution. The second case includes all remaining values of p.

Case 1. $p = C\sqrt{d}$, C constant.

Let the number of clones satisfied by Majority Vote be sn, for a random variable s, and let the number of unsatisfied clauses be xn for a random variable x. Clearly the expectation of x is $\frac{s^2}{4d'}n$, where $2d'n$ is the total number of clones. We evaluate the probability that x deviates by at least $\delta\sqrt{d}$ from its expectation conditioned on the assumption that $2d' = 2d$. Since $p = C\sqrt{d}$, whp $s = d - \alpha\sqrt{d}$. In fact, $\mathbb{E}[s] = \mathbb{E}[\min(X, Y)]$, where X, Y are defined as in part (1) of the proposition. We need a tighter estimation, therefore, rather than using Chernoff bound we resort to a more straightforward method.

We count the total number of formulas with $2dn$ clones, and the number of those of them having sn unsatisfied literals and xn unsatisfied clauses. The first number, N equals the number of matchings of a complete graph, see, e.g., [16]

$$N = \left(\frac{2dn}{e}\right)^{dn} \frac{e^{\sqrt{2dn}}}{(4e)^{1/4}}(1 + o(1)).$$

The second number M is the number of ways to complete the following process: choose a matching of xn edges in the complete graph with sn vertices (i.e., choose xn unsatisfied clauses among pairs of unsatisfied literals), then choose matches for the remaining $(s - 2x)n$ unsatisfied literals among $(2d - s)n$ satisfied ones, and, finally, choose a matching among the $(2d - 2s + 2x)n$ remaining satisfied literals. Thus, $M = M_1 \cdot M_2 \cdot M_3$, where

$$M_1 = \frac{(sn)!}{2^{xn}((s - 2x)n)!(xn)!}, \qquad M_2 = \frac{((2d - s)n)!}{((2d - 2s + 2x)n)!},$$

$$M_3 = \left(\frac{(2d - 2s + 2x)n}{e}\right)^{(d-s+x)n} \frac{e^{\sqrt{(2d-2s+2x)n}}}{(4e)^{1/4}}(1 + o(1)).$$

Using Stirling's approximation up to a factor of $1 + o(1)$ we can represent the probability we are looking for, that is, the fraction $\frac{M}{N}$, as $\frac{M}{N} = n^A \cdot e^{Bn} \cdot e^{o(n)}$, where B is a constant, while A may depend on n. As is easily seen, $A = 0$. A straightforward (although somewhat tedious) calculation shows that $B < -16\delta^2(1 + o(1))$.

Thus,

$$\text{Prob}[|x - \mathbb{E}[x]| \geq \delta\sqrt{d}] \leq e^{-16\delta^2 n}.$$

Case 2. $p = \Omega(\sqrt{d})$.

In this case the result follows from Chernoff bound.

Remark 2. Analytically, $\beta(d, p)$ can be expressed as follows

$$\beta(d, p) = e^{-\mu_2} \int_0^{\mu_1} e^{-t} I_0(2\sqrt{\mu_2 t}) dt.$$

Table 2 below contains values of $\beta(d, p)$, that is, the probability that majority vote assigns values corresponding to the all-ones assignment, and the number of unsatisfied clauses left by majority vote. These numbers depend on two parameters, d the density, and p the parameter regulating the number of clauses satisfied by the all-ones assignment. Recall that for the uniform distribution whp $p = \Theta(\sqrt{d})$, more precise estimations give $0.34\sqrt{d} \leq p \leq 0.51\sqrt{d}$ [8].

Table 2. Values of $\beta(d, p)$ and $\text{num}(d, p)$. Each entry has the form $\beta(d, p)/\text{num}(d, p)$. The values in the empty cells are too small and cannot be reliably approximated.

d / p	$0.34\sqrt{d}$	$0.42\sqrt{d}$	$0.51\sqrt{d}$	$\frac{1}{16}d$	$\frac{1}{8}d$	$\frac{1}{4}d$
25	0.284/5.661	0.235/5.321	0.191/4.916	0.302/5.761	0.133/4.182	0.010/0.230
49	0.277/11.395	0.229/10.914	0.186/10.343	0.219/10.796	0.054/7.040	0.0005/0.138
64	0.275/15.012	0.227/14.460	0.184/13.806	0.184/13.806	0.032/8.752	
100	0.272/23.745	0.225/23.052	0.182/22.233	0.126/20.765	0.010/12.950	
144	0.270/34.479	0.223/33.645	0.180/32.660	0.083/28.978	0.0025/18.298	
256	0.268/61.947	0.221/60.830	0.179/59.514	0.031/49.453		

5.3 Satisfiability of Random Max-2-SAT

We now combine Propositions 9, 10, and Proposition 5 to estimate the maximal expected number of clauses that can be simultaneously satisfied in $\varphi \in \Phi(n, dn)$.

Proposition 11. *Let* $r(d) = \beta^2(d)d$ *be the fraction of clauses unsatisfied by Majority Vote on* $\varphi \in \Phi(n, dn)$, *and let* $p = p(d)$ *be the value of* p *such that*

$$r(d) = \frac{1}{4}d - p + \frac{1}{3}p\beta(d, p)(2 - \beta(d, p)).$$

Then whp the maximum number of simultaneously satisfiable clauses of $\varphi \in \Phi(n, dn)$ *belongs to the interval* $[p(d) - \delta, p(d) + \delta]$, $\delta(d) = 0.2\sqrt{d}$.

Proof. Let $\overline{C}_p$ be the event that $\varphi \in \Phi(n, dn)$ has a p'-satisfying assignment, for some $p' \geq p$. Let also M_p denote the event that Majority Vote finds a p'-satisfying assignment, $p' \geq p$, and $\overline{M}_p$ the event that it does not happen. By Proposition 9 $\mathcal{P}_d^u(M_{d-(r(d)+\varepsilon)})$ is exponentially small for any $\varepsilon > 0$. On the other hand, by Proposition 5 for $p \geq p(d) + \delta$

$$\mathcal{P}_d^u(M_{d-(r(d)+\varepsilon)} \wedge \overline{C}_p)$$
$$= \mathcal{P}_d^u(\overline{C}_p)\mathcal{P}_{d,p}^{u*}(M_{d-(r(d)+\varepsilon)}) = \mathcal{P}_d^u(\overline{C}_p)(1 - \mathcal{P}_{d,p}^{u*}(\overline{M}_{d-(r(d)+\varepsilon)}))$$
$$\geq \mathcal{P}_d^u(\overline{C}_p)\left(1 - \left(\frac{d}{4} - p\right)n \cdot \exp\left[n \log\left(1 + e^{-\frac{16}{9}\frac{p^2}{d}}\right)\right]\mathcal{P}_{d,p'}^{pl}(\overline{M}_{d-(r(d)+\varepsilon)})\right)$$

for some $p' \geq p$. Since $\mathcal{P}_{d,p'}^{pl}(\overline{M}_{d-(r(d)+\varepsilon)}) \leq \mathcal{P}_{d,p}^{pl}(\overline{M}_{d-(r(d)+\varepsilon)})$ we assume $p' = p$. By [8] whp $p(d) > \frac{1}{3}\sqrt{d}$. Therefore,

$$\exp\left[n \log\left(1 + e^{-\frac{16}{9}\frac{p^2}{d}}\right)\right] \leq e^{0.599n};$$

and if n is sufficiently large

$$\left(\frac{d}{4} - p\right)n \cdot \exp\left[n \log\left(1 + e^{-\frac{16}{9}\frac{p^2}{d}}\right)\right] \leq e^{0.6n}.$$

On the other hand, since $r(d) = \text{num}(d, p(d))$, if we prove that $\text{num}(d, p(d)) > \text{num}(d, p(d) + \delta) + \delta$, we get

$$\mathcal{P}_{d,p}^{pl}(\overline{M}_{d-(r(d)+\varepsilon)}) = \text{Prob}[x(d, p(d) + \delta, n) > (r(d) + \varepsilon)n]$$
$$\leq \text{Prob}[|x(d, p(d) + \delta, n) - \text{num}(d, p(d)+\delta)n| > \delta n] < e^{-16\frac{\delta^2}{d}n}.$$

Thus,
$$\mathcal{P}_d^u(M_{d-(r(d)+\varepsilon)} \wedge \overline{C}_p) \geq \mathcal{P}_d^u(\overline{C}_p)(1 - e^{0.6n} \cdot e^{-0.64n}),$$

and as it is exponentially small, $\mathcal{P}_d^u(\overline{C}_p)$ is exponentially small.

The lower bound is similar.

It remains to show that $\text{num}(d, p(d)) > \text{num}(d, p(d) + \delta) + \delta$. Let $g(x) = x\beta(d, x)(2 - \beta(d, x))$. It is not hard to see that $g(x)$ decreases on $[0, \frac{d}{4}]$. Thus,

$$\text{num}(d, p(d) + \delta) + \delta = \frac{d}{4} - p(d) - \delta + \frac{1}{3}g(p(d) + \delta) + \delta$$
$$< \frac{d}{4} - p(d) + \frac{1}{3}g(p(d)) = \text{num}(d, p(d)).$$

This completes the proof.

The values of $p(d)$ for some values of d are given in Table 3. It seems that the fraction of satisfiable clauses should be very close to $\frac{3}{4}d + \frac{1}{2}\sqrt{d}$.

Table 3. Values of $p(d)$ depending on d

d	5^2	6^2	7^2	8^2	9^2	10^2	11^2	12^2	13^2	14^2	15^2	16^2	17^2	18^2
$p(d)$	2.572	3.069	3.565	4.061	4.558	5.054	5.550	6.047	6.543	7.039	7.535	8.032	8.528	9.025
$p(d)/\sqrt{d}$	0.514	0.512	0.509	0.507	0.506	0.505	0.504	0.504	0.503	0.503	0.502	0.502	0.502	0.501

References

1. Achlioptas, D., Naor, A., Peres, Y.: On the maximum satisfiability of random formulas. J. ACM 54(2) (2007)
2. Achlioptas, D., Peres, Y.: The threshold for random k-SAT is $2^k \ln 2 - o(k)$. J. AMS 17, 947–973 (2004)
3. Bellare, M., Goldreich, O., Sudan, M.: Free bits, PCPs, and nonapproximability-towards tight results. SIAM J. Comput. 27(3), 804–915 (1998)
4. Bollobás, B.: Random Graphs. Cambridge Studies in Advanced Mathematics, vol. 73. Cambridge University Press (2001)
5. Coja-Oghlan, A., Krivelevich, M., Vilenchik, D.: Why almost all k-CNF formulas are easy, http://www.wisdom.weizmann.ac.il/vilenchi/papers/2007/UniformSAT.pdf
6. Coja-Oghlan, A., Krivelevich, M., Vilenchik, D.: Why almost all satisfiable k-CNF formulas are easy. In: DTMC, pp. 89–102 (2007)
7. Coja-Oghlan, A., Krivelevich, M., Vilenchik, D.: Why almost all k-colorable graphs are easy to color. Theory Comput. Syst. 46(3), 523–565 (2010)
8. Coppersmith, D., Gamarnik, D., Hajiaghayi, M.T., Sorkin, G.B.: Random MAX SAT, random MAX CUT, and their phase transitions. In: SODA, pp. 364–373 (2003)
9. Feige, U.: Relations between average case complexity and approximation complexity. In: IEEE Conference on Computational Complexity, p. 5 (2002)
10. Flaxman, A.: A spectral technique for random satisfiable 3CNF formulas. In: SODA, pp. 357–363 (2003)
11. Friedgut, E.: Sharp thresholds of graph properties, and the k-SAT problem. J. Amer. Math. Soc. 12(4), 1017–1054 (1999)
12. Hajiaghayi, M., Kim, J.H.: Tight bounds for random MAX 2-SAT, http://research.microsoft.com/en-us/um/redmond/groups/theory/jehkim/a-yspapers/max2sat.pdf
13. Håstad, J.: Some optimal inapproximability results. J. ACM 48(4), 798–859 (2001)
14. Interian, Y.: Approximation algorithm for random MAX-kSAT. In: Hoos, H.H., Mitchell, D.G. (eds.) SAT 2004. LNCS, vol. 3542, pp. 173–182. Springer, Heidelberg (2005)
15. Kim, J.H.: The Poisson cloning model for random graphs, random directed graphs and random k-SAT problems. In: Chwa, K.-Y., Munro, J.I. (eds.) COCOON 2004. LNCS, vol. 3106, p. 2. Springer, Heidelberg (2004)
16. Knuth, D.E.: The Art of Computer Programming. Sorting and Searching, vol. III. Addison-Wesley (1973)
17. Krivelevich, M., Vilenchik, D.: Solving random satisfiable 3CNF formulas in expected polynomial time. In: SODA, pp. 454–463 (2006)
18. Mitchell, D.G., Selman, B., Levesque, H.J.: Hard and easy distributions of sat problems. In: AAAI, pp. 459–465 (1992)
19. de la Vega, W.F., Karpinski, M.: 1.0957-approximation algorithm for random MAX-3SAT. RAIRO - Operations Research 41(1), 95–103 (2007)

Hypergraph Acyclicity and Propositional Model Counting[*]

Florent Capelli[1], Arnaud Durand[1,2], and Stefan Mengel[3]

[1] IMJ UMR 7586 - Logique, Université Paris Diderot, France
[2] LSV UMR 8643, ENS Cachan, France
[3] Laboratoire d'Informatique, LIX UMR 7161, Ecole Polytechnique, France

Abstract. We show that the propositional model counting problem #SAT for CNF-formulas with hypergraphs that allow a disjoint branches decomposition can be solved in polynomial time. We show that this class of hypergraphs is incomparable to hypergraphs of bounded incidence cliquewidth which were the biggest class of hypergraphs for which #SAT was known to be solvable in polynomial time so far. Furthermore, we present a polynomial time algorithm that computes a disjoint branches decomposition of a given hypergraph if it exists and rejects otherwise. Finally, we show that some slight extensions of the class of hypergraphs with disjoint branches decompositions lead to intractable #SAT, leaving open how to generalize the counting result of this paper.

1 Introduction

Proposition model counting (#SAT) is the problem of counting satisfying assignments (models) to a CNF-formula. It is the canonical #**P**-hard counting problem and is important due to its applications in Artificial Intelligence. Unfortunately, #SAT is extremely hard to solve: Even on restricted classes of formulas like monotome 2CNF-formulas or Horn 2CNF-formulas it is **NP**-hard to approximate within a factor of $2^{n^{1-\epsilon}}$ for any $\epsilon > 0$ [13]. Fortunately, this is not the end of the story: While syntactical restrictions on the types of allowed clauses do not lead to tractable counting, there is a growing body of work that successfully applies so-called *structural* restrictions to #SAT, see e.g. [7,14,12,16]. In this line of work one does not restrict the individual clauses of CNF-formulas but instead the interaction between the variables in the formula. This is done by assigning graphs or hypergraphs to formulas and then restricting the class of (hyper)graphs that are allowed for instances (see Section 2 for details). In this paper we present a new class of hypergraphs, such with disjoint branches decompositions [5], for which #SAT is tractable.

Having a disjoint branches decomposition is a so-called acyclicity notion for hypergraphs. Unlike for graphs, there are several resonable ways of defining acyclicity for hypergraphs [6] which have been very successful in database theory.

---

[*] Partially supported by DFG grants BU 1371/2-2 and BU 1371/3-1.

Mostly three "degrees of acyclicity" have been studied: α-acyclicity, β-acyclicity and γ-acyclicity, where the α-acyclic hypergraphs form the most general and the γ-acyclic hypergraphs the least general class. Prior to this paper it was known that #SAT for CNF-formulas with α-acyclic hypergraphs was #**P**-hard [14], while it is tractable for γ-acyclic hypergraphs as the latter have incidence cliquewidth bounded by 3 [9] and thus the results of [16] apply.

To understand the influence of hypergraph acyclicity on the complexity of #SAT, the next natural step is thus analyzing the intermediate case of β-acyclic hypergraphs. For this class it is known that SAT is tractable [11], unlike for α-acyclic hypergraphs. Unfortunately, the algorithm in [11] is based on a resolution-like method and it is not clear whether one can obtain tractability for counting from the method used for decision. In fact, most classical decision results based on tractability of resolution (such as for $2-\text{SAT}$) or unit propagation (Horn$-$SAT) do not extend to counting as the respective counting problems are hard (see e.g. [13]).

Unfortunately, #SAT for CNF-formulas with β-acyclic hypergraphs has turned out to be a stubborn problem whose complexity could so far not be determined despite considerable effort by us and others [15]. A natural approach which we follow in this paper is thus trying to understand slightly more restrictive notions of acyclicity. We focus here on hypergraphs with disjoint branches decompositions, a notion which was introduced by Duris [5] and which lies strictly between β-acyclicity and γ-acyclicity. We show that for CNF-formulas whose hypergraphs have a disjoint branches decompositions we can solve #SAT in polynomial time. We also show that hypergraphs with disjoint branches decompositions are incomparable to hypergraphs with bounded incidence cliquewidth which so far were the biggest class of hypergraphs for which #SAT was known to be tractable. Thus our results give a new class of tractable instances for #SAT, pushing back the known tractability frontier for this problem.

Our main contribution is twofold: Most importantly, we present the promised counting algorithm for CNF-formulas whose hypergraphs have a disjoint branches decomposition in Section 3. Secondly, we present in Section 4 a polynomial time algorithm that checks if a hypergraph has a disjoint branches decomposition and if so also constructs it. On the one hand, this gives some confidence that hypergraphs with disjoint branches decompositions form a well-behaved class as it can be decided in polynomial time. On the other hand, the counting algorithm will depend on knowing a decomposition, so its computation is an essential part of the counting procedure. Finally, in Section 5 we then turn to generalizing the results of this paper, unfortunately showing only negative results. We consider some natural looking extensions of hypergraphs with disjoint branches and show that #SAT is intractable on these classes under standard complexity theoretic assumptions.

Due to lack of space some of the proofs are omitted and can be found in the full version of this paper.

2   Preliminaries and Notation

2.1   Hypergraphs and Graphs Associated to CNF-Formulas

The *primal graph* of a CNF-formula F has as vertices the variables of F and two vertices are connected by an edge if they appear in a common clause of F. The *incidence graph* of F is defined as the bipartite graph which has as vertices the variables and the clauses of F and two vertices u and v are connected by an edge if u is a variable and v is a clause such that u appears in v.

A (finite) hypergraph $\mathcal{H}$ is a pair (V, E) where V is a finite set and $E \subseteq \mathcal{P}(V)$. A subhypergraph $\mathcal{H}' = (V', E')$ of $\mathcal{H}$ is a hypergraph with $V' \subseteq V$ and $E' \subseteq \{e \cap V' \mid e \in E, e \cap V' \neq \emptyset\}$. A path between two vertices $u, v \in V$ is a sequence $e_1, \ldots, e_k$ such that $u \in e_1$, $v \in e_k$ and for every $i = 1, \ldots, k-1$ we have $e_i \cap e_{i+1} \neq \emptyset$. A hypergraph $\mathcal{H}$ is called *connected* if there is a path between every pair of vertices of $\mathcal{H}$. A (connected) *component* of $\mathcal{H}$ is defined to be a maximal connected subhypergraph of $\mathcal{H}$.

To a CNF-formula F we associate a hypergraph $\mathcal{H} = (V, E)$ where V is the variable set of F and the hyperedge set E contains for each clause of F an edge containing the variables of the clause.

Graph Decompositions. We will not recall basic graph decompositions such as tree-width and clique-width (see e.g. [9,7]). A class of CNF-formulas is defined to be of bounded (signed) incidence clique-width, if their (signed) incidence graphs are of bounded clique-width. A set $X \subseteq V$ of vertices of a graph is called a *module*, if every $v \in V \backslash X$ has the same set of neighbours and non-neighbours in X. If X is a module of a graph G, the graph $G' = (V', E')$ obtained after contraction of X is defined by $V' := (V \backslash X) \cup \{x\}$ where x is a new vertex not in V and $E' := (E \cap (V \backslash X)^2) \cup \{ux : u \notin X$ and $\exists v \in X$ s.t. $uv \in E\})$. A class of CNF-formulas is of bounded *modular incidence treewidth* if their incidence graphs are of bounded tree-width after contracting all modules.

Acyclicity in hypergraphs. In contrast to the graph setting, there are several non equivalent notions of acyclicity for hypergraphs [6]. Most of these notions have many equivalent definitions , but we will restrict ourselves to definitions using the notion of join trees.

Definition 1. *A* join tree *of a hypergraph* $\mathcal{H} = (V, E)$ *is a pair* $(\mathcal{T}, \lambda)$ *where* $\mathcal{T} = (N, T)$ *is a tree and* λ *is a bijection between* N *and* E *such that:*

- *for each* $e \in E$, *there is a* $t \in N$ *such that* $\lambda(t) = e$, *and*
- *for each* $v \in V$, *the set* $\{t \in N \mid v \in \lambda(t)\}$ *is a connected subtree of* $\mathcal{T}$.

The second condition in Definition 1 is often called the *connectedness condition*. We will follow the usual convention of identifying an edge $e \in E$ with the vertex $\lambda(e)$. We call a join tree $(\mathcal{T}, \lambda)$ a *join path* if the underlying tree $\mathcal{T}$ is a path.

A hypergraph is defined to be α-*acyclic* if it has a join tree [6]. This is the most general acyclicity notion for hypergraphs commonly considered. However,

Table 1. Known complexity results for structural restrictions of #SAT

class	lower bound	upper bound
primal treewidth		**FPT** [14]
incidence treewidth		**FPT** [14]
modular incidence treewidth	**W**[1]-hard [12]	**XP** [12]
signed incidence cliquewidth		**FPT** [7]
incidence cliquewidth	**W**[1]-hard [11]	**XP** [16]
γ-acyclic		**FP** [9,16]
α-acyclic	#**P**-hard [14]	#**P**
disjoint branches		**FP** (this paper)

α-acyclicity is not closed under taking subhypergraphs. To remedy this situation, one considers the restricted notion of β-*acyclicity* where a hypergraph is defined to be β-acyclic if it is α-acyclic and all of its subhypergraphs are also all α-acyclic.

Definition 2. *A* disjoint branches decomposition *of a hypergraph $\mathcal{H}$ is a join tree $(\mathcal{T}, \lambda)$ such that for every two nodes t and t' appearing on different branches of $\mathcal{T}$ we have $\lambda(t) \cap \lambda(t') = \emptyset$.*

Disjoint branches decompositions were introduced in [5]. A hypergraph is called γ-acyclic if and only if it has a disjoint branches decomposition for any choice of hyperedge as a root. Furthermore, hypergraphs with disjoint branches decompositions are β-acyclic.

2.2 Known Complexity Results and Comparisons between Classes

Known complexity results for the restrictions of #SAT can be found in Table 1; for definitions of the appearing complexity classes see e.g. [8].

The acyclicity notions and classes defined by bounding the introduced width measures form a hierarchy for inclusion (see Figure 1). Most proofs of inclusion can be found in [6,5,9,12] and the references therein.

We now show that the hypergraphs with disjoint branches decompositions have unbounded cliquewidth. In fact we will even show this for hypergraphs with join paths. Since join paths do not branch, these hypergraphs are a subclass of the hypergraphs with disjoint branches decompositions.

We will use the following characterization of hypergraphs with join paths.

Lemma 1. *A hypergraph $\mathcal{H} = (V, E)$ has a join path if and only if there exists an order $<_E$ on the edge set E of $\mathcal{H}$ such that for all $e, f, g \in E$ such that $e <_E f <_E g$, if $v \in e \cap g$ then $v \in f$.*

Definition 3. *Let $G = (X, Y, E)$ be a bipartite graph. A* strong ordering *$(<_X, <_Y)$ of G is a pair of orderings on X and Y such that for all $x, x' \in X$ and $y, y' \in Y$, such that $x <_X x'$ and $y <_Y y'$, if $(x, y) \in E$ and $(x', y') \in E$, then $(x, y') \in E$ and $(x', y) \in E$. G is called a* bipartite permutation graph *if it admits a strong ordering.*

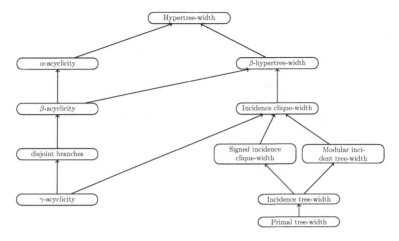

Fig. 1. A hierarchy of inclusion of graph and hypergraph classes. Classes not connected by a directed path are incomparable

Brandstädt and Lozin showed the following property of bipartite permutation graphs.

Lemma 2 ([2]). *Bipartite permutation graphs have unbounded cliquewidth.*

It turns out that hypergraphs with a bipartite permutation incidence graph admit a join path.

Proposition 1. *Every hypergraph $\mathcal{H}$ without empty hyperedges, whose incidence graph $\mathcal{H}_I$ is a bipartite permutation graph, has a join path.*

Proof. Let $(<_V, <_E)$ be a strong ordering of $\mathcal{H}_I = (V, E, A)$. We claim that for all $e <_E f <_E g$, if $v \in e \cap g$ then $v \in f$. Indeed, as f is not empty, there exists $w \in f$. If $w = v$, there is nothing to prove. Otherwise, suppose that $w <_V v$. Then, by definition of strong orderings, as $(f, w) \in A$ and $(g, v) \in A$, we have $(f, v) \in A$. Thus $v \in f$. The case $v <_V w$ follows symetrically: $(f, w) \in A$ and $(e, v) \in A$ implies that $(f, v) \in A$ so $v \in f$. Thus the ordering $<_E$ has the property of Lemma 1 and it follows that $\mathcal{H}$ has a join path.   □

By combining Lemma 2 and Proposition 1 we get:

Corollary 1. *The class of CNF-formulas with join paths has unbounded incidence cliquewidth. The same is thus true for CNF-formulas with disjoint branches decompositions.*

2.3   Representation of #SAT by Constraint Satisfaction Problems

It will be convenient to describe our counting algorithm in the framework of constraint satisfaction problems in negative representation [4]. We will discuss below how this representation relates to #SAT.

Let D be a finite set called domain. A constraint $C = (R, \bar{x})$ is a pair where $R \subseteq D^r$ is a relation and $\bar{x} = (x_{i_1}, \ldots, x_{i_r})$ is a list of variables of length r. An instance Φ of the (uniform) constraint satisfaction problem, is a set of constraints. We denote by $\mathsf{var}(\Phi)$ the set X of variables of Φ. The instance Φ is satisfied by an assignment $a : \mathsf{var}(\Phi) \to D$ if for all $(R, \bar{x}) \in \Phi$ we have $a(\bar{x}) = (a(x_{i_1}), \ldots, a(x_{i_r}))$ is in the relation R. We denote this by $a \models \Phi$.

The associated counting problem, #CSP, is, given an instance Φ, to compute

$$|\{a \mid a \models \Phi\}|,$$

i.e., the number of satisfying assignments of Φ.

No hypothesis is made above on the arity of relations which is not a priori bounded and may differ for different relations. So, it may be more succinct to represent each relation R by listing the tuples in its complement $R^c := D^r \backslash R$. Consequently, we define the counting constraint satisfaction problem in negative representation, #CSP$_{\mathrm{neg}}$, that is to compute, given Φ where each relation R is encoded by listing the tuples in R^c, the number of satisfying assignments of Φ.

The relation to #SAT: It is natural to represent CNF-formulas by a Boolean CSP$_{\mathrm{neg}}$-instance. Indeed, as we do not bound the length k of clauses, it is more realistic to represent each associated constraint relation not by its set of $2^k - 1$ models (as common in the area of constraint satisfaction) but by its complement containing the unique counter model of the clause.

In the other direction one can easily encode every #CSP$_{\mathrm{neg}}$-instance by a CNF-formula (see also [3]): In a first step encode all domain elements in binary, introducing vertex modules in the hypergraph. Then we encode every tuple in every relation by a clause that disallows the respective tuple. Observe that after the contraction of some modules, the CNF-formula has the same hypergraph as the original #CSP$_{\mathrm{neg}}$-instance. Since the class of hypergraphs with disjoint branches decompositions is stable under introducing or contracting of modules, it follows that #SAT and #CSP$_{\mathrm{neg}}$ are equivalent for our considerations.

3   Counting Solutions of Disjoint Branches Queries

In this section we will show that #CSP$_{\mathrm{neg}}$—and thus also #SAT—restricted to hypergraphs with a disjoint branches decomposition can be solved in polynomial time. It will be convenient to work with inputs in a disjunctive form: Let #DCSP be the problem of computing, given a finite domain D and a set of constraints $C_1, \ldots, C_m$ with variables X, the number of assignments $a : X \to D$ such that at least one constraint is satified. The following is easy to see.

Observation 1. *#CSP$_{\mathrm{neg}}$ is polynomial time reducible to #DCSP.*

Note that, as discussed in Section 2.3, we may assume that the domain of all relations is $\{0, 1\}$, so we restrict ourselves to this case.

Let us introduce some notation: Let X, Y be two sets of variables and let $a : X \to \{0, 1\}$ and $b : Y \to \{0, 1\}$ be two assignments. We call a and b *consistent,* symbol $a \sim b$, if they agree on their common variables $X \cap Y$. If, in addition, $X \subseteq Y$, we write $a \subseteq b$. Finally, if a and b have disjoint domains, i.e., $X \cap Y = \emptyset$, we denote by $a \oplus b$ the assignment on $X \cup Y$ defined in the obvious way.

For an assignment $a \colon X \mapsto \{0,1\}$ and a set of variables $Y \subseteq X$ the restriction of a onto Y is denoted by $a|_Y$.

Let ϕ be an instance and let ψ be a subformula of ϕ. Let X a subset of the variables such that $\mathsf{var}(\psi) \subseteq X \subseteq \mathsf{var}(\phi)$ and let a be a partial assignment of variables in $\mathsf{var}(\phi)$. We denote by $Sol_X(\psi, a) = \{b : X \to \{0,1\}, b \models \psi, a \sim b\}$ and by $S_X(\psi, a) = |Sol_X(\psi, a)|$. The number of solutions of ϕ is then $S_{\mathsf{var}(\phi)}(\phi, \emptyset)$, where $\emptyset$ denotes the empty partial assignment. We show that by computing a polynomial number of values $S_X(\psi, a)$ in polynomial time, we can compute $S_{\mathsf{var}(\phi)}(\phi, \emptyset)$. To this end we state several lemmas that will allow us a recursive computation.

The first lemma shows how the disjointness naturally appears when we want to count solutions:

Lemma 3. *Let ϕ_1 and ϕ_2 be two instances and $X \supseteq \mathsf{var}(\phi_1) \cup \mathsf{var}(\phi_2)$, then*

$$S_X(\phi_1 \vee \phi_2, a) = S_X(\phi_1, a) + S_X(\phi_2, a) - |Sol_X(\phi_1, a) \cap Sol_X(\phi_2, a)|.$$

The next lemma will allow us to efficiently compute $Sol_X(\phi_1 \vee \phi_2, a)$ recursively.

Lemma 4. *Let $X_1 = \mathsf{var}(\phi_1)$ and $X_2 = \mathsf{var}(\phi_2)$. Assume that $X_1 \cap X_2 = \emptyset$ and let $X = X_1 \cup X_2$. Let a be a partial assignment of the variables of X and $a_1 = a|_{X_1}$, $a_2 = a|_{X_2}$. Then $Sol_X(\phi_1, a) \cap Sol_X(\phi_2, a) = \{b_1 \oplus b_2 \mid b_i : X_i \to \{0,1\}, b_i \models \phi_i, a_i \subseteq b_i\}$ and $|Sol_X(\phi_1, a) \cap Sol_X(\phi_2, a)| = S_{X_1}(\phi_1, a_1) S_{X_2}(\phi_2, a_2)$.*

We now show how we can add variables that do not appear in ϕ.

Lemma 5. *Let $Y \subseteq X$ and $a : X_0 \to \{0,1\}$ for $X_0 \subseteq X$ with $\mathsf{var}(\phi) \subseteq X_0 \cup Y$. Then*

$$S_X(\phi, a) = 2^{|X \setminus (Y \cup X_0)|} S_Y(\phi, a|_Y).$$

The next corollary lets us handle the disjunction of more than two terms. The proof is by induction with Lemma 4.

Corollary 2. *Let $\phi_1, \ldots, \phi_k$ be formulas with $X_i = \mathsf{var}(\phi_i)$ and $X_i \cap X_j = \emptyset$ for every combination $i, j \in [k], i \neq j$. Let X be a set such that $X_1 \cup \ldots \cup X_k \subseteq X$, let $Y := X_0 \cup X_1 \cup \ldots \cup X_k$ and let $a : X_0 \to \{0,1\}$ for $X_0 \subseteq X$, $a_i = a|_{X_i}$. Then*

$$S_X\left(\bigvee_{j=1}^{k} \phi_j, a\right) = 2^{|X \setminus Y|} \sum_{i=1}^{k} S_{X_i}(\phi_i, a_i) \prod_{j=1}^{i-1} (2^{|X_j \setminus X_0|} - S_{X_j}(\phi_j, a_j)) \prod_{j=i+1}^{k} 2^{|X_j \setminus X_0|}$$

Lemma 6. $|Sol_X(R, a) \cap Sol_X(\phi, a)| = \sum_{b \in Sol_X(R, a)} S_X(\phi, b)$

We now finally show the main result of this section.

Theorem 1. *There is a polynomial time algorithm that, given an instance $\phi = \bigvee_{i=1}^{m} R_i$ and a disjoint branches decomposition of the hypergraph of ϕ, computes the number of satisfying assignments of ϕ.*

Proof. Let $\phi = R_1 \vee \ldots \vee R_m$ be an instance with hypergraph $\mathcal{H}$. Let $r_i := |R_i|$ and let $r := \sum_{i=1}^{m} r_i$. Let furthermore $(\mathcal{T}, \lambda)$ be a disjoint branches decomposition of $\mathcal{H}$. For a vertex t of $\mathcal{T}$, we denote by $\mathcal{T}_t$ the subtree of $\mathcal{T}$ rooted in t and by ϕ_t the associated subinstance. Furthermore, R_t is defined to be the relation associated to t. Finally, we denote be V_t the set of variables of ϕ_t.

We will give a polynomial time algorithm that computes inductively from the leaves to the root of $\mathcal{T}$ certain values $S_X(\psi, a)$ where ψ is a subinstance of ϕ. Our goal is to compute $S_{V_r}(\phi_r, \emptyset)$ where r is the root of $\mathcal{T}$, since this value is the number of solutions of ϕ. More precisely, for a given t in the tree, we compute $S_{V_t}(\phi_t, \emptyset)$ and for all ancestors u of t and all $b \in R_u$, we compute $S_{V_t}(\phi_t, b|_{\text{var}(R_t)})$. Since there are m vertices in $\mathcal{T}$ and at most $r + 1$ values to compute for each vertex, we will to compute at most $m(r + 1)$ different values. We will show how to compute these values in polynomial time to get a polynomial time algorithm overall.

Let t be a vertex of $\mathcal{T}$, u one of his ancestors and $b \in R_u$. If t is a leaf, then ϕ_t consists only of the relation R_t. Then $Sol_{V_t}(\phi_t, \emptyset) = R_t$ and $Sol_{V_t}(\phi_t, b|_{\text{var}(R_t)}) = \{a \in R_t \mid a \sim b\}$ so the computations can be done efficiently.

Now assume that t has children $t_1, \ldots, t_k$. We set $V_i := V_{t_i}$, $\phi_i := \phi_{t_i}$ and $R_i := R_{t_i}$. Observe that by Lemma 3 we have

$$S_{V_t}(\phi_t, \emptyset) = S_{V_t}(R_t, \emptyset) + S_{V_t}(\bigvee_{j=1}^{k} \phi_j, \emptyset) - |Sol_{V_t}(R_t, \emptyset) \cap Sol_{V_t}(\bigvee_{j=1}^{k} \phi_j, \emptyset)|.$$

As the variables of the ϕ_i are disjoint, Corollary 2 lets us compute $S_{V_t}(\bigvee_{j=1}^{k} \phi_j, \emptyset)$ in time $O(k)$ if the values of $Sol_{V_i}(\phi_i, \emptyset)$ are precomputed, which is the case by induction.

In addition, $S_{V_t}(R_t, \emptyset) = 2^{|V_t \setminus \text{var}(R_t)|} |R_t|$ since a solution of R_t on variables V_t is a solution of R_t on $\text{var}(R_t)$ and any assignment of the other variables.

Finally, we have $|Sol_{V_t}(R_t, \emptyset) \cap Sol_{V_t}(\bigvee_{j=1}^{k} \phi_j, \emptyset)| = \sum_{a \in R_t} S_{V_t}(\bigvee_{j=1}^{k} \phi_j, a)$ by Lemma 6. By Corollary 2, one can compute for each a the value $S_{V_t}(\bigvee_{j=1}^{k} \phi_j, a)$ in time $O(k)$ if the values of $S_{V_i}(\phi_i, a|_{V_i})$ are precomputed. But since the domain of a is $\text{var}(R_t)$, we have $a|_{V_i} = a|_{\text{var}(R_t) \cap V_i} = a|_{\text{var}(R_i)}$ by connectedness of the variables in the join tree $\mathcal{T}$. Thus $S_{V_i}(\phi_i, a|_{V_i}) = S_{V_i}(\phi_i, a|_{\text{var}(R_i)})$ which is precomputed by hypothesis.

Let $b' = b|_{\text{var}(R_t)}$. We compute $S_{V_t}(\phi_t, b')$ in the following way, similarly to before. We start with Lemma 3 to get

$$S_{V_t}(\phi_t, b') = S_{V_t}(R_t, b') + S_{V_t}(\bigvee_{j=1}^{k} \phi_j, b') - |Sol_{V_t}(R_t, b') \cap Sol_{V_t}(\bigvee_{j=1}^{k} \phi_j, b')|.$$

Again, by Corollary 2, one can compute $S_{V_t}(\bigvee_{j=1}^{k} \phi_j, b')$ in time $O(k)$ if $S_{V_i}(\phi_i, b'|_{V_i})$ are known. But as the domain of b' is $\text{var}(R_t)$, $b'|_{V_i} = b'|_{\text{var}(R_i)}$

by connectedness of the variables in $\mathcal{T}$. So $S_{V_i}(\phi_i, b'|_{V_i})$ is precomputed since u is also an ancestor of t_i.

Moreover, $S_{V_t}(R_t, b') = \sum_{a \in R_t, b' \subseteq a} 2^{|V_t \setminus \mathsf{var}(R_t)|}$ which can be computed in time $O(|R_t|)$.

Finally, by Lemma 6, we have

$$|Sol_{V_t}(R_t, b') \cap Sol_{V_t}(\bigvee_{j=1}^{k} \phi_j, b')| = \sum_{a \in R_t, b' \subseteq a} S_{V_t}(\bigvee_{j=1}^{k} \phi_j, a).$$

And again, by Corollary 2, we can compute $S_{V_t}(\bigvee_{j=1}^{k} \phi_j, a)$ in time $O(k)$ if $S_{V_i}(\phi_i, a|_{V_i})$ is precomputed. For the same reasons as above, $a|_{V_i} = a|_{\mathsf{var}(R_i)}$, thus these values were already computed by induction.

To conclude, we have seen how to compute the $S_{V_t}(\phi_t, \emptyset)$ and $S_{V_t}(\phi_t, b|_{\mathsf{var}(R_t)})$ for each $b \in R_u$ where u is an ancestor of t with $O(k \cdot r)$ arithmetic operations. Thus we can compute $S_{V_r}(\phi_r, \emptyset)$ in polynomial time. $\qquad\square$

4 Computing Disjoint Branches Decompositions

In this section we will show how to compute disjoint branches decompositions of hypergraphs in polynomial time. We will first introduce PQF-trees, the datastructure that our algorithm relies on, then consider some structural properties of hypergraphs with disjoint branches decompositions and finally describe the algorithm itself, relying on objects we call A-separators.

4.1 PQF-Trees

PQ-trees are a data structure introduced by Booth and Lueker [1] originally to check matrices for the so-called consecutive ones property. This problem can be reformulated as follows in our setting: Given a hypergraph $\mathcal{H} = (V, E)$, is there an ordering $\ell = e_1 \ldots e_m$ of the edges such that if $v \in e_i \cap e_j$, then for all $i \leq k \leq j$, $v \in e_k$? We encode ordering of edges by lists. We call such a list *consistent* for $\mathcal{H}$. Note that the notion of consistent lists matches exactly our notion of join paths.

A PQ-tree is a compact way of representing all the consistent lists for a hypergraph. We introduce a generalization of this data structure which we call PQF-trees.

Definition 4. *Let $\mathcal{H} = (V, E)$ be a hypergraph. A PQF-tree for $\mathcal{H}$ is defined to be an ordered tree with leaf set E such that*

- *the internal nodes are labeled with P, Q or F,*
- *the P-nodes and F-nodes have at least two children, and*
- *the Q-nodes have at least 3 children.*

A PQF-tree without F-nodes is called a PQ-tree.

PQF-trees will be used to encode sets of permutations of the edge set of a hypergraph that have certain properties. We write these permutations simply as (ordered) lists. To this end, we define some notation for lists and sets of lists. The concatenation of two ordered lists ℓ_1, ℓ_2 will be denoted by $\ell_1\ell_2$. If L_1, L_2 are two sets of lists, we denote by $L_1 L_2$ the set $\{\ell_1\ell_2 \mid \ell_1 \in L_1, \ell_2 \in L_2\}$.

Definition 5. *The* frontiers $\mathcal{F}(T)$ *of a PQF-tree T for $\mathcal{H} = (V, E)$ are a set of ordered list of the elements of E defined inductively by*

- *if T is a leaf e, then $\mathcal{F}(T) = \{e\}$,*
- *if T is rooted in t, having children $t_1, \ldots, t_k$, then*
  - *if t is an F-node then $\mathcal{F}(T) = \mathcal{F}(T_1) \ldots \mathcal{F}(T_k)$,*
  - *if t is a Q node then $\mathcal{F}(T) = (\mathcal{F}(T_1) \ldots \mathcal{F}(T_k)) \cup (\mathcal{F}(T_k) \ldots \mathcal{F}(T_1))$,*
  - *if t is a P-node then $\mathcal{F}(T) = \bigcup_{\sigma \in \mathcal{S}_k} \mathcal{F}(T_{\sigma(1)}) \ldots \mathcal{F}(T_{\sigma(k)})$ where $\mathcal{S}_k$ is the set of permutations of $[k]$,*

  where T_i is the subtree of T rooted in t_i.

If for all $\ell \in \mathcal{F}(T)$, ℓ is a consistent list for $\mathcal{H}$, we say that T is consistent for $\mathcal{H}$.

We recall the main theorem of [1], which allows to compute all possible join paths of a hypergraph in polynomial time.

Theorem 2 ([1]). *Given a hypergraph $\mathcal{H} = (V, E)$, one can compute in time $O(|E||V|)$ a PQ-tree T such that $\mathcal{F}(T)$ is exactly the set of consistent lists for $\mathcal{H}$.*

In order to compute disjoint branches decompositions, we will need to compute join paths with additional restrictions. This is the reason for the introduction of F-nodes. It will be convenient to not have F-nodes that are children of other F-nodes and thus we introduce the following normal form for PQF-trees.

Definition 6. *A PQF-tree T is said to be in* normal form *if there is no F-node in T having an F-node as a child.*

Clearly, if an F-node t has a child u in T which is also an F-node, then we can remove u from T and connect its children to t without changing $\mathcal{F}(T)$. Thus we may always assume that all PQF-trees we encounter are in normal form.

We will in the remainder of this section use certain subtrees of PQF-trees which we call PQF-subtrees. These will be trees rooted in a vertex t of a PQF-tree, but they will not necessarily contain all descendants of t. Instead, we allow to "cut off" certain trees that are rooted by children of t. We now give a formal definition of PQF-subtrees. As usual, the *subtree rooted in t* is defined to be the tree induced by t and all its descendants.

Definition 7. *Let T be a PQF-tree, and let t be a vertex of T. A subgraph S of T is said to be a PQF-subtree rooted in t if*

- *t is a leaf and S consists of the graph containing only t,*
- *t is a P-node and S is the subtree rooted in t, or*

- t is a Q-node or an F-node with children $t_1, \ldots, t_k$ and there exists i, j such that $1 \leq i < j \leq k$ and S is the graph containing t and $T_i, \ldots, T_k$, the subtrees rooted in $t_i, \ldots, t_j$.

We will now show that PQF-subtrees allow us to filter the frontier of a PQF-tree for certain lists that we will be interested in later. Remember that the *depth* of a node in a tree is its distance from the root.

Lemma 7. *Let T be a consistent PQF-tree for (V, E) in normal form. Let $V' \subseteq V$ and $A = \{e \in E \mid V' \subseteq e\}$. Then there exists a PQF-subtree $T_{V'}$ of T such that the labels of the leaves of $T_{V'}$ are exactly A.*

During the construction of disjoint branches decompositions later, we will put restrictions on the position of some edges in join paths. To do so we will use the algorithm of the following proposition.

Proposition 2. *There is a polynomial time algorithm* **Force** *that, given a PQF-tree T and a PQF-subtree S of T, computes in polynomial time a PQF-tree $T' = \mathrm{Force}(T, S)$ such that $\mathcal{F}(T') = \{\ell_1 \ell_2 \in \mathcal{F}(T) \mid \ell_2 \in \mathcal{F}(S)\}$. If this set is empty, the algorithm rejects.*

Corollary 3. *Let $\mathcal{H} = (V, E)$ be a hypergraph and T a consistent PQF-tree for $\mathcal{H}$. Let $V' \subseteq V$ and $A = \{e \in E \mid e \cap V' \neq \emptyset\}$. Suppose that for all $e, f \in A$, $e \cap V' \subseteq f \cap V'$ or $f \cap V' \subseteq e \cap V'$. Then we can compute in polynomial time a PQF-tree T' such that $\mathcal{F}(T') = \{e_1 \ldots e_m \in \mathcal{F}(T) \mid \forall i < j, e_i, e_j \in A \Rightarrow e_i \cap V' \subseteq e_j \cap V'\}$.*

Proof. We want to compute T' such that the frontiers of T' are the frontiers of T in which the edges of A appear in increasing order with respect to inclusion relative to V'. The set $\{e \cap V' \mid e \in A\}$ is ordered by inclusion, thus it has a smallest element V_1 and a biggest element V_2. Obviously, $V_1 \subseteq V_2 \subseteq V'$. Furthermore, $A = \{e \in E \mid V_1 \in e\}$ and $A_2 = \{e \in A \mid V_2 \subseteq e\}$ is not empty.

First use Lemma 7 to find a PQF-subtree S of T whose leaves are exactly $\{e \mid V_1 \subseteq e\} = A$. Then use Lemma 7 again to find a PQF-subtree R of S whose leaves are exactly A_2. Now use the procedure $\mathrm{Force}(S, R)$ to compute S' as in Proposition 2 and let T' be the tree where we replace S by S' in T. As finding the right subtrees in T can easily be done in polynomial time by finding a least common ancestor and Force is a polynomial time, it is clear that one can compute T' as well. We now show that T' has the desired properties.

To this end, let $\ell \in \mathcal{F}(T')$. By definition of T', we have that ℓ is also in $\mathcal{F}(T)$. In addition, ℓ is of the form $\ell_1 \ell_A \ell_2$ with $\ell_A \in \mathcal{F}(S')$. By definition of Force, ℓ_A is of the form $\sigma_1 \sigma_2$ with $\sigma_2 \in \mathcal{F}(R)$, that is, consisting only of the edges in A_2, which are maximal for the inclusion. Let $g \in A_2$. Let $e, f \in A$ with e appearing before f in ℓ_A. If $e \cap V'$ is not included in $f \cap V'$, then there exists a $v' \in V_2$ such that $v' \in e$, $v' \notin f$ and $v' \in g$ since $V_2 \subseteq g$. That would lead to an inconsistent list which is a contradiction.

Reciprocally, let $\ell \in \{e_1 \ldots e_m \in \mathcal{F}(T) \mid \forall i < j, e_i, e_j \in A \Rightarrow e_i \cap V' \subseteq e_j \cap V'\}$. ℓ is of the form $\ell_1 \ell_A \ell_2$ with $\ell_A \in \mathcal{F}(S)$. As it is organized by inclusion relative to V', the elements of A_2 should all lie at the end of ℓ_A. Thus $\ell_A \in \{\sigma_1\sigma_2 \mid \sigma_2 \in \mathcal{F}(R)\} = \mathcal{F}(S')$ by Proposition 2. It follows that $\ell \in \mathcal{F}(T')$. $\qquad\square$

4.2 Db-Rootable Hypergraphs and Separators

In this section we will prove several structural properties of hypergraphs with disjoint branches decompositions which we will use in the algorithm in the next section.

Definition 8. *Let $\mathcal{H} = (V, E)$ be a hypergraph. For $e \in E$, we say that $\mathcal{H}$ is db-rootable* in e *if there is a disjoint branches decomposition of $\mathcal{H}$ rooted in e.*

The algorithm for the construction of disjoint branches decompositions will delete edges of hypergraphs. To this end we introduce the following notation.

Definition 9. *For a hypergraph $\mathcal{H} = (V, E)$ and an edge $e \in E$, we denote $H \setminus e$ the hypergraph $(V_e, E \setminus \{e\})$ where $V_e := \bigcup_{e' \in E \setminus \{e\}} e'$. For a set $A = \{e_1, \ldots, e_k\} \subseteq E$, we define $\mathcal{H} \setminus A$ to be the hypergraph $((\mathcal{H} \setminus e_1) \setminus \ldots) \setminus e_k$.*

We make the following observation which will simplify our arguments later.

Observation 2. *Let $\mathcal{H} = (V, E)$ be a hypergraph and $e \in E$. Then $\mathcal{H}$ is db-rootable in e if and only if for every connected component $C = (V_C, E_C)$ of $\mathcal{H} \setminus e$ the hypergraph $C' := (V_C \cup e, E_C \cup \{e\})$ is db-rootable in e.*

Lemma 8. *Let $\mathcal{H}$ be a hypergraph $\mathcal{H}$ with a disjoint branches decomposition T that is rooted in e. Let and v_1, v_2 be two vertices that appear in different trees T_1 and T_2 of the forest $T \setminus \{e\}$. Then v_1 and v_2 lie in different connected components of $\mathcal{H} \setminus e$.*

Lemma 9. *If $\mathcal{H}$ is db-rootable in e and $\mathcal{H} \setminus e = (V', E')$ has one connected component then there exists $e' \in E'$ such that $e \cap V' \subseteq e'$.*

By Observation 2 we may deal with the components of $\mathcal{H} \setminus e$ for a hypergraph $\mathcal{H}$ and an edge e independently. Thus we will in this section always assume that $\mathcal{H} \setminus e$ just has a single component. We will consider restricted join paths that we call A-separators. In the following, all join paths will be denoted as ordered lists of edges, which corresponds to the notation in Section 4.1.

Definition 10. *Let $\mathcal{H} = (V, E)$ be a hypergraph and let $\mathcal{P} = a_1 \ldots a_m$ be a join path of $A \subseteq E$. We call $\mathcal{P}$ an A-separator of $\mathcal{H}$ if for all connected components $C = (V_C, E_C)$ of $\mathcal{H} \setminus A$ we have that if $a_j \cap V_C \neq \emptyset$, then for all $i \leq j$, $a_i \cap V_C \subseteq a_j \cap V_C$.*

Theorem 3. *There is a polynomial time algorithm* ComputeSeparator$(\mathcal{H}, A)$ *that, given a hypergraph $\mathcal{H} = (V, E)$ and a set $A \subseteq E$, computes an A-separator of $\mathcal{H}$ if it exists and rejects otherwise.*

Proof. We will iterate the algorithm described in Corollary 3. We first compute a PQ-tree T_0 such that $\mathcal{F}(T_0)$ is the set of all join paths for A using Theorem 2. Then, for each component $C = (V_C, E_C)$ of $\mathcal{H} \setminus A$, we iteratively do the following: If there are edges a_i and a_j such that $a_i \cap V_C \not\subseteq a_j \cap V_C$ and $a_j \cap V_C \not\subseteq a_i \cap V_C$, then $\mathcal{H}$ cannot have an A-separator and we reject. Otherwise, we can use the algorithm of Corollary 3 on the PQF-tree we have computed so far to construct a PQF-tree whose frontiers respect the order condition for the edges imposed by C or rejects.

An easy induction shows that if this algorithm does not reject at any point, then the computed PQF-tree T has as frontiers all join paths that satisfy the order conditions of Definition 10. Thus we can choose one of these join paths arbitrarily as the desired A-separator. □

Definition 11. *Let $\mathcal{H} = (V, E)$ be a hypergraph and let $A \subseteq E$. We call an A-separator $\mathcal{P} = a_1 \ldots a_m$ of $\mathcal{H}$ a strong A-separator if for all connected components $C = (V_C, E_C)$ of $\mathcal{H} \setminus A$, $C' = (V_C, E_C \cup \{a_{l_C}\})$ is db-rootable in a_{l_C} where $l_C = \max\{i \mid a_i \cap V_C \neq \emptyset\}$.*

Proposition 3. *Let $\mathcal{H} = (V, E)$ be a hypergraph and $A \subseteq E$. If there exists a strong A-separator of $\mathcal{H}$ then all A-separators of $\mathcal{H}$ are strong.*

Theorem 4. *Let $\mathcal{H} = (V, E)$ be a hypergraph, $e \in E$, and $\mathcal{H}' = (V', E') = \mathcal{H} \setminus e$. Let furthermore $A_e := \{e' \in E' \mid e \cap V' \subseteq e'\}$. Assume that $\mathcal{H}'$ has only one connected component. Then $\mathcal{H}$ is db-rootable in e if and only if there exists a strong A_e-separator of $\mathcal{H}'$.*

Proof. Suppose $\mathcal{H}$ is db-rootable in e. Let $\mathcal{T}$ be a disjoint branches decomposition of $\mathcal{H}$ rooted in e. By Lemma 9, A_e is not empty. Let $v \in e \cap V'$. We know that v is contained in all edges $e' \in A_e$. Thus, by disjointness, the edges in A_e are on the same branch of $\mathcal{T}$. Moreover, we claim that A_e is connected in $\mathcal{T}$. To see this, suppose that $b \in E$ is between $a, c \in A_e$ on this branch. Then by connectedness, $e \cap V' \subseteq b$, so $b \in A_e$. Consequently, A_e is connected and thus forms a path. Let $\mathcal{P} = a_1 \ldots a_k$ be this path in $\mathcal{T}$ in the direction from the root to the leaves of the tree. We claim that $\mathcal{P}$ is a strong A_e-separator.

To this end, let $C = (V_C, E_C)$ be a connected component of $\mathcal{H}' \setminus A_e$. We consider the forest obtained by removing A_e in $\mathcal{T}$. By Lemma 8, vertices in different trees of this forest are in different connected component of $\mathcal{H}' \setminus A_e$ as well. Thus there is a tree $\mathcal{T}_C$ that contains all the edges in E_C. Let a_{l_C} be the edge of A_e to which the root of $\mathcal{T}_C$ is connected. If $j > l_C$, then we claim that $a_j \cap V_C = \emptyset$. Assume this were not the case, then $\mathcal{T}_C$ and the subtree of a_{l_C} containing a_j were not disjoint which is a contradiction to $\mathcal{T}$ being a disjoint branches decomposition. Thus $a_j \cap V_C = \emptyset$. Now consider $i < j \leq l_C$ and let $v \in a_i \cap V_C$. We have $v \in a_{l_C} \cap V_C$ by the connectedness condition since v appears in $\mathcal{T}_C$ and, again by connectedness the connectedness condition, $v \in a_j \cap V_C$. Thus, $a_i \cap V_C \subseteq a_j \cap V_C$. It follows that $\mathcal{P}$ is an A_e-separator. As $C \cup \{a_{l_C}\}$ is db-rootable in a_{l_C} using the tree $\mathcal{T}_C$, we have that $\mathcal{P}$ is a strong A_e-separator.

Assume now that there is a strong A_e-separator $\mathcal{P} = a_1 \ldots a_k$. For a connected component C of $\mathcal{H}' \setminus A_e$, let $l_C := \max\{i \mid a_i \cap V_C \neq \emptyset\}$. By definition, there exists a disjoint branches decomposition $\mathcal{T}_C$ for $C \cup \{a_{l_C}\}$ rooted in a_{l_C}. We construct a disjoint branches decomposition for $\mathcal{H}$ as follows:

- We root the path $ea_1 \ldots a_k$ in e.
- For each component C of $\mathcal{H}' \setminus A_e$, we connect the root of $\mathcal{T}_C$ to a_{l_C}.

We claim that the resulting tree $\mathcal{T}$ is a disjoint branches decomposition. We first show that the branches of $\mathcal{T}$ are disjoint. Indeed, if a, b are edges in two different branches then two cases can occur: Either a and b are in two different connected component of $\mathcal{H}' \setminus A_e$. But then they are disjoint, because they lie in different components of $\mathcal{H}$ by construction. Otherwise, let a be in a connected component C of $\mathcal{H}' \setminus A_e$ and let b be in A_e. But as b is on a different branch as a, it follows that b comes after a_{l_C} on $\mathcal{P}$. Thus $b \cap V_C = \emptyset$ and it follows that $a \cap b = \emptyset$ since $a \subseteq V_C$. Thus the branches of $\mathcal{T}$ are disjoint.

Now we show that $\mathcal{T}$ is a join tree, i.e., it satisfies the connectedness condition for all vertices. For a connected component C of $\mathcal{H}' \setminus A_e$, for a vertex $v \in V_C$, its connectedness is ensured along $\mathcal{P}$ since $\mathcal{P}$ is a join path and in $\mathcal{T}_C$ since $\mathcal{T}_C$ is a join tree. Furthermore, by construction $v \in a_{l_C}$ so the edges containing v are connected in $\mathcal{T}$. For a vertex v, which does not appear in any V_C, that is, which only appears in A_e, its connectedness is ensured by the fact that $\mathcal{P}$ is a join path for A_e. □

Corollary 4. *There is a polynomial time algorithm* ComputeDB$(\mathcal{H}, e)$ *that, given a hypergraph* $\mathcal{H} = (V, E)$ *and and edge* $e \in E$, *returns a disjoint branches decomposition of* $\mathcal{H}$ *rooted in* e *if it exists and rejects otherwise.*

A rough analysis of the runtime of the algorithm gives an upper bound of $O(|E|^3|V|)$. We believe firmly this can be improved by using smart data structures.

5   Some Negative Results on Generalizations

We now discuss several approaches to generalizing the counting algorithm for hypergraphs with disjoint branches to more general classes of hypergraphs. Unfortunately, all these results will be negative as we will show hardness results for all extensions we consider. We still feel these results are worthwhile because they might help in guiding future research.

We call a hypergraph $\mathcal{H}$ *undirected path acyclic* if there is a join tree $(\mathcal{T}, \lambda)$ of $\mathcal{H}$ such that for every $v \in V$ the edge set $\{e \in E \mid v \in e\}$ forms an undirected path in $\mathcal{T}$. Undirected path acyclicity is a seemingly natural generalization of disjoint branches acyclicity by allowing undirected paths instead of directed paths. Unfortunately, undirected path acyclicity does not not allow tractable SAT and thus it is not a good generalization in our setting.

Theorem 5. SAT *on undirected path acyclic CNF-formulas is* **NP**-*complete.*

Algorithm 1. The algorithm `ComputeDB` of Corollary 4.

`ComputeDB`$(\mathcal{H} = (V, E), e) =$
if $|E| = 1$ **then return** the tree with the only vertex e
else
    for each connected component $\mathcal{H}_i = (V_i, E_i)$ of $\mathcal{H} \setminus e$ **do**
        $A_e \leftarrow \{e' \in E_i \mid e \cap V_i \subseteq e'\}$
        if $A_e = \emptyset$ **then** Reject.
        $\mathcal{P} \leftarrow$ `ComputeSeparator`$(\mathcal{H}_i, A_e)$
        $\mathcal{T}_i \leftarrow \mathcal{P}$
        for each connected component $C = (V_C, E_C)$ of $\mathcal{H}_i \setminus A_e$ **do**
            $l_C \leftarrow \max\{j \mid a_j \cap V_C \neq \emptyset\}$ (where $\mathcal{P} = a_1 \dots a_k$)
            $C' \leftarrow (V_C \cup a_{l_C}, E_C \cup \{a_{l_C}\})$
            $\mathcal{T}_C \leftarrow$ `ComputeDB`(C', a_{l_C})
            connect $\mathcal{T}_C$ to $\mathcal{T}_i$ in a_{l_C}
    return the tree rooted in e having $\mathcal{T}_1, \dots, \mathcal{T}_p$ as children

One also can generalize disjoint branches decompositions by allowing limited intersections between branches. Unfortunately, this approach leads to hard counting problems even if we only allow the intersection to contain one variable.

Theorem 6. *#SAT is #**P**-hard for CNF-formulas that have a join tree in which the branches may may have a pairwise intersection containing one variable.*

6 Conclusion

We have presented a new structural class of tractable #SAT instances, those whose hypergraphs admit a disjoint branches decomposition. To this end, we also presented an algorithm that computes the decompositions.

Several questions remain, the most obvious open problem certainly being the complexity of #SAT on β-acyclic hypergraphs. Can one show a #**P**-completeness result or a polynomial time algorithm for this case?

We would like to turn the disjoint branches property into a hypergraph width measure such that #SAT—or even SAT—for the hypergraphs for which this width measure is bounded is tractable? Can we construct this measure to even allow fixed-parameter tractability? Note that it is known that the parameterization by incidence cliquewidth does not allow fixed-parameter tractability [11].

More generally, we feel that it is very desirable to understand the tractability frontier for SAT and #SAT with respect to structural restrictions better overall. Is there a width measure that generalizes both hypergraphs with disjoint branches and incidence cliquewidth that leads to tractable #SAT? A natural candidate would be β-hypertree width (see Figure 1). Are there other classes of hypergraphs incomparable to those studied so far that give large structural classes of tractable #SAT-instances? Note that there is a similar line in the area of constraint satisfaction (see e.g. [10] for an overview) that has been very successful but unfortunately does not apply directly.

References

1. Booth, K., Lueker, G.: Testing for the consecutive ones property, interval graphs, and graph planarity using PQ-tree algorithms. Journal of Computer and System Sciences 13(3), 335–379 (1976), http://www.sciencedirect.com/science/article/pii/S0022000076800451
2. Brandstädt, A., Lozin, V.V.: On the linear structure and clique-width of bipartite permutation graphs. Ars Combinatoria 67(1), 273–281 (2003)
3. Brault-Baron, J.: A Negative Conjunctive Query is Easy if and only if it is Beta-Acyclic. In: Computer Science Logic, 26th International Workshop/21st Annual Conference of the EACSL, pp. 137–151 (2012)
4. Cohen, D.A., Green, M.J., Houghton, C.: Constraint representations and structural tractability. In: Gent, I.P. (ed.) CP 2009. LNCS, vol. 5732, pp. 289–303. Springer, Heidelberg (2009), http://dl.acm.org/citation.cfm?id=1788994.1789021
5. Duris, D.: Some characterizations of γ and β-acyclicity of hypergraphs. Inf. Process. Lett. 112(16), 617–620 (2012)
6. Fagin, R.: Degrees of acyclicity for hypergraphs and relational database schemes. Journal of the ACM 30(3), 514–550 (1983)
7. Fischer, E., Makowsky, J.A., Ravve, E.V.: Counting truth assignments of formulas of bounded tree-width or clique-width. Discrete Applied Mathematics 156(4), 511–529 (2008)
8. Flum, J., Grohe, M.: Parameterized Complexity Theory. Springer-Verlag New York Inc. (2006)
9. Gottlob, G., Pichler, R.: Hypergraphs in Model Checking: Acyclicity and Hypertree-Width versus Clique-Width. SIAM Journal on Computing 33(2) (2004)
10. Miklós, Z.: Understanding Tractable Decompositions for Constraint Satisfaction. Ph.D. thesis. University of Oxford (2008)
11. Ordyniak, S., Paulusma, D., Szeider, S.: Satisfiability of acyclic and almost acyclic CNF formulas. Theoretical Computer Science 481, 85–99 (2013)
12. Paulusma, D., Slivovsky, F., Szeider, S.: Model Counting for CNF Formulas of Bounded Modular Treewidth. In: 30th International Symposium on Theoretical Aspects of Computer Science, STACS 2013, pp. 55–66 (2013)
13. Roth, D.: On the hardness of approximate reasoning. Artificial Intelligence 82(1-2), 273–302 (1996), http://www.sciencedirect.com/science/article/pii/0004370294000921
14. Samer, M., Szeider, S.: Algorithms for propositional model counting. Journal of Discrete Algorithms 8(1), 50–64 (2010)
15. Slivovsky, F.: Personal communication (2014)
16. Slivovsky, F., Szeider, S.: Model Counting for Formulas of Bounded Clique-Width. In: Cai, L., Cheng, S.-W., Lam, T.-W. (eds.) ISAAC 2013. LNCS, vol. 8283, pp. 677–687. Springer, Heidelberg (2013)

Automatic Evaluation of Reductions between NP-Complete Problems*

Carles Creus, Pau Fernández, and Guillem Godoy

Universitat Politècnica de Catalunya, Department of Software
ccreuslopez@gmail.com, ggodoy@lsi.upc.edu, pau.fernandez@upc.edu

Abstract. We implement an online judge for evaluating correctness of reductions between NP-complete problems. The site has a list of exercises asking for a reduction between two given problems. Internally, the reduction is evaluated by means of a set of tests. Each test consists of an input of the first problem and gives rise to an input of the second problem through the reduction. The correctness of the reduction, that is, the preservation of the answer between both problems, is checked by applying additional reductions to SAT and using a state-of-the-art SAT solver. In order to represent the reductions, we have defined a new programming language called REDNP. On one side, REDNP has specific features for describing typical concepts that frequently appear in reductions, like graphs and clauses. On the other side, we impose severe limitations to REDNP in order to avoid malicious submissions, like the ones with an embedded SAT solver.

Keywords: SAT application, Reductions, NP-completeness, Self-learning.

1 Introduction

Nowadays, there is an increasing interest in offering college-level courses online. Several websites like Khan Academy [1], Coursera [2], Udacity [3] and edX [4], provide online courses on numerous topics. The users/students have access to videos and texts explaining several subjects, as well as tools for automated evaluation by means of exercises. In the specific context of computer science, online judges for testing correctness of programs are used in several academic domains as a self-learning tool for students, as well as an objective method in exams for scoring their programming skills (see, e.g., [5]). This kind of judges are also used in online contests organised by several sites, like TopCoder [6] and Codeforces [7].

For the last two years we have developed a specific online judge for the subject of Theory of Computation [8], located at http://racso.lsi.upc.edu/juez. The site offers exercises about deterministic finite automata, context-free grammars, push-down automata, and reductions between undecidable problems. Users

---

* The authors were supported by an FPU grant (first author) and the FORMALISM project (TIN2007-66523) from the Spanish Ministry of Education and Science.

C. Sinz and U. Egly (Eds.): SAT 2014, LNCS 8561, pp. 415–421, 2014.

can submit their solutions, the judge evaluates them, and offers a counterexample when the submission is rejected. This is very useful to make students understand why their solutions are wrong, and to keep them motivated during the learning process. We have used the judge in the classroom, not only as a support tool for the students, but also as an evaluation method on exams. This has had a marked effect on the motivation and involvement of the students: during a fifteen-week course, each student has solved more than 150 exercises in average, with more than 680 submissions. This means that each exercise needed over 4 submissions to be accepted by the judge, and the students were motivated enough to perform new attempts to reach an acceptance verdict. These results have motivated us to extend the judge to evaluate other kinds of exercises related to the subject. In particular, we have developed an evaluator of reductions between NP-complete problems. In this paper we explain the main aspects of this new evaluator. In Section 2, we explain our approach for checking the reductions, based on the use of SAT solvers, and the need for a specialised programming language to describe the reductions. This language is described in Section 3. In Section 4 we analyse the performance of the judge.

2 Approach

We start by recalling the well-known concept of reduction [9]. Given two decisional problems P_1, P_2, a reduction $P_1 \leq P_2$ is a program R that transforms a valid input of P_1 into a valid input of P_2 preserving the answer between the respective problems, i.e., that satisfies $P_1(I) \Leftrightarrow P_2(R(I))$ for any valid input I of P_1. In the specific setting of reductions for proving NP-hardness, R must be a polynomial time algorithm. Also, recall that any NP-complete problem can be reduced to SAT in polynomial time.

Currently, our website has a list of exercises of reductions between NP-complete problems. Each exercise describes two problems P_1, P_2, and the user must provide a program R, which is intended to be a polynomial time reduction $P_1 \leq P_2$. Internally, the judge has a set of tests, where each test is a valid input I_1 of P_1, and uses two aditional reductions $R_1 : P_1 \leq$ SAT, $R_2 : P_2 \leq$ SAT to transform the respective $I_1, I_2 = R(I_1)$ into instances of SAT $S_1 = R_1(I_1)$, $S_2 = R_2(I_2)$. Next, it checks whether S_1, S_2 are logically equivalent using a state-of-the-art SAT solver. In the negative case, the judge offers I_1 and the corresponding I_2 to the user as a counterexample to the correctness of R.

The previous approach has two main drawbacks. On the one side, passing a finite set of tests does not guarantee the preservation of the answer. Nevertheless, the same objection arises for programming judges, and experience shows that a good set of tests often suffices to avoid false positives (except for those malicious submissions that fail deliberately on a particular case). Typically, creating a good set of tests requires to anticipate wrong solutions to the exercise that can be apparently correct, and produce one or more tests that make those wrong reductions fail. Also, it is convenient to cover a large number of cases by generating several input instances systematically, perhaps randomly. On the other side, since the involved problems in the reduction are hard, we need to use small tests I_1. In fact, these tests must be very small since, otherwise, the composition of reductions might produce instances S_2 that are too complex for the SAT solver. However, an exercise $P_1 \leq P_2$ checked with small tests could be passed by a malicious submission that solves $P_1(I_1)$ directly (by, e.g., using an adequate backtracking that works fine for small inputs, or with a direct translation to SAT combined with the use of a simple embedded SAT solver), and produces a forged I_2 accordingly. Thus, an exponential time transformation might be wrongly accepted.

We have designed a concrete programming language specially suited for describing reductions between NP-complete problems called REDNP. The transformations can be described succinctly due to the management of the input and the output based on a predefined structure for each problem and the use of strings for naming basic objects. REDNP also has features for describing typical concepts that frequently appear in reductions, like graphs and clauses. In addition, we impose severe limitations to REDNP, like the use of structured intermediate memory, in order to avoid malicious submissions (this should not be an obstacle for describing reductions, since many natural reductions between NP-complete problems are just projections [10]). Also, we limit the maximum execution time and the value of integers (this makes it difficult to use integers as bitvectors representing sets of true/false values for performing some kind of exhaustive search).

3   REDNP

REDNP is a simple C-like programming language with the classical if-else, for, while instructions, and expressions with the usual integer and boolean operators. In our case, the assignment =, and the pre- and post- increment and decrement are also instructions. We also admit a foreach instruction. Except for two special variables in and out, all the variables have simple types int and string. This is to avoid the use of intermediate structured data for the computation of the reduction. For the same reason, strings can be concatenated and assigned, but their positions cannot be accessed with an index. Thus, strings should be used only to name objects in the result of the reduction. The string operator for concatenation is +, and integer values are automatically converted to strings when they are operated or assigned. In string literals like "s{i}{0} iff s{i-1}{0} and not n{i}", the expressions inside curly braces are evaluated, making it easier to describe indexed names. The two special variables in and out mentioned above refer to the input and output of the program, respectively. The contents of in can be accessed but cannot be modified, and the contents of out can be assigned but cannot be accessed.

For each exercise, the types of in and out are fixed. As an example, consider the exercise asking for a reduction VERTEX-COVER≤DOMINATING-SET. These problems receive a natural number k and an undirected graph G as input. In the first case, the question is whether there exists a choice of k vertices of G such that each edge of G contains at least one chosen vertex. In the second case, the question is whether there exists a choice of k vertices such that any other vertex is adjacent to a chosen one. A typical correct reduction consists in preserving the same k, removing all isolated vertices, and for each edge (u, v), adding a new vertex new_{uv} and two new edges (u, new_{uv}) and (v, new_{uv}). Let's see how we represent this reduction in REDNP. For this exercise, we have the types:

```
in: struct {                        out: struct {
  k: int                              k: int
  nvertices: int                      edges: array of array [2] of string
  edges: array of array [2] of int    vertices: array of string
}                                   }
```

It is implicit that the vertices are numbered from 1 to `in.nvertices`, and hence, each cell of `in.edges` contains an array with 2 different values inside such interval. Note that the edges described by `in` are pairs of int's, while the edges described by `out` are pairs of `string`'s. This is also the case for all the exercises of our site: names of objects are always int's inside `in`, and `string`'s inside `out`. This is because, usually, reductions need to traverse all input objects in a simple way, and naming them from 1 to a certain value eases this task. Moreover, reductions usually have to name the output objects meaningfully, using names that keep some relation with input objects. Indexed names in the output where the indices refer to input objects is a typical case, like the new_{uv} vertex. The field `out.vertices` is optional and does not need to be filled by the reduction, but it is provided for the cases where the user considers necessary to notify the existence of isolated vertices, since they are not present in any edge. A possible solution to this exercise described in REDNP is:

```
out.k = in.k;
foreach (edge; in.edges) {
  u = edge[0]; v = edge[1];
  out.edges.push = u, v;
  out.edges.push = u, "new{u}{v}";
  out.edges.push = v, "new{u}{v}";
}
```

Note that arrays of non-fixed size in the output can be extended using `.push` to add a new element. Also, note that values of several positions of arrays can be simultaneously assigned by separating the list of expressions by commas.

As we have mentioned in Section 2, the problem setter must provide additional reductions to SAT in order to validate the submissions. In particular, a reduction from DOMINATING-SET to SAT must be provided for the previous exercise, and this is also done using REDNP. Since this reduction works considering the input elements represented by integers, an internal process transforms the generated graph by converting strings to integers again. The following program could be the reduction provided by the problem setter, as well as a solution for the exercise DOMINATING-SET$\leq$SAT:

```
for (i=1; i<=in.numnodes; ++i) {
  s = "n{i}";
  foreach (node; in.adjacents[i]) s = s + " or n{node}";
  insertsat(s);
}
insertsat("s{0}{0}");
for (j=1; j<=in.k; ++j) insertsat("not s{0}{j}");
for (i=1; i<=in.numnodes; ++i) {
  insertsat("s{i}{0} iff s{i-1}{0} and not n{i}");
  for (j=1; j<=in.k; ++j)
    insertsat("s{i}{j} iff (s{i-1}{j} and not n{i}) or " +
                         "(s{i-1}{j-1} and n{i})");
}
insertsat("s{in.numnodes}{in.k}");
```

The previous program appends clauses to out using insertsat, an additional feature provided by REDNP that allows to transform any given propositional formula into a list of clauses using the Tseitin transformation.

4 Performance

As mentioned in Section 2, it may be hard to solve the resulting SAT instance $S_2 = R_2(R(I_1))$ compared to the cost of solving I_1 and $S_1 = R_1(I_1)$. In this section we explain some experiments[1] that confirm this statement, and justify the need for considering only simple I_1's as tests for the judge. MINISAT [11] has been the final choice used by the judge since PICOSAT [12] gave lower performance.

As a first experiment, we consider the family of tests $\langle k, \mathcal{K}_n \rangle$ as input for the exercise VERTEX-COVER≤DOMINATING-SET, where $\mathcal{K}_n$ represents the complete graph of size n. Figure 1 shows the time needed by MINISAT to solve S_1 and S_2 for n's chosen among $\{12, 14, 16, 18, 20\}$ and all possible k's. The critical k with respect to performance is the maximum number of vertices that cannot cover $\mathcal{K}_n$, that is $k = n - 2$. All the S_1's are solved very fast, whereas for $n = 20$ and $k = 18$ the solver needs about 50 seconds to solve S_2.

We have conducted additional experiments with similar results. For instance, when using matrix-like graphs with size $n \times n$ as input for the same exercise, S_1 is solved quickly and S_2 takes significantly longer for the critical k. In the exercise CLIQUE≤INDUCED-SUBGRAPH, we have modified the test set $\langle k, \mathcal{K}_n \rangle$ removing n edges from each graph to guarantee that there is no clique of size $n/2 + 1$. We observed that, in this case, solving S_2 was either easy or very difficult, with an abrupt transition at $k \approx n/2$. Another interesting example is the reduction of HAMILTONIAN-CIRCUIT from directed to undirected graphs. In this case, with a relatively small matrix-like graph of 4×4, the composition of reductions produces an S_2 which is very difficult to solve.

These experiments show that very small instances give rise to SAT instances that are difficult for state-of-the-art SAT solvers after applying the composition

---

[1] The data is available at http://www.lsi.upc.edu/~ggodoy/publications.html

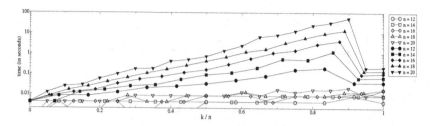

Fig. 1. Performance of MiniSat for VERTEX-COVER≤DOMINATING-SET (continuous lines; dotted lines correspond to solving VERTEX-COVER) with input $\langle k, \mathcal{K}_n \rangle$

of reductions to them. This seems to happen when the initial small instance belongs to a parameterized family of conceptually difficult instances. This means that the SAT solver is not able to detect the backdoor variables in such cases. SAT instances generated in this way could be included in the benchmarks for the SAT competitions, since they have a practical interest.

5   Conclusions

We have developed an online judge for automatic evaluation of exercises on reductions between NP-complete problems. To this end, we have designed the programming language REDNP, which facilitates the task of describing such reductions, imposing limits to the algorithmic descriptions in order to avoid malicious submissions. This last feature was mandatory since, due to the hardness of the testing process, the tests used by the judge must be very simple.

Several questions arise from this work. On the one side, it could be interesting to design an alternative non-imperative language for the description of reductions, since some students might find more natural to express reductions in a declarative way. Nevertheless, one should decide how to restrict the language in order to avoid non-polynomial time reductions. On the other side, variants of our approach using different kinds of solvers (e.g., SMT solvers) should be experimented, since they might offer better performance and ease of use for reductions that need to express arithmetic properties.

References

1. Khan, S.: Khan Academy (2006), http://www.khanacademy.org
2. Ng, A., Koller, D.: Coursera (2012), http://www.coursera.org
3. Thrun, S., Stavens, D., Sokolsky, M.: Udacity (2012), http://www.udacity.com
4. Agarwal, A.: edX (2012), http://www.edx.org
5. García, C., Revilla, M.A.: UVa online judge (1997), http://uva.onlinejudge.org
6. Hughes, J.: TopCoder (2000), http://www.topcoder.com
7. Mirzayanov, M.: Codeforces (2010), http://www.codeforces.com
8. Hopcroft, J.E., Motwani, R., Ullman, J.D.: Introduction to Automata Theory, Languages, and Computation, 3rd edn. Addison-Wesley (2006)

9. Garey, M.R., Johnson, D.S.: Computers and Intractability: A Guide to the Theory of NP-Completeness. W. H. Freeman & Co., New York (1979)
10. Skyum, S., Valiant, L.G.: A complexity theory based on boolean algebra. Journal of the ACM 32(2), 484–502 (1985)
11. Sörensson, N., Eén, N.: MiniSat 2.1 and MiniSat++ 1.0. Tech. rep., SAT Race 2008 Editions (2008)
12. Biere, A.: PicoSAT essentials. Journal on Satisfiability, Boolean Modeling and Computation 4(2-4), 75–97 (2008)

DRAT-trim: Efficient Checking and Trimming Using Expressive Clausal Proofs*

Nathan Wetzler, Marijn J.H. Heule, and Warren A. Hunt, Jr.

The University of Texas at Austin

Abstract. The DRAT-trim tool is a satisfiability proof checker based on the new DRAT proof format. Unlike its predecessor, DRUP-trim, all presently known SAT solving and preprocessing techniques can be validated using DRAT-trim. Checking time of a proof is comparable to the running time of the proof-producing solver. Memory usage is also similar to solving memory consumption, which overcomes a major hurdle of resolution-based proof checkers. The DRAT-trim tool can emit trimmed formulas, optimized proofs, and new TraceCheck$^+$ dependency graphs. We describe the output that is produced, what optimizations have been made to check RAT clauses, and potential applications of the tool.

1 Introduction

The DRAT-trim satisfiability (SAT) proof checker and trimming utility is designed to validate proofs of unsatisfiability. DRAT-trim uses the new DRAT proof format which combines the previous DRUP and RAT formats [9,10] and allows all presently known SAT techniques to be expressed. The tool maintains the efficiency of its predecessor, DRUP-trim, while integrating stronger forms of clause redundancy.

SAT solvers, as well as SMT and QBF solvers, have documented bugs [6,7]. Jävisalo et. al. [12] described a subtle error in the blocked clause addition [14] routine of Lingeling [5], one of the best SAT solvers available. Despite this bug, Lingeling was claimed to be "experimentally correct" on millions of satisfiability benchmarks and was used by industry for over a year and half before the bug was discovered. The process inspired a proof system [12] designed to detect this and similar errors.

The DRAT-trim proof checker is based on this proof system and addresses the main limitation of contemporary proof checkers: they cannot validate blocked clause addition and other techniques such as extended resolution [19], extended learning [1], and bounded variable addition [15]. DRAT proofs are easy to emit, require relatively little space on disk, and can be used to check all known solving and preprocessing techniques. DRAT-trim validates unsatisfiability results in a time comparable with solving time and uses far less physical memory than previous checkers.

* The authors are supported by DARPA contract number N66001-10-2-4087 and by the National Science Foundation under Grant No. CCF-1153558.

C. Sinz and U. Egly (Eds.): SAT 2014, LNCS 8561, pp. 422–429, 2014.

Proof checking is just one function of the DRAT-trim tool; trimmed formulas, optimized proofs, and TraceCheck$^+$ dependency graphs can also be emitted. These extra results have a variety of applications. Trimmed formulas and dependency graphs can be used as input to MUS extraction tools, similar to the relationship of a preprocessor to a solver. Reduced proofs can speed up proof playback, a feature that may be leveraged by a mechanically-verified proof checking tool [21]. DRAT-trim is the first tool to emit these additional results for proofs generated by techniques such as blocked clause addition and extended resolution.

In the remainder of this tool paper, some of the technical details associated with producing proofs in the DRAT format are described in Section 2. The input formats are detailed in Section 3. Output formats and applications of the tool are presented in Section 4, and implementation details are discussed in Section 5. Finally, conclusions are drawn in Section 6.

2 Redundancy Properties

DRAT-trim is designed to accept proofs in a specific format, which will be described in the next section. The proof checking algorithms are based on clause redundancy properties presented in earlier work [9,10] and are briefly explained below. This brief paper does not contain preliminaries on CNF formulas, resolution, and unit propagation. Readers who are not familiar with these concepts are referred to our earlier work [11] which uses the same notation as this system description. This material is presented for completeness as the definition of the main clause redundancy property has changed slightly from earlier publications [10].

Definition 1 (Asymmetric Literal Addition (ALA) [11]). *Let C be a clause occurring in a CNF formula $F \cup \{C\}$. The clause* ALA(F, C) *is the unique clause obtained by applying the following extension rule until fixpoint:*

$$C := C \cup \{\bar{l}\} \text{ if } \exists\, l_1, \ldots, l_k \in C, (l_1 \vee \ldots \vee l_k \vee l) \in F$$

A clause C has property AT (*Asymmetric Tautology*) with respect to CNF formula F if ALA(F, C) is a tautology. The property AT is also known as Reverse Unit Propagation (RUP) [20]. A clause C has property RAT (*Resolution Asymmetric Tautology*) with respect to CNF formula F if there exists a literal $l \in C$ such that for all $D \in F$ with $\bar{l} \in D$, it holds that $(D \setminus \{\bar{l}\}) \cup C$ has property AT with respect to F. Due to the monotonicity of asymmetric literal addition, a clause C with property AT with respect to formula F, has property RAT with respect to F. If a clause $C \notin F$ has property RAT with respect to F, then F and $F \cup \{C\}$ are satisfiability equivalent [10].

Example 1. Consider the CNF formula $F = (\bar{a} \vee b) \wedge (a \vee c) \wedge (b \vee \bar{c})$. The clause (a) has property RAT with respect to F, because $(a \vee b)$ has property AT with respect to F. Therefore, (a) can added to F, while preserving unsatisfiability. The clause (b) has property AT (and hence property RAT) with respect to F.

3   Input

DRAT-trim requires two input files: a formula and a proof. Formulas must be in DIMACS CNF format—the conventional input format for SAT solvers. Proofs must be expressed in the new DRAT (**D**eletion **R**esolution **A**symmetric **T**autology) notation, which is a generalization of the DRUP (**D**eletion **R**everse **U**nit **P**ropagation) format [9]. The DRAT format has the advantage that all presently-known techniques are expressible in this notation [10].

A mathematical proof of a theorem can be constructed from smaller theorems, or lemmas. We use this terminology to describe clausal proofs. Thus, clausal proofs are sequences of "lemma" additions and "lemma" or input clause deletions. More specifically, each line of the proof file is either a lemma (a sequence of literals terminated by 0) or a deletion instruction (with a "d " prefix). Proof files terminate with the empty clause: a line containing only a zero. Let F be a CNF formula and P be a DRAT proof for F. The number of lines in a proof P is denoted by $|P|$. For each $i \in \{0, \ldots, |P|\}$, a CNF formula is defined F_P^i below. L_i refers to the lemma on line i of P.

$$F_P^i := \begin{cases} F & \text{if } i = 0 \\ F_P^{i-1} \setminus \{L_i\} & \text{if the prefix of } L_i \text{ is "d "} \\ F_P^{i-1} \cup \{L_i\} & \text{otherwise} \end{cases}$$

Lemma addition steps are validated using both RUP and RAT checks. The RUP check for lemma L_i in proof P for CNF formula F succeeds if L_i has the property AT with respect to F_P^{i-1}. Let l_i denote the first literal in lemma L_i. The RAT check for lemma L_i in proof P for CNF formula F succeeds if and only if L_i has the property RAT on literal l_i with respect to F_P^{i-1}.

Note that the DRUP and DRAT formats are syntactically the same, but DRAT-trim checks each line of the proof for a stronger form of clause redundancy, RAT, when the RUP check has failed. Furthermore, a tool that emits DRUP proofs is compatible with DRAT-trim. Fig. 1 shows an example DIMACS CNF file by Rivest [17] and a DRAT proof file.

Emitting a DRAT proof from a conflict-driven clause-learning (CDCL) [16] solver and preprocessors is relatively easy. CDCL solvers maintain a database containing the original clauses and lemmas. Creating a DRAT proof is typically as simple as printing each modification to the database to a file. Several state-of-the-art SAT solvers support emitting DRUP proofs, the format used for the SAT Competition 2013 [3]. For example, Glucose 3.0 [2] supports emitting proofs in the DRAT format with "-certified" and "-certified-output=" options.

4   Output and Applications

The default output of the DRAT-trim tool is a message indicating the validity of a proof with respect to an input formula. This information can be very useful while developing SAT solvers, especially since the DRAT format supports checking all existing techniques used in state-of-the-art SAT solvers [10].

CNF formula   DRAT proof

```
p cnf 4 10
  1   2 -3  0                -1  0
 -1  -2  3  0        d  -1  2  4  0
 -1  -2 -3  0                 2  0
  2   3 -4  0                    0
 -2  -3  4  0
 -1  -3 -4  0
  1   3  4  0
 -1   2  4  0
  1  -2 -4  0
 -1   2 -4  0
```

Fig. 1. DRAT-trim accepts two files as input: a formula in DIMACS CNF format (left), and a proof in the new DRAT format (right). Each line in the proof is either a lemma or a deletion step identified by the prefix "d". Spacing in both examples is to improve readability. If the RUP check fails, the DRAT-trim tool expects that a lemma has RAT on its first literal, as in the RAT format [10]. No check is performed for clause deletion steps, although the deleted clause needs to be present.

Additionally, SAT competitions can benefit from having a fast proof checker with an easy-to-produce input format. As mentioned in Section 1, SAT solvers have been shown to be buggy, even during competitions. DRUP-trim, the predecessor of DRAT-trim, was used to check the results of the unsatisfiability tracks of the SAT Competition 2013 [3]. Some techniques, such as bounded variable addition [15], cannot be validated using only RUP checks, and hence cannot be checked by DRUP-trim. The new DRAT-trim tool now supports all these techniques. Aside from checking the validity of proofs, DRAT-trim can optionally produce three outputs: a *trimmed formula*, an *optimized proof*, and a *dependency graph* in the new TraceCheck$^+$ format. These output files are described below and illustrated in Fig. 2.

Trimmed Formula. The trimmed formula produced by DRAT-trim is a subset of the input formula in DIMACS format, and the remaining clauses appear in the same order as the input file. However, the order of the literals in each clause may have changed. Trimming a formula can be a useful preprocessing step in extracting a Minimal Unsatisfiable Subset (MUS). Since DRAT-trim supports validating techniques that are based on (generalizations of) extended resolution, one can use these techniques to improve MUS extraction tools.

Optimized Proof. The optimized proof contains lemmas as well as deletion information in the DRAT format. Lemmas appearing in the optimized proof will be an ordered subset of the lemmas in the input proof. The first literal of each lemma is the same as the first literal of that lemma in the input proof, however the order of all other literals for the lemma may be permuted. The output proof is called "optimized" because it contains extra deletion information that is obtained during the backward checking process, described in Section 5. The optimized proof file may be larger than the input proof in size (as in Fig. 2); however, the additional deletion information helps reduce computation time because fewer clauses are active during each check. A smaller proof is very useful for potentially slower, mechanically-verified solvers. The optimized proof from one round of

trimmed formula	optimized DRAT proof	TraceCheck+ file

```
 p cnf 4 8
   1   2 -3  0
  -1  -2   3  0
   2   3 -4  0
  -2  -3   4  0
  -1  -3 -4  0
   1   3   4  0
  -1   2   4  0
   1  -2 -4  0
```

```
      -1   0
 d   -1  -2   3  0
 d   -1  -3 -4  0
 d   -1   2   4  0
       2   0
 d    1   2 -3  0
 d    2   3 -4  0
       0
```

```
 1    1   2 -3   0   0
 2   -1  -2   3   0   0
 4    2   3  -4   0   0
 5   -2  -3   4   0   0
 6   -1  -3  -4   0   0
 7    1   3   4   0   0
 8   -1   2   4   0   0
 9    1  -2  -4   0   0
11   -1   0   2   6   8   0
12    2   0   1   4   7  11   0
13    0   5   7   9  11  12   0
```

Fig. 2. Three optional output files from DRAT-trim for the input formula and proof from Fig. 1. A trimmed formula (left) is an ordered subset of the input formula. An optimized proof (middle) is an ordered subset of the input proof with extra deletion instructions. A TraceCheck+ file (right) is a dependency graph that includes the formula and proof. Each line begins with a clause identifier (bold), then contains the literals of the original clause or lemma, and ends with a list of clause identifiers (bold).

checking is often much smaller than the original, but it is still far from minimal. One can iteratively apply the checking process to further optimize a proof by submitting the reduced output proof of one round of checking to the tool for another round of checking and trimming.

Dependency Graph. DRUP-trim, the predecessor to this tool, can produce a resolution graph of a proof in the TraceCheck [18,13,4] format. DRAT-trim can validate techniques that cannot be checked with resolution, and we designed a new format that is backward-compatible with TraceCheck, allowing expression of all presently-known solving techniques. Resolution graphs in the TraceCheck format begin each line with a clause identifier, followed by the literals of the clause, a zero, the identifiers of antecedents, finally, followed by another zero. The new TraceCheck+ format uses this syntax as well. The formats only differ in the expressing the reasons for a lemma's redundancy. If the RUP check succeeds, the reasons are the antecedents as in the TraceCheck format. If a RAT check is needed, the reasons are the clauses required to let the RAT check succeed.

Dependency graphs have many potential uses. The dependency graph may be supplied to a MUS extractor as input, avoiding the need to recompute clause dependencies. Another use is for solver debugging: a dependency graph gives a step-by-step account of why each clause can be added to a clause database.

5   Methodology and Optimizations

DRAT-trim uses backward checking [8] to track and limit dependencies between clauses and lemmas; this means that lemmas are checked in reverse of the order they occur in the original proof. As each lemma is checked, any clauses or lemmas that were crucial to the check are marked. Marked clauses are preferred over unmarked clauses during unit propagation [9]. The process of marking not only keeps the trimmed formula and optimized proof small, but also reduces the number of lemmas that need to be validated, lowering the run time of the tool.

Clauses are marked by a conflict analysis routine that runs after unit propagation derives a conflict. When a clause is marked, the timestamp and location of the clause are stored as deletion information. The first time a clause is marked (and is determined to be necessary to the proof) during backward checking is the last time a clause is used during forward checking or optimized proof emission. Therefore, as each clause is marked, optimized deletion information is stored. During post-processing, marked lemmas are printed in order and deletion information sharing the same timestamp is printed afterwards.

During a RAT check, DRAT-trim needs access to all clauses containing the negation of the resolution literal. One could build a literal-to-clause lookup table of the original formula and update it after each lemma addition and deletion step. However, these updates can be expensive and the lookup table potentially doubles the memory usage of the tool. Since most lemmas emitted by state-of-the-art SAT solvers can be validated using the RUP check, such a lookup table has been omitted. Instead, when a RUP check failed, the currently active formula is scanned to find all clauses containing the complement of the resolution literal.

6 Performance and Conclusion

The DRAT-trim tool outperforms its RAT checker [10] and DRUP-trim [9] predecessors. To illustrate the performance gain, consider the rbcl_xits_09_unknown benchmark, one of the harder benchmarks [10]. The solving time with Coprocessor and Glucose 3.0 is 95 seconds and the proof checking time with DRAT-trim is 91 seconds, compared to 1096 seconds for a previous RAT proof checker. This particular benchmark can only be solved with bounded variable addition, necessitating a RAT/DRAT proof. On the application suite of SAT Competition 2009, DRAT-trim is 2% faster than DRUP-trim on average, within a range of 85% faster to 13% slower. Thus, the addition of RAT checking in DRAT-trim does not have a noticeable impact on DRUP proof checking.

DRAT-trim[1] and its associated DRAT proof format can be used to validate all contemporary SAT pre-processing and solving techniques with similar time and memory consumption to solving. Resolution-based proof checkers are not subsumed by this effort, but the performance of DRAT-trim appears better on large, industrial-scale examples. Our tool can optionally emit trimmed formulas, optimized proofs, and dependency graphs, and we look forward to seeing creative utilization of these resources by the SAT community. We believe that all SAT solvers should emit proofs in the DRAT format so that their results can be validated. Not only does this provide implementers with a convenient way of debugging, but it gives users the confidence to find new applications for satisfiability solvers.

---

[1] DRAT-trim is available at http://cs.utexas.edu/~marijn/drat-trim/.

References

1. Audemard, G., Katsirelos, G., Simon, L.: A restriction of extended resolution for clause learning SAT solvers. In: Fox, M., Poole, D. (eds.) Proceedings of the 24th AAAI Conference on Artificial Intelligence (AAAI). AAAI Press (2010)
2. Audemard, G., Simon, L.: Glucose's home page (2014), http://www.labri.fr/perso/lsimon/glucose/ (accessed: January 21, 2014)
3. Balint, A., Belov, A., Heule, M., Järvisalo, M.: SAT Competition 2013 (2013), http://www.satcompetition.org/2013/ (accessed: January 21, 2014)
4. Biere, A.: PicoSAT essentials. Journal on Satisfiability, Boolean Modeling and Computation (JSAT) 4, 75–97 (2008)
5. Biere, A.: Lingeling, Plingeling, and Treengeling (2014), http://fmv.jku.at/lingeling/ (accessed: January 27, 2014)
6. Brummayer, R., Biere, A.: Fuzzing and delta-debugging SMT solvers. In: International Workshop on Satisfiability Modulo Theories (SMT), pp. 1–5. ACM (2009)
7. Brummayer, R., Lonsing, F., Biere, A.: Automated testing and debugging of SAT and QBF solvers. In: Strichman, O., Szeider, S. (eds.) SAT 2010. LNCS, vol. 6175, pp. 44–57. Springer, Heidelberg (2010)
8. Goldberg, E.I., Novikov, Y.: Verification of proofs of unsatisfiability for CNF formulas. In: Design, Automation and Test in Europe Conference and Exhibition (DATE), pp. 10886–10891. IEEE (2003)
9. Heule, M.J.H., Hunt Jr., W.A., Wetzler, N.: Trimming while checking clausal proofs. In: Formal Methods in Computer-Aided Design (FMCAD), pp. 181–188. IEEE (2013)
10. Heule, M.J.H., Hunt Jr., W.A., Wetzler, N.: Verifying refutations with extended resolution. In: Bonacina, M.P. (ed.) CADE 2013. LNCS (LNAI), vol. 7898, pp. 345–359. Springer, Heidelberg (2013)
11. Heule, M.J.H., Järvisalo, M., Biere, A.: Clause elimination procedures for CNF formulas. In: Fermüller, C.G., Voronkov, A. (eds.) LPAR-17. LNCS, vol. 6397, pp. 357–371. Springer, Heidelberg (2010)
12. Järvisalo, M., Heule, M.J.H., Biere, A.: Inprocessing rules. In: Gramlich, B., Miller, D., Sattler, U. (eds.) IJCAR 2012. LNCS (LNAI), vol. 7364, pp. 355–370. Springer, Heidelberg (2012)
13. Jussila, T., Sinz, C., Biere, A.: Extended resolution proofs for symbolic SAT solving with quantification. In: Biere, A., Gomes, C.P. (eds.) SAT 2006. LNCS, vol. 4121, pp. 54–60. Springer, Heidelberg (2006)
14. Kullmann, O.: On a generalization of extended resolution. Discrete Applied Mathematics 96-97, 149–176 (1999)
15. Manthey, N., Heule, M.J.H., Biere, A.: Automated reencoding of boolean formulas. In: Biere, A., Nahir, A., Vos, T. (eds.) HVC 2012. LNCS, vol. 7857, pp. 102–117. Springer, Heidelberg (2013)
16. Marques-Silva, J.P., Lynce, I., Malik, S.: Conflict-Driven Clause Learning SAT Solvers. In: Handbook of Satisfiability, Frontiers in Artificial Intelligence and Applications, ch. 4, vol. 185, pp. 131–153. IOS Press (February 2009)
17. Rivest, R.L.: Partial-match retrieval algorithms. SIAM J. Comput. 5(1), 19–50 (1976)
18. Sinz, C., Biere, A.: Extended resolution proofs for conjoining BDDs. In: Grigoriev, D., Harrison, J., Hirsch, E.A. (eds.) CSR 2006. LNCS, vol. 3967, pp. 600–611. Springer, Heidelberg (2006)

19. Tseitin, G.S.: On the complexity of derivation in propositional calculus. In: Siekmann, J., Wrightson, G. (eds.) Automation of Reasoning 2, pp. 466–483. Springer (1983)

20. Van Gelder, A.: Verifying RUP proofs of propositional unsatisfiability. In: International Symposium on Artificial Intelligence and Mathematics (ISAIM) (2008)

21. Wetzler, N., Heule, M.J.H., Hunt Jr., W.A.: Mechanical verification of SAT refutations with extended resolution. In: Blazy, S., Paulin-Mohring, C., Pichardie, D. (eds.) ITP 2013. LNCS, vol. 7998, pp. 229–244. Springer, Heidelberg (2013)

MPIDepQBF: Towards Parallel QBF Solving without Knowledge Sharing*

Charles Jordan[1], Lukasz Kaiser[2,**], Florian Lonsing[3], and Martina Seidl[4]

[1] Division of Computer Science, Hokkaido University, Japan
[2] LIAFA, CNRS & Université Paris Diderot
[3] Knowledge-Based Systems Group, TU Wien, Austria
[4] Institute for Formal Models and Verification, JKU Linz, Austria

Abstract. Inspired by recent work on parallel SAT solving, we present a lightweight approach for solving *quantified Boolean formulas (QBFs)* in parallel. In particular, our approach uses a sequential state-of-the-art QBF solver to evaluate subformulas in working processes. It abstains from globally exchanging information between the workers, but keeps learnt information only locally. To this end, we equipped the state-of-the-art QBF solver DepQBF with assumption-based reasoning and integrated it in our novel solver MPIDepQBF as backend solver. Extensive experiments on standard computers as well as on the supercomputer Tsubame show the impact of our approach.

1 Introduction

Recently, there has been much progress in solvers for *quantified Boolean formulas* (QBF) [4,7]. The quest for QBF solvers is motivated by the vision that QBF solvers become powerful general purpose reasoning engines for PSPACE problems in the same way as SAT solvers are for problems in NP. Then, many interesting application problems that have compact QBF encodings but (assuming $NP \neq PSPACE$) no compact SAT encodings could be handled by QBF solvers [1].

Most of the recent advances in QBF solving are realized within sequential systems (e.g., [8,10,11,14]), thus not taking advantage of the parallel computing resources provided by modern computer architectures. Although some dedicated parallel QBF solver implementations have been presented [6,12,13,18], to the best of our knowledge none is actively maintained and publicly available. Additionally, the usual focus on sharing information between different working processes can be influential to the solver's overall performance [17] due to restrictions of bandwidth.

In this paper, we propose solving QBFs without global information sharing and present MPIDepQBF[1], a parallel solver for quantified Boolean formulas based

---

* Partially supported by the Austrian Science Fund (FWF) under grants S11408-N23 and S11409-N23, by the Wiener Wissenschafts-, Forschungs- und Technologiefonds (WWTF) under grant ICT10-018, and by the Japan Society for the Promotion of Science (JSPS) as KAKENHI No. 25106501.
** Currently at Google Inc.
[1] Open-source, available via SVN (rev. 1928) at http://toss.sf.net/develop.html

C. Sinz and U. Egly (Eds.): SAT 2014, LNCS 8561, pp. 430–437, 2014.
© Springer International Publishing Switzerland 2014

on a new version of the sequential solver DepQBF. The rest of the paper is organized as follows. First, we review related work in Section 2. Then we describe the basic algorithm of our approach in Section 3. In Section 4, we evaluate the performance of our solver on standard benchmark instances using Tsubame2.5[2]. In Section 5 we conclude with an outlook to future work.

2 Related Work

In the year 2000, work on parallel QBF solvers began with PQSolve, a parallel version of QSolve [6], which implements techniques like quantifier inversion or trivial truth. However, much progress has since been made in QBF solving, in particular clause and cube sharing have been introduced. More recently, PAQuBE [13], QMiraXT [12], and the approach of Da Mota et al. [18] have been presented. However, to the best of our knowledge, none of these tools is publicly available and all seem to be no longer developed.

QMiraXT uses a shared memory architecture for sharing clauses which increases flexibility, but has scalability issues. Da Mota et al. split non-prenex non-CNF formulas into QBFs with free variables which are rewritten to propositional formulas by the workers. The master collects these propositional formulas from which the final result is obtained. PAQuBE was developed as a parallel version of QuBE and supports sharing learnt information between workers. It is the solver most closely related to our approach. However, our solver uses a different approach to generate subproblems. For quantified constraint satisfaction problems (QCSP), a problem-partition approach has been presented by Vautard [20] et al. where parallelism seems to be very beneficial. Inspired by the recent success of Treengeling [2] in SAT, we allow each worker short (but increasing) timeouts to solve particular subproblems. If a worker solves its subproblem, the result is combined with previous results to determine what portion of the search tree is completed. Otherwise, we either set more variables and distribute the resulting problems or allow the worker to continue with a longer timeout (depending on the number of free workers, subproblems not yet assigned to workers and parameters). Each worker retains information learnt from its previous subproblems.

3 The Architecture of MPIDepQBF

MPIDepQBF is a QBF solver that accepts input in the standard qdimacs format, i.e., QBFs $\psi = Q_1 B_1 Q_2 B_2 \ldots Q_n B_n.\phi$ in prenex conjunctive normal form (PCNF), where the formula ϕ is a conjunction of clauses. The quantifier prefix $Q_1 B_1 Q_2 B_2 \ldots Q_n B_n$ is a sequence of quantified blocks B_i of variables where $Q_i \in \{\forall, \exists\}$ and adjacent blocks are quantified differently, i.e. $Q_i \neq Q_{i+1}$.

MPIDepQBF has one master process to coordinate arbitrarily many worker processes via MPI. The workers use the sequential QBF solver DepQBF [14], which we extended to allow solving under assumptions. No other modifications

---

[2] We are grateful to the ELC project for providing access to Tsubame.

were necessary. The workers obtain the input formula only once, at the start. This is the only information shared, otherwise they are completely agnostic of the global solving process. When a worker process is idle, the master process supplies it with assumptions and a timeout. Then, the worker process tries to solve the formula under these assumptions. The result is communicated to the master process which adapts the time limit or selects another set of assumptions, as described below. If the final value of the formula has been determined, the worker processes are stopped. Details on master and worker processes follow.

3.1   Master Process

The master process generates a stream of subproblems, represented by partial variable assignments and timeouts, and distributes them to the worker processes. First, starting from the formula $Q_1 B_1 Q_2 B_2 \ldots Q_n B_n.\phi$, we sort each B_i according to the number of occurrences of each variable in ϕ and concatenate these sorted lists. This determines the order in which variables will be set.

Next we build a search tree. The tree has 3 kinds of leaves: sat, unsat, and open, where each open leaf contains a variable assignment and a timeout. We start with a fully-balanced binary tree that has only open leaves with the starting timeout (by default 0.1s). The number of leaves is initially the highest power of 2 smaller than the number of available MPI worker processes times a busy-factor (we use 1.25 by default). Each open leaf contains an assignment of the variables: We assign the variables one-by-one in the order determined in the first step. For each open leaf, we send the subproblem (determined by the timeout and the assignment) to a free worker process.

When a worker returns a result, the master process merges it with the current search tree. If the result is sat or unsat, then the open leaf gets replaced by the result, and the tree is simplified as shown in Procedure simplify. If the result is a timeout, then what happens depends on how many free MPI worker processes are available. If the MPI queue is full (except for the one process that just returned), then we multiply the timeout by a timeout-factor (1.4 by default) and send the same case back to the worker process with the new, longer timeout. If there are fewer open leaves than the number of MPI worker processes times the busy-factor, then we replace the open leaf by, again, a fully-balanced binary sub-tree with assignments that prolong the previous one. This is repeated until the whole search tree is simplified to sat or unsat, when the problem is solved.

3.2   Worker Processes: Search-Based Solving under Assumptions

A worker process runs an instance of the QBF solver DepQBF which is initialized with the complete input formula. We extended the API of DepQBF to allow for assumptions as input. These assumptions can be regarded as assignments to variables which are fixed in the current run. All the learnt information is shared over different runs. Note that this information is only kept locally, so no exchange between the different worker processes is realized. QBF solving with

---

Procedure simplify(t)

if $t = Branch(\exists, v, t_1, t_2)$ *and* simplify(t_1) = *sat* **then** sat;
if $t = Branch(\exists, v, t_1, t_2)$ *and* simplify(t_2) = *sat* **then** sat;
if $t = Branch(\exists, v, t_1, t_2)$ *and* simplify(t_1) = simplify(t_2) = *unsat* **then** unsat;
if $t = Branch(\forall, v, t_1, t_2)$ *and* simplify(t_1) = *unsat* **then** unsat;
if $t = Branch(\forall, v, t_1, t_2)$ *and* simplify(t_2) = *unsat* **then** unsat;
if $t = Branch(\forall, v, t_1, t_2)$ *and* simplify(t_1) = simplify(t_2) = *sat* **then** sat;
else if $t = Branch(Q, v, t_1, t_2)$ **then** $Branch(Q, v,$ simplify(t_1), simplify(t_2));

---

assumptions has been applied in the context of an incremental approach to QBF-based bounded model checking of partial designs [16]. In this section, we describe the handling of assumptions in DepQBF. Assumptions are used in MPIDepQBF to split the search space and thus generate subproblems for the worker processes.

DepQBF [14] is a search-based QBF solver with conflict-driven clause learning and solution-driven cube learning [9,21]. In this approach, called $QCDCL$ [15], backtracking search is combined with the dynamic generation of new clauses and cubes. Given a QBF, variables are assigned successively until either all clauses of the QBF are satisfied or one clause is falsified under the current assignment. Depending on the assignment, clauses are derived by Q-resolution [5] and cubes are derived by the model generation rule and term resolution [9]. The newly derived clauses and cubes are added to the formula as part of separate sets of learnt clauses and learnt cubes to prune the search-space. The parallel variant MPIDepQBF of DepQBF applies QCDCL where the given PCNF is solved under a set of predefined variable assignments, called *assumptions*.

A set of *assumptions* $A := \{l_1, \ldots, l_n\}$ is a set of literals of variables such that $var(l_i) \in B_1$ for all literals $l_i \in A$. The variables of literals in A are from the first block B_1 of ψ. Each literal $l_i \in A$ represents an assignment to the variable $var(l_i)$. Positive literals $l_i = var(l_i)$ and negative literals $l_i = \neg var(l_i)$ represent the assignment of *true* and *false* to the variable $var(l_i)$, respectively. The *PCNF* $\psi[A]$ *under the assumptions* A is obtained from ψ by deleting the clauses which are satisfied under the assignments represented by A, deleting literals from clauses in ψ which are falsified, and deleting superfluous quantifiers from the quantifier prefix of ψ.

In MPIDepQBF, subproblems for the worker processes are generated by applying the definition of assumptions recursively to the formula $\psi[A]$ under some set A of assumptions. If A assigns all variables from the first block B_1 of ψ, then the quantifier prefix of $\psi[A]$ has the form $Q_2 B_2 \ldots Q_n B_n$. Since B_2 is now the leftmost block in $\psi[A]$, variables from B_2 can be assigned in some other set A' of assumptions with respect to $\psi[A]$.

The following properties follow from the semantics of QBF. Given an instance ψ in PCNF, if $Q_1 = \exists$ and $\psi[A]$ is satisfiable then ψ is also satisfiable since the variables of the literals in A are all from the first block B_1. If $Q_1 = \forall$ and $\psi[A]$ is unsatisfiable then ψ is also unsatisfiable. These properties are checked in Procedure simplify presented in Section 3.1.

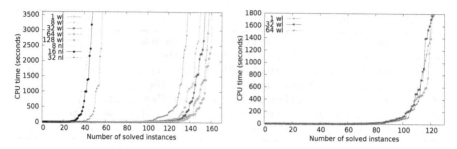

Fig. 1. Cactus plots on eval12r2-bloqqer (with (wl)/without (nl) learning) and eval12r2

Similar to SAT solving under assumptions, assumptions in QCDCL-based QBF solving can be modeled on a syntactic level by adding unit clauses for existential literals in A and unit cubes for universal literals. This is the "clause-based multiple instances (CM)" approach in the terminology of [19].

Alternatively, assumption based reasoning can be realized on the semantic level. Here, the variables of the literals in A are assigned as special *decision variables* (also called *branching variables*) in ψ before the solver makes any assignments to variables not in A. The backtracking procedure of QCDCL has to be modified to guarantee that the assignments represented by A are never retracted. This is the "literal-based single instance (LS)" approach [19]. As in SAT solving, the advantage of the LS approach compared to CM is that all the learnt clauses and cubes can be kept across different calls of the solver with different sets of assumptions. Since assumptions are assigned as decision variables, the learning procedure of QCDCL can never generate clauses and cubes by resolving on variables in A. Because of this advantage, we implemented the semantic LS approach in DepQBF. For the application in MPIDepQBF, we applied a sophisticated analysis of variable dependencies based on the *standard dependency scheme* implemented in DepQBF [14].

4   Evaluation

We evaluated the performance of MPIDepQBF as the number of cores is increased. To this end, we used the eval2012r2 and eval12r2-bloqqer benchmark sets[3], where the latter results from the former by applying the preprocessor Bloqqer [3]. We run our experiments on a small portion of the supercomputer Tsubame. In particular, we used the 'V' queue running on qemu virtual machines with eight physical 2.93 GHz Xeon 5670 cores and 30 GB memory per node.

The left-hand side of Figure 1 shows the performance of MPIDepQBF with various numbers of cores on instances from eval12r2-bloqqer with (wl) and without learning (nl) using a timeout of one hour. Whereas with one core only 139 formulas are solved, with 128 cores 160 formulas can be solved. A detailed

---

[3] See http://www.kr.tuwien.ac.at/events/qbfgallery2013/results.html

# cores (x)	# solved both x/128	avg time (s) x cores	avg time (s) 128 cores
1	137	168.11	62.26
8	148	180.64	64.03
16	149	154.44	76.26
32	151	163.74	79.46
64	155	122.96	98.47

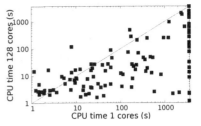

Fig. 2. Runtime comparison: (a) runtimes of x vs. 128 cores for the formulas solved with both x and 128 cores, (b) scatter plot for runtimes of 1 vs. 128 cores

comparison of the runtimes with different numbers of cores is shown in Figure 2. Besides the number of solved formulas as well as the average runtime for solving these formulas, we also show the average runtime for solving exactly the same set of formulas when using 128 cores. In additional experiments, we experienced that disabling learning drastically decreases the number of solved formulas. Although workers in MPIDepQBF do not share learnt information, this demonstrates the importance of each worker retaining its learnt information.

The right-hand side of Figure 1 shows performance on non-preprocessed instances. Scaling is limited or non-existent, suggesting the importance of preprocessing for MPIDepQBF. In particular, preprocessing breaks very long clauses into shorter clauses, which is important for memory efficiency in MPIDepQBF.

Besides the experiments on Tsubame, we also evaluated MPIDepQBF on a 12-core 2.4 GHz Intel Xeon 5645 with 96 GB RAM. The results were similar to those above. In particular, with one core 127, with two cores 130 formulas, with four cores 137, with eight cores 139 and with twelve cores 140 formulas could be solved within a timelimit of 600 seconds.

5 Conclusion

We presented MPIDepQBF, a parallel QBF solver based on a search-space splitting approach. A master process generates subproblems and distributes them to arbitrarily-many worker processes, as described above in Section 3. Workers do not exchange information with each other, but keep learnt information locally. To this end, we extended DepQBF with support for assumption-based reasoning. Initial experiments show that more processing power results in a gain of performance, especially when dozens of cores are available. For us, this is surprising given that individual workers do not share learnt information with other workers and gives several possibilities for future development using sophisticated information sharing. In addition, the memory-efficient nature of DepQBF makes usage of many-core, memory-constrained coprocessors promising. More experiments are required to systematically characterize and understand the impact of clause and cube sharing as done in previous works (e.g., [17]).

Our main motivation for working on parallel QBF solving is the desire to solve hard QBF instances stemming from applications. Our hope is that larger systems and parallel solvers will allow us to solve instances beyond the reach of current solvers. That is, while we have challenging instances and access to large systems, we did not have solvers that can utilize these systems.

MPIDepQBF is available at http://toss.sf.net/develop.html via SVN.

References

1. Benedetti, M., Mangassarian, H.: QBF-based formal verification: Experience and perspectives. Journal on Satisfiability, Boolean Modeling and Computation 5(1-4), 133–191 (2008)
2. Biere, A.: Lingeling, Plingeling and Treengeling Entering the SAT Competition 2013. In: Proc. of the SAT Competition 2013. Dep. of Computer Science Series of Publications B, University of Helsinki, vol. B-2013-1, pp. 51–52 (2013)
3. Biere, A., Lonsing, F., Seidl, M.: Blocked Clause Elimination for QBF. In: Bjørner, N., Sofronie-Stokkermans, V. (eds.) CADE 2011. LNCS (LNAI), vol. 6803, pp. 101–115. Springer, Heidelberg (2011)
4. Büning, H.K., Bubeck, U.: Theory of Quantified Boolean Formulas. In: Handbook of Satisfiability, pp. 735–760. IOS Press (2009)
5. Büning, H.K., Karpinski, M., Flögel, A.: Resolution for Quantified Boolean Formulas. Information and Computation 117(1), 12–18 (1995)
6. Feldmann, R., Monien, B., Schamberger, S.: A Distributed Algorithm to Evaluate Quantified Boolean Formulae. In: Proc. of the 17th Nat. Conference on Artificial Intelligence (AAAI 2000), pp. 285–290. AAAI (2000)
7. Giunchiglia, E., Marin, P., Narizzano, M.: Reasoning with Quantified Boolean Formulas. In: Handbook of Satisfiability, pp. 761–780. IOS Press (2009)
8. Giunchiglia, E., Marin, P., Narizzano, M.: QuBE7.0. Journal on Satisfiability, Boolean Modeling and Computation 7(2-3), 83–88 (2010)
9. Giunchiglia, E., Narizzano, M., Tacchella, A.: Clause/Term Resolution and Learning in the Evaluation of Quantified Boolean Formulas. Journal of Artificial Intelligence Research 26(1), 371–416 (2006)
10. Goultiaeva, A., Bacchus, F.: Recovering and Utilizing Partial Duality in QBF. In: Järvisalo, M., Van Gelder, A. (eds.) SAT 2013. LNCS, vol. 7962, pp. 83–99. Springer, Heidelberg (2013)
11. Janota, M., Klieber, W., Marques-Silva, J., Clarke, E.: Solving QBF with Counterexample Guided Refinement. In: Cimatti, A., Sebastiani, R. (eds.) SAT 2012. LNCS, vol. 7317, pp. 114–128. Springer, Heidelberg (2012)
12. Lewis, M., Schubert, T., Becker, B.: QMiraXT – a multithreaded QBF solver. In: Methoden und Beschreibungssprachen zur Modellierung und Verifikation von Schaltungen und Systemen, MBMV (2009)
13. Lewis, M., Schubert, T., Becker, B., Marin, P., Narizzano, M., Giunchiglia, E.: Parallel QBF Solving with Advanced Knowledge Sharing. Fundamenta Informaticae 107(2-3), 139–166 (2011)
14. Lonsing, F., Biere, A.: DepQBF: A dependency-aware QBF solver. Journal on Satisfiability, Boolean Modeling and Computation 7, 71–76 (2010)
15. Lonsing, F., Egly, U., Van Gelder, A.: Efficient Clause Learning for Quantified Boolean Formulas via QBF Pseudo Unit Propagation. In: Järvisalo, M., Van Gelder, A. (eds.) SAT 2013. LNCS, vol. 7962, pp. 100–115. Springer, Heidelberg (2013)

16. Marin, P., Miller, C., Lewis, M., Becker, B.: Verification of partial designs using incremental QBF solving. In: Proc. of the Int. Conf on Design, Automation & Test in Europe (DATE 2012), pp. 623–628. IEEE (2012)

17. Marin, P., Narizzano, M., Giunchiglia, E., Lewis, M.D.T., Schubert, T., Becker, B.: Comparison of knowledge sharing strategies in a parallel QBF solver. In: Proc. of the Int. Conf. on High Performance Computing & Simulation (HPCS 2009), pp. 161–167. IEEE (2009)

18. Mota, B.D., Nicolas, P., Stéphan, I.: A new parallel architecture for QBF tools. In: Proc. of the Int. Conf. on High Performance Computing and Simulation (HPCS 2010), pp. 324–330. IEEE (2010)

19. Nadel, A., Ryvchin, V.: Efficient SAT solving under assumptions. In: Cimatti, A., Sebastiani, R. (eds.) SAT 2012. LNCS, vol. 7317, pp. 242–255. Springer, Heidelberg (2012)

20. Vautard, J., Lallouet, A., Hamadi, Y.: A parallel solving algorithm for quantified constraints problems. In: Proc. of the 22nd IEEE International Conference on Tools with Artificial Intelligence (ICTAI 2010), pp. 271–274. IEEE Computer Society (2010)

21. Zhang, L., Malik, S.: Towards a Symmetric Treatment of Satisfaction and Conflicts in Quantified Boolean Formula Evaluation. In: Van Hentenryck, P. (ed.) CP 2002. LNCS, vol. 2470, pp. 200–215. Springer, Heidelberg (2002)

Open-WBO: A Modular MaxSAT Solver[*,**]

Ruben Martins[1], Vasco Manquinho[2], and Inês Lynce[2]

[1] University of Oxford, Department of Computer Science, United Kingdom
ruben.martins@cs.ox.ac.uk
[2] INESC-ID / Instituto Superior Técnico, Universidade de Lisboa, Portugal
{vmm,ines}@sat.inesc-id.pt

Abstract. This paper presents OPEN-WBO, a new MaxSAT solver. OPEN-WBO has two main features. First, it is an open-source solver that can be easily modified and extended. Most MaxSAT solvers are not available in open-source, making it hard to extend and improve current MaxSAT algorithms. Second, OPEN-WBO may use any MiniSAT-like solver as the underlying SAT solver. As many other MaxSAT solvers, OPEN-WBO relies on successive calls to a SAT solver. Even though new techniques are proposed for SAT solvers every year, for many MaxSAT solvers it is hard to change the underlying SAT solver. With OPEN-WBO, advances in SAT technology will result in a free improvement in the performance of the solver. In addition, the paper uses OPEN-WBO to evaluate the impact of using different SAT solvers in the performance of MaxSAT algorithms.

1 Introduction

Maximum Satisfiability (MaxSAT) formulations are currently used for solving many different real-world problems [20,1,16]. This results from the recent improvements of MaxSAT algorithms, which are now able to solve much larger instances than before. These improvements result mainly from (i) an increased performance of the underlying SAT solver and (ii) novel techniques and algorithms proposed for MaxSAT solving.

Currently, MaxSAT solvers more suited for industrial instances may follow different algorithmic approaches, although the common feature is that MaxSAT algorithms rely on successive calls to a SAT solver. In this paper, we present OPEN-WBO, an open-source modular MaxSAT solver that enables an easy replacement of the underlying SAT solver for any MiniSAT-like solver. The OPEN-WBO architecture also allows an easy implementation and extension of different MaxSAT algorithms. Another contribution of the paper is an evaluation of different state of the art SAT solvers in solving MaxSAT. We provide the results of OPEN-WBO for linear search and unsatisfiability-based algorithm when using different underlying SAT solvers.

In what follows we assume that the reader is familiar with MaxSAT [21,26].

---

[*] Supported by the ERC project 280053.

[**] Partially supported by FCT grants ASPEN (PTDC/EIA-CCO/110921/2009), POLARIS (PTDC/EIA-CCO/123051/2010), and INESC-ID's multiannual PIDDAC funding PEst-OE/EEI/LA0021/2011.

C. Sinz and U. Egly (Eds.): SAT 2014, LNCS 8561, pp. 438–445, 2014.

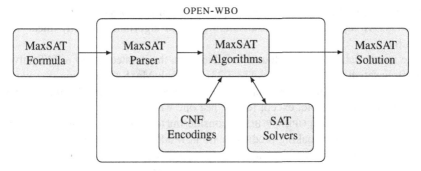

Fig. 1. Overview of the architecture of OPEN-WBO

2   Solver Description

OPEN-WBO is implemented in C++ and extends the interface of MINISAT [12] for solving MaxSAT formulas. OPEN-WBO is open-source and is available for download at `http://sat.inesc-id.pt/open-wbo`. Figure 1 provides an overview of the OPEN-WBO architecture. The main components are the following: (i) MaxSAT Parser, (ii) MaxSAT Algorithms, (iii) CNF Encodings, and (iv) SAT Solvers. Each of the components of OPEN-WBO is briefly described in this section.

MaxSAT Parser. OPEN-WBO can take as input any MaxSAT formula. The parser reads the MaxSAT formula and stores the hard and soft clauses into two different data structures. Each hard and soft clauses may have private fields that describe properties of those clauses. Currently, to each soft clause we associate an integer weight, a set of relaxation variables, and a selector variable. (In the case of unweighted MaxSAT formulas, weight 1 is associated with each soft clause.) The integer weight is used to store the cost of unsatisfying the soft clause. The set of relaxation variables keeps the fresh variables that have been added to soft clauses during the relaxation procedure of the MaxSAT algorithm. Selector variables are fresh variables that are used to extract an unsatisfiable subformula [5]. For that we first associate each selector variable to the corresponding soft clause. Next, the SAT solver is called with a set of assumptions, where the assumptions correspond to the negation of the selector variables. Assumptions are always picked as the first decision literals by the SAT solver. If the SAT solver returns unsatisfiable, then the solver is able to infer a conflict clause that only contains assumptions. As a result, the subset of soft clauses that contains those assumption variables corresponds to the soft clauses in an unsatisfiable subformula. (Note that unsatisfiability-based algorithms do not need to relax hard clauses. Therefore, we only need to extract the soft clauses from the unsatisfiable subformula.)

If a MaxSAT algorithm requires an additional property of the hard or soft clauses, this can be easily added to the respective data structure. Hence, the proposed architecture allows the working formula to be easily rebuilt at each iteration of the MaxSAT algorithm.

MaxSAT Algorithms. The current implementation of OPEN-WBO has two orthogonal MaxSAT algorithms: (i) unsatisfiability-based algorithm, and (ii) linear search algorithm.

The unsatisfiability-based algorithm is based on the iterated use of a SAT solver to identify unsatisfiable subformulas [13,3,23]. At each MaxSAT iteration, the working formula is relaxed until the formula becomes satisfiable. Therefore, all calls to the SAT solver return unsatisfiable, except for the last one which returns satisfiable. The basic algorithm is improved in OPEN-WBO by considering techniques that have been recently proposed. Symmetry breaking predicates are now added to the formula to break symmetries that arise from having multiple relaxation variables [2]. For weighted partial MaxSAT instances, this algorithm is further improved by considering the weight-based partitioning scheme [25] or the diversity-based heuristic [2].

The linear search algorithm [19,17] starts by adding a new relaxation variable to each soft clause and solving the resulting formula with a SAT solver. Whenever a model is found, a new constraint on the relaxation variables is added such that models with a greater or equal value are excluded. The algorithm terminates when the SAT solver returns unsatisfiable. Therefore, all calls to the SAT solver return satisfiable, except for the last one which returns unsatisfiable. The basic algorithm is improved in OPEN-WBO by considering lexicographic optimization for weighted partial MaxSAT instances where the optimality criterion is lexicographical [24].

CNF Encodings. Most MaxSAT algorithms require the encoding of constraints that are not originally expressed in CNF, such as: (i) at-most-one constraints ($x_1 + \ldots + x_n \leq 1$), (ii) cardinality constraints ($x_1 + \ldots + x_n \leq k$), and (iii) pseudo-Boolean constraints ($a_1 x_1 + \ldots + a_n x_n \leq k$). OPEN-WBO uses the following encodings for these different constraints:

- *Ladder* encoding [4,14] (at-most-one constraints): for a constraint with n literals, this encoding creates $O(n)$ clauses with $O(n)$ auxiliary variables.
- *Cardinality Networks* encoding [6] (cardinality constraints): for a constraint with n literals and with right-hand side k, this encoding creates $O(nlog^2 k)$ clauses with $O(nlog^2 k)$ auxiliary variables.
- *Sequential* encoding [15] (pseudo-Boolean constraints): for a constraint with n literals and with right-hand side k, this encoding creates $O(nk)$ clauses with $O(nk)$ auxiliary variables.

These encodings are compact and maintain generalized arc consistency by unit propagation. For the cardinality and the pseudo-Boolean constraints, the user as a programmer has two options: (i) to encode the constraint into CNF or (ii) to update the right-hand side value of the encoding. If the encoding was already built and the MaxSAT algorithm found a smaller right-hand side value, then the programmer may update the encoding instead of rebuilding it [6]. In this case, all learned clauses from the previous SAT call may be kept in the next call to the SAT solver. With these encodings, a programmer may implement MaxSAT algorithms that are based on successive calls to a SAT solver.

Table 1. Number of instances solved by different SAT solvers in the SAT Competition 2013 in the Application SAT+UNSAT track (out of $150 + 150 = 300$ instances)

	#SAT	#UNSAT	Total
ZENN	113	95	208
SINN	120	86	206
glue_bit	102	102	204
Glucose 2.3	103	98	201
GluH	99	97	196
Glueminisat	100	96	196
Glucored	93	95	188

SAT Solvers. OPEN-WBO can use any MiniSAT-like SAT solver, i.e. any SAT solver that extends and uses the same interface as MINISAT [12]. When compiling OPEN-WBO, the user may choose which MiniSAT-like solver is going to be used. Currently, OPEN-WBO includes the following SAT solvers: (i) MiniSAT2.0 [12], (ii) Mini-SAT2.2 [12], (iii) Glucose2.3 [8,9], (iv) Glucose3.0 [8,7], (v) ZENN [33], (vi) SINN [32], (vii) Glueminisat [27], (viii) GluH [29], (ix) glue_bit [11], and (x) GlucoRed [31]. Note that Glucose3.0 was modified to support the optimizations on assumptions [7] for MaxSAT. MiniSAT2.2 extends MiniSAT2.0 by using Luby restarts [22] and phase saving [30]. The other solvers differ mainly in the strategy employed for deleting learned clauses and for restarting. These strategies are usually based on the Literal Block Distance (LBD) measure [8] or in variations of LBD. For details on the SAT solvers see the proceedings of the SAT Competition 2013 [10].

Table 1 shows the number of instances solved by the different SAT solvers that are being used in OPEN-WBO and also participated in the Application SAT+UNSAT track of the SAT Competition 2013. For a better understanding of the performance of the solvers, we split the number of solved instances into unsatisfiable (#UNSAT) and satisfiable (#SAT). Note that MiniSAT2.0, MiniSAT2.2, and Glucose3.0 are not included in Table 1 since they did not participate in the SAT Competition 2013.

Overall, ZENN was the best performing MiniSAT-like solver in the SAT Competition 2013. The best solver for satisfiable instances was SINN, whereas the best solver for unsatisfiable instances was glue_bit. The difference between the number of instances solved by MiniSAT-like solvers is small. For example, ZENN only solved 12 more instances than Glueminisat. On the other hand, GlucoRed was the solver with the worst performance. GlucoRed uses two concurrent threads, which are called the solver and the reducer. The solver uses the SAT solver as usual, while the reducer attempts to strengthen the clauses derived by the solver. In practice, GlucoRed only has half of the CPU time for the SAT solver since the SAT Competition 2013 enforces a CPU time limit. Hence, it is expected to solve less instances. Nevertheless, we included this solver in our tool since it may solve instances that are not solved by other solvers.

Note that OPEN-WBO is not restricted to these ten solvers and so any MiniSAT-like solver can be easily plugged in. This allows OPEN-WBO to take advantage of the constant improvement in SAT solver technology.

Table 2. Impact of different SAT solvers in MaxSAT algorithms

(a) Unsat-based algorithm

	ms	pms	wpms	Total
VBS	42	381	331	754
Glucose 3.0	41	365	313	719
Glucose 2.3	40	343	315	698
GluH	40	341	315	696
ZENN	40	341	315	696
Minisat 2.2	41	340	313	694
SINN	41	335	318	694
Glueminisat	39	329	315	683
Minisat 2.0	38	330	310	678
GlucoRed	38	293	295	626
glue_bit	40	286	294	620

(b) Linear search algorithm

	ms	pms	wpms	Total
VBS	10	535	246	791
Glucose 2.3	5	525	242	772
Glucose 3.0	6	521	242	769
ZENN	6	511	242	759
Glueminisat	6	509	238	753
glue_bit	7	510	232	749
GluH	5	505	235	745
GlucoRed	3	509	225	737
SINN	7	484	239	730
Minisat 2.2	7	468	235	710
Minisat 2.0	4	454	233	691

3 Experimental Results

All experiments were run on the unweighted MaxSAT (ms, 55 instances), partial Max-SAT (pms, 697 instances) and weighted partial MaxSAT (wpms, 396 instances) instances from the industrial category of the MaxSAT evaluation of 2013. The evaluation was performed on two AMD Opteron 6276 processors (2.3 GHz) running Fedora 18 with a timeout of 1,800 CPU seconds and a memory limit of 16 GB.

Impact of Different SAT Solvers. Table 2 compares the performance of MaxSAT algorithms when using different SAT solvers. On the left, we present the impact of different SAT solvers in the unsatisfiability-based algorithm, and on the right, the impact of different SAT solvers in the linear search algorithm. For each algorithm, we grouped SAT solvers that performed similarly. The table also includes the Virtual Best Solver (VBS), i.e. the number of instances that were solved by at least one of the SAT solvers.

The performance of SAT solvers in MaxSAT algorithms is very different from the ranking of the SAT solvers at the SAT Competition 2013. This is particularly noticeable for the unsatisfiability-based algorithm. Most SAT solvers have a performance similar to MiniSAT2.2, which is the baseline solver for all SAT solvers reported in this paper. Therefore, the remaining solvers are expected to improve the performance of MiniSAT 2.2. However, this is not the case for MaxSAT when using the unsatisfiability-based algorithm. It has been observed that the LBD measure is affected when using assumptions [7]. Since each assumption has its own decision level, the LBD measure will be similar to the clause size when learning clauses that contain a large number of assumptions. With the exception of MINISAT, all other solvers use the LBD measure. This may explain why the performance of most SAT solvers is similar to MiniSAT2.2. Furthermore, assumptions may also affect other heuristics. For example, glue_bit was the best performing MiniSAT-like solver for unsatisfiable instances in the SAT Competition 2013, but is one of the worst solvers when using the unsatisfiability-based algorithm, since it uses a restart strategy based on the depth of the search which is greatly affected by the large number of assumptions. On the other hand, Glucose3.0 was the best performing solver for the unsatisfiability-based algorithm since it is the only solver with optimizations when using assumptions [7]. The VBS solves 35 more instances than any individual solver.

For the linear search algorithm, all SAT solvers outperform MINISAT. The linear search algorithm does not use assumptions on the SAT calls. Hence, the heuristics of SAT solvers are not affected. SINN was the best performing MiniSAT-like solver for satisfiable instances in the SAT Competition 2013, but it is one of the worst solvers in Table 2b. This is due to the formula in the last call of the linear search algorithm being unsatisfiable. Even though SINN performs well for satisfiable instances, it does poorly on unsatisfiable instances. Glucose2.3 was the best solver for the linear search algorithm with results similar to Glucose3.0. The VBS solves only 19 more instances than any individual solver. It seems that the performance of the linear search algorithm does not depend much on the SAT solver performance.

State-of-the-art MaxSAT Solvers. We have compared the performance of OPEN-WBO (with Glucose3.0) against QMAXSAT [17] and WPM1 [3] (MaxSAT Evaluation 2013 versions) in order to show that OPEN-WBO is competitive.

QMAXSAT uses a linear search algorithm similar to the one implemented in OPEN-WBO and is particularly effective for solving industrial pms instances. QMAXSAT solved 641 instances (6 ms, 545 pms, 90 wpms). When compared to the OPEN-WBO linear search algorithm, QMAXSAT solved the same number of ms instances, 24 more pms instances, and 152 less wpms instances. The performance gains on the pms instances are mostly due to a new cardinality encoding that has been recently proposed [28]. On the other hand, QMAXSAT is much less efficient than OPEN-WBO when solvingwpms instances for not using lexicographic optimization.

WPM1 uses an unsatisfiability-based algorithm similar to the one implemented in OPEN-WBO and is particularly effective for solving weighted partial MaxSAT instances. WPM1 solved 772 instances (22 ms, 405 pms, 345 wpms). When compared to OPEN-WBO, WPM1 solved 19 less ms instances, 40 more pms instances, and 32 more wpms instances. The current version of WPM1 uses a SMT solver instead of a SAT solver, and is able to keep some learned clauses between iterations of the MaxSAT algorithm. This may explain the performance gains of WPM1 when compared to OPEN-WBO for the pms and wpms instances. On the other hand, OPEN-WBO is more efficient for solving ms instances. These instances have a very large number of soft clauses, which results in a considerable overhead to the current version of WPM1.

4   Conclusions

In this paper we presented OPEN-WBO, an extensible and modular open-source MaxSAT solver. OPEN-WBO implements state of the art linear search and unsatisfiability-based algorithms and can use any MiniSAT-like SAT solver. An experimental evaluation was conducted to show the performance of OPEN-WBO when using different SAT solvers.

References

1. Achá, R.A., Nieuwenhuis, R.: Curriculum-based course timetabling with SAT and MaxSAT. Annals of Operations Research, 1–21 (2012)
2. Ansótegui, C., Bonet, M.L., Gabàs, J., Levy, J.: Improving SAT-Based Weighted MaxSAT Solvers. In: Milano, M. (ed.) CP 2012. LNCS, vol. 7514, pp. 86–101. Springer, Heidelberg (2012)
3. Ansótegui, C., Bonet, M.L., Levy, J.: Solving (Weighted) Partial MaxSAT through Satisfiability Testing. In: Kullmann [18], pp. 427–440
4. Ansótegui, C., Manyà, F.: Mapping Problems with Finite-Domain Variables into Problems with Boolean Variables. In: Hoos, H.H., Mitchell, D.G. (eds.) SAT 2004. LNCS, vol. 3542, pp. 1–15. Springer, Heidelberg (2005)
5. Asín, R., Nieuwenhuis, R., Oliveras, A., Rodríguez-Carbonell, E.: Practical algorithms for unsatisfiability proof and core generation in SAT solvers. AI Communications 23(2-3), 145–157 (2010)
6. Asín, R., Nieuwenhuis, R., Oliveras, A., Rodríguez-Carbonell, E.: Cardinality Networks: a theoretical and empirical study. Constraints 16(2), 195–221 (2011)
7. Audemard, G., Lagniez, J.M., Simon, L.: Improving Glucose for Incremental SAT Solving with Assumptions: Application to MUS Extraction. In: Järvisalo, M., Van Gelder, A. (eds.) SAT 2013. LNCS, vol. 7962, pp. 309–317. Springer, Heidelberg (2013)
8. Audemard, G., Simon, L.: Predicting Learnt Clauses Quality in Modern SAT Solvers. In: Boutilier, C. (ed.) International Joint Conference on Artificial Intelligence, pp. 399–404 (2009)
9. Audemard, G., Simon, L.: Glucose 2.3 in the SAT 2013 Competition. In: Proceedings of SAT Competition 2013: Solver and Benchmark Descriptions [10], pp. 42–43
10. Balint, A., Belov, A., Heule, M., Järvisalo, M.: Proceedings of SAT Competition 2013: Solver and Benchmark Descriptions. Tech. rep., Department of Computer Science Series of Publications B, vol. B-2013-1, University of Helsinki, Helsinki (2013)
11. Chen, J.: Solvers with a Bit-Encoding Phase Selection Policy and a Decision-Depth-Sensitive Restart Policy. In: Proceedings of SAT Competition 2013: Solver and Benchmark Descriptions [10], pp. 44–45
12. Eén, N., Sörensson, N.: An Extensible SAT-solver. In: Giunchiglia, E., Tacchella, A. (eds.) SAT 2003. LNCS, vol. 2919, pp. 502–518. Springer, Heidelberg (2004)
13. Fu, Z., Malik, S.: On Solving the Partial MAX-SAT Problem. In: Biere, A., Gomes, C.P. (eds.) SAT 2006. LNCS, vol. 4121, pp. 252–265. Springer, Heidelberg (2006)
14. Gent, I.P., Nightingale, P.: A new encoding of All Different into SAT. In: International Workshop on Modelling and Reformulating Constraint Satisfaction Problems (2004)
15. Hölldobler, S., Manthey, N., Steinke, P.: A Compact Encoding of Pseudo-Boolean Constraints into SAT. In: Glimm, B., Krüger, A. (eds.) KI 2012. LNCS, vol. 7526, pp. 107–118. Springer, Heidelberg (2012)
16. Janota, M., Lynce, I., Manquinho, V., Marques-Silva, J.: PackUp: Tools for Package Upgradability Solving. Journal on Satisfiability, Boolean Modeling and Computation 8(1/2), 89–94 (2012)
17. Koshimura, M., Zhang, T., Fujita, H., Hasegawa, R.: QMaxSAT: A Partial Max-SAT Solver. Journal on Satisfiability, Boolean Modeling and Computation 8, 95–100 (2012)
18. Kullmann, O. (ed.): SAT 2009. LNCS, vol. 5584. Springer, Heidelberg (2009)
19. Le Berre, D., Parrain, A.: The Sat4j library, release 2.2. Journal on Satisfiability, Boolean Modeling and Computation 7(2-3), 59–66 (2010)
20. Le Berre, D., Rapicault, P.: Dependency Management for the Eclipse Ecosystem: Eclipse P2, Metadata and Resolution. In: International Workshop on Open Component Ecosystems, pp. 21–30. ACM (2009)

21. Li, C.M., Manyà, F.: MaxSAT, Hard and Soft Constraints. In: Handbook of Satisfiability, pp. 613–631. IOS Press (2009)
22. Luby, M., Sinclair, A., Zuckerman, D.: Optimal Speedup of Las Vegas Algorithms. Information Processing Letters 47(4), 173–180 (1993)
23. Manquinho, V., Marques-Silva, J., Planes, J.: Algorithms for Weighted Boolean Optimization. In: Kullmann [18], pp. 495–508
24. Marques-Silva, J., Argelich, J., Graça, A., Lynce, I.: Boolean lexicographic optimization: algorithms & applications. Annals of Mathematics and Artificial Intelligence 62(3-4), 317–343 (2011)
25. Martins, R., Manquinho, V.M., Lynce, I.: On Partitioning for Maximum Satisfiability. In: Raedt, L.D., Bessière, C., Dubois, D., Doherty, P., Frasconi, P., Heintz, F., Lucas, P.J.F. (eds.) European Conference on Artificial Intelligence. Frontiers in Artificial Intelligence and Applications, vol. 242, pp. 913–914. IOS Press (2012)
26. Morgado, A., Heras, F., Liffiton, M., Planes, J., Marques-Silva, J.: Iterative and core-guided MaxSAT solving: A survey and assessment. Constraints 18(4), 478–534 (2013)
27. Nabeshima, H., Iwanuma, K., Inoue, K.: GLUEMINISAT2.2.7. In: Proceedings of SAT Competition 2013: Solver and Benchmark Descriptions [10], pp. 46–47
28. Ogawa, T., Liu, Y., Hasegawa, R., Koshimura, M., Fujita, H.: Modulo Based CNF Encoding of Cardinality Constraints and Its Application to MaxSAT Solvers. In: International Conference on Tools with Artificial Intelligence, pp. 9–17. IEEE (2013)
29. Oh, C.: gluH: Modified Version of glucose 2.1. In: Proceedings of SAT Competition 2013: Solver and Benchmark Descriptions [10], p. 48
30. Pipatsrisawat, K., Darwiche, A.: A Lightweight Component Caching Scheme for Satisfiability Solvers. In: Marques-Silva, J., Sakallah, K.A. (eds.) SAT 2007. LNCS, vol. 4501, pp. 294–299. Springer, Heidelberg (2007)
31. Wieringa, S.: GlucoRed. In: Proceedings of SAT Competition 2013: Solver and Benchmark Descriptions [10], pp. 40–41
32. Yasumoto, T., Okugawa, T.: SINNminisat. In: Proceedings of SAT Competition 2013 : Solver and Benchmark Descriptions [10], p. 85
33. Yasumoto, T., Okugawa, T.: ZENN. In: Proceedings of SAT Competition 2013: Solver and Benchmark Descriptions [10], p. 95

Author Index